Heteroaromatic Nitrogen Compounds

THE AZOLES

Heteroaromatic Nitrogen Compounds

THE AZOLES

K. Schofield, M. R. Grimmett and B. R. T. Keene

CAMBRIDGE UNIVERSITY PRESS
Cambridge
London · New York · Melbourne

CAMBRIDGE UNIVERSITY PRESS
Cambridge, New York, Melbourne, Madrid, Cape Town,
Singapore, São Paulo, Delhi, Tokyo, Mexico City

Cambridge University Press
The Edinburgh Building, Cambridge CB2 8RU, UK

Published in the United States of America by Cambridge University Press, New York

www.cambridge.org
Information on this title: www.cambridge.org/9780521275217

First published 1976
First paperback edition 2011

A catalogue record for this publication is available from the British Library

Library of Congress Catalogue Card Number: 74-17504

ISBN 978-0-521-20519-1 Hardback
ISBN 978-0-521-27521-7 Paperback

Contents

Preface

In 1967 one of us published an account of the chemistry of the pyrroles and pyridines in which these two families of heterocyclic compounds were regarded as the prototypes of heteroaromatic nitrogen systems in general (K. Schofield: Heteroaromatic Nitrogen Compounds: Pyrroles and Pyridines, pp. viii + 434, Butterworth, London, 1967). It was intended to give a similar treatment of the other major nitrogen systems, and the present volume fulfils that aim for the five-membered ring systems, the azoles.

The chemistry of the azoles, classified and viewed from the same point of view as that adopted in the original volume, is dealt with comprehensively, though not exhaustively, and the book takes account of work published up to a very recent date. Appendix 3 contains references not only to early work overlooked in the preparation of the main text, but also to new work which has appeared since that preparation was completed; the available literature to the end of September 1974 has been scrutinised.

It is hoped that the book will serve as a guide to those working in the field, and also as a work of reference.

K. S.
M. R. G.
September 1975   B. R. T. K.

1   Introduction

Azoles are five-membered heterocyclic compounds containing in their rings one or more hetero-atoms, at least one of which must be nitrogen; the rings contain the maximum number of non-cumulative double-bonds consistent with the normal valencies of the ring atoms. The simplest azole, pyrrole, was dealt with in a previous publication† and this volume is concerned with the azoles formally derived from pyrrole by replacement of one or more -CH: groups by one or more N-atoms. They contain nitrogen atoms of two electronic types: the pyrrolic, electron-releasing type, and the pyridinic, electron-attracting type (Schofield†, chapter 2). The five parent members of the group are pyrazole (1), imidazole (2), 1,2,3-triazole (3), 1,2,4-triazole (4), and tetrazole (5).

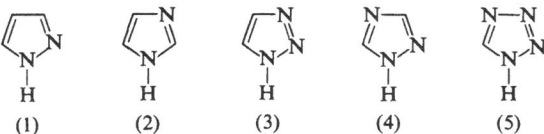

(1)   (2)   (3)   (4)   (5)

The first mononuclear pyrazoles were described by Knorr[1a] in 1883, and it was Knorr and Blank[1b] who introduced the name 'pyrazole', from the relationship to pyrrole, and the correct cyclic formulation. Buchner[2a-c] first prepared pyrazole itself, ascribed to it structure (1), and established some of its fundamental properties. A point of the first importance was established by Knorr[1c] in the following way: 3-methyl-1-phenylpyrazole, the structure of which was fixed by its method of synthesis, was nitrated to give 3-methyl-1-nitrophenylpyrazole (see p. 63). Reduction gave the amine which was oxidised with acidic permanganate to 3-methylpyrazole. Similar oxidation of 5-methyl-1-phenylpyrazole, or of 1-aminophenyl-5-methylpyrazole derived from 5-methyl-1-phenylpyrazole-3-carboxylic acid, should have given 5-methyl-

†   K. Schofield: *Heteroaromatic Nitrogen Compounds: Pyrroles and Pyridines*, Butterworth, London, 1967 (USA: Plenum, New York, 1968)

pyrazole. In fact, the supposed 3- and 5-methyl isomers were identical. Later, Buchner and v. der Heide[2d] showed that only two *C*-phenyl-pyrazoles and two pyrazole-*C*-carboxylic acids could be prepared. An explanation of the equivalence of the 3- and 5-positions in pyrazole by use of a symmetrical structure (6) would not account for the aromatic properties of the compound nor for its properties as a secondary amine (see p. 74), and Knorr[1c] recognised the existence of tautomerism in *N*-unsubstituted pyrazoles.

Tautomerism in the azoles makes ambiguous the position of the imino-hydrogen atom. Consequently a problem of nomenclature arises with *N*-unsubstituted derivatives. Thus, structures (7) and (8) do not refer to preparatively separable entities, a situation indicated by the name 3(5)-methylpyrazole. *The simpler form, 3-methylpyrazole (and for (9) and (10), 4-methylimidazole) will be used here, and the same simplification will be used in related cases, the tautomeric implications*

(6)   (7)   (8)   (9)   (10)   (11)

(12)   (13)   (14)   (15)   (16)   (17)

of the name being implicit. Derivatives of the triazoles and tetrazoles substituted on nitrogen will be related to the tautomeric pairs (11-12), (13-14), and (15-16) by use of the forms 1*H*-1,2,3- and 2*H*-1,2,3-triazole, 1*H*-1,2,4- and 4*H*-1,2,4-triazole, and 1*H*-1,2,3,4- and 2*H*-1,2,3,4-tetrazole[3].

Although allantoin, an imidazole derivative, was known in the eighteenth century, and lophine, a compound holding a central place in the history of the series, was discovered in 1845[4], it was not until the 1880s that satisfactory structures for imidazoles emerged. Debus[5], who first synthesised imidazole from glyoxal and ammonia (cf. [6a]), called it 'glyoxaline', a name indicating its source and still used. The name 'imidazole', indicating that the compound is a member of the azole series, was introduced by Hantzsch[7a].

Fischer and Troschke[8a] first postulated a cyclic structure for lophine (2,4,5-triphenylimidazole), but left the position of the imino-hydrogen atom and of the nuclear double-bonds open. Mainly on synthetic grounds Radziszewski[9a,d] urged the structure (17), R = Ph, for lophine, and also postulated the analogous expression (17), R = H, for imidazole. Finally Japp argued against Radziszewski's structures on the grounds that they would not account for the 'stability' of the compounds and, in the case of imidazole, for the formation of a tribromo-derivative and a silver derivative, and for the characteristics of the compound as a secondary base. Japp[10] correctly formulated imidazole as (2) and lophine as its 2,4,5-triphenyl-derivative.† Evidence in support of Japp's structure was presented by Marckwald[12a,b] and by Bamberger[13a,b] who degraded benzimidazole to imidazole.

As in the case of pyrazole, it was soon shown that in imidazoles unsubstituted on nitrogen the 4- and 5-positions are equivalent. Syntheses which should have given one or other of such isomers gave identical compounds[14a,b] and Jänecke[15] postulated that the hydrogen atom of the imino-group vibrated between the two nitrogen atoms. Further proof of tautomerism came from alkylation studies by Pyman[16a] (see p. 77).

v. Pechmann[17a] prepared the first mononuclear 1,2,3-triazoles by the action of acid on the oxidation products of osazones. These compounds were derivatives of 2-phenyl-2*H*-1,2,3-triazole and because of their origin were called osotriazoles. The parent compound was prepared from 2-phenyl-2*H*-1,2,3-triazole-4-carboxylic acid by nitration, reduction to the amine, oxidation with alkaline permanganate to 1,2,3-triazole-4-carboxylic acid, and decarboxylation of the latter[17f]. Formally the product would be 2*H*-1,2,3-triazole, but other syntheses which might have been expected to yield 1*H*-1,2,3-triazole gave a compound identical with v. Pechmann's. Thus, Bladin[18h] oxidised 5-methylbenzo-1,2,3-triazole to the triazole-4,5-dicarboxylic acid and decarboxylated this, whilst Dimroth[19a,b] submitted 1-phenyl-1*H*-1,2,3-triazole-5-carboxylic acid and 1-phenyl-1*H*-1,2,3-triazole-4-carboxylic acid to the sequence of reactions used by v. Pechmann with 2-phenyl-2*H*-1,2,3-triazole-4-carboxylic acid. The identity of v. Pechmann's with Dimroth's 1,2,3-triazole-4-carboxylic acid, and the formation of one and the same triazole in all these experiments, clearly demonstrates the tautomerism of *N*-unsubstituted 1,2,3-triazoles.

Bladin[18a,c,d] who synthesised the first derivatives of 1,2,4-triazole, correctly represented their cyclic structures, and named the then unknown parent compound 'triazole', but deduced the wrong orientations of substituents in his products. The orientations were corrected by the work of Andreocci[20a], Bamberger and de Gruyter[13c] and Widman[21]. 1,2,4-Triazole itself was prepared almost simultaneously by Andreocci [20b] and by Bladin[18e] through the reactions shown below (Bladin

---

† Radziszewski[9b] synthesised 2-methylimidazole and prepared from it a compound thought to be a tribromo-derivative, which evidence he took to support his imidazole structure. Later[11] the 'tribromo-derivative' was shown to be 4,5-dibromo-2-methylimidazole hydrobromide.

mistakenly took his 1-phenyl-1*H*-1,2,4-triazole-3-carboxylic acid to be the 5-carboxylic acid).

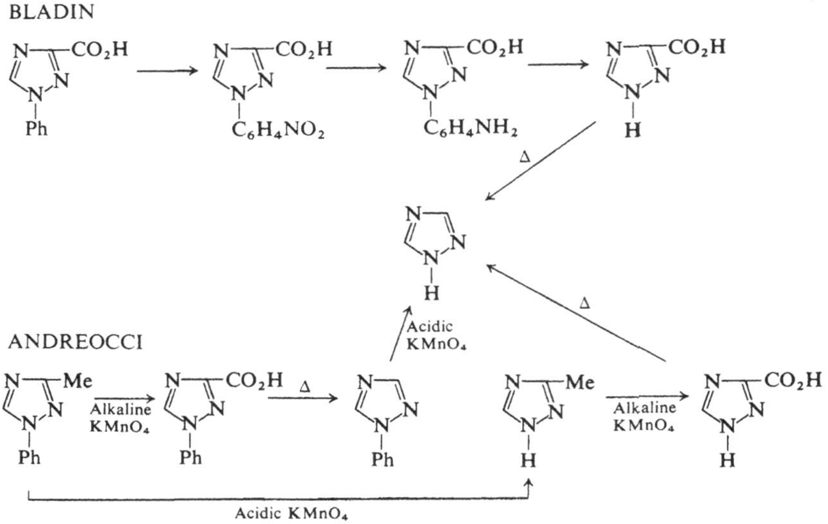

BLADIN

N—CO₂H ... N—CO₂H ... N—CO₂H ... N—CO₂H
Ph ... C₆H₄NO₂ ... C₆H₄NH₂ ... H

(Δ)

N
H

Acidic KMnO₄

ANDREOCCI

N—Me →(Alkaline KMnO₄) N—CO₂H (Δ)→ N ... N—Me →(Alkaline KMnO₄) N—CO₂H
Ph ... Ph ... Ph ... H ... H

Acidic KMnO₄

The later demonstration that the triazole obtained by oxidising 4-phenyl-4*H*-1,2,4-triazole was identical with the compound prepared as above and by several other methods[22b,c,e,g] uncovered the tautomeric character of 1,2,4-triazole.

It was also Bladin[18a,b,c] who obtained the first tetrazoles and since he used the same starting material for preparing 1,2,4-triazoles and the tetrazole derivative, he again mistook the orientation of his product[13c, 20a,21,23b], describing 5-cyano-2-phenyl-2*H*-1,2,3,4-tetrazole as 5-cyano-1-phenyl-1*H*-1,2,3,4-tetrazole. The preparation of 2-phenyl-2*H*-1,2,3,4-tetrazole[18b,23b] and of the quite different 1-phenyl-1*H*-1,2,3,4-tetrazole[24], and the conversion of these compounds (or derivatives)

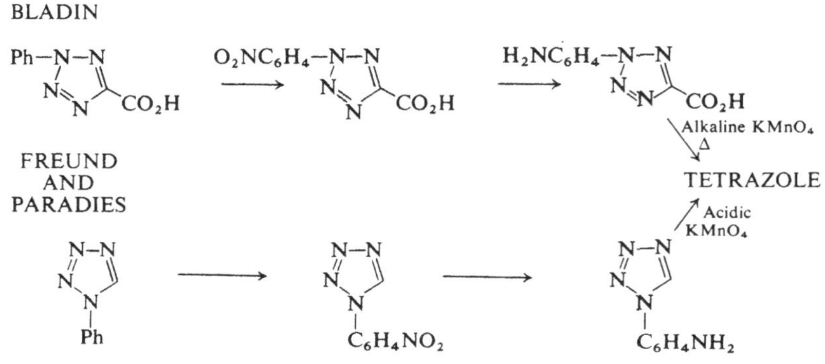

BLADIN

Ph—N—N → O₂NC₆H₄—N—N → H₂NC₆H₄—N—N
CO₂H ... CO₂H ... CO₂H

Alkaline KMnO₄ Δ

FREUND
AND
PARADIES

TETRAZOLE

Acidic KMnO₄

N—N → N—N → N—N
Ph ... C₆H₄NO₂ ... C₆H₄NH₂

into one and the same tetrazole[18f,24c] implied that *N*-unsubstituted tetrazoles are tautomeric substances[23a]. Other syntheses of tetrazole produce one and the same compound[17c,23a,25c].

Early workers on azoles correctly deduced the cyclic structures of these compounds, and from their 'stabilities' (i.e. lack of reactivity towards some reagents), ability to undergo substitution, and properties of their functional groups, recognised their aromatic character. The usual problems of valency in aromatic structures remained. Broadly, the problem was taken only to the stage of adopting for the azoles v. Baeyer's formulation of pyrrole (Schofield, p.6.), leading to the structures (1)-(5). Some other alternatives, as in the case of pyrazole[1c,2a,b,c,d,26] were considered, but the only general treatment of the series was that of Bamberger (Schofield, p.6) which could, for example, account for the increase in basicity of pyrazole and imidazole over that of pyrrole. The theory of the 'aromatic sextet' gives an equivalent account of the facts[27].

The chemical demonstrations of tautomerism in the azole series, described above, and later confirmation, as for example by alkylation studies, contributed nothing to our knowledge of the nature of the tautomerism. Later physico-chemical studies (see chapter 2) have provided many data on this point.

Although the natural occurrence of imidazoles has long been known, it was only relatively recently that a pyrazole derivative was found in nature; β-(pyrazolyl-*N*)alanine was isolated from water-melon seeds[28]. The alkaloid withasomnine is also a pyrazole derivative[29].

2 Physical properties

2.1 General properties. Dipole moments

The molecular geometry of some of the azoles and their derivatives has been studied by X-ray and related methods. Pyrazole itself has been studied by both X-ray[30] and neutron diffraction[31]. In pyrazole crystals hydrogen bonding creates chains in a figure of eight spiral. The neutron diffraction experiments showed the two crystallographically independent molecules to be of identical dimensions within experimental error. The bond angles, and the bond lengths, corrected for rigid body motion, are shown in the following diagrams[31], the second of which includes (in parentheses) the corresponding results for pyridine (Schofield, p.15) and pyrrole[32]. The similarities are close. Within experimental error both of the C–N bonds and both of the C–C bonds in the planar pyrazole molecule are of the same lengths. Dichloro-[33] and dibromo-tetrapyrazolenickel(II)[34] yield results consistent with these, and results for some derivatives of antipyrine are available[35].

Imidazole has been examined by X-ray methods[36,37]. The results from the low-temperature study[37,38] are illustrated. Again, the molecule is planar. In 4,5-di-t-butylimidazole the dimensions are generally similar, though the C-4–C-5 bond is slightly stretched[38]. Amongst derivatives of imidazole which have been examined are histidine hydrochloride[39,41], parabanic acid[42], and 4-acetylamino-2-bromo-1-methyl-5-isopropylimidazole[43]. The imidazolium ion is practically a regular pentagon. An X-ray study[44a,b] of a supercooled melt of 4-methylimidazole revealed polymer structures with linear N–H- -N bonds about 300 pm long.

X-Ray studies of crystals show 1,2,4-triazole to exist in the solid state as planar 1*H*-1,2,4-triazole[45,46]. (The vapour-phase microwave spectrum shows the same tautomer to be present[47].) The results of the low-temperature (–160°C) X-ray study[46], which are consistent with results found for other triazoles, are illustrated. Hydrogen bonding (N–H- -N,

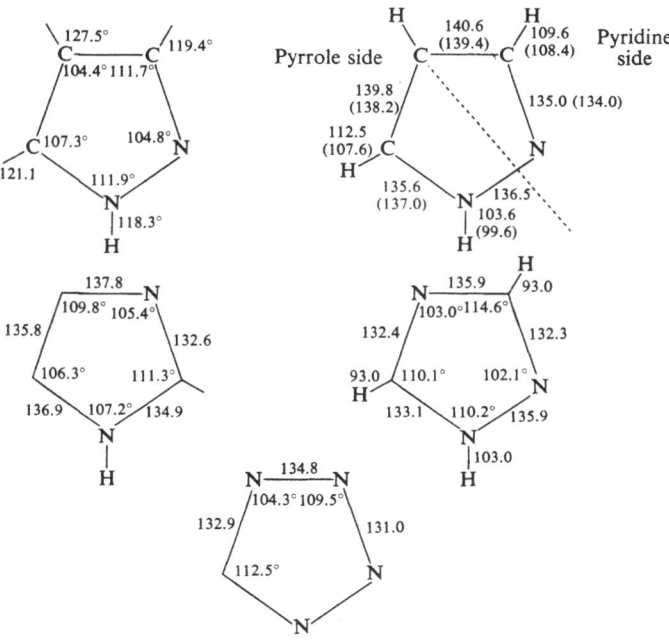

282 pm; C-H- -N, 330 pm) produces infinite corrugated sheets of molecules in the crystals. 4-Amino-3-hydrazino-5-mercapto-4*H*-1,2,4-triazole has also been studied[48].

The tetrazole series has been neglected. X-Ray studies of 1,3-dimethyl-5-iminotetrazole[49a], 5-amino-2-methyltetrazole[49b], and sodium tetrazole monohydrate[50] have been reported. The results for the anion in the last compound are illustrated.

The heats of combustion of a number of derivatives of tetrazole and 1,2,4-triazole have been determined. Some of the compounds did not burn properly, and some of the tetrazoles detonated[51]. The resonance energies of the azoles deduced from heats of combustion[44d] were reported earlier (Schofield, p.19).

The pyridine nitrogen atoms of the azoles influence profoundly the physical properties of these compounds by comparison with those of pyrrole. Thus, whilst pyrrole is only slightly soluble in water, pyrazole and imidazole are very soluble (one part in one, and greater than one part in one of water at 20 °C, respectively)[52c]. 1,2,4-Triazole is, similarly, readily soluble in polar solvents but only slightly soluble in non-polar solvents. This drastic change is due to the ability of the pyridinic nitrogen atom to join in hydrogen bonding with water, a function forbidden to the pyrrolic nitrogen atom; hydrogen bond formation in aqueous solution thus compensates for the hydrogen bonds broken during dissolution of the crystals. Non-polar solvents are not able to overcome intermolecular hydrogen bonds, and the solubilities of azoles in them are small.

There are further consequences. In the vapour state pyrazole[2a,b,c], imidazole[53,54a], and 1,2,4-triazole[20b] are monomeric, and cryoscopic methods show the molecular weights of pyrazole and imidazole to be roughly normal over a wide concentration in water. Dioxan is also able to disrupt the intermolecular hydrogen bonds of a solute, and azoles dissolved in it are only slightly associated[52b,c]. In contrast, in non-polar solvents pyrazole, imidazole, 4-methylimidazole, and 1,2,4-triazole maintain their intermolecular hydrogen bonds[2a,b,c,52c]; the apparent molecular weight of pyrazole in 0.0056 molal benzene solution is 66.7, in 0.5825 molal solution it is 166.2, and the other compounds show similar trends. As would be expected, replacement of the imino-hydrogen atom by an alkyl group destroys association in both pyrazoles[55] and imidazoles[56] by making hydrogen bonding impossible. That imidazole is considerably more associated than pyrazole is due to the formation by imidazole of linear polymeric forms, whilst pyrazole gives the strainless trimer[55d], as shown here, and a similar dimer (see p. 16). As mentioned above, a supercooled melt of 4-methylimidazole contains neighbouring molecules joined by N-H--N bonds of length 300 pm[44b].

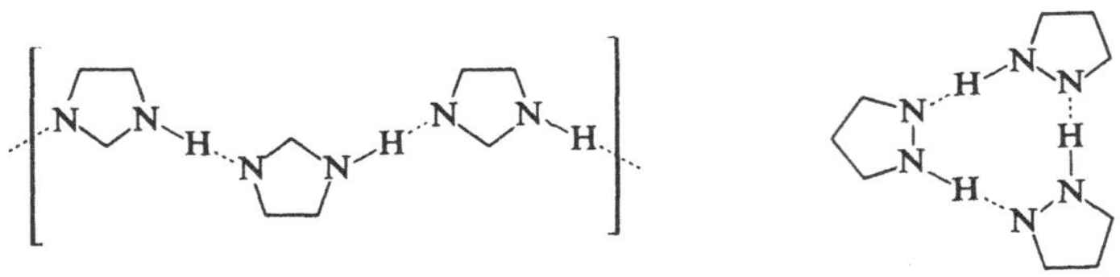

Hydrogen bonding shows its influence in the melting and boiling points of azoles (table A.1). *N*-Substituted derivatives boil at lower temperatures than corresponding *N*-unsubstituted derivatives. The effect of *N*-substitution on melting points is similar. The boiling points of the parent azoles are strikingly higher than that of pyrrole. Imidazole and 1,2,4-triazole are particularly noteworthy; that hydrogen bonding is an important factor is shown by comparing imidazole with 1-methylimidazole, but even so the boiling point of 1-methylimidazole itself is surprisingly high[52c].

The dipole moments of a number of azoles have been measured. Some results for the simpler compounds are collected in table 1, and others include those for halogeno-, amino-, and hydroxy-pyrazoles[107], imidazole-2-aldehydes[108], and various 1,2,4-triazoles[109]. The relatively low value for pyrazole in solution in benzene was thought to be due to dimeric association (see p. 16); dioxan can evidently disrupt association, and in it pyrazole shows a dipole moment little different from that of 1-methylpyrazole. Gas-phase, microwave spectroscopic

Table 1. Dipole moments

$p/10^{-30}$ Cm		$p/10^{-30}$ Cm	
Pyrazole[105][a]	7.386	1,2,3-Triazole[101][b]	5.90
Pyrazole	5.24[101][b]; 7.39[52c][d]	1,2,4-Triazole[101][d]	10.57; 9.074[47][a]
1-Methylpyrazole[c]	7.643	1-Ph-1*H*-1,2,3-Triazole [101][b]	13.59
3-Methylpyrazole [101][b]	4.77	2-Ph-2*H*-1,2,3-Triazole [101][b]	3.23
1-Ethylpyrazole[c]	7.923	1-Ph-1*H*-1,2,4-Triazole [101][b]	9.61
1-Isopropylpyrazole[c]	8.206	4-Ph-4*H*-1,2,4-Triazole [101][b]	18.78
Imidazole	12.81[101][b] 13.3[102][d]	Tetrazole[101][d]	17.05
1-Methylimidazole [52c][be]	12.11	1-Methyl-1*H*-1,2,3,4-tetrazole[101][b]	17.95
1-Propylimidazole [103][bf]	13.74	1-Ethyl-1*H*-1,2,3,4-tetrazole[104][b]	18.21
1-Phenylimidazole [103][bf]	10.47	2-Ethyl-2*H*-1,2,3,4-tetrazole[104][b]	8.84

(a) Gas-phase microwave spectrum. (b) In benzene. (c) In benzene at 25.0 °C. Values for solutions in cyclohexane, dioxan, carbon tetrachloride, toluene, trichloroethylene, ether, chloroform, and chlorobenzene were also reported. The extrapolated gas-phase values were 7.873, 8.139, and 8.340 × 10^{-30} Cm for 1-methyl-, 1-ethyl-, and 1-isopropyl-pyrazole, respectively[106]. (d) In dioxan. (e) 20 °C. (f) 25 °C.

studies[105] show the pyrazole molecule to be planar with no symmetry axis so that the imino-hydrogen atom is not shared by the two nitrogen atoms and is not oscillating rapidly between them. The dipole moment of the isolated pyrazole molecule agrees well with that for pyrazole in dioxan.

The dipole moment of imidazole in solution decreases with dilution to a value close to that for 1-methylimidazole, which is but little dependent on concentration[52c].

The value for 1,2,3-triazole is suggested to indicate the predominating tautomer to be 2*H*-1,2,3-triazole, being closer to that of pyrazole than to that of imidazole[101]. Comparison with the isomeric *N*-alkyl compounds shows tetrazole to be predominantly 1*H*-1,2,3,4-tetrazole[104], a conclusion supported by calculation[110]. Microwave spectroscopic studies of 1,2 4-triazole[47] show it to exist in the gas phase as 1*H*-1,2,4-triazole (see p. 6 for the solid state).

The performance of molecular orbital theories in treating the dipole moments of azoles is mentioned below (p. 29).

Combining the fact of association in the azoles with a resonance description of the hydrogen bond (N–H...N⇔N...H–N), Hunter[55c] concluded that these compounds exemplified 'mesohydric tautomerism'. The status of this theory is doubtful, for the resonance description of the hydrogen bond is not now accepted.

From refractivity studies, v. Auwers[63g] concluded that pyrazoles are equilibrium mixtures of tautomers, the position of equilibrium varying with the substituents present. The data indicated that in some cases one tautomer was so highly favoured that the pyrazole was essentially homogeneous. Thus, comparison of methyl 1-methyl-3-phenyl-pyrazole-5-carboxylate with methyl 1-ethyl-5-phenylpyrazole-3-carboxylate suggested that in quinoline solution (18) existed almost to the exclusion of its tautomer. Similarly, (19) is predominant, rather than its tautomer.

Certain derivatives excepted, azapyrroles show great stability to acids, bases, oxidising agents, and heat. Some tetrazoles are exceptions in the last respect. Many tetrazoles melt without decomposition but others explode when heated. Tetrazole itself explodes above its melting point, and some metallic derivatives of tetrazoles are explosive. Addition of nitrogen-containing functions to the tetrazole ring increases thermal instability. Aqueous solutions containing more than about 2% of tetra-zolediazonium cation are said to explode at $0\,^{\circ}C$[25a,b].

Extensive information about the chromatographic identification of pyrazoles and of imidazoles[1576] is available.

2.2   Spectroscopic properties

2.2.1   Ultraviolet spectroscopy

The ultraviolet absorption of the azoles in the vacuum ultraviolet has not been measured. In solution (table A.2) the parent azoles all show absorption maxima close to that of pyrrole at 210 nm, although the absorption intensities are weakened. Because of the limitations of the quartz-prism instrument the data recorded in table A.2 for the region below 220 nm must be accepted with caution. At higher wavelengths only low-intensity end absorption is found, as with the imidazoles[112]. The significance of the weak long-wavelength absorption is not clear, and $n \rightarrow \pi^*$ bands have not been reported. The suggestion that the weak absorption may be due to impurities[113] has been borne out for imidazole[114d]. Although pyrazole and imidazole show no long-wavelength peaks, conversion into their anions causes a rise in absorbance at 230-250 nm sufficient to permit its use for determining ionisation ratios (see p. 25) [115].

Alkyl groups produce small bathochromic shifts, and aryl groups generate new maxima in the 250 nm region. In the pyrazole series 4-alkyl groups cause larger bathochromic shifts than 3- or 5-alkyl groups [116], and in the 1,2,4-triazole series the order of effectiveness of aryl groups is 3>1>4[117]. In arylpyrazoles[97,118,119,125b], and aryl-1,2,4-triazoles[117] suitably situated methyl groups cause intensity changes suggestive of steric hindrance. In contrast, methyl groups substituted into 4-nitropyrazole cause bathochromic shifts and produce none of the usual indications of steric inhibition of conjugation[120]. The spectrum of 1-phenylpyrazole shows weak absorption at 323 nm which, because it disappears on protonation, has been ascribed to a n→π* transition[121].

The spectra of nitro-imidazoles show features which are altered characteristically by substituents and by pH, and which are useful for purposes of orientation[122].

As regards nuclear tautomerism in the azoles ultraviolet absorption spectra are not often informative; thus, for 1,2,3-triazole they are equivocal[123].

Ultraviolet spectra are more helpful in the cases of tautomerism in which both the nucleus and a substituent are involved. The selective absorption of imidazole-4-aldehyde has been referred to resonance forms such as (20) (which are improbable in the cation), and tautomerisation to a hydroxymethylene form is excluded[124].

As regards the amino-azoles in which amine–imine tautomerism is formally possible, ultraviolet spectroscopic evidence is very incomplete. All 5-aminotetrazoles capable of tautomerism of this kind have λ_{max} 218-232 nm in ethanol, whereas the 'fixed' models (21) R = H and Me, have λ_{max} 266-7 nm. Thus, 5-aminotetrazoles probably exist as amines [155,157]. Other evidence[123] is unacceptable[157] because 'fixed' models were not used.

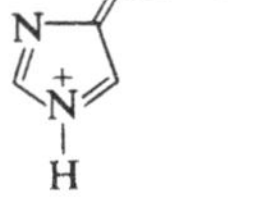

(20) 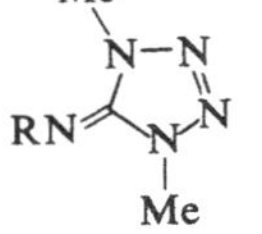   (21)

More evidence is available concerning the hydroxyazoles. The situation with 1-substituted pyrazol-3- and -5-ones has been thoroughly examined, and the main conclusions seem now to be clear (table 2). In the first case two forms are possible, (22) and (23), and in the second case three forms, (24), (25) and (26), sufficiently identified as NH-, CH- and HO-forms. Ultraviolet spectroscopic evidence[130a] supports other evidence (pp. 17, 22) in showing that in non-polar solvents 3-hydroxy-1,5-dimethyl- and 3-hydroxy-5-methyl-1-phenyl-pyrazole exist as HO-forms (as (23)) whereas in water NH- (22) and HO-forms (23) are present in comparable amounts. Similarly, a study of 'fixed' models of (24), (25) and (26) shows that in cyclohexane solution compounds in which R = Me or Ph exist predominantly as the CH-forms (25) with some of the NH-form also present, but with very little HO-form. On the other hand, in aqueous buffer solution the NH-form (24) predominates, with some of the HO-form (26), but little of the CH-form (25). The absorption spectra of many 1-substituted (especially 1-aryl) 4-arylazopyrazol-5-ones have been discussed in terms of the structure (29), strongly supported by other evidence (p. 17)[114c,132a,b].

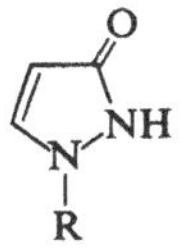

(22) 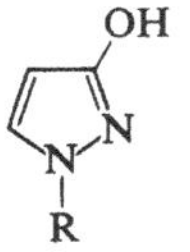   (23)

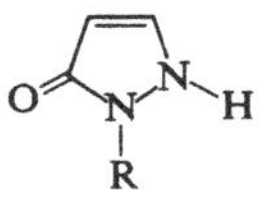

(24) 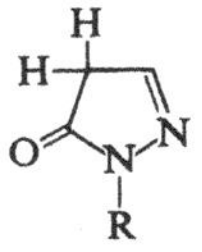   (25) 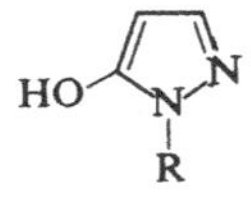   (26)

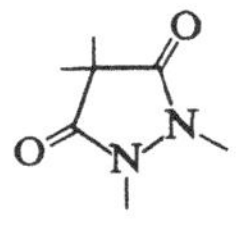

(27) 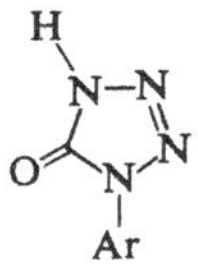   (28) 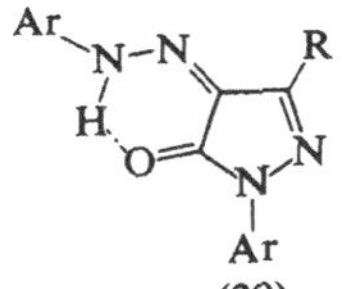   (29)

For *N*-unsubstituted 'hydroxypyrazoles' the tautomeric possibilities are more complicated. Ultraviolet and infrared spectroscopy and measurements of ionisation constants have been used in examining some of these compounds. The main features of the situation seem now to be clear, although different workers have drawn conclusions differing in detail. Comprehensive results are available for '5-hydroxy-3-methyl-', '5-hydroxy-4-methyl-' and '5-hydroxy-3,4-dimethyl-pyrazole', having been obtained by correct comparisons with 'fixed models'[114g,130b]. Thus, in aqueous solution of '5-hydroxy-3-methylpyrazole' the form analogous to (24), R = H, is dominant, whilst in cyclohexane that analogous to (25), R = H, is dominant. Table 2 (based on ref.114g) summarises the most recent conclusions based on evidence of several kinds, including ultraviolet spectroscopy. These results are more comprehensive, and more soundly based than those of earlier workers[158].

Table 2. The tautomerism of the pyrazolones [114g] [a]

Compound	C_6H_{12}	CCl_4	$CHCl_3$	Dioxan	MeCN	MeOH	H_2O	Pyridine	Solid
3-Me-5-one	85 (25) 15 (26)		(25)		(26)	40 (24) 60 (26)	90 (24) 10 (26)	(26)	(26)↔(24)
4-Me-5-one	85 (25) 15 (26)		(25)			30 (24) 70 (26)	85 (24) 15 (26)	(26)	(26)↔(24)
3,4-Me2-5-one	85 (25) 15 (26)		(25)			65 (24) 35 (26)	(24)	(26)	(26)↔(24)
1,3-Me2-5-one	(25)	(25)	(25)			75 (24) 25 (26)	(24)	(26) and trace of (25)	(26)↔(24)
1,4-Me2-5-one			(25)<(24)[b]			80 (24) 20 (26)	(24)	(26)	
1,3,4-Me3-5-one			(25)>(24)[b]			(24)	(24)	(26)(24) and trace of (25)	
1,5-Me2-3-one	(23)	(23)	(23)			10 (22) 90 (23)	70 (22) 30 (23)	(23)	(23)
1,4-Me2-3-one			(23)			10 (22) 90 (23)	60 (22) 40 (23)	(23)	
1,4,5-Me3-3-one			(23)			30 (22) 70 (23)	(22)	(23)	
1-Ph-5-one			(25)	55 (25) 45 (26)	(25) (26)				(26)↔(24)
3-Me-1-Ph-5-one	(25)	(25)	(25)	90 (25) 10 (26)	(25) (26)		(24)		(26)↔(24)
3,4-Me2-1-Ph-5-one	90 (25) 10 (24)		(24) (25)[b]			(24)	(24)		(26)↔(24)
1-Ph-3-one	(23)	(23)	(23)						(23)
5-Me-1-Ph-3-one	(23)	(23)	(23)			(22) (23)	(22) (23)		(23)

(a) In each case the proportions (%) of tautomeric forms are given and (in parenthesis) the number of the structure.
(b) With increased concentration the equilibrium is moved towards form (24).

With pyrazol-3,5-diones the situation is not altogether clear. 1-Phenyl-pyrazol-3,5-diones have been represented as NH-forms with one carbonyl and one enol group, but their ultraviolet spectra resemble those of 4,4-disubstituted derivatives[159a]. Most frequently however, compounds of this series seem to be of the type (27); one or two substituents at C-4 have little effect on the ultraviolet spectra[160].

Ultraviolet spectroscopic evidence supports other results (p. 18) for the assignment of the 'dioxo' form to 1-phenylurazole (1-phenyl-1,2,4-triazol-3,5-dione)[161], and of NH-forms (28) to some 1-aryl-5-hydroxy-tetrazoles[156].

Little is known about the tautomerism of the azole *N*-oxides. '2,4,5-Triphenylimidazole 3-oxide' has been examined spectroscopically along with the 'fixed' models, 1-methyl-2,4,5-triphenylimidazole 3-oxide and 1-methoxy-2,4,5-triphenylimidazole, and ionisation constants have also been recorded[142]. The ultraviolet spectroscopic results show the parent compound to predominantly 1-hydroxy-2,4,5-triphenylimidazole in ethanol, chloroform, acetonitrile, and cyclohexane, whilst in aqueous ethanol there is a move toward the oxide form.

Ultraviolet spectroscopic data support other evidence (p. 18) for ascribing thione structures to 2-mercapto-imidazoles[162], mercapto-1,2,4-triazoles[163], and mercapto-tetrazoles[148f].

2.2.2   Infrared spectra

Vibration spectra have played an important role in the study of the azoles. Total or extensive vibrational assignments have been made for pyrazole and some methyl derivatives[164], imidazole, 1-methylimidazole, and imidazole-d[165-7a], 1,2,3-triazole[138,144,168b,c], and the tetrazolate anion[169]. Bands characteristic of the tetrazole ring have been recognised[148b,176,177c]. Alkyl-[170b], 3,4,5-tri-substituted-[171], 1-phenyl-[121], and 5-amino-pyrazoles[172b], 4-methyl-[167c], nitro-[173], and amino-imidazoles[174], and numerous 1,2,4-triazoles[175b] have also been studied. Table 3 lists some assignments.

The Raman spectrum of pyrazole indicates the equivalence of the two nitrogen atoms and, for the imidazole cation, C_{2v} symmetry[39,178-9]. The spectrum in solution confirms the existence of 4- and 5-methyl-1*H*-imidazole, but shows only one form in the solid state[167c]. The infra-red spectrum of 1,2,3-triazole favours the 1*H*-1,2,3-triazole form[168b].

The infrared characteristics of the NH-bond in azoles have attracted much attention. Strong hydrogen bonding leads to absorption at frequencies much lower than usual[175b], with features of great theoretical interest[130d]. Some features of the NH-bond have been clarified by the examination of pyrazole and imidazole in argon matrices at 20 K[180]. Proton-tunnelling in the azole hydrogen bond has also been discussed[167b,175b,181]. The occurrence of hydrogen bonding

Table 3. Infrared absorption of azole nuclei[a]

Substance	Ring skeletal/cm⁻¹			Ring breathing and C-H in-plane deformation/cm⁻¹			C-H out-of-plane deformation/cm⁻¹		
Pyrazoles[b]	1608-1597	1555-1553	1484-1462		1130-1310	1037-974	935	790	
Imidazoles	~1550	~1490	~1325	~1260	~1140	~1060	~940	~840	~760
1,2,3-Triazole	~1520	~1450	~1410	~1250	~1100	~1050	~990	~850-700	
Substituted 1,2,3-triazoles				1290-1150	1130-1105	1090-1010	1000-900		
Tetrazoles	1640-1615	1450-1410	1400-1335	1300-1260	1200-1110	1095-1035	990-900	810-715	

(a) This table[176] differs somewhat from that given in ref. 130c, from which some of the data come. See also refs. 39, 138, 144 and 148b. A comprehensive tabulation is given in ref.130d.

(b) For 3-, 4-, and 1-alkylpyrazoles see ref. 170a.

causes the spectra of azoles in solution to be dependent on concentration. For monomeric pyrazole in carbon tetrachloride[182] the NH-stretching band is at 3485 ± 10 cm^{-1}, and for 3,5-dimethylpyrazole in the same solvent[164] νNH = 3484 cm^{-1} and βNH = 1147 cm^{-1}.

In the concentration range 10^{-4}-1 M an equilibrium exists between pyrazole and its cyclic hydrogen bonded dimer and trimer[183a], as Hückel had suggested (see p. 8). For 3,5-dimethyl-[183c,184] and 3,5-diethyl-pyrazole[183c] the situation is similar, but with 3,5-diphenyl-pyrazole it is different[183c]; with it a cyclic dimer and a series of linear polymers exist, but a cyclic trimer is not formed because of steric hindrance. 3,5-Diphenylpyrazole thus lies between pyrazole and imidazole [183b]. For imidazole the concentration-dependence of the infrared spectrum in carbon tetrachloride shows the N–H stretching band at 3485 ± 10 cm^{-1} to be due to linear oligomers. An earlier suggestion [185b] that imidazole exists as an ionic structure formed by transfer of a proton from one molecule to a second, thus producing a cation and an anion, is not tenable[44b]. Substituted imidazoles in which intra-molecular hydrogen bonding is possible show no intermolecular hydrogen bonding[186]. The association of 4-methylimidazole has also been studied by infrared spectroscopy[187].

Studies of the triazoles in carbon tetrachloride also reveal association [168a,188]. The infrared spectrum of dilute solutions shows that here, as in the vapour phase (see p. 9) 1,2,3-triazole exists as the unsymmetrical 1H-1,2,3-triazole[168b].

The infrared characteristics of N-acylazoles (table 4) are interesting [185a,189b,191]. The carbonyl stretching bands generally appear at frequencies slightly higher than those associated with aliphatic saturated ketones (1765-1725 cm^{-1}), and much higher than those for disubstituted amides (1650 cm^{-1}). This illustrates the small importance of ring nitrogen–acyl group conjugation

a consequence of participation of the nitrogen atom in the aromatic structure of the ring. The carbonyl stretching frequency increases with the number of ring nitrogen atoms. Intense bands at 1370-1420 cm^{-1} parallel those shown by methyl ketones, and bands at 1250-1330 cm^{-1}, have been ascribed to C-N stretching[191].

A systematic study of C-acyl-azoles has not been made. Like the ultraviolet spectrum (see p. 11), the infrared spectrum of imidazole-4-aldehyde supports the aldehyde, rather than the tautomeric hydroxy-methylene formulation[124].

The infrared spectra of amino-azoles have been examined neither extensively nor systematically. For some aminopyrazoles[192c,193a] the evidence indicates an amino-structure, as it does also for 3-amino-1-

Table 4. Infrared characteristics of *N*-acylazoles

Compound	Carbonyl stretching cm^{-1}	MeCO absorption cm^{-1}	C–N stretching cm^{-1}
1-Acetylpyrrole[185a] [a]	1732	1325	1305
3,5-Dimethyl-1-propionylpyrazole[189]	1722		
1-Acetylimidazole[185a] [b]	1747	1294	1258
1-Acetyl-1*H*-1,2,3-triazole[190f] [c]	1762		
2-Acetyl-2*H*-1,2,3-triazole	1780		
1-Acetyl-4-methyl-1*H*-1,2,3-triazole	1761		
2-Acetyl-4-methyl-2*H*-1,2,3-triazole	1771		
N-Acetyl-1,2,4-triazole[185a] [b]	1765	1282	1245 1212
N-Acetyltetrazole[185a] [b]	1779	1205	1198

(a) Liquid film. (b) KBr disc. (c) For 1,2,3-triazoles the results refer to solutions in carbon tetrachloride. Other examples have been recorded, and some for solutions in chloroform.

phenylpyrazol-5-one[193b] and 4-amino-antipyrine[194], and for amino-1,2,4-triazoles[191]. The same is true for 5-aminotetrazoles, which have been more fully studied[148b,155,195d]. Infrared spectroscopy is said to indicate that the position of equilibrium between amide and imide forms of 5-acylaminotetrazoles depends on the phase, the nature of the acyl group, and the position of any *N*-methyl group which may be present[195a,b,d].

'Hydroxyazoles' have been extensively examined[131,193c]. Infrared spectroscopy contributed to the results summarised in table 2[114g]. In chloroform or tetrachloroethylene and in Nujol mulls the infrared spectrum of 3-hydroxy-5-methyl-1-phenylpyrazole resembles that of 3-methoxy-5-methyl-1-phenylpyrazole more than that of 2,5-dimethyl-1-phenylpyrazol-3-one. Thus, the HO-form predominates, and the same is true for the 1-methyl series[130a]. In the solid state the distinction between HO- and NH-forms is not easily made. In this connection the near infrared spectra have been examined (4000-6000 cm^{-1})[196]. The carbonyl group in 'fixed' models of 1-substituted pyrazol-5-ones related to (24) absorbs at ~1655 cm^{-1} in chloroform solution, but rather higher in carbon tetrachloride. 'Fixed' models of (25) absorb at 1694–1698 cm^{-1} with a 1-methyl group, and at 1705–1712 cm^{-1} with a 1-phenyl group. For compounds of these types, for 1-substituted 5-alkoxypyrazoles[130a], and for various tautomeric forms of 1-aryl-3- and -4-substituted pyrazol-5-ones[197] assignments have been made.

The infrared spectra of 4-arylazo-1-phenylpyrazol-5-ones have been argued to support CH-forms[198-9], hydrazone forms (29)[199,200] and internally hydrogen bonded azo-enol forms[201]. Recently, infrared spectroscopic evidence has been interpreted to support the hydrazone form (29)[202]. In consonance with conclusions drawn from ultraviolet spectroscopy (p. 12) and n.m.r. spectroscopy (p. 22), infrared results have been interpreted by later workers also as supporting the chelated structure (29)[114c,132a]. For solutions in chloroform ν(C=O) lies between 1655 and 1668 cm^{-1}.

4-Hydroxypyrazoles have been much less studied, but available evidence relating to solutions and solid state shows them to be hydroxy-compounds [193c].

Infrared data are available for a number of *N*-unsubstituted 'hydroxypyrazoles'[203-5], in which, as has been pointed out already, the tautomeric possibilities are more complicated. Some of the results[114g,130b, 158] have been outlined in the discussion of ultraviolet spectra (p. 12), and incorporated in table 2.

Some members of the '3,5-dihydroxypyrazole' series with 1-substituents have been formulated as CH-compounds with one carbonyl and one enol group[159b], but most members of this series are diones (27), showing two carbonyl bands in their infrared spectra[160,206-7]. Substituents which take part in hydrogen bonding stabilise enol forms[160,206].

Much less is known about other 'hydroxyazoles'. Infrared spectroscopy points to carbonyl forms for '2-hydroxyimidazoles in the solid state' [208b], '3-hydroxy-1,2,4-triazoles'[209,210], 1-phenylurazole[161], and '1-aryl-5-hydroxytetrazoles'[156].

Infrared spectroscopy, in conjunction with n.m.r. spectroscopy, establishes the 1*H*-1,2,4-triazolin-5-thione structure for *N*-unsubstituted 1,2,4-triazolthiones[211].

Infrared spectra have been recorded for halogeno-[212] and nitro-imidazoles[213], imidazolium salts[167d], and 1,2,3-triazole-1-oxides [168d].

2.2.3 *N.m.r. spectroscopy*

Proton magnetic resonance spectroscopy in particular has been widely studied in the pyrazole series (see especially ref. 114k) and to a smaller extent with the other azoles. For *N*-unsubstituted azoles rapid exchange of the NH-proton leads to the observation of averaged CH-proton signals, a fact which accounts for the way in which many of the results in table A.3 are there reported. Couplings are commonly obscured by poor resolutions (as in solutions in trifluoroacetic acid[216]) or by quadrupole coupling of adjacent nitrogen atoms. NH-Proton signals have been observed much broadened at very low fields[226a].

Self-association in pyrazoles[227] has been studied using p.m.r. spectroscopy. In dilute solutions it is agreed that monomers exist, but there is disagreement as to whether polymers are cyclic or open-chain [114k]. Similar studies have been reported for imidazoles[219,226b,228] and 1,2,4-triazoles[175b].

Azoles unsubstituted on nitrogen give averaged signals because of proton exchange. The mechanism of exchange is not known. The problems surrounding the structures of *N*-unsubstituted azoles were discussed by Elguero *et al.*[114k] who correctly pointed out that up to that time work purporting to measure the proportions of tautomers present in solutions of a particular azole suffered from the shortcoming that the existence of tautomers was a mere assumption and the results could also be interpreted as indicating the presence of only one form in which the proton was not localised[44c]. Recent p.m.r. spectroscopic results give *a posteriori* support to the assumptions of earlier workers who used various physical methods, as for example refractivity measurements in the work of v. Auwers (§2.1), purportedly to measure tautomeric ratios or to demonstrate the dominance of one or other of the possible forms.

Especially interesting in this connection is the work of Creagh and Truitt[229] who recorded the 60 MHz-spectra of 1,2,4-triazole at various temperatures with the results shown in table 5. For molten 1,2,4-triazole the NH-singlet is sharp because of rapid exchange. The solution spectrum differs very little except that the NH-signal is broadened. As the temperature of the solution is lowered H-exchange is slowed down and the CH-signal broadens and below 0°C splits to a doublet. It is concluded that at -34°C 1-*H*-1,2,4-triazole is present and the down-field peak of the doublet is assigned to H-3 on this unsymmetrical tautomer. At the same

Table 5

Compound	Solvent	Temp./ °C	P.p.m. down-field from $SiMe_4$ CH	NH	Me
1,2,4-Triazole	(Neat)	125	7.85	13.9 (sharp)	
	$(CD_3)_2SO^a$	37	8.25	13.9 (broad)	
	$[(CD_3)_2N]_3PO^b$	37	8.17	15.1	
		10	8.03 (broad)	15.9	
		0	8.95, 8.59	15.10	
		-10	8.72, 7.87	15.15	
		-34	8.85, 7.92	15.25	
1-Methyl-1*H*-1,2,4-triazole	$[(CD_3)_2N]_3PO^b$	37	8.70^c, 8.01^d		3.64
		-40	8.93, 8.02		3.64
4-Methyl-4*H*-1,2,4-triazole	$[(CD_3)_2N]_3PO^b$	37	8.20		3.64
		-40	8.34		

(a) 3.8% solution. (b) 4.3% solution. (c) H-5. (d) H-3.

time the NH-signal moves downfield with increased intermolecular hydrogen bonding. NH-Multiplicity is not observed because of broadening caused by quadrupole effects. If the splitting at -34 °C is assumed to be the maximum, the mean half-life of the proton at N-1 at 10 °C is about 4×10^{-3} s.

In a related study[230] the spectra of solutions of azoles in $(CD_3)_2CO$ were examined. Temperature variation revealed an equilibrium (not present when $CDCl_3$, MeOD or MeCN were the solvent) written as follows. Here 'Azole$_x$' is the form involved in the

$$\text{Azole}_x + \text{Acetone} \rightleftharpoons x \, [\text{Azole–Acetone}]$$

intermolecular exchange phenomena which make protons equivalent. Lower temperatures favoured the association with acetone, which was regarded as arising from interaction of a lone-pair of electrons on a nitrogen atom with the electrophilic carbon atom of the acetone. For 1,2,4-triazole the unsymmetrical structure was involved, whilst for 1,2,3-triazole both possible forms took part equally. Similarly both forms of tetrazole were involved, the 1*H*-form being the more important. Various substituted pyrazoles and 1,2,4-triazoles behaved similarly, but even at -90 °C the phenomenon did not appear with 3,5-dimethyl-4-nitropyrazole, 3,5-dimethyl-, and 3-bromo-5-methyl-1,2,3-triazole. When solutions of 4-nitropyrazole were examined at successively lower temperatures the averaged signal of the CH-protons broadened and split, and at -98 °C signals at $\tau = 1.53$ and $\tau = 0.94$, assigned to H-3 and H-5 respectively, were apparent. Similarly, 3,5-dimethyl-4-nitropyrazole, which does not show the association phenomenon, gave two methyl signals ($\tau = 7.44$ and $\tau = 7.58$) at -98 °C.

Comparison of the spectra of (30) and (31) with that of 3(5)-methyl-4-phenylpyrazole showed that in deuteriochloroform the tautomer (32) predominated[98,233]. The use of 1,3-dimethyl-4-phenyl- and 1,5-dimethyl-4-phenyl-pyrazole as models led to the same conclusion, and study of 1,3- and 1,5-dimethylpyrazole showed 3(5)-methylpyrazole to be dominantly the 5-methyl tautomer[98]. This work has been criticised on the basis of a study of the spectra of *N*-substituted pyrazoles and imidazoles, it being suggested that the effects of *N*-substitution may not have been allowed for correctly[114j]. For 1,3,5-tri-substituted pyrazoles an additive relationship can be used to express the chemical shift of

H-4. Several pyrazoles unsubstituted on nitrogen give H-4 shifts intermediate between those calculated for the two possible tautomeric forms.
In other cases, notably those of 1*H*-3-phenyl-, 1*H*-5-methyl-3-phenyl-,
and 1*H*-5-methoxycarbonyl-3-phenylpyrazole, the H-4 shifts suggested the
named tautomers to be present; these conclusions[215] agree with those
of v. Auwers (§2.1).

The spectra of tetrazoles show systematic shifts of the C-H signal,
depending on the orientation of substituents at N-1 or N-2. Comparison
with the spectrum of tetrazole itself, in dimethylformamide, supports
the view, based on dipole moments (§2.1), that tetrazole is mainly 1*H*-
1,2,3,4-tetrazole[231]. The chemical shifts of the *C*-methyl protons in
1,5-dimethyl-1,2,3,4- and 2,5-dimethyl-1,2,3,4-tetrazole are too similar to
help in deciding what is the dominant tautomer in *C*-methyltetrazole.
The temperature and concentration dependence of the N-H resonance of
the latter compound in liquid sulphur dioxide suggest that it exists as a
hydrogen bonded dimer[95].

The effect of protonation is illustrated by various results collected in
table A.3. Protonation of pyrazole leaves H-3 and H-5 equivalent, and
the spectra are useful for orientation purposes[59k]. The protonation of
pyrazole and of imidazole,[150o,219] and the quaternisation of 1-methyl-
imidazole produce symmetrical cations[232]. Generally, protonation of
N-methyl-imidazoles and -1,2,4-triazoles gives the stabilised amidinium-
like ions[216].

That tautomerism in imidazoles is due to prototropy, and not to
mesohydric tautomerism or ion-pair formation (see p. 10) is shown by
the following work. If imidazole is dissolved in concentrated sulphuric
acid, proton loss from nitrogen is slowed down and spin-spin coupling
involving protons on nitrogen becomes observable and, as already mentioned, the imidazolium is seen to be symmetrical[150o]. When 2-deutero-
4(5)-bromo-imidazole is dissolved in D_2SO_4, and 1,2-dideutero-4(5)-
bromo-imidazole in H_2SO_4, the spectra reveal the situation shown here
[150n,219]. Thus, in these imidazoles the 1,4-structures are preferred.
The same is true for the 2-methyl homologues of these compounds.

Table A.3 illustrates the effects of various substituents on azole p.m.r.

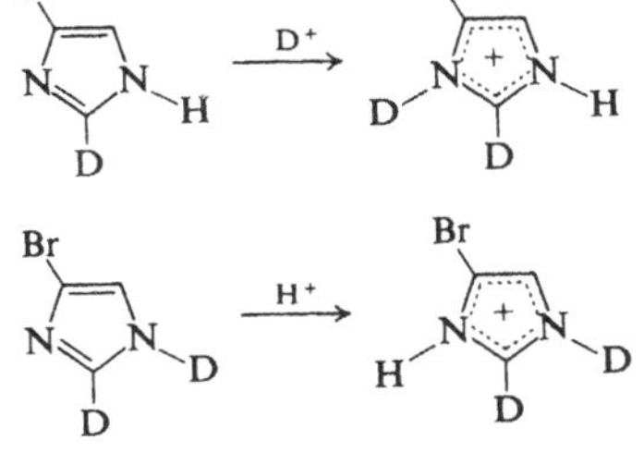

spectra. The phenylpyrazoles fall into two groups: one group of compounds give phenyl signals which are essentially singlets, the other give signals which are multiplets. The first situation arises if there is another substituent in the pyrazole ring adjacent to the phenyl group[114k], or if the phenyl group contains an *ortho*-substituent[217], that is if steric hindrance to coplanarity is present. The production of multiplicity has been attributed to deshielding caused by the diamagnetic anisotropy of the pyrazole ring[121,129,215] and additionally to the resonance effect[119].

With the assumption that, because of steric hindrance, acetylation of 3-t-butylpyrazole must give 1-acetyl-1*H*-3-t-butylpyrazole, and also by studying a *N*-acylpyrazole, 4,4-dimethyl-4,5,6,7-tetrahydro-7-ketopyrazolo [2,3-*a*]pyridine, of unambiguous structure, coupling constants have been derived ($J_{3,5} \sim$ 0.6-0.7; $J_{3,4} \sim$ 1.6; $J_{4,5} \sim$ 2.9 Hz) which should assist structure determination of *N*-acetyl derivatives of unsymmetrical pyrazoles[218].

P.m.r. spectra can be used for orienting substituents in imidazoles; the method has been used for halogen-, arylazo-[234], and methyl-nitro-imidazoles[213]. Similar remarks apply to *N*-acetyl-[190f], *N*-alkyl-[235], and *N*-aryl-1,2,3-triazoles[114f], and *N*-acetyl-1,2,4-triazoles[223a].

P.m.r. spectroscopy has been used to a considerable degree in the study of ring-substituent tautomerism in the azoles. As already mentioned, in conjunction with other techniques it has contributed to our knowledge of the tautomerism of aminotetrazoles[157], pyrazolones[114g,130a,b, 197a], arylazopyrazolones[114c,132a,202,236-8], 1-aryl-3-hydroxypyrazoles[131], and 1,2,4-triazolinthiones[211].

In the case of 2-methylimidazole and a number of other aromatic compounds consideration of the ring current and its determination from chemical shifts led to the conclusion that the ring current is a sensitive test for the detection of aromatic character (Schofield, p.145) but is not related to resonance energy or reactivity[220].

^{13}C-N.m.r. spectroscopy is beginning to be applied to the azoles. Chemical shifts, and in some cases coupling constants, have been reported for pyrazole[239-242], imidazole[239-241], their anions and cations[239], 1,2,3-triazole[240], 1,2,4-triazole[240,241], tetrazole[240], 1-methyl-[242], 1-phenyl-[242], 3-methyl-[240], and 3-methyl-1-phenyl-pyrazole [242], and 4-methylimidazole and its cation[243]. Natural abundance ^{13}C-n.m.r. spectroscopy has also been applied to the problem of tautomerism in the pyrazolones[244].

2.3   Acidity constants. Complexes with metal ions and other substances

2.3.1   *Acidity constants*

Pyrrole is a very weak base (pK_a = -3.80; Schofield, p.60). Protonation, which occurs on carbon, interferes with the aromatic structure. Pyrrole is acidic, though weakly (Schofield, p. 61) (pK_a = 17.51)[115], because acceptance by the ring of electrons from the nitrogen atom weakens the N-H bond, and ionisation produces a resonance-stabilised anion.

The introduction of one pyridinic nitrogen atom into the pyrrole ring should produce bases stronger than pyrrole because the new nitrogen atom possesses a lone-pair of electrons which can bind a proton without detriment to the aromatic structure. At the same time, the introduction of a pyridinic nitrogen atom should increase the acidity of the N-H link by making the ring better able to accept electronic charge from the pyrrolic nitrogen atom. More pyridinic nitrogen atoms can only augment this last effect, but cannot further increase basic strength because they will be in competition for the available electronic charge and thus will be mutually base-weakening.

Examination of table A.4 shows that pyrazole, imidazole, and the two triazoles are all markedly stronger bases than pyrrole. Imidazole is in fact a surprisingly strong base, presumably because of the symmetry of its mesomeric cation (33). This is additionally stabilised by comparison with the free base in the same way that the cation of 4-aminopyridine is stabilised with respect to pyridine (Schofield, p.153)[149a]. It has already been noted that n.m.r. evidence shows protonation of *N*-methyl-1,2,4-triazoles to give the stabilised amidinium-like ions (see p. 21), another illustration of the same phenomenon.

(33)

That pyrazole should be so much weaker than imidazole is surprising, but the fact illustrates a common characteristic of bases in which two nitrogen atoms stand adjacent. A suggestion[149a] that pyrazole is relatively weak because of the occurrence of the free base in the hydrogen bonded dimeric form (see above) is unlikely because 1-methylpyrazole, in which such bonding is impossible, is even weaker.

The second pyridinic nitrogen atom in the triazoles causes a decrease in basic strength, as compared with that of pyrazole. The effect is more marked in 1,2,3- than in 1,2,4-triazole. This is not surprising for in the former the base-weakening effect of the second pyridinic nitrogen atom is felt equally by both of the other nitrogen atoms. This is not so in 1,2,4-triazole.

The large differences in basicities amongst the azoles are due to large differences in ΔH^O[248a].

Imidazole forms well defined salts[84a,274c]. Those of pyrazole[2] and the other weak bases[7b,80b,275], are frequently hygroscopic.

1,2,3-Triazole hydrochloride[275] and 1-methyltetrazole hydrochloride [80b] were prepared in anhydrous solution. A curious feature of the behaviour of 1,2,4-triazole nitrate is that with sodium hydroxide or ethoxide it does not give triazole but triazole-sodium nitrate $C_2H_3N_3$-$NaNO_3$[7b]. Whilst 4-bromo-1-methyl-1H-1,2,3-triazole and 5-bromo-1-methyl-1H-1,2,3-triazole gave hydrochlorides, 4-bromo-2-methyl-2H-1,2,3-triazole did not; generally salts were obtained where an imidazolium-like cation could arise, but not where the cation would be pyrazolium-like[276b].

C-Alkyl groups are generally base-strengthening, but this is not always so with N-alkyl groups (see below). It is interesting that whilst 1-methyl-1H-1,2,3-triazole is slightly more basic than 1,2,3-triazole, 2-methyl-2H-1,2,3-triazole is weaker. This also is presumably due to the ability of 1-methyl-1H-1,2,3-triazole and 1,2,3-triazole to form imidazolium-like cations, whilst 2-methyl-2H-1,2,3-triazole must give a pyrazolium-like cation.

Early figures (table A.4) for some N-alkyltetrazoles, derived from the effect of the hydrochlorides of these compounds upon ester hydrolysis, indicate basic strengths of about the same magnitude as that of aniline (pK_a = 4.62 at 25 °C)[80f,277], and need to be re-examined.

Qualitatively substituents other than alkyl groups can be seen (table A.4) to have the expected effects upon basic strengths. The effect of methyl groups in imidazoles was early noticed to be substantially additive[256], and this is more widely true in this series[140]. This fact makes determinations of pK_a useful in orientation; thus with diazomethane 2,4-dinitroimidazole gives a product (pK_a = –7.47) (see p. 80) which from consideration of the pK_a values predicted for the two possible isomers must be 1-methyl-2,4-dinitroimidazole[140]. The corresponding situation in the pyrazole series has been extensively studied [114l]. Except where effects regarded as steric in origin were present substituent effects could be expressed by a Hammett relationship using σ_m (see p. 26); corrected σ-values were obtained by reference to a value of ρ = 6.6. Correlation for N-alkylpyrazoles was not quite so good. In the pyrazole series, N-alkyl groups are base-weakening (ΔpK_a = 0.45 ± 0.15), and if the statistical factor mentioned below is taken into account, the effect is seen to be larger than the pK_a values suggest[114l].

Nitro-groups are noteworthy for their very marked base-weakening influence. 1-Methyl-4-nitroimidazole is a weaker base than 4(5)-nitroimidazole. Because the imidazolium ion contains two equivalent ionisable protons, this factor of itself would produce a dissociation constant twice that of the 1-methylimidazolium cation, resulting in a pK_a difference of 0.3 units. It has been argued[141] that, in fact, imidazole and 1-methyl-imidazole have almost equal strengths, so that the methyl group is base-weakening to about that extent. In solutions of 4(5)-nitroimidazole the structure (34), X = NO$_2$, Y = H, predominates (see p. 27) and in the cation the two protons are no longer equivalent, so that in comparing (34), X = NO$_2$, Y = H, with 1-methyl-4-nitroimidazole the statistical factor is not applicable, and the base-weakening effect of the 1-methyl group is evident. The decrease in basicity is presumably due to decreased solvation. The much greater basicity of 1-methyl-5-nitroimidazole than of 4-nitroimidazole is noteworthy. It arises from the fact that the basicity of the former compound is due to the nitrogen atom remote from the nitro-group, whilst with 4-nitroimidazole, since the dominant tautomer is (34), X = NO$_2$, Y = H, (see p. 27), the basicity is that of the nitrogen atom nearer to the nitro-group[141].

The use of σ_m and σ_0 values of -0.34 and 0.42, respectively, for the -NH- group, in conjunction with the Hammett relationship for pyridines, permits the prediction with reasonable accuracy of pK_a values for imidazoles and pyrazoles[278]. That the method fails badly with 5-nitro-imidazole is perhaps not surprising in view of what has just been said about the tautomerism of this compound. Extension of the method to triazoles and tetrazoles was not very successful. Correlation of pK_a values for 2-substituted imidazoles by the Hammett equation is best when σ_m constants are used[279a]. With 1-substituted imidazoles correlation is best with σ_I, though not significantly better than with σ_m, whilst for 4(5)-substituted imidazoles σ_m is best and σ_I good[280] (see also ref. 266).

The Hammett treatment of 1,2,4-triazoles as bases is not so satisfactory as when they act as acids (see below)[268a,270].

The protonation of azoles has been examined with respect to various acidity functions (table A.4). Some alkylpyrazoles behave like Hammett bases[247], as do some pyrazolones despite their being protonated on oxygen[114g]. The protonation of some antipyrines is best correlated with H_A, of others with H_0 [253].

The introduction of a nitrogen atom into pyrrole to give pyrazole or imidazole increases the acidity by 3.30 and 3.34 units, respectively[115]. There is a linear correlation between the acidities of the azoles and the number of nuclear nitrogen atoms[248a] (though extrapolation of the correlation to the case of pyrrole gave pK_a ~ 19, a value substantially different from that later measured[115]). The correlation arises from the roughly constant value of ΔS^0 = -50.2 J K^{-1} mol^{-1} for imidazole, the triazoles, and tetrazole, and the accompanying roughly linear variation of ΔH^0. The two pK_a values of a series of imidazoles and benzimidazoles are linearly related[255b,281b] by the equation

$$\text{p}K_a \text{ (as base) = 0.94 p}K_a \text{ (as acid) + 7.43}$$

As can be seen from table A.4, 1,2,3- and 1,2,4-triazole are about as acidic as β-keto-esters or 1,3-diketones, or hypoiodous acid (pK_a = 10).

Acidity is an outstanding feature of the tetrazoles, the parent being as strong an acid as acetic acid, and 5-halogenotetrazoles as strong as chloracetic acid. The dihalogeno-1,2,4-triazoles are as strongly acidic as tetrazole[148a], being surpassed only by 3,5-bis-perfluoroalkyl-1,2,4-triazoles[271].

Electron-attracting substituents increase the acid strength and 2,4,5-tribromo-imidazole is soluble in sodium carbonate solution[133]. Pyrazole (and 1,2,3-triazole) aldehydes form salts[77a,c], and nitro-groups are markedly acid-strengthening.

The use of σ_m constants gives a good Hammett relationship ($\rho = 6$) with the values of pK_a for the 1,2,4-triazoles acting as acids, and also for 5-hydroxy-1,2,4-triazoles ($\rho = 7.62$)[270]. The same substituent constants correlate well the pK_a values of 5-substituted tetrazoles acting as acids[279a].

Azoles form many characteristic salts with metals. Ammoniacal silver solutions precipitate the silver salts of pyrazole[2], imidazole[6d], 1,2,3-triazole[275], and tetrazole[17c,18g]. (Regarding the supposed chloromercuri-salts of azoles see §3.8.)

1,2,4-Triazole gives a blue cupric salt[7b,22h,24b]. Like that of pyrrole, the potassium salt of pyrazole must be prepared in anhydrous circumstances, preferably using potassium in benzene[52a]. The sodium salt of 1,2,4-triazole is precipitated by adding ether to the alcoholic solution of the base and sodium ethoxide[282]. In contrast, the sodium and barium salts of tetrazole can, of course, be prepared in water[25c]. Imidazole and 1,2,4-triazole do not form ammonium salts by evaporation of their solutions in liquid ammonia, but sodium, magnesium, calcium and other salts can be prepared in this way. Tetrazole with gaseous ammonia forms its ammonium salt[283]. Pyrazole[56] and imidazole [284] react with ethyl magnesium bromide in ether to give ethane and solutions of their Grignard reagents. Derivatives of imidazole and 1,2,4-triazole with thallium(I) and thallium(III) can be prepared from the bases and thallium(I) hydroxide or dimethylthallium(III) hydroxide[301]. 5-Aminotetrazole readily gives salts[25].

Measurements of ionisation constants of azoles and their derivatives have provided useful information about tautomerism in the series. The general principles relating to the use of 'fixed' models have been outlined in Schofield (p.152). Some tautomeric equilibrium constants so obtained for the forms (34) and (35) are given in table 6. The relative basic strengths of 4(5)-phenylimidazole and 1-methyl-4-phenylimidazole show that in the former compound the tautomer 4-phenylimidazole (34), X = Ph, Y = H, predominates[141]. This and the examples tabulated suggest that the relative stabilities of the tautomeric forms of an imidazole carrying a 4(5)-substituent capable of conjugating with the nucleus are but little influenced by the consequent differences in

(34)   (35)

Table 6. Tautomeric equilibrium constants

Y	X	$K_T = [(34)]/[(35)]$	
H	NO_2	400^a	500^b
H	Cl	447^b	
Cl	NO_2	186^b	

(a) Ref. 141.   (b) Ref. 140.

π-electronic energy. With electron-attracting substituents the greater base-weakening effect upon the nearer nitrogen atom leaves the remote nitrogen more strongly basic (in the conjugate base), and it is for this reason the preferred seat of the proton[141].

The Hammett correlation of imidazole pK_a values with σ_m constants (see p. 25) has been used to discuss imidazole tautomerism, and leads to the same conclusion; namely, that in 4(5)-substituted imidazoles an electron-attracting substituent causes the 4-substituted form to predominate, whilst an electron-releasing substituent causes the 5-substituted form to predominate[279a].

A study has been made[279b] of the ionisation constants (as acids) of a series of 5-substituted tetrazoles in terms of the extended Hammett equation, using suitable model systems (1-methyl-4-substituted and 1-methyl-5-substituted imidazolium ions, 2-substituted imidazolium ions, and substituted imidazoles) to permit the calculation of constants relating to the two possible nuclear-tautomeric forms. It led to the conclusion that, regardless of the nature of the 5-substituent, the dominant nuclear tautomer in aqueous solution was that with the imino-hydrogen atom at N-2. This conclusion is the reverse of that reached for tetrazole itself (in organic solvents) from dipole moments (§2.1) and n.m.r. spectroscopy (§2.2.3).

Some use has been made of acidity constants in the study of tautomerism in azoles involving both the nucleus and a substituent.

An estimate of the tautomeric equilibrium between 1-substituted pyrazol-2*H*-5-ones of the type (24) and 1-substituted 5-hydroxypyrazoles of the type (26) (see table 2) can be made because they have a common cation[130a]. In both the 1-methyl and the 1-phenyl series the data suggest that the NH-forms predominate. Comparison of the results of using this method of studying tautomerism with those obtained from ultraviolet spectroscopy shows the former not to work for 1,3-dimethyl-

and 1,3,4-trimethyl-pyrazol-5-one, and for 1,5-dimethyl- and 1,4,5-trimethyl-pyrazol-3-one. Evidently replacing hydrogen by methyl in the model compounds produces effects which are not negligible[114g].

1-Phenylurazole is predominantly present in aqueous solution in the di-carbonyl form with about 10% of the form enolised at C-3 (see table A.4)[161].

Acidity constants (table A.4) show 1-methyl-5-methylamino-1*H*-tetrazole to be present in aqueous solution overwhelmingly in that form (pK_T = 9.02)[157].

2.3.2 Complexes with metal ions and other substances

The co-ordinating ability conferred upon azoles by the presence in their structures of pyridinic nitrogen atoms has been extensively studied.

Pyrazole and its *C*-methyl derivatives give tetragonal complexes NiX_2B_2 (where B is a molecule of the pyrazole base), though the complex formed from nickel bromide and 3,5-dimethylpyrazole is tetrahedral[285]. Some hexakis(pyrazole)nickel(II) complexes are known [286b,287a]. Complexes of 3(5)-methylpyrazole with salts of a number of bivalent metals seem to involve co-ordination to the pyridinic nitrogen atom of the form 5-methylpyrazole[287b,c]. Metal complexes of 4-aryl-azopyrazol-3- and -5-ones have attracted attention because of their interest as dye-stuffs[288].

The pyridinic nitrogen atom makes imidazole able to co-ordinate with ions such as Zn^{2+}, Ni^{2+}, Cu^{2+} and Cd^{2+}, a property which has been much studied because of its biological significance. As would be expected, co-ordinating ability is also found in 1-methylimidazole[254,273,289-91].

Co-ordination compounds of 1,2,4-triazole and 1-aryl-1,2,4-triazoles are also known[302]. Tetrazole, 5-aryltetrazoles and 5-aminotetrazole also form complexes with divalent cations, but in these anions of the tetrazoles are involved[169,286a,292,293].

The effect upon the acidity constants of complexing with metal ions has been examined[258,294-5].

1,5-Pentamethylenetetrazole (metrazole) forms complexes with many metal ions[296a,297a,b].

With iodine[297b,c] and 'π-acids' such as tetracyanoethylene and trinitrobenzene[298] metrazole gives charge-transfer complexes, whilst with iodide, bromide or chloride it complexes at nitrogen[296b,299].

Sodium chloride and related salts dissolve appreciably in molten 1,2,4-triazole and are readily recovered from the solutions. Cuprous chloride and transition metal halides react to form compounds of unknown structure[300].

2.4   Theoretical studies

A number of MO-theoretical studies have been carried out of some of
the ground-state properties, electronic and n.m.r. spectroscopic properties,
and basicities discussed in previous sections of this chapter. The HMO
method (Schofield, p.22) used in early studies[303-5] was used mainly
in connection with dipole moments. The method is not now of much
interest, but was applied fairly recently using parameters based on ultraviolet spectroscopy[59*l*]. Another early study used the VESCF method
to calculate dipole moments[306].

More advanced MO methods have subsequently been applied to
calculate ground-state and spectroscopic properties. Heats of combustion
of several azoles have been successfully calculated[307b]. Applied to
azoles the EHT method suggests the σ-framework to be markedly
polarised, and the π-framework to be polarised independently; the
method gives dipole moments which are much too large[308a]. The
CNDO/2 method gives dipole moments in excellent agreement with
experiment for imidazole, 1,2,3-triazole, and tetrazole, but for pyrazole
the agreement is not so good[309,310]. The method has been applied
to alkylpyrazoles[309].

Several workers have examined the electronic spectra of the azoles
using calculations of the Pariser–Parr–Pople type[311-13], and even more
interest has centred on n.m.r. spectroscopy. Using EHT calculations a
good correlation was found between total electron densities at carbon
atoms and proton chemical shifts and ^{13}C-chemical shifts[239,240,
308a,b] but the method did not perform well for spin–spin coupling
constants[240]. Other methods which have been applied to related azines
were not used with azoles for lack of results[310].

MO-theoretical treatments of the acidity constants of azoles have also
been examined[309,313]. π- and total-electron densities, but not lone-
pair character, on N-2 in pyrazole and alkylpyrazoles correlated fairly well
with basicities[309]. The Pariser–Parr–Pople method has been used to
estimate the energy change in the π-bonds, ΔE_π, between the base and
its cation, and this quantity was found to correlate well with pK_a values
for pyrazole, imidazole, and their methyl derivatives (the methyl group
being treated as exerting a purely inductive effect); the correlation did
not extend to the triazoles, unless the two-centre Coulomb core integrals
were drastically altered[314b]. Accordingly the validity of ΔE_π as a
measure of pK_a was regarded as doubtful, and the CNDO/2 method was
used to calculate ΔE_{total} ($\Delta E_{total} = \Delta E_\pi + \Delta E_\sigma$) and ΔE_T ($\Delta E_T =
\Delta E_e + \Delta E_R$; ΔE_R being the change in internuclear replusion). ΔE_T
correlated the pK_as of pyrazole and imidazole, and of the triazoles, but
none of the quantities correlated all of these acidity constants taken
together (perhaps because of entropy changes). ΔE_e did not indicate

unequivocally the relative pK_as of the triazoles, and q_π^N failed to predict the correct order for imidazole and pyrazole. ΔE_T predicted 1*H*-1,2,4- and 4*H*-1,2,4-triazole to be equally stable, and 1*H*-1,2,3- to be more stable than 2*H*-1,2,3-triazole, and also that the most stable triazole cations were those in which the protons on nitrogen were most separated. ΔE_T was a more reliable index of basicity than ΔE or ΔE_π[314c].

Tautomerism in the azoles has also been studied by other workers using MO methods. They suggest that 1*H*-1,2,3,4-tetrazole is the preferred tautomer[310]. Calculations of heats of combustion in the pyrazole, imidazole, triazole and tetrazole series suggest that in substituent–nucleus tautomerism the tautomer with the mobile proton on nitrogen should be more stable than that with it on carbon, and that the amino-forms of amines and the carbonyl forms of hydroxy-compounds will be preferred [315]. As regards the former point theoretical studies of pyrazolone tautomerism are more in line with experiment; their main conclusion was that in the vapour phase or in an inert solvent the CH-form would be the most stable, followed by the NH-form. Change to more polar solvents would favour the NH-form (see table 2)[316].

3 Behaviour in substitution reactions: electrophilic substitution

Pyrrole is extremely reactive towards electrophilic reagents. Entry into its nucleus of one or more pyridinic nitrogen atoms must reduce the availability of electrons and with it the ease of electrophilic substitution. Azoles might be expected, then, to behave like pyrroles containing electron-attracting substituents, and the deactivation will increase with the increasing number of nuclear nitrogen atoms. This situation will be seen in the relatively small number of electrophilic substitutions which have been effected with azoles as compared with pyrroles, and in the more severe conditions which are usually necessary in these reactions.

By the same token, nucleophilic substitutions should be of some importance in azole chemistry, though they are of no consequence in the pyrrole series.

3.1 Acylation and carboxylation

Pyrazole and its 1-acyl derivatives are not formylated by the Vilsmeier method, but a number of 1-methyl- and 1-phenyl-pyrazoles give the 4-aldehydes[59d,f,317,318k]. The reaction works with 5-amino-1-phenyl-pyrazoles. Intermediates of the type (36) have been isolated; they are rapidly converted into the aldehydes by water[59g].

Whilst 1-unsubstituted pyrazoles react at nitrogen with acyl chlorides (see below), both 1-aryl- and 1-alkyl-pyrazoles undergo 4-acylation when heated with acetyl or benzoyl chloride[60b,100e,126b,d,319].

3,5-Dimethylpyrazole is not acetylated by the Friedel–Crafts method [320a], but 1-arylpyrazoles give the 4-aroyl derivatives[62a,126d,320a, 321l], sometimes in poor yield[59i]. Despite some failures with 1-alkyl-pyrazoles[59i,62a,321l], 1,3,5-trimethylpyrazole gave 40% of the 4-acetyl derivative by Friedel–Crafts acetylation in carbon tetrachloride[126d], and 1,3-dialkyl-5-chloropyrazoles have been successfully aroylated in s-tetrachloroethane[322].

1-Phenylpyrazole did not react with acetyl or benzoyl chloride in the presence of aluminium chloride, but 3,5-dimethyl-1-phenylpyrazole gave the 4-acetyl derivative[59i,126d]. Attempts to benzoylate 1-phenylpyrazole in a Friedel–Crafts reaction carried out in carbon tetrachloride gave only a very small yield of di-(4-*N*-pyrazolylphenyl)ketone, arising from reaction of the solvent with the phenyl groups[59i]. Both 1-alkyl- and 1-aryl-pyrazoles are 4-acylated by solutions of acetic anhydride or acyl chlorides in sulphuric acid[323a].

The simplest method reported for *C*-aroylation of imidazoles is exemplified by the reaction of 1-methylimidazole with benzoyl chloride in the presence of triethylamine in acetonitrile; 2-benzoyl-1-methylimidazole is formed in good yield[324p].

4-Methyl-2-phenylimidazole does not react with *N*-methylformanilide and phosphoryl chloride[325]. The failure of Friedel–Crafts acylation with *N*-unsubstituted imidazoles[320a,c], as with pyrazoles, is not surprising; markedly basic compounds such as these must be deactivated by the presence of Lewis acids. Imidazoles can be *N*-acylated in the right conditions, but other reactions also occur (see §§3.13.5 and 6.4).

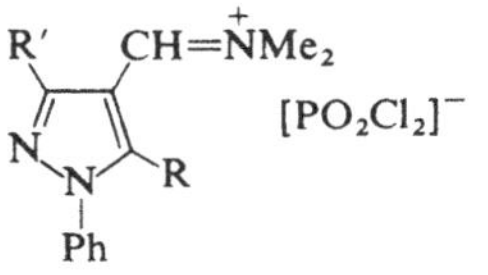

(36) 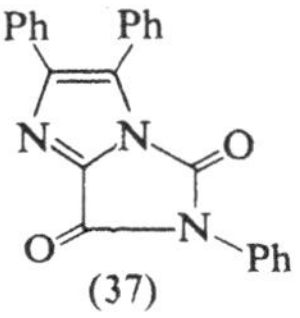   (37)

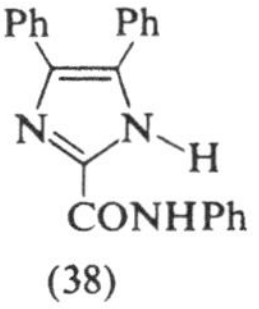

(38) 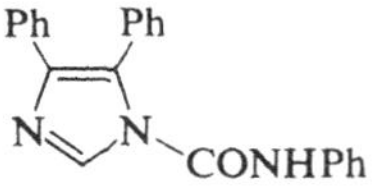   (39)

Some imidazoles and pyrazoles are *N*-acylated by aryl isocyanates (§3.12), but with 4,5-diphenylimidazole reaction at both nitrogen and carbon occurs. Thus, phenyl isocyanate at 80 °C gives (37) and (38), in boiling nitrobenzene (38) only, and at room temperature (39). 2-Methyl-4-phenylimidazole gives 2-methyl-4-phenylimidazole-5-carboxylic acid anilide[208c].

1,3-Diphenylimidazolium perchlorate is converted by potassium t-butoxide into an ylide which reacts at C-2 with phenyl isothiocyanate [326].

An example of the rare process of electrophilic substitution into a 1,2,4-triazole is the formation of 3-aroyl- and 3,5-di-aroyl-4-phenyl-4*H*-1,2,4-triazole when 4-phenyl-4*H*-1,2,4-triazole is heated with aroyl chlorides[327].

'Hydroxypyrazoles' are more favourably situated with regard to electrophilic substitution. A hydroxyl group would be highly activating, and with decreased basic strength there is less tendency to deactivation by salt formation. However, the tautomerism of the hydroxyazoles does confuse our judgement of the true character of substitution reactions in these compounds, and whilst the older literature refers to reactions of the enolic forms, or of the methylene group in the tautomeric pyrazolone structures, nothing is known of the mechanisms of these reactions.

With acyl chlorides and alkali, 3-methylpyrazol-5-one and its 1-phenyl derivative undergo *O,N*-di- or *N*-mono-acylation[62a,321*l*], exemplifying a pattern common in this series (and with 4-hydroxypyrazoles[76a,328]), but *C*-acylation at C-4 is also common with *N*-substituted pyrazolones [318g,321h,329e]. 1-Alkylpyrazol-3-ones are *O*-acylated by acyl chlorides [330a,c,e]. Reaction of many pyrazolones at C-4 has been recorded, using acids[329e,331], amides[332a,333b,334,335b c], amidines[335b,c, 336-7], esters[333b,338], and orthoesters[67c,339b]. Most commonly the products are of the type (40), (40) itself being readily obtained from 3-methyl-1-phenylpyrazol-5-one and formic acid[335e,g,340-1], and analogously phthalic anhydride gives a product of the phthalein type[342-3].

As well as compounds of type (40) the reaction with amides can give 4-aminomethylene (or tautomeric) derivatives (:C(NH$_2$)R$\rightleftharpoons$·C(:NH)R at C-4) (see §3.4), which are easily hydrolysed to ketones. Amidines give similar derivatives. 3-Methyl-1-phenylpyrazol-5-one with ethyl isoformani-lide[318e,339e], aromatic isonitriles[344a] and formamidines[344b] gives the Schiff bases of the 4-formyl derivative. Oxalic esters in the presence of ethoxide give the expected pyrazolone-4-glyoxylic esters[333a,338].

The Vilsmeier reaction has been used with *N*-substituted pyrazolones [348]. 3-Methyl-1-phenylpyrazol-5-one reacts at C-4 and the initial Vilsmeier product is hydrolysed by alkali to the aldehyde, but with acid gives (40)[345]. With isocyanates or isothiocyanates 5-methyl-1-phenyl-pyrazol-3-one forms the anilides at C-4[346a].

In antipyrine (41) and its analogues reaction at nitrogen is excluded, and at oxygen can only produce quaternary salts. This appears to happen with benzoyl chloride, giving (42)[1f] but nuclear acylation at C-4 occurs readily. Thus, dimethylformamide and phosphoryl chloride give the aldehyde[347] and other acylations occur with or without a Friedel-Crafts catalyst[324d,348,349,350c,351]. '3,5-Dihydroxypyrazoles' (pyrazolidinediones) are acylated on nitrogen and at C-4[206,321a,i,352]. 352].

(40)   (41)   (42)

Imidazol-2-ones are easily acetylated and the mono- and di-acetyl derivatives have been regarded, without proof, as *N*-acetylated compounds. However, under Friedel–Crafts conditions *C*-acylation occurs[353-4].

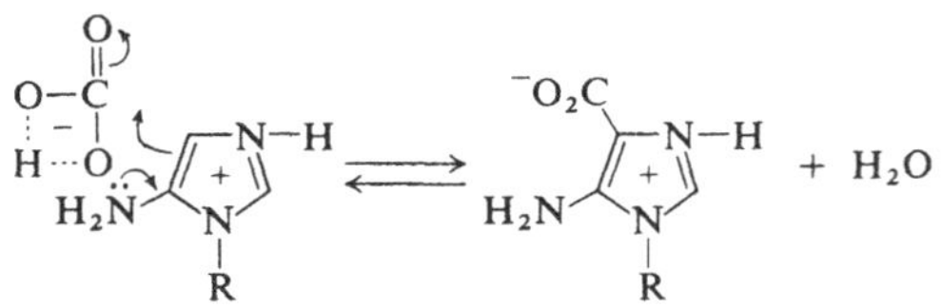

No work seems to have been reported on the carboxylation of azoles, apart from the interesting results for 5-amino-imidazole and some of its 1-substituted derivatives. In aqueous solution this compound and 5-amino-imidazole-4-carboxylic acid are in equilibrium; in acid decarboxylation occurs, and in presence of an excess of bicarbonate carboxylation occurs. Potassium carbonate and carbon dioxide are ineffective, and the two reactions are regarded as involving the system illustrated. 5-Amino-pyrazoles and (less successfully) 5-amino-1,2,3-triazoles can be carboxy-lated similarly[355].

3.2 Alkylation and substituted alkylation

C-Alkylations are uncommon among the reactions of the azoles. Only recently was the 4-alkylation of 1-substituted pyrazoles by the Friedel–Crafts method described[323a].

Pyrazole, 3,5-di-, and 3,4,5-tri-methylpyrazole react with methyl acetylenedicarboxylate at a nitrogen atom, as do some imidazoles (p. 107). 4-Methylimidazole was mistakenly supposed to give (43); in fact reaction occurs at nitrogen (p. 109).

N-Unsubstituted azoles and their salts commonly undergo *N*-alkylation, and alkyl migration from nitrogen to carbon is known (pp. 204, 244).

Whilst imidazol-2-ones are alkylated on the nitrogen atom (p. 91), pyrazolones give *O*-, *N*-, and *C*-alkyl derivatives according to circumstances. Unambiguous *C*-alkylation is seen in the formation of 3-methyl-

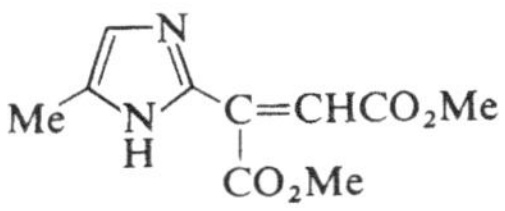

(43)

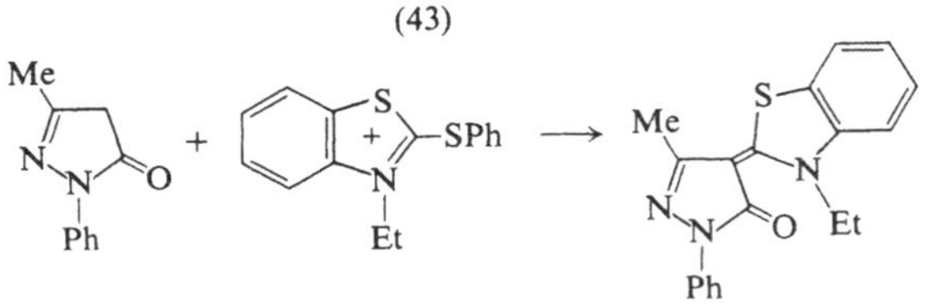

(44)

1-phenyl-4-triarylmethylpyrazol-5-ones from triarylcarbinols and 3-methyl-1-phenylpyrazol-5-one [357a,b]. Some reactions resulting in *C*-alkylations of antipyrine may involve initial reactions at nitrogen, followed by rearrangement (p. 101).

1,2-Diaryl- or -dialkyl-pyrazol-3,5-diones are readily mono- or dialkylated at C-4 by alkyl halides in presence of sodium hydroxide or alkoxides (see p. 89) [339f,358-9].

Aminomethylations of pyrazoles by the Mannich reaction are mentioned in §3.4.

3.3 Arylation

Electrophilic arylation of unactivated azoles has not been described. However, pyrazolones are reactive enough to be the source of useful merocyanines such as (44) in reactions of the kind illustrated [318d,f, 339a]. 4-Chloroquinaldine reacts similarly with pyrazolones [360]. Various reactions closely related to these are discussed in §3.4.

1-Benzoyloxyquinolinium and 2-benzoyloxyisoquinolinium chlorides (from benzoyl chloride and the corresponding *N*-oxides) have been used to introduce 2-quinolyl and 1-isoquinolyl at C-4 in 3-methyl-1-phenyl-pyrazol-5-one [361] and in antipyrine [361-2]. Toluene-*p*-sulphonyl chloride cannot be used in place of benzoyl chloride, but 1-methoxyquinolinium methosulphate takes part in the reaction [362].

3.4 Carbonyl reactions

In contrast to pyrrole, *N*-unsubstituted pyrazoles give *N*-hydroxymethyl compounds with formalin [74a,363b,364-5]. *C*-Hydroxymethylation occurs when 1,3,5-trimethyl-[77a,363,366] and 1-hydroxymethyl-3,5-dimethyl-pyrazole [364] react with paraformaldehyde and hydrochloric acid or hydrochloric acid with piperidine. As well as the 4-hydroxymethyl derivatives, 1,3,5-trimethylpyrazole and its analogues give 4,4′-dipyrazolyl-methanes [323h]. 1-Benzylpyrazole is said not to react with formaldehyde [363b]. 3,5-Dimethylpyrazole reacts at nitrogen with formaldehyde but its *N*-cyanoethyl derivative (p. 203) gives the dipyrazolylmethane from which the *N*-substituents can be removed by pyrolysis [126e]. 5-Amino-1-phenylpyrazole with paraformaldehyde in sulphuric acid gives the product analogous to Tröger's base. In formic acid the 4,4′-dipyrazolyl-methane linked by methylene between the amino-group, and a related product, are formed [367].

Imidazoles behave differently; usually those unsubstituted on nitrogen react with formalin at C-4 (or C-5), whilst *N*-substituted imidazoles react at C-2. This generalisation holds for 4-methyl-[368], 4-bromo-[369], 1-methyl-[369-372,375], 5-chloro-1-methyl-, 5-bromo-1-methyl-[370,373b,

374b], 1-benzyl-[363a,372,375], 1-methoxymethyl-[372], 1,5-dimethyl-[369], and 1-aryl-imidazole[376], and other examples[377]. However, 1,4-dimethylimidazole gives the 5-hydroxymethyl derivative[369], 4,5-dibromo-1-methyl-[370], 4-methyl-2-phenyl-[325], and 4-chloro-1-methyl-imidazole[369], do not react with formaldehyde, and imidazole itself gives no crystalline product[369]. 1,2,4-Triazole and its 1-benzyl derivative undergo *C*-hydroxymethylation[363c].

Under some circumstances, reaction of *N*-substituted pyrazoles with formaldehyde and hydrochloric acid effects 4-chloromethylation. Examples of compounds which react in this way are 1-phenylpyrazole[59a], di-(1,5-diphenylpyrazolyl)[59b], and 1,3,5-triphenylpyrazole[323h].

The related Mannich reaction has met with limited success, occurring at nitrogen in *N*-unsubstituted pyrazoles, and in other cases giving only *C*-hydroxymethyl compounds[74a]. No Mannich reaction is observed with 2-ethyl- and 2-methyl-4,5-diphenyl-imidazole[365]. However, histamine and histidine with formaldehyde and hydrochloric acid give (45)[378]. Some of these results are not entirely in line with those of a more general study[379]; this showed that under the usual acidic conditions of the Mannich reaction *N*-substitution occurred. This *N*-substitution is reversible in base so that in basic conditions *C*-substitution products accumulate. In basic conditions all positions can be substituted; thus, 2-methyl-imidazole gave 1,4,5-tris-, 4,5-bis- and 4-mono-substituted products, 4,5-dimethylimidazole reacted at C-2, and 2,4-dimethylimidazole at C-5. *N*-Substituted imidazoles did not react, and it was concluded that imidazole anions are the reactive entities.

1,2,3-Triazole probably reacted at nitrogen with formaldehyde and *p*-aminobenzoic acid[380], whilst 3-chloro-5-ethyl-1,2,4-triazole does not react even in this way[365].

Not much is known about the reactions of azoles lacking activating substituents with aldehydes other than formaldehyde, or with ketones. With chloral, 4-methylimidazole gives the expected carbinol, reaction occurring at the 5-position[381a,b], but the product from 5-chloro-1-methylimidazole is of uncertain nature[373b,374b]. In contrast, 1-alkyl-imidazoles do not react with chloral, and mixed results have been obtained with other aldehydes, reaction when successful occurring at C-2[372]. Imidazole itself gave tractable products only with 1,4-dicarbonyl compounds, e.g. (46) from hexan-2,5-dione[372].

The reaction in which 1,3,4-trimethylimidazolium cation catalyses the

(45) R = H or CO_2H (46)

benzoin condensation in alkaline solution probably involves reaction of benzaldehyde at C-2 in a zwitterion formed from the quaternary cation (*cf.* §3.5).

As regards the triazoles it seems only to have been reported that *N*-substituted 1,2,4-triazoles give resins with aldehydes[318b,339d]. 1,2,4-Triazole does not give the indophenin reaction[83a].

With diethylamine and trioxymethylene, 3-amino-4-ethoxycarbonyl-imidazole gives the 2-substituted Mannich base[382].

3-Methyl-1-phenylpyrazol-5-one[383e] and antipyrine[383a] are 4-hydroxymethylated by formaldehyde. 3-Methyl-1-phenylpyrazol-5-one [384], 5-methyl-1-phenylpyrazol-3-one[385], and antipyrine[386-7,388b] have been used successfully in the Mannich reaction. Antipyrine gives thiomethyl compounds with formaldehyde and thiols[389]. 1,2-Diphenyl-pyrazol-3,5-dione is readily aminomethylated[206].

Whilst imidazol-2-thione gives only the *N*-hydroxymethyl compound in the Mannich reaction, 4-methylimidazol-2-thione gives the 4-dimethyl-aminomethyl compound[143].

5-Amino-3-methyl-1-phenylpyrazole is said to react with benzaldehyde at C-4, giving (47)[330j], whereas 4-amino-3-methyl-1-phenylpyrazole merely gives Schiff bases[330k]. 4-Aryl-2-phenylhydrazino-imidazoles give (48) with salicaldehyde[390a].

(47) (48) (49)

Numerous examples are known of reactions between pyrazol-3-[330a, c,e] and pyrazol-5-ones and carbonyl compounds other than formaldehyde. Chloral reacts normally, giving the carbinols, with antipyrine[391] and isoantipyrine[335g]. With benzaldehyde, pyrazol-5-one gives (49)[392h] and 3-methyl-1-phenylpyrazol-5-one behaves similarly with aromatic aldehydes[392h,k,393]. Many other aromatic aldehydes have been used [388b,394] and products of the type (51), ·CHAr· in place of ·C(Me$_2$)·, have been obtained[383d,395]. Antipyrine and related compounds give products of the type (50)[383b,392k]. Under the same conditions, 3-antipyrine does not react with aldehydes[330a]. Pyrazolone aldehydes have been condensed with pyrazolones[335a,385].

1-Aryl-3-methylpyrazol-5-ones give with acetone the 4-isopropylidene derivatives or products of the type (51)[392k,396]. The reactions with acetone and acetophenone to give the alkylidene derivatives are favoured by acidic conditions, and to give the compounds of type (51) by alkaline conditions[397].

In obviously close relationship to these various reactions are those of (52) and (53) with 3-methyl-1-phenylpyrazol-5-one to give (54) and (55), respectively[318d,f,339a,c]. Arylhydrazones of 5-formylbarbituric acids react like aldehydes with pyrazolones[335d].

(50)

(51)

(52)

(53)

(54)

(55)

3-Methyl-1-phenylpyrazol-5-one gives, with amides, 4-aminomethylene (or tautomeric) derivatives[335b,c,d,e] as mentioned above (p. 33).

Imidazol-4-ones with a free 5-position readily give benzylidene derivatives, but 2-methylimidazol-4-one reacts also at the methyl group [398]. Since the original description of the reactions using benzaldehyde [399], numerous aldehydes have been condensed with hydantoin and 2-thiohydantoin.

3.5 Deuteration and protiation

A number of deuterated pyrazoles can be prepared by direct substitution because of the different reactivities of the various nuclear positions. The processes are summarised below[400]. The proton on nitrogen is most easily exchanged, followed by that at C-4; those at C-3 and C-5 are only replaced in the presence of base (see below for mechanism).

4-Phenyl-3-trimethylsilyl- and 3,4-di(trimethylsilyl)-pyrazole undergo protodesilylation with acids[190e]. The acidolysis of 4-acetoxymercuri-1-phenylpyrazole occurs readily[59a]. See also the reactions of sulphonic acids (p. 251).

The proton attached to nitrogen in imidazole is rapidly exchanged for deuterium in deuterium oxide, and the proton at C-2 is exchanged at

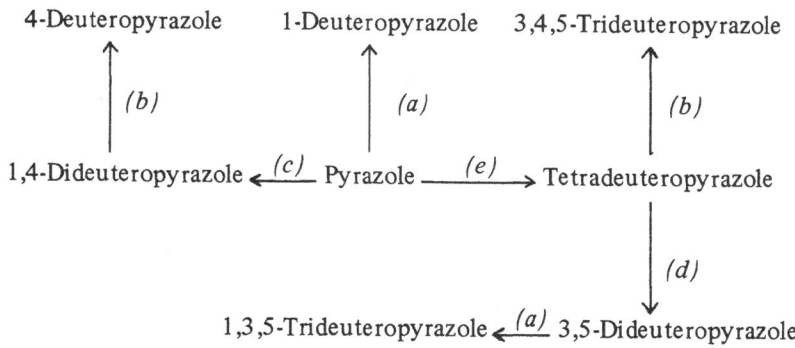

(a) 3 Exchanges for 3 h with D_2O. (b) 1 Exchange for 3 h with H_2O. (c) 2 Exchanges of 12 h in sealed tube at 200 °C with D_2O. (d) 2 Exchanges of 12 h in sealed tube at 200 °C with H_2O. (e) 2 Exchanges of 12 h in sealed tube at 200 °C with 1M NaOD.

room temperature in a reaction with $t_{1/2} \sim 700$ minutes at 37 °C[150p, 219]; the reaction is essentially complete in 2 h at 150 °C, with or without the addition of base[232,401], but exchange at C-2 does not occur in deuterosulphuric acid during a day at 37 °C, nor in deutero-acetic acid during 5 days[150p]. Complete exchange of the protons of imidazole occurs in 4 h at 250 °C in deuterium oxide[401]†.

4-Bromoimidazole in 7.5 h at 100 °C gives the 1,2-dideuterated compound which, crystallised from water, gives 4-bromo-2-deuteroimid-azole[219]. 4-Nitroimidazole when heated for 13.5 h at 100 °C in deuterium oxide is deuterated at C-2 and C-4 in the ratio 5 : 3 in the presence of sodium deuteroxide exchange at these positions is complete in 12 h at room temperature[150p]. Two exchanges in D_2O–NaOD at 100 °C for 4-5 h converts 4-methylimidazole into 2,5-dideutero-4-methylimidazole[402]. At 37 °C exchange in 4-methylimidazole is slow, and in acid exchange occurs at C-5 and not at C-2[150p].

Like imidazole, 1-methylimidazole undergoes 2-deuteration in deuter-ium oxide at room temperature, and in 4 h at 250 °C gives 2,4,5-trideutero-1-methylimidazole. The rate of the reverse reaction at C-2 in the trideutero-compound is essentially independent of pH, but it drops rapidly to zero in acidic solution. A mechanism of 2-substitution involv-ing an ylide derived from the conjugate acid is thought to be involved

† The apparent change in orientation of deuteration with pH observed by Gillespie *et al.*[401] was based on the mistaken conclusion that at 250 °C in deuterium oxide the product was 4,5-dideuteroimidazole. It was, in fact the 2,4,5-trideutero-compound, but the deuterium atom at C-2 was lost through too long contact with water during working up (private communi-cation from Dr J. H. Ridd).

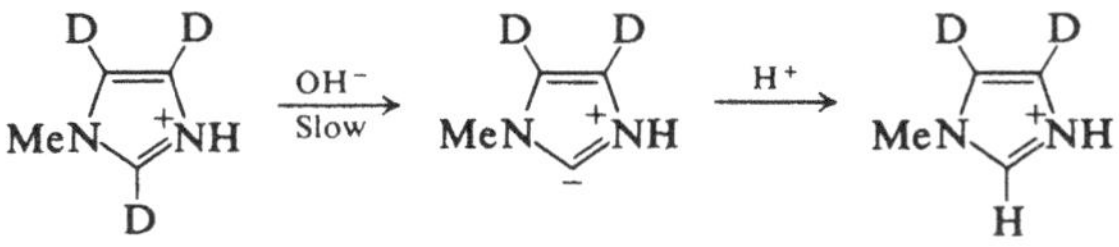

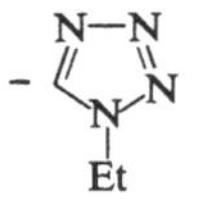

(56)

[403]. The involvement of the conjugate acid was also suggested from experiments with 1-benzylimidazole[150p].

The *N*-methyl-nitro-imidazoles behave surprisingly. With deuterium oxide at 100 °C for 12.5 h 1-methyl-4-nitro-imidazole reacts more at C-5 than at C-2 (exchange being 90% complete at C-5 and 50% complete at C-2), but under the same conditions exchange is complete at C-2 in 1-methyl-5-nitro-imidazole but only 10% complete at C-4[150p].

The kinetics of deuteration of imidazole and pyrazole have recently been studied[404-5]. Rates of 2-deuteration of imidazole at 65 °C and 70 °C, and of 4(5)-deuteration at 180 °C and 190 °C were determined in D_2O at various pD values[404]. Analysis of the rate profile for the dependence of rate upon acidity (coupled with the fact that 2-deutero-1-methylimidazole is protiated readily at C-2 at room temperature whereas imidazole needs a higher temperature, suggesting that the imidazole anion is not the reactive species (see p. 39)[403] led to the conclusion that exchange at C-2 involved two parallel processes; rate determining proton abstraction from the imidazolium cation by D_2O and by OD⁻ to give the ylide at C-2, followed by deuteration there. As regards exchange at C-4(5), the difference in the rate profile requires that in addition to two processes analogous to those mentioned (but with ylide formation occurring at C-4(5)) a third is involved in which proton abstraction from C-4(5) in the imidazole molecule is effected by OD⁻. Pathways for exchange at high pD values involving σ-intermediates, though not excluded by the kinetics, were regarded as less likely because of the behaviour of pyrazole.

That behaviour[405] showed marked differences from the behaviour of imidazole. 4-Deuteration and 3(5)-deuteration were followed in sealed tubes at 200 °C and higher temperatures. 4-Deuteration evidently involved general acid catalysed production of σ-intermediates (Wheland intermediates) from pyrazole itself, and from pyrazole anion. In contrast, the rate of deuteration at C-3(5) was independent of pD and of buffer catalysis; evidently C-3(5)-deuteration depends on ylide formation at C-3(5) from pyrazolium cations by the action of OD⁻.

The authors of these studies of hydrogen isotope exchange in pyrazole and imidazole point out that two types of mechanism are to be discerned in heteroaromatic systems: one involves base-catalysed proton removal from the site of exchange, and the second involves acid-catalysed formation of a Wheland intermediate. The relative reactivities of the various species arising from the heteroaromatic compound are different in the two mechanisms: in the ylide mechanism the sequence is conjugate acid > neutral molecule > conjugate base (unreactive), whilst in the mechanism proceeding through a Wheland intermediate the sequence is conjugate base > neutral molecule > conjugate acid.

More recent studies of H-D-exchange in pyrazoles have been concerned with reactions at lower temperatures and higher acidities than those just described. Qualitatively, the following situation was observed; in D_2SO_4 exchange in 1-methylpyrazole occurred initially at C-4 and then simultaneously at C-3 and C-5, in 1,2-dimethylpyrazolium only at C-4, in 1,3-dimethylpyrazole at C-4 and then at C-5, in 1,5-dimethylpyrazole at C-4 and then at C-3, in 1,2,3-trimethylpyrazolium only at C-4, in 1,4-dimethylpyrazole simultaneously at C-3 and C-5, and in 1,3,4- and 1,4,5-trimethylpyrazole smoothly at the free position. Exchange did not occur in 1,2,4-trimethyl- and 1,2,3,4-tetramethylpyrazolium. Thus, for all the methylpyrazoles studied with unsubstituted 4-positions, exchange occurred there preferentially. Rate profiles (70–90 °C) for 4-exchange revealed a change from reaction via the free base to reaction via the cation at H_0 = -5 to -6, a conclusion confirmed by the behaviour of the quaternary cations. At higher temperatures (140–220 °C) exchange occurred at C-3 and C-5 on the free base, with no mechanistic change-over. The methyl substituents invariably increased the rate of exchange, but their quantitative effect depended on the position at which reaction was occurring, on the relative orientation of the methyl groups, and on the species reacting. There was no significant difference in the effects of methyl groups at C-3 or C-5 upon exchange at C-4[470].

Similar studies have been made with 3,5-dimethylpyrazole and 1,2,3,5-tetramethylpyrazolium. For the former the change from reaction via free base or by cation occurs at D_0 = -4.5, and the free base is about 10^9 times as reactive as the cation[471].

The other azoles have been less thoroughly studied. 1-Ethyltetrazole is 5-deuterated in deuteromethanol with catalysis by methoxide or piperidine. The anion (56) is probably involved[406a].

The base-catalysed deuteration of quaternary azolium ions (table 7), like that of quaternary pyridinium salts (Schofield, p.164), goes through ylides.† The third cation in table 7, 1,4-diethyltetrazolium, undergoes base-catalysed exchange even in 2M DCl at room temperature, and a number of triazolium salts react in the range pH 5–8[406b]. In 1,3-

---

† For examples other than those discussed here see p. 145.

dimethyltetrazolium, in addition to complete exchange at C-5, exchange in the 3-methyl group occurs ($t_{1/2}$ = 130 ± 10 min)[407a]. Comparison of the imidazolium with the pyrazolium cation shows the accelerating effect of a nitrogen atom on each side of the potential carbanion. Their effect is greatly augmented by the additional nitrogen atoms in 1,4-diethyltetrazolium, but in 2,3-diphenyltetrazolium, where the formal positive charge is not adjacent to the carbanion (and where strictly, an ylide cannot be formed), the four nitrogen atoms are much less effective. Deuterium exchange in 1,3-dibenzylimidazolium[150p] and 1,3,4-trimethylimidazolium[243] cations has also been reported. In the second case base-catalysed rates were studied, and compared with those for related thiazolium and oxazolium cations.

Table 7. Base-catalysed deuteration of quaternary azolium ions

Cation	Product	Conditions[a]	$t_{1/2}$/s	Relative rate
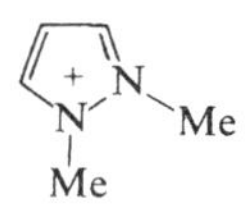	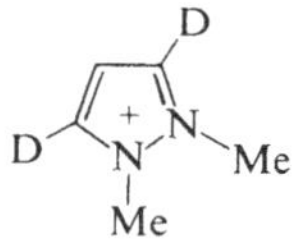	pD 12.95[406b]	840	1
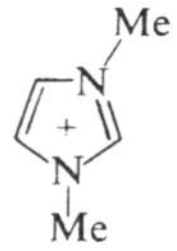	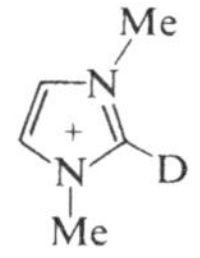	pD 8.92[406b]	270	3×10^4
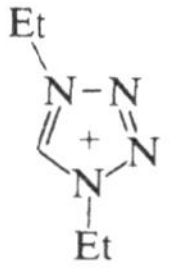	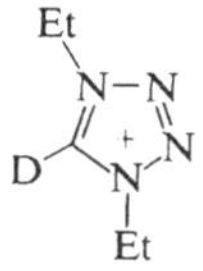	2.04M DCl-D_2O [406b]	306	(5×10^{13})[b]
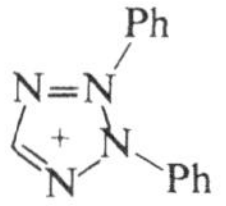	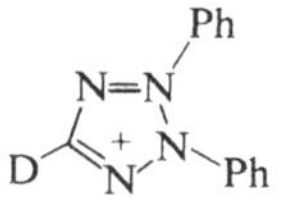	pD 8.85[406b]	372	3×10^4
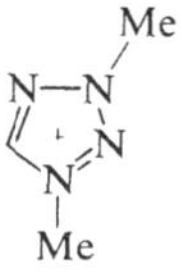	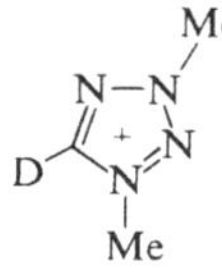	pD 11.15 (60°C) [407a]	3600 ± 300	

(a) At ~ 31 °C unless otherwise stated.
(b) The relative rates were calculated assuming the reactions to be first order in substrate concentration and in [OH⁻]. For the 1,4-diethyltetrazolium ion other bases seem to be important so a true relative rate could not be calculated.

3.6   Diazo-coupling

Pyrazole has not been observed to couple with diazonium salts[13d]. However, the recent observation of coupling between 3,5-dimethylpyrazole and diazotised 3-aminopyrazole, albeit in poor yield[419b], makes it improbable that the failure is general.

Imidazoles with a free NH-group and either a hydrogen atom or a carboxyl group at C-2, C-4 or C-5 couple readily in alkaline solution. Early workers, who observed this reaction with imidazole, 2-methyl-, and 2-phenyl-imidazole, also noted that coupling does not occur with 1-methyl-imidazole, and believed the substitution to occur on a nitrogen atom[54b, 408-9]. Pauly obtained from histidine and diazotised arsanilic acid the product of bis-coupling, and suggested the arylazo-imidazoles to be C-substituted compounds[410a]. Fargher and Pyman improved the coupling reaction by carrying it out in sodium carbonate solution, and proved the main product from imidazole and diazotised aniline to be 2-benzeneazo-imidazole. (This orientation is supported by n.m.r. studies [234] for the 2-(5-tetrazolylazo)imidazole.) Some tris-coupling also occurred, and with p-bromobenzenediazonium cation some of the 4-isomer resulted as well as 2-p-bromobenzeneazo-imidazole. 4-Methyl-imidazole gave equal amounts of the 2- and 4-isomers, and some 2,4-bis-compound. 2,4,5-Trimethylimidazole does not react[274] (*cf.* tetra-methylpyrrole, Schofield, p. 76). 4,5-Tetramethyleneimidazole coupled at C-2[411a]. Imidazole-4-carboxylic acid and its 5-methyl and 2-alkyl derivatives couple readily, but their esters and anilides do not. Acids, esters, and anilides are all unreactive in acid solution[16c,274d,381b]. Nitroimidazoles do not couple[274d]. Imidazole-4,5-dicarboxylic acid and its 2-substituted derivatives couple with loss of one carboxyl group [54b,274b,c,408-9]. An *ortho*-substituent in the diazonium component is said to lower the yield in coupling reactions[412d]. The coupling reaction, sometimes called the Pauly reaction, is used to detect histamine and histidine in paper chromatography[137a,413]. Diazotised 4-amino-imidazoles[414a] and 5-aminotetrazole[234] have been used in the coupling reaction.

A kinetic study of the coupling of imidazole at C-2 with diazotised sulphanilic acid, over the range of pH 7-11, showed that the imidazole anion is the active reagent in this bimolecular reaction[415]. In the coupling reaction using 2,4,5-trideuteroimidazole the kinetic isotope effect is negligible[416].

Coupling of aminopyrazoles is exemplified by the reaction of 5-amino-3-methyl-1-phenylpyrazole with diazotised aniline[330b], and by the similar reactions of 5-amino-1-phenylpyrazole and its p-aminobenzene-sulphonyl derivative[417b]. 3-Amino-5-hydroxy- and 5-amino-3-hydroxy-1-phenylpyrazole couple readily[192a]. As regards the amino-imidazoles, it seems only to have been reported that the 4-amino and 4-acetylamino compounds give a blue and a red colour, respectively, in the 'diazo test' [418b].

Many coupling reactions of pyrazolones have been described. Generally, coupling occurs if in the enolic form of the structure of the pyrazolone the situation is as represented in (57); R and R' are substituents or hydrogen[329b,420]. Thus, for example, pyrazolone itself[421a], 3-methylpyrazol-5-one[419b,421c], and 1-phenylpyrazol-5-ones[132a,b, 329a,392k] couple, but antipyrines[321g], 4-methyl-1-phenylpyrazol-5-one[422], and 4-methyl-1-phenylpyrazol-3-one[422] do not (see also ref. 329). Some reported exceptions such as 3,4-dimethylpyrazolone[421c] and other 3,4-di-substituted pyrazolones[420] seemed to have been removed by the identification of the products as diazoamino compounds [420], but more recently pyrazol-5-ones with hydrogen or phenyl at N-1, methyl or phenyl at C-3, and alkyl or phenyl at C-4 have been shown to couple at C-4[201]. The rule relating ability to couple with structures of the kind (57) is clearly permissive but not proscriptive.

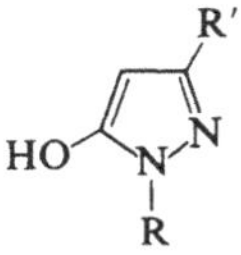

(57) 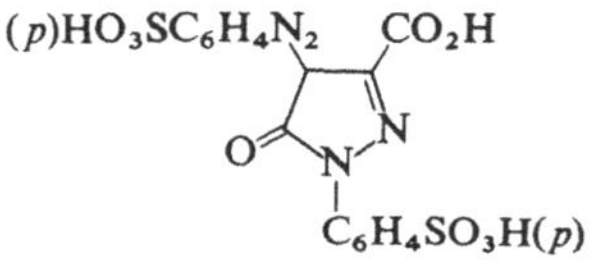   (58)

Coupling does occur when the substituent adjacent to the oxygen function is easily replaced by the entering azo-group. Thus, bis-pyrazolones of the type (51), bridged by ·CHR· and ·CHAr·, couple with the elimination of aldehydes, as do the corresponding alkylidene and arylidene pyrazolones[329b,395]. Similarly, 3-methyl-1-phenyl-4-triphenylmethyl-pyrazol-5-one couples with elimination of the triphenylmethyl group [357c], and 1,3-disubstituted 4-acylpyrazol-5-ones with loss of the acyl groups[318g]. Pyrazolone-sulphonic acids can couple with displacement of the sulpho-group[427]. These examples should be compared with those in the pyrrole series (Schofield, p.76).

Many azo-derivatives of pyrazolones (mostly of 1-arylpyrazol-5-ones) have been prepared because of their valuable properties as dyestuffs[132b, 423]. The first compound important for this purpose was tartrazine, the tri-sodium salt of tartrazinic acid (58). First prepared by a ring-synthetic reaction by Ziegler in 1884, it was described in the scientific literature [424] in 1887. Knorr recognised these compounds as pyrazolones and this conclusion was confirmed by Anschütz[425a] who also prepared its ethyl ester by the coupling reaction.

1-Substituted pyrazol-3-ones (22) and (23) also couple readily[131,

288,321f,330g] (but see ref. 323c). Some 1-aryl-4-hydroxypyrazoles have been coupled with diazotised aniline[76a], and pyrazol-3,5-diones with a free 4-position react readily[321a,426d].

Imidazol-4-ones couple at C-5[398], and 1,2,3-triazolones also react [19e,276a,428b,429b]. 1,2,4-Triazol-3-one and its 5-carboxylic acid are said to give red colours in the coupling reaction[430a].

3.7 Halogenation

Halogenation is the most studied of electrophilic substitutions in the azole series. The results are complicated and mechanistically interesting.

3.7.1 Chlorination

Chlorine water effects substitution at C-4 in pyrazole[392g], 3,5-dimethyl-pyrazole[431c], and pyrazole-3,5-dicarboxylic acid[392g]. 4-Chlorination of pyrazole and 3-methylpyrazole has also been effected with chlorine in carbon tetrachloride, but in the hot pyrazole gives higher substituted compounds and the trichloroterpyrazolyl (59)[77d,432f]. Pyrazole with 1 equivalent of hypochlorous acid in aqueous acetic acid gives 4-chloro-pyrazole, but 3 equivalents of the reagent produce (59). As well as 4-chloropyrazole, aqueous chlorine produces (60) from pyrazole. Chlorination occurs more easily in acetic acid solution and 3-methylpyrazole gives

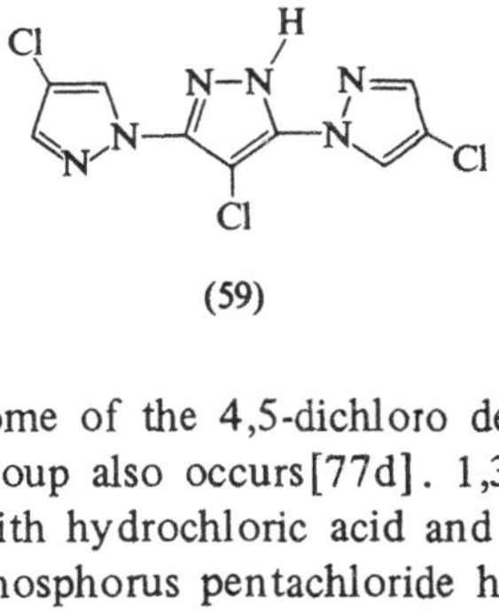

(59) 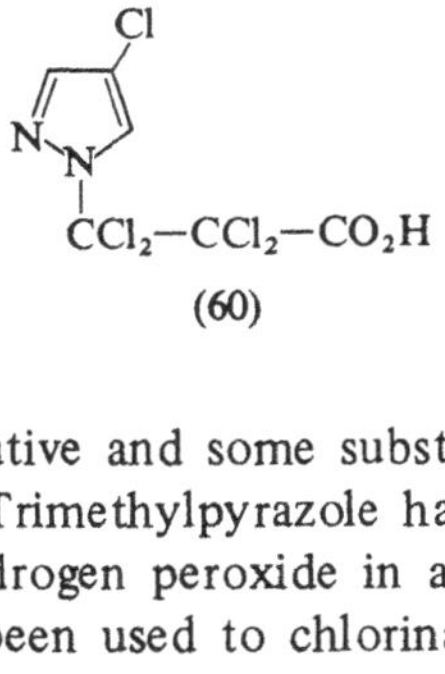   (60) 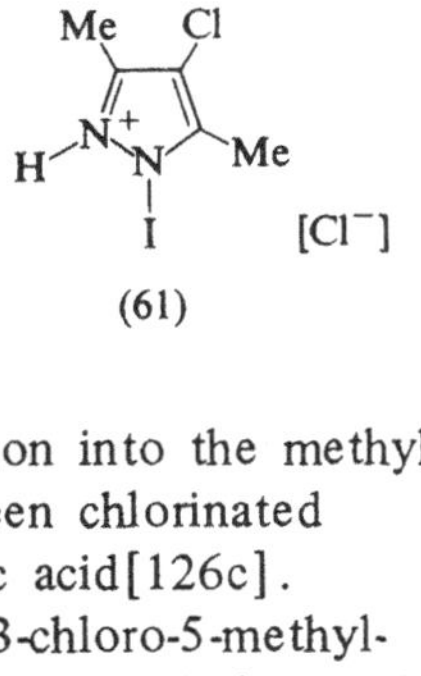   (61)

some of the 4,5-dichloro derivative and some substitution into the methyl group also occurs[77d]. 1,3,5-Trimethylpyrazole has been chlorinated with hydrochloric acid and hydrogen peroxide in acetic acid[126c]. Phosphorus pentachloride has been used to chlorinate 3-chloro-5-methyl-pyrazole[321j,330d,426a]. The formation of 3,4-dichloro-5-methylpyrazole when a solution of 3-chloro-5-methylpyrazole in nitric acid was evaporated and the residue heated with sulphuric acid (in an attempted nitration) was ascribed to the partial decomposition of the starting material, with liberation of chlorine[426a].

Sulphuryl chloride chlorinates pyrazole[433] and 1,3-, 1,5- and 3,5-dimethylpyrazole[100d] at C-4. Sulphur monochloride with 1,3,5-tri-methylpyrazole gives the 4-chloro compound and di-4-(1,3,5-trimethyl-pyrazolyl) sulphide[126c]. Work in the pyrrole series suggests that these may not be purely electrophilic chlorinations (Schofield, p.78).

Chlorine in chloroform converts 4-iodo-3,5-dimethylpyrazole into (61), and with 4-bromopyrazole in carbon tetrachloride gives 4-chloropyrazole and (59)[77d], but such replacements of groups other than hydrogen are rarer than in the pyrrole series (Schofield, p.78).

With silver pyrazole, chlorine in chloroform or carbon tetrachloride rapidly gives 4-chloropyrazole, the formation of some 4-chloro-1-(4-chloro-3-pyrazolyl)pyrazole being unavoidable. Silver 4-chloropyrazole in this reaction did not give the dichloropyrazole, but the di- and trinuclear (59) pyrazoles were formed (contrast bromination and iodination). Silver 3-chloropyrazole gave 3,4-dichloropyrazole (72%), and trichloropyrazole could be obtained similarly [432f]. Silver salts of 3-methyl- and 3,5-dimethyl-pyrazole gave the 4-chloro compounds, and chlorine displaced iodine from the silver salt of 4-iodo-3,5-dimethylpyrazole [432g].

It is surprising that the chlorination of imidazole has only been reported so recently. With chlorine, imidazole gave only 'undesired products containing carbonyl groups' [434], or (as with sulphuryl chloride) very poor yields of undescribed chlorinated products [212]. However, the use of sodium hypochlorite under carefully defined conditions gave 4,5-dichloroimidazole, and 2-ethyl- and 2-methylimidazole behaved similarly, whilst 4-bromoimidazole gave 4-bromo-5-chloroimidazole. From a large-scale preparation of 4,5-dichloroimidazole small amounts of 2,4,5-trichloroimidazole were isolated [434]. Imidazole can also be chlorinated in boiling chloroform with N-chlorosuccinimide or N-chlorophthalimide. Use of the latter produces 4-chloro- (13%) and 4,5-dichloro-imidazole (25%). 2-Methylimidazole behaves similarly and 2,4-dimethylimidazole gives 60% of 5-chloro-2,4-dimethylimidazole [212].

1,2,3-Triazole is not affected by chlorine water, and hypochlorite gives no recognisable product. Such is the case with 2-methyl-2H-1,2,3-triazole, but with chlorine water or hypochlorite 1-methyl-1H- and 4-methyl-1,2,3-triazole give 4-chloro-1-methyl- and 5-chloro-4-methyl-1,2,3-triazole, respectively [77a,c].

Strikingly different is the behaviour of 1,2,4-triazole. With chlorine in aqueous alkali this gives a N,C-dichloro compound, also obtained from 3-chloro-1,2,4-triazole and formulated as 1,3-dichloro-1H-1,2,4-triazole. 3-Methyl- similarly gave 1,3-dichloro-5-methyl-1H-1,2,4-triazole, and N-chloro compounds were also obtained from 3,5-dichloro-, 3-phenyl-, and 3,5-diphenyl-1,2,4-triazole [268b].

In the chlorination of arylazoles there is, of course, the possibility of the aryl group competing successfully with the azole nucleus for the electrophile. Most of the arylazoles which have been halogenated have been N-aryl compounds. However, 3-phenylpyrazole, 1-methoxycarbonyl-3-phenylpyrazole and its 5-methyl and -phenyl analogues are all chlorinated at C-4 with chlorine or sulphuryl chloride [100i]. With hydrochloric acid and hydrogen peroxide 4-chlorination of 3,5-diphenylpyrazole occurs[126e].

Several examples are known of the chlorination of 1-phenylpyrazole and its derivatives occurring at C-4[330d,435]; they include 5-chloro-1-phenyl- reacting with phosphorus pentachloride and 3,5-dichloro-1-phenyl-pyrazole reacting with phosphorus pentachloride or chlorine in chloroform[330i], 5-chloro-1-phenylpyrazole-3-carboxylic acid reacting with chlorine in acetic acid[330i], and 1-phenyl- and 3,5-dimethyl-1-phenyl-pyrazole[63b], and 3-methyl-1-(2,4-dinitrophenyl)pyrazole[436] reacting with sulphuryl chloride. As indicated above (p. 45), sulphuryl chloride may not function as an electrophilic reagent.

In these examples chlorination occurs in the pyrazole ring and not in the phenyl ring, even when the latter contains no de-activating group. This situation should be compared with the situation revealed in brominations (p. 51).

The consequences of chlorinating 1-aryl-4-bromopyrazoles in nitric acid are discussed below (p. 52).

Reaction of 1-phenyl-1*H*-1,2,3,4-tetrazole with mercuric acetate followed by chlorine gives the 5-chloro compound[83d].

The chlorination of azoles containing tautomerisable substituents will now be considered. Few examples of the chlorination of amino-azoles have been reported. 3-Acetamidopyrazole reacts at C-4 with sulphuryl chloride[172f]. Chlorination of 5-amino-3-methyl-1-phenylpyrazole occurs at C-4, but the reaction of the 5-acetamido compound is cleaner[330b] (*cf.* bromination, p. 53). Halogenation of 3-anilino-5-phenylpyrazole and related compounds gives 4-halogeno-3-(dihalogenophenylamino)-5-phenyl-pyrazoles[437].

1-Benzyl-3-methylpyrazol-5-one with phosphorus pentachloride gives the 4,4-dichloro compound[428d].

As would be expected from what has gone before, in the chlorination of arylpyrazolones the heterocyclic ring is the one to be attacked. 1-Arylpyrazol-5-ones with chlorine, hypochlorite, or phosphorus penta-chloride are chlorinated[439] or dichlorinated[330g,392e,k,439]. With chlorine in acetic acid 3,4-dimethyl-1-phenylpyrazol-5-one is chlorinated at C-4[439].

1-Phenylpyrazol-3-one is chlorinated at C-4 with sulphuryl chloride [131], and the 5-methyl homologue reacts similarly with phosphorus pentachloride or with chlorine in chloroform[330a].

3-Ethoxycarbonyl-4-hydroxy-1-nitrophenylpyrazole, its 5-chloro and 5-bromo derivatives, and the corresponding 3-benzoyl compounds give the 5,5-dichloro compounds with chlorine in chloroform or acetic acid. When the phenyl group contains both nitro and chloro substituents the 5-monochloro derivative can be isolated[440b,c].

The first substitution products in the chlorination of antipyrines are the 4-chloro compounds[324c,441c,442].

Chlorination of 1,2-diarylpyrazol-3,5-diones at C-4 is easily effected [443c].

With hypochlorite hydantoins are *N*-chlorinated[438].

3.7.2 Bromination†

Bromine water effects 4-substitution with pyrazole[2b,c,392g], 3,5-dimethylpyrazole[431c], and pyrazole-3,5-dicarboxylic acid[392g]. Bromine in acetic acid and in aqueous alkali similarly substitutes 3-methylpyrazole-5-carboxylic acid and its amide[426c], respectively. 1-Cyanoethyl-3,5-dimethylpyrazole readily gives the 4-bromo compound[126e]. Early workers[2f] remarked on their inability to introduce more than one bromine atom into pyrazole even at 150°C. A reported dibromination of 1,3-dimethylpyrazole[443d] could not be repeated, and 1-methylpyrazole gave only 4-bromo-1-methylpyrazole[77f].

In view of these reports it is of interest that whilst treatment of 3-trifluoromethylpyrazole with bromine (1 equivalent) in carbon tetrachloride with sodium acetate gave the 4-bromo compound, the use of bromine (2 equivalents) in aqueous sodium acetate gave 3,4-dibromo-5-trifluoromethylpyrazole[444].

Multiple bromination of *N*-unsubstituted pyrazoles was achieved using a catalyst[77f]. Thus, pyrazole in carbon tetrachloride with bromine and iron gave 3,4-di- and 3,4,5-tri-bromopyrazole.

Early workers noticed the formation of orange-red 'perbromides' in some cases[321j,330d,431c]. With bromine in ligroin 3-chloro-5-methylpyrazole gave the perbromide $C_4H_4ClBr_3N$, converted by alkali into 4-bromo-3-chloro-5-methylpyrazole. Several of these perbromides have been more closely examined[77f]. If in the starting pyrazole the 4-position was free, then in the perbromide it was brominated. Pyrazoles already substituted at C-4 can give perbromides lacking nuclear bromine. The perbromides are formulated as (62), R′ = Br or a previously present substituent. Such perbromides, from *N*-unsubstituted pyrazoles, are very unstable, readily losing hydrogen bromide to give 1-bromopyrazoles. That from 4-nitropyrazole has been isolated. 1-Bromopyrazoles may be important in the process of *C*-bromination, not because of $N \rightarrow C$ rearrangement, but by acting as a source of the powerfully electrophilic brominium ion. Substituents might change this behaviour; the formation of (63) when 4-methylpyrazole is boiled with bromine in chloroform could be due to the dissociation of 1-bromo-4-methylpyrazole into bromide ion and the electrophile (64). Other products of dehydrogenative condensation are also formed, but not when electron-attracting groups are present. Bromination in the presence of iron can cause substitution

† For brominations by ylide mechanisms see p. 143.

(62)　　　　　(63)　　　　(64)　　　　(65)

at positions other than C-4; thus, 4-methylpyrazole-3-carboxylic acid[77f] and 4-bromo-3-methylpyrazole[114a] are brominated at C-5. With bromine and iron, or with hypobromous acid, 4-chloropyrazole gives 3,5-dibromo-4-chloropyrazole[77d].

Bromination of 4-bromo-1,5-dimethylpyrazole in nitric acid gives 3,4-dibromo-1,5-dimethylpyrazole[114a].

Contrary to an earlier report[77f], 3,4,5-trimethylpyrazole does react with bromine or *N*-bromosuccinimide in chloroform. The product has been shown to have the structure (65)[445], a result which must raise doubts about the formulation as *N*-bromopyrazoles of some of the compounds mentioned above.

In recent work brominations have been carried out in organic solvents. Some examples are illustrated here[114a].

R = H,　R′ = Me,　R″ = Br
R = Me,　R′ = Me,　R″ = Br *or* CO_2Et　} $Br_2/CHCl_3$
R = Me,　R′ = CO_2Et, R″ = Me

R = Me,　R′ = Me,　R″ = H *or* CO_2H　} $Br_2/AcOH$
R = Me,　R′ = H,　R″ = Me

Replacement of groups other than hydrogen is not common, but with bromine water 3-methyl-1,5-diphenylpyrazole-4-sulphonic acid gives 4-bromo-3-methyl-1,5-diphenylpyrazole[446]. 3,4-Di(trimethylsilyl)pyrazole reacts with bromine, one or both of the silicon groups being replaced, that at C-4 going first[190e].

Silver pyrazoles behave towards bromination very much as they do in chlorination. However, silver 4-chloro- and 4-bromo- and 4-iodo-pyrazole can all be 3-brominated (though not very efficiently). Silver 4-iodo-pyrazole gives 4-bromopyrazole as well as 3-bromo-4-iodopyrazole[432f]. Silver 4-methylpyrazole with bromine gives (63)[432g].

5-Bromopyrazoles have been obtained by applying the Hunsdiecker reaction to the silver salts of 1-substituted pyrazole-5-carboxylic acids [59h,114a,447]. The reaction also works with 1-phenylpyrazole-4-carboxylic acid, but not with 1-substituted pyrazole-3-carboxylic acids [447].

Imidazole is brominated more readily than pyrazole, giving, with bromine in water or ether, 2,4,5-tribromoimidazole[6b]. With bromine in chloroform at -10 °C 4-methylimidazole gives the dibromo compound and 34% of 4-bromo-5-methylimidazole[16a,412b] (see ref. 234), but even under these conditions imidazole gives mainly the tribromo derivative with only a trace of 4,5-dibromoimidazole (concerning orientation see ref. 234). Complete *C*-bromination of the empty nuclear positions occurs readily with 2-methyl-[9b,11], 2-iodo-, 2-iodo-4-methyl-[410b], 4-methyl-5-nitro-[368a], 4-ethoxycarbonyl-[374a], 4-aminocarbonyl-[374f], and 4-ethoxycarbonyl-5-methyl-imidazole[412b]. *N*-Bromosuccinimide in aprotic solvents also tribrominates imidazole[448b].

1-Alkylimidazoles also are so readily tribrominated that a lower degree of substitution is not readily achieved[54d,374b]. Conversion of 1-ethyl-2-methyl- into 4,5-dibromo-1-ethyl-2-methyl-imidazole by bromine in dilute sulphuric acid, and of 5-chloro-1-ethyl-2-methyl- into 4-bromo-5-chloro-1-ethyl-2-methyl-imidazole by bromine in chloroform or carbon disulphide goes through 'polybromides'[54c,d]. 1,5-Dimethyl- is more readily brominated than is 1,4-dimethyl-imidazole[16a,c,412b]. Cyanogen bromide converts 1,4-dimethyl- into 2-bromo-1,4-dimethyl-imidazole but does not react with 1-ethyl-2-methylimidazole[449].

With bromine in chloroform imidazole is said to give a complex $[(C_3H_4N_2)_2Br^+]Br_3^-$, in which two imidazole nuclei are joined to a bromine cation through nitrogen atoms, and 1-methylimidazole behaves similarly[450].

The isolation of some ammonium bromide from the bromination of imidazole was an early indication that halogens could effect ring-opening [374a], and the ring-degradation which accompanies the tribromination of imidazole with *N*-bromosuccinimide in aqueous media has been examined. In aqueous buffers or acids the formation of tribromoimidazole is suppressed and ring-degradation is enhanced by increasing acidity. Tribromoimidazole is itself degraded by *N*-bromosuccinimide as are various 4-substituted imidazoles[448b]. The reactions are initiated by electrophilic bromination and are discussed in §6.2.

Monobromination of 1,2-dimethylimidazole can be achieved by lithiation, followed by reaction with diethylbromamide, which produces 5-bromo-1,2-dimethylimidazole[451].

Bromine water or hypobromite converts 1,2,3-triazole into the 4,5-dibromo compound, and bromine water produces the same product from 1,2,3-triazole-4-aldehyde and -4-carboxylic acid, but not from the -4,5-dicarboxylic acid. 1-Methyl- and 4-methyl-1,2,3-triazole are readily monobrominated, but 2-methyl-2*H*-1,2,3-triazole needs the help of an iron catalyst, when it gives 4,5-dibromo-2-methyl-2*H*-1,2,3-triazole. In contrast to 1-methyl-1*H*-1,2,3-triazole, 4-formyl-1-methyl-, 4-formyl-1-phenyl-, and 1-phenyl-1*H*-1,2,3-triazole could not be brominated at room temperature

[77a,c] (concerning phenyltriazoles in general, see below).

1,2,4-Triazole can be brominated in aqueous alkali, one equivalent of bromine giving 4,5-dibromo-1,2,4-triazole. Several 3-monosubstituted compounds have been similarly brominated[268c,f].

Kinetic studies of bromination have been reported only in the pyrazole series. Studies were made using solutions of bromine in aqueous potassium bromide, and the results were extrapolated to refer to aqueous solution. It was concluded that molecular bromine reacted with the free bases in the cases of pyrazole, 1-methylpyrazole and 3,5-dimethylpyrazole, in a conventional two-stage reaction. Substitution occurred quantitatively at C-4, the second order rate constants at 25 °C being 3.8 × 10^5, 8.0 × 10^5, and 1.4 × 10^9 1 mol^{-1} s^{-1} for pyrazole, 1-methylpyrazole, and 3,5-dimethylpyrazole respectively. In the first two cases the rate constants are comparable with those for phenol and anisole, and in the third case the value is within an order of magnitude of that for reaction upon encounter. For pyrazole and 3,5-dimethylpyrazole finite kinetic isotope effects were observed with k_H/k_D values of 1.39 ([HBr] = 0.369 mol 1^{-1}) and 1.08 ([HBr] = 1.480 mol 1^{-1}), respectively[472].

4-Substitution occurs when 1-phenylpyrazole[60b,251] and its derivatives containing carboxyl, halogen, or methyl groups at C-3 or C-5, or even carboxyl groups at both C-3 and C-5 are brominated in chloroform or acetic acid[60e,321b,c,d,e,330i]. C-4 is, expectedly, the reactive position in 5-chloro-3-methyl-1-(p-nitrophenyl)pyrazole[321c], and even in 1-(p-biphenyl)pyrazole[251]. 4-Bromo-3-methyl-1-phenylpyrazole-5- and 4-bromo-5-methyl-1-phenylpyrazole-3-carboxylic acid could not be further brominated[59h], and neither could 4-bromo-1-p-bromophenyl-3-methyl-pyrazole in neutral solution. However, the last compound was further brominated at C-5 by bromine in 8M-nitric acid[59h]. The dibromo compound from 1-phenylpyrazole[60b] is 4-bromo-1-(p-bromophenyl)-pyrazole[447], and the derived tribromo compound is 4,5-dibromo-1-(p-bromophenyl)pyrazole[114e]. Similarly, the dibromo compound from 3-methyl-1-phenylpyrazole is not the 4,5-dibromo compound, as originally thought[321c], but 4-bromo-3-methyl-1-(p-bromophenyl)pyrazole[59h], and 4-bromo-3-methyl-1-(p-nitrophenyl)pyrazole with bromine in acetic acid gives the 4,5-dibromo compound[114e].

It has been claimed that, whereas bromination of 1-phenylpyrazole in neutral solution gives 4-bromo-1-phenylpyrazole, bromination with bromine and silver sulphate in concentrated sulphuric acid gives 1-(p-bromophenyl)pyrazole[251]. The change in orientation was ascribed to bromination in sulphuric acid proceeding through the cation of the substrate, whereas in other conditions the neutral molecule was presumably involved. A detailed re-examination of the case showed that bromine in neutral or acidic organic solvents produces 4-bromo-1-phenyl-pyrazole, and bromine in concentrated sulphuric acid also gave this product mainly, together with a small proportion of 4-bromo-1-(*p*-bromophenyl)pyrazole. Further, the 1 : 1 complex of 1-phenylpyrazole with aluminium chloride when treated with bromine in carbon tetrachloride in the cold gave 4-bromo-1-phenylpyrazole, and in the hot a mixture of this with 4-bromo-1-(*p*-bromophenyl)pyrazole. The sole formation of 1-(*p*-bromophenyl)pyrazole by use of bromine and silver sulphate in sulphuric acid could not be confirmed, but rather, a mixture of this with 4-bromo-1-(*p*-bromophenyl)pyrazole was obtained. The observed differences in orientation were, therefore, attributed not to protonation of the base, but to the differing properties of molecular bromine and brominium ion as electrophilic reagents[59i]. The case should be compared with that of nitration (p. 63).

4-Acetoxymercuri- and 4-chloromercuri-1-phenylpyrazole both react with bromine to give 4-bromo-1-phenylpyrazole[59a], and 3-methyl-1,5-diphenylpyrazole-4-sulphonic acid gives 4-bromo-3-methyl-1,5-diphenyl-pyrazole[446].

3-Methyl-1-(2,4-dinitrophenyl)pyrazole is brominated at C-4, and the product reacts with chlorine and nitric acid to give 4-bromo-4-chloro-3-methyl-1-(2,4-dinitrophenyl)pyrazol-5-one; the same product results from the action of bromine and nitric acid on 4-chloro-3-methyl-1-(2,4-dinitrophenyl)pyrazole. Electrophilic attack of halogen at C-4 is followed by nucleophilic attack of water at C-5 and oxidation by nitric acid[114e, 436].

Ar = 2,4-(NO$_2$)$_2$C$_6$H$_3$ –

4-Arylimidazoles are brominated at C-5 and di-substituted at C-5 and C-2[453b]. 4,5-Diphenylimidazole gives, with bromine in chloroform, the 2-bromo compound[454].

Bromine with silver sulphate and concentrated sulphuric acid converts 1-phenyl-1*H*-1,2,3-[455] and 2-phenyl-2*H*-1,2,3-triazole[456b] into the *p*-bromophenyl compounds. Bromination in acetic acid of 4-methyl-2-phenyl- and 4,5-dimethyl-2-phenyl-2*H*-1,2,3-triazole proceeds similarly, but under other conditions the methyl groups are brominated[457]. In contrast to these cases, hypobromite converts 4-phenyl- into 4-bromo-5-phenyl-1,2,3-triazole[458c].

Few examples of the bromination of amino-azoles have been described.
With bromine in acetic acid 3-aminopyrazole reacts at C-4[172f], as does
5-amino-3-methyl-1-phenylpyrazole[330b]. In contrast, with bromine
4-amino-3-methyl-1-phenylpyrazole is believed to lose its amino group
and give a bromopyrazolone[330k].

3,5-Diaminopyrazole is brominated at C-4[459].

Pyrazolones unsubstituted on nitrogen and having a methyl or bromo
group at C-4 have been brominated in chloroform and are said to give
N-bromo compounds. Thus, 3-methylpyrazol-5-one gave first 4-bromo-3-
methylpyrazol-5-one, and then a compound formulated as 1,4-dibromo-3-
methylpyrazol-5-one[77e]. 3,4-Dimethylpyrazol-5-one gave a similar
compound with *N*-bromosuccinimide, whilst 1,3,4-trimethyl- and 1-acetyl-
2,3-dimethyl-pyrazol-5-one could not be brominated[460b]. Earlier
workers[421c,463] had regarded the product from 4-bromo-3-methyl-
pyrazol-5-one as the 4,4-dibromo compound (66), R = Me, R′ = Br, and
this formulation has been supported by infrared spectroscopic evidence
[464], as have the similar ones (66), R = R′ = Me, and R = H, R′ = Me.

Bromination of 1-acetyl-3-methyl- and 1,3-dimethyl-pyrazol-5-one
occurs at C-4[460b], and 1-aminocarbonyl-3-methylpyrazol-5-one gives
the 4-bromo and 4,4-dibromo compounds[465]. The dibromo derivative
has also been obtained from 1-benzyl-3-methylpyrazol-5-one[428d].

(66)

Aliphatic substituents at C-4 in hydantoins are usually brominated in
preference to C-4[438].

Bromination of 1,2,4-triazol-5-ones at C-3 with bromine in aqueous
alkali goes only with *N*-substituted compounds. *N*-Unsubstituted
compounds apparently react at nitrogen, and ring-opening follows[461].

1-Phenylpyrazol-3-ones are brominated at C-4[131,330a], and 1-aryl-
pyrazol-5-ones are similarly brominated[392k,439,460] or dibrominated
[392k,439,460a].

With bromine in acetic acid, 3,4-dimethyl-1-phenylpyrazol-5-one gave
the 4-bromo compound. The bromination stops half-way until water is
added to hydrolyse the hydrobromide of the starting material formed.
The same phenomenon was observed in the 4,4-dibromination of
3-methyl-1-phenylpyrazol-5-one[439].

The first two stages in the bromination of 3-methyl-1-phenylpyrazol-
5-one are thus 4-mono- and 4,4-di-bromination. Only after this does
bromination in the phenyl ring occur, at the *para*-position and, in the
case of the corresponding sulphonic acid, with replacement of the -SO$_3$H
group[392e,462].

Antipyrines are brominated at C-4, usually with bromine in chloroform or acetic acid[321k,324c,330h,392c,k,442a,466] although hydrobromic acid and hydrogen peroxide have been used[442b]. 3-Antipyrine[321k, 330e] and isoantipyrine[330d] are also brominated at C-4.

Antipyrine can also be 4-brominated with *N*-bromosuccinimide, but substitution in the 3-methyl group also occurs[467b]. The same reagent replaces carboxyl and formyl groups at C-4 in antipyrines, and again brominates the 3-methyl group if any but the mildest conditions are used[467a].

In his original experiments on the bromination of antipyrine, Knorr [392k] observed the initial formation of a compound which he took to be the addition compound (67). With water it gave 4-bromo-antipyrine. Similar observations in brominations were made by all the subsequent workers quoted above. Subsequently the initial product was shown to be 4-bromoantipyrine hydrobromide (68)[439,441a]. The structures of other reported bromides and perbromides are in doubt[439,441b].

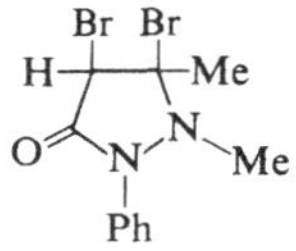

(67) 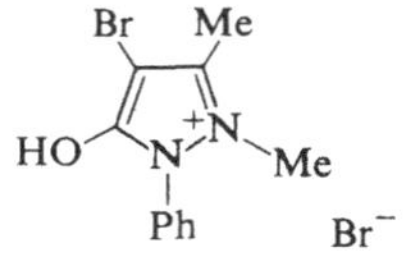   (68)

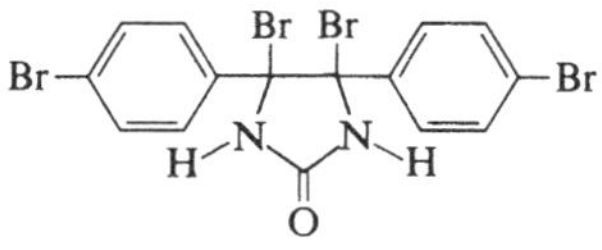

(69) 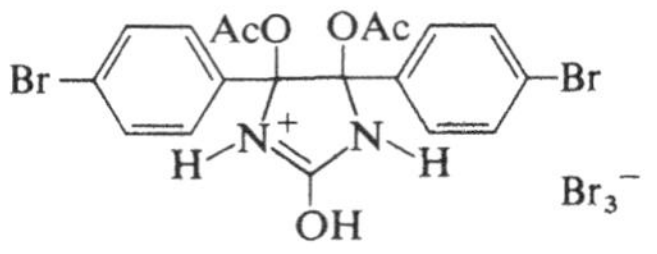   (70)

Relatively little work has been done on imidazolones having aryl groups. Biltz[353c,d] examined the reactions of 4,5-diarylimidazol-2-ones with bromine in acetic acid. The final result was ring-opening, but Biltz believed the initial step to be 4,5-addition of bromine; he formulated the products from 4,5-di-(*p*-bromophenyl)imidazol-2-one as (69). More recently this product was re-formulated[468] as (70). From 4,5-diphenyl-imidazol-2-one the 4,5-di-(*p*-bromophenyl) compound is evidently the first product (see p. 157).

Bromination of 1,2-diarylpyrazol-3,5-diones at C-4 is easily effected when C-4 is unsubstituted or substituted by alkyl or aryl groups[426d, 443a,b]. Similarly 4-phenyl- and 1-methyl-4-phenyl-hydantoin are brominated at C-4[14c]. In other cases a side-chain at C-4 may be brominated, or C-4 bromination may be followed by easy loss of hydrogen bromide between the nucleus and a side-chain[469].

3.7.3 Iodination

Pyrazole does not react with iodine although silver pyrazole is converted into 4-iodopyrazole[2f] (see below). Pyrazole gives this product with potassium tri-iodide and sodium acetate in boiling water. 1- and 3-methylpyrazole[77b], and 3,5-dimethylpyrazole[431c] behave similarly, whilst 4-methylpyrazole gives 3-iodo-4-methylpyrazole[77b]. The iodination of 3- is much faster than that of 4-methylpyrazole[77b].

Iron does not catalyse the multiple iodination of pyrazole, but nitric acid, silver nitrate, and iodine together convert pyrazole into 1,3,4-triiodopyrazole, which with sulphur dioxide gives 3,4-di-iodopyrazole. Repetition of the process produces 3,4,5-tri-iodopyrazole[77b]. With iodine in carbon tetrachloride, 4-methylpyrazole gives a deep-red oil, believed to be 1-iodo-4-methylpyrazolium iodide. With alkali the latter gives 3-iodo-4-methylpyrazole, but the red intermediate from 3,4-dimethylpyrazole gives 1-iodo-3,4-dimethylpyrazole. Presumably the second methyl group discourages the dissociation to I^+. However, with hot hydrochloric acid and sulphur dioxide, or in boiling carbon tetrachloride, 1-iodo-3,4-dimethylpyrazole gives 5-iodo-3,4-dimethylpyrazole. With iodine in carbon tetrachloride, 5-iodo-3,4-dimethyl- and 3,4,5-trimethyl-pyrazole give the *N*-iodo derivatives[77b].

Other workers found the method of multiply iodinating 4-iodopyrazole with iodine and silver nitrate to be preparatively unsatisfactory; yields were poor and large quantities of reagents were needed. 3,4-Di- and 3,4,5-tri-iodopyrazole could be prepared satisfactorily from pyrazole or 4-iodopyrazole with iodine and ammonia. Under these conditions 3-methylpyrazole gave 4,5-di-iodo-3-methylpyrazole. With iodine monochloride and hydrochloric acid at room temperature pyrazole gave 4-iodopyrazole, and at 100 °C a poor yield of tri-iodopyrazole. Iodine and caustic soda, a satisfactory reagent with imidazole (see below) gave with pyrazole only *N*-iodopyrazole[473].

The method of iodinating using pyrazole silver salts, mentioned above, has been extended. From silver pyrazole itself, 95% of 4-iodopyrazole is produced, 3-iodination occurs with the silver salts of the 4-halogenopyrazoles, and silver 3-chloropyrazole can be iodinated at C-4. With iodine bromide, silver pyrazole gives 4-iodopyrazole exclusively[432f]. The silver salts of the methylpyrazoles are also readily iodinated[432g].

Iodine in aqueous alkali converts imidazole into 2,4,5-tri-iodoimidazole, with some of the 2,4-di-iodo compound, whilst 4-methyl- gives 2-iodo- and 2,5-di-iodo-4-methylimidazole[410b,c]. Imidazole-4-carboxylic acid [452b] and 2,5-dibromoimidazole[410b] give, respectively, 2,4,5-triiodoimidazole and 2,5-dibromo-4-iodoimidazole. 1,4(or 5?)-Dimethylimidazole does not react with iodine[410c]. In the iodination of imidazole initial reaction occurs at C-4[234] (see p. 56).

Pauly[410d] showed that 2,4,5-tri-iodo-, 4,5-di-iodo-2-methyl- and 2,4,5-trimethyl-imidazole gave their 1-iodo derivatives with iodine in sodium hydroxide solution. A related observation may be that of the formation of a red 'di-iodide' from 5-chloro-1-ethyl-2-methylimidazole [54d]. Pauly[410b] also drew an incorrect parallel between iodination and diazo-coupling which were held to proceed initially at C-2, in contrast to nitration and sulphonation which occur at C-4 (see pp. 62, 71).

4-Methyl-1,2,3-triazole is readily iodinated in boiling chloroform, whilst 1-methyl-1*H*-1,2,3-triazole gives the 4-iodo compound with iodine in nitric acid; under these last conditions 2-methyl-2*H*-1,2,3-triazole is di-iodinated[77a,c].

1,2,3-Triazole with iodine or hypoiodite gives a *N*-iodo derivative. Other *N*-halogeno compounds arise when 4-iodo-5-methyl-1,2,3-triazole reacts with iodine, when 4-methyl-1,2,3-triazole reacts with hypobromite solution, and when 4,5-dimethyl-1,2,3-triazole is treated with hypohalite solutions. With two molecular equivalents of a hypohalite solution, 4-methyl-1,2,3-triazole gives the *N*,5-dihalogeno derivative. *N*-Bromo-4-methyl-1,2,3-triazole is orange-red, and quickly changes to the colourless 5-bromo-4-methyl-1,2,3-triazole[77a,c]. 3,5-Disubstituted 1,2,4-triazoles give *N*-bromo derivatives[268f].

It is appropriate to consider at this stage quantitative studies which throw light on the mechanism of halogenation of the azoles. The first of these, due to Ridd[474a], demonstrated that in aqueous solution at pH = 7 the iodination of imidazole followed the equation

$$\text{Rate} = k[C_3H_3N_2^-][H_2OI^+] + k'[C_3H_3N_2^-][C_3H_4N_2I^+].$$

The first term corresponds to an uncatalysed, and the second to a catalysed reaction, both involving the imidazole anion, and the second being a special case of general buffer catalysis of iodination. The equation refers to monoiodination, but even with a large excess of imidazole the 2,4-di-iodo compound is the main product. The reaction is first order in $[I_3^-]$ and the monoiodinating step is rate-determining, being followed by a rapid second iodination of the iodo-imidazole anion. The possibility that the uncatalysed reaction might be a $N \to C$ rearrangement, and that catalysis by imidazole may involve assisted proton loss in the transition state was not favoured. The involvement of imidazole anions accounts for the failure of *N*-substituted imidazoles to react. Iodination of 2,4,5-trideutero-imidazole showed a large kinetic isotope effect[416] (k_H/k_D = 4.5) for catalysed and uncatalysed reactions. The effect was present for 4,5-dideutero- but not for 2-deutero-imidazole, and the relative rates for 2-deutero-, 4,5-dideutero-, and 2,4,5-trideutero-imidazole prove the initial substitution to occur at C-4.

Recent work on aromatic iodination modifies to some extent our view of these results[474b]. It seems probable that the initial step involves molecular iodine, perhaps as is shown here. The kinetics and isotope effects show that in the rate-determining step a proton is lost from a complex of the imidazole anion and the iodine cation, but reveal nothing about the earlier steps. In the above sequence the parent imidazole molecule might, in fact, be involved in the first step.

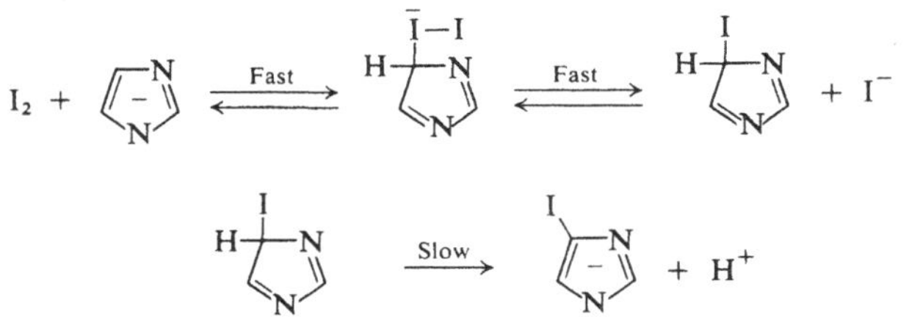

In this light, iodination and diazo-coupling are seen to differ in important respects.

Other workers have examined the iodination of imidazole and of histidine and several related compounds. The first point of attack in histidine was C-4, and for all the compounds the kinetics were consistent with the attack of molecular iodine upon the anionised imidazole nuclei, followed by proton abstraction from the σ-complex in the rate-determining step. The anions were estimated to be more reactive than the neutral molecules by a factor $\ll 10^8$. With histidine and its derivatives hydrogen bonding between the side-chain and the nucleus will de-activate the anion. Phosphate ions appear to exercise a special function in catalysing iodination. The very rapid reaction of homohistidine was attributed to the ability of the amino group in the side-chain to assist deprotonation of the σ-complex. Histidine derivatives may exhibit the same phenomenon, but to a smaller degree. The rate equations for histidine and some of its derivatives contain terms which are zero order in iodinating agent; these represent the rate-determining formation of a reactive species, possibly of the form (71). Accordingly, in the zero order reaction the orientation of iodination may be different (at C-2 rather than C-4)[475a].

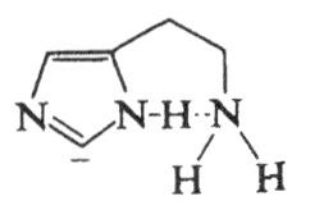

(71)

Di-iodohistidine reacts with iodine through its singly and doubly charged anions, and the rate equation containing these terms does not refer to the formation of a tri-iodo compound, but to oxidative degradation[475a,b], (§6.2). These conclusions are subject to the same qualification regarding mechanisms as those relating to the iodination of imidazole (see p. 57).

The catalytic effects of various bases on the iodination of imidazole in borate buffers at fixed ionic strength show a better correlation with the nucleophilicities of the bases than with their basicities[475c].

In the iodination of the Ni^{2+} complex of imidazole the catalytic term is negligible, as would be expected since base catalysis is eliminated by the nickel–nitrogen co-ordination[476]. The iodination has been concluded to involve molecular iodine reacting with both $Ni(C_3H_4N_2)^{2+}$ and with $Ni(C_3H_3N_2)^+$.

Because 1-methylpyrazole, like pyrazole, can be iodinated at C-4, it has been suggested[477] that the neutral molecule is involved in the latter case. However, results of a study of the kinetics of iodination in aqueous solution were interpreted as showing that the anion was the reactive species[314a]. The rate expression was of the same form as that for imidazole, and contained 'catalytic' and 'non-catalytic' terms,

$$- d[I_3^-]/dt = (k_0 + k_{cat}[B])/[I^-]^2 [H^+]$$

Comparison of the 'non-catalytic' constants, k_0, for imidazole and pyrazole, was said to show the imidazole anion to be 1.3 times more reactive than the pyrazole anion. However, those constants are themselves composite, being made up of the forward and reverse rate constants for the step in which iodine is supposed to react with the anion and the equilibrium constants of the pre-equilibria. In addition, the mechanism needs to be re-written in terms similar to those mentioned above for the modified mechanism of imidazole iodination. Where the last step is rate-determining the idea of relative reactivity cannot have its normal meaning.

The 1-alkylpyrazoles show similar kinetics. In phosphate buffers the rate law is of the same form as that for aniline, with catalytic and non-catalytic terms. Because of the probable nature of these reactions the attempt to use these results in conjunction with those for pyrazole to calculate relative reactivities of free base and anion in this series is not justified[478].

5-Amino-3-methyl-1-phenylpyrazole is iodinated at C-4[330b]. Attempts to iodinate derivatives of 4-aminopyrazole failed[473].

Some 4-arylimidazoles have been mono- (at C-5?) and di-iodinated (at C-5 and C-2?)[452b].

1-Arylpyrazol-5-ones are iodinated at C-4 with iodine in alkali[439]. The same orientation is found for the iodination of antipyrines with iodine in organic solvents[321k,324c].

The 4-iodination of antipyrine has been achieved through the 4-chloromercuri compound[479].

3.8   Metalation, halogen–metal exchange

The lithiation of pyrazoles produces some interesting reagents[74e,f].
With butyl or phenyl lithium, pyrazole gives first the *N*-lithio compound.
A second equivalent of the reagent causes metalation, so that on
carboxylation the reaction solution gives mainly pyrazole, but also small
proportions (7–9%) of pyrazole-3-carboxylic acid[77g]. 1-Methylpyrazole
is metalated at C-5 by phenyl or butyl lithium[77g,480b,481], the second
reagent being more effective (66%) than the first (39%). Two equivalents
of reagent do not effect di-metalation. 1-Benzylpyrazole is also lithiated
at C-5, as is 1,3-dimethylpyrazole, but 1,5-di- and 1,3,5-tri-methylpyrazole
do not react at the nucleus[77g]. In fact, 1,3,5-trimethyl- and 5-methoxy-
1-methyl-3-phenylpyrazole are lithiated at the methyl groups[482]. Butyl
lithium metalates 1-phenylpyrazole at C-5, and substitution also occurs at
C-2′·in the phenyl group. Excess of the reagent causes di- and tri-
metalation[480b,481]. Other 1-phenylpyrazoles containing substituents at
C-3 and C-5 are also lithiated at C-2′[482].

Lithium or butyl lithium metalate 1-methyl-, 1-methoxymethyl-, and
1-benzyl-imidazole at C-2[372,480a,483]. 1-Phenylimidazole reacts
similarly, and with excess of butyl lithium is 2,2′-di-metalated[480a].

5-Methyl-1-phenyl- and 1,5-diphenyl-1*H*-1,2,3-triazole are lithiated by
butyl lithium at C-4, and the methyl compound in that substituent also.
1-Phenyl-, 4-methyl-1-phenyl-, and 1,4-diphenyl-1*H*-1,2,3-triazole are all
lithiated at C-5. Ring opening can follow lithiation (p. 162)[484].

Early workers observed[1c,2c] the formation from mercuric chloride
and pyrazole and some of its homologues (both *N*- and *C*-substituted)
of compounds regarded as 'double salts'. These were reported for
pyrazole, 3,5-dimethyl- and 1,3,5-trimethyl-pyrazole. The composition
recorded for that from pyrazole $(C_3H_3N_2.HgCl)$ represented it as being
formed by loss of a hydrogen atom, but the product from 3-methyl-
pyrazole was formulated as $(C_4H_6N_2)_2.3HgCl_2$, and in other cases
compositions were not recorded. Later workers[485] represented the
product from 5-chloro-3-methyl-1-phenylpyrazole and mercuric acetate
as the 4-acetoxymercuri compound, but without any particular
justification.

The reactions of several pyrazoles with mercuric chloride were more
recently re-examined by Russian workers[323b]. Whilst 4-benzyl-3,5-
dimethyl-1-phenylpyrazole gave a 1 : 1 complex from which alkali
regenerated starting material, and 4-chloro-1-phenylpyrazole did not react,
several other compounds were 4-chloromercurated. These were 1,3,5-
trimethyl-, 1-ethyl-3,5-dimethyl-, 1-phenyl-, 5-chloro-3-methyl-1-phenyl-
and 5-chloro-3-methyl-1-phenethylpyrazole. The products from the first
two compounds were associated with a molecular equivalent of mercuric
chloride, whilst those from the second two were not, and the amount of

'complexed' mercuric salt was thought to be related to the basicity of the product. Experiments with *N*-unsubstituted compounds resulted in the isolation of a product only from 4-methyl-3,5-dipropylpyrazole, which gave the 1-chloromercuri compound.

The early-obtained products from pyrazole homologues substituted at nitrogen but not at C-4 are therefore probably 4-chloromercuri compounds. Those from pyrazole and compounds unsubstituted on nitrogen may also be so, but need re-examination.

1-Phenylpyrazole gives 4-acetoxymercuri-1-phenylpyrazole [59a], and 3-phenylpyrazole behaves similarly [486a]. The latter compound thus provides the only case at present clearly established of nuclear mercuration in a *N*-unsubstituted pyrazole.

1-Phenyltetrazole is mercurated at C-5 [83d].

The above mercurations are normal electrophilic substitutions. 1,3-Diphenylimidazolium perchlorate gives with potassium t-butoxide an ylide, which with mercuric chloride reacts at C-2 [326].

Little is known about the mercuration of pyrazolones. 3-Phenylpyrazol-5-one gives a product formulated with one mercuric substituent at C-4 and two in the phenyl ring [485]. With mercuric chloride in methanol, 3-methyl-1-phenylpyrazol-5-one gives a compound represented as (72), together with the corresponding compound disubstituted in the phenyl ring. Antipyrine gave an analogous compound written as (73) [485]. With mercuric acetate in acetic acid or with mercuric chloramide, antipyrine gave the normal product of 4-substitution. More vigorous treatment led also to substitution in the phenyl ring [479].

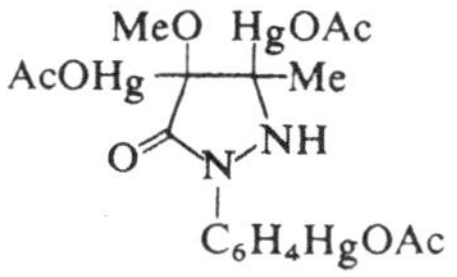

(72) 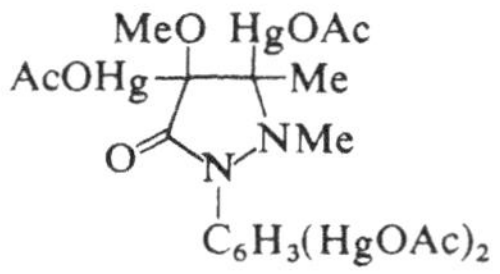   (73)

Halogen–metal interchange has been very little investigated in the azole series. 4-Bromo-1-methylpyrazole does not react with lithium, but with butyl and phenyl lithium some success has been achieved [77g]. 1,3-Dialkyl-4-bromo-5-chloropyrazoles exchange their bromine atoms for lithium with butyl lithium [322]. Other results are summarised in table 8. It will be seen that in some cases the bromo compounds underwent metalation, rather than or as well as halogen–metal interchange.

3.9   Nitration

The stability to acids of the pyrazoles shows in the methods used to nitrate them. Heated with a mixture of fuming nitric and sulphuric acids,

Table 8. Halogen–metal interchange in pyrazoles[77g]

Pyrazole	Reagent	Products of carboxylation		
		A[a]	B[b]	Carboxylic acids
4-Br	BuLi	51%		pyrazole-4- and 4-bromopyrazole-3-carboxylic acid
4-Br	2BuLi	23%		pyrazole-4-carboxylic acid (72%)
4-Br	PhLi	Sole product		
4-Br	2PhLi	52%		4-bromopyrazole-3-carboxylic acid (35%)
4-Br-1-Me	BuLi		37%	1-methylpyrazole-4-carboxylic acid (52%)
4-Br-1-Me	PhLi	28%	29%	Traces of 4-bromo-1-methylpyrazole-5- and 1-methylpyrazole-4-carboxylic acid
4-Br-1,5-Me$_2$	BuLi		+	1,5-dimethylpyrazole-4-carboxylic acid
4-Br-1,3,5-Me$_3$	2BuLi		+	1,3,5-trimethylpyrazole-4-carboxylic acid
3,4,5-Br$_3$	4BuLi			3,5-dibromopyrazole-4-carboxylic acid

(a) Starting material.
(b) De-brominated analogue of starting compound, formed by inefficient carboxylation.

pyrazole gives 4-nitropyrazole in good yield[1c,2f,74b]. 3-Methyl-[1c], 3,5-dimethyl-[431a], 1,3,5-trimethyl-[1c], and 3-chloro-5-methyl-pyrazole [426a], and 3-methylpyrazole-5-carboxylic acid[426b] are also nitrated at C-4. 1-Methylpyrazole is similarly nitrated[487c,488], as is 3-t-butyl-pyrazole[489b].

4-Nitropyrazole cannot be nitrated further[489b], but with mixed acid 3-nitro- and 3-methyl-5-nitro-pyrazole are further nitrated at C-4, and 4-ethyl-3-nitropyrazole gives 4-ethyl-3,5-dinitropyrazole[489b]. 4-Bromo-1-methylpyrazole gives 4-bromo-3,5-dinitropyrazole[490b]. These results show the belief that nitration could only be effected at C-4 to be unfounded. Acetyl nitrate nitrates 1,4-dimethyl- and 4-ethyl-1-methyl-pyrazole at C-3[491].

The nitrates of pyrazole and several of its derivatives, whether substituted at C-4 or not, when treated with acetic anhydride give *N*-nitro compounds[74c]. Using solutions of acetyl nitrate, or by adding nitric acid and then acetic anhydride to a solution of the pyrazole in acetic acid, a wide range of *N*-nitropyrazoles has been prepared. Thus, several 4-alkylpyrazoles gave 4-alkyl-1-nitropyrazoles, and 4-nitropyrazole gave 1,4-dinitropyrazole; 3-t-butyl- and 3-nitro-pyrazole gave the 1,5-disubstituted compounds. With 1 equivalent of acetyl nitrate 3-methylpyrazole gave 3-methyl- and 5-methyl-1-nitropyrazole, but with a large excess of the reagent only the 1,3-compound was obtained[489a,b].

The *N*-nitro compounds from pyrazole and 3,5-dialkylpyrazoles are converted by concentrated sulphuric acid into the 4-nitropyrazoles[74c, 489a]. 4-Iodo-1-nitropyrazole gives only 4-iodopyrazole, but the other 4-halogeno compounds give free halogens and 'nitrous fumes'[74c]. The production of 4- from *N*-nitropyrazoles by reaction with strong acid is clearly a dissociation followed by normal nitration. The thermal rearrangement of *N*-nitro- to 3-nitropyrazoles is a process of an entirely different kind, and is discussed in §9.11.

Nitration of imidazoles with hot mixed acid or nitric acid proceeds at C-4(5) and not at all if these positions are blocked, or if a carboxyl group is present[274c]. This statement is based on the behaviour of imidazole[408,492], 4,5-dimethyl-[274c], and 4,5-dibromo-imidazole [374a], a number of alkyl-[16c,274a,c,368a], halogeno-[374a], alkyl-halogeno-imidazoles[373c], 1,2,5-trimethyl-[496] and 1,2-dimethyl-4-isopropylimidazole[493], some 1,2-dialkyl-4- and -5-chloroimidazoles [494a,495a,b], and imidazole-4- and -4,5-di-carboxylic acid[274c]. 1-Methyl-[497] and 1,2-dimethyl-imidazole[498] give their respective 4- and 5-nitro derivatives in the rough ratios of 5 : 2 and 2 : 1. Generally the conditions used to nitrate imidazoles have been severe.

Evaporation of a solution of 1,2,4-triazole in nitric acid gives only the nitrate[7b,430c]. Chloro-1,2,4-triazole is unchanged, and bromo- and iodo-1,2,4-triazole give mainly 1,2,4-triazole nitrate[430c].

Table 9. Nitration of imidazole in sulphuric acid at 25 °C

H_2SO_4/%	83.7	89.6	93.8	98.8	1% SO_3
Yield of 4-nitroimidazole/%	46	26	19	19	90

Kinetic studies have been made of the nitration in sulphuric acid of both pyrazole and imidazole[128]. Whilst the 4-nitration of pyrazole was quantitative, that of imidazole gave yields dependent on the acidity, as shown in table 9. Ring-opening occurred, giving ammonia and, presumably, products of oxidation. It was, however, possible to separate the nitration process kinetically, and it and the nitration of pyrazole were shown to be of the usual second order kinetic form. For pyrazole the plot of $\log_{10}k_2$ *v.* %H_2SO_4 (90–99% range) resembled that for the nitration of quinoline, though the data for pyrazole referred to 80.8 °C and those for quinoline to 25 °C. The conclusion derived, that nitration involved the pyrazolium cation, was supported by consideration of the Arrhenius parameters. At lower acidities the slope of the plot for pyrazole was considerably smaller than that for quinoline and the cation nitration may be limited to acidities above 90% sulphuric acid. The corresponding plot for imidazole also indicates that the cation is being

Table 10. Comparison of nitration rates with encounter rates for the free bases present in 98% sulphuric acid at 25 °C

Compound	pK_a	$\log_{10}k_2$ (calc.)[a,b]	$\log_{10}k_2$ (obs.)[b]	k_2 (obs.)/k_2 (calc.)
Quinoline	4.94	–5.88	–2.02	7×10^3
Pyrazole	2.53	–3.47	–4.88[c]	4×10^{-2}
Imidazole	7.03	–7.94	–3.42	3×10^4

(a) Calculated from an estimated encounter rate of 8.5×10^8 1 mol^{-1} s^{-1} by assuming the protonation of the bases to follow H_0.
(b) k_2 in 1 mol^{-1} s^{-1}.
(c) Extrapolated from kinetic results for higher temperatures.

nitrated but the side reactions (which appear to be alternative reaction paths for a common intermediate formed in the rate-determining step) complicate the situation. Knowledge of the ionisation constants for these bases makes it possible to calculate the encounter rate constants for the free bases, and comparison with the experimental rate constant (see table 10) shows clearly that with imidazole (and quinoline) the cation must be involved, but is equivocal for pyrazole. Partial rate factors of 2.1×10^{-10} and 3.0×10^{-9} were calculated for the nitration in 98% sulphuric acid at 25 °C of the four position of the pyrazolium and imidazolium cation, respectively. The general significance of these values in indicating the very low reactivity of these cations is obvious, but too precise an interpretation should not be attempted. Partial rate factors only have significance if the rate of nitration of the compound in question depends upon the medium in the same way as does the rate of nitration of benzene. It is not clear in the case of pyrazole what is happening at lower acidities, and the case of imidazolium is complicated by the dependence of the yield of 4-nitro-imidazole upon acidity.

The kinetics of nitration in sulphuric acid (75–98%) of 1,3,5-trimethyl-pyrazole (20.9 °C) and 1,2,3,5-tetramethylpyrazolium cation show the trimethyl compound to react as the cation over the acidity range studied. The reactivities of the tetramethyl quaternary cation and of the trimethyl cation were very closely similar, and in 98% sulphuric acid both were very roughly 10^4 times as reactive as the pyrazole cation[471].

Kinetic studies of the nitration of 1-arylpyrazoles and azole *N*-oxides are mentioned below.

Like the halogenation (p. 46), the nitration of arylpyrazoles (table A.5) shows an interesting dependence on the reagent used. With 1-aryl-pyrazoles the dependence is broadly that nitration with nitric acid in acetic anhydride causes substitution at C-4 of the pyrazole ring whilst nitration in mixed acid causes substitution in the phenyl ring. A similar result, affecting a C-phenyl group, is seen in the case of 1-(*p*-nitrophenyl)-5-phenylpyrazole which with nitric acid in acetic anhydride gives the

4-nitro compound, but with mixed acids, 1,5-di-(*p*-nitrophenyl)pyrazole. These differences might be a consequence of nitration occurring on the free base in nitric acid–acetic anhydride, and on the cation in mixed acids[251a,499a]. However, it has been suggested[129] that such an explanation is not reasonable in such cases as that of the 1-(*p*-nitrophenyl) compound mentioned, which, it is thought, would provide measurable amounts of the free base even in solutions in mixed acids. The alternative explanation, that specific interaction occurs between the pyrazole ring and the nitrating agent in nitric acid–acetic anhydride, as represented here, (the nitrating agent being shown as protonated acetyl nitrate) is offered[129]. The observation that nitration of 1-phenylpyrazole with nitronium fluoborate gave only 4-nitro-1-phenylpyrazole was adduced to support this view[129].

Kinetic studies of the dependence upon acidity of the rate of nitration of some 1-arylpyrazoles in sulphuric acid at 25 °C have partly clarified these problems; all of the compounds examined reacted as their cations so that the direction of substitution into the aryl ring is a consequence of the de-activation of the pyrazole ring caused by protonation[217]. 1-Phenylimidazole (see below) also reacted as the cation (relative rate, 1.9×10^{-5}, $f_p = 11 \times 10^{-5}$: see table 11).

A striking feature of these results (table 11) is the high degree of de-activation which the protonated pyrazole ring causes in the 1-aryl groups, coupled with the attendant almost quantitative *para*-orientation. The partial rate factors for the *para*-positions (f_p) are very small, being comparable with f_m associated with nitro- and trimethylammonio-substituents[217]. Both the quaternary 1-methyl-2-phenylpyrazolium cation and 5-methyl-1-phenylpyrazolium cation were markedly less reactive than 1-phenylpyrazolium, a consequence of the steric hindrance to coplanarity of the two nuclei caused by the methyl groups.

For nitrations in acetic anhydride good kinetics could not be observed, and the yields of mono-nitro compounds were low (40–75%). Surprisingly, the major product formed from 5-methyl-1-phenylpyrazole was 5-methyl-3-nitro-1-phenylpyrazole, a smaller proportion of 5-methyl-4-nitro-1-phenylpyrazole also being formed. This is the only reported example of 3-nitration accompanying 4-nitration of a compound unsubstituted at C-4, and it, combined with the non-quantitative character of these reactions makes it unwise on present knowledge to ascribe 4-nitration to a simple free base reaction. The possible role of initial

Table 11. Nitration of 1-arylpyrazoles in sulphuric acid at 25 °C[217]

R^1	R^2	R^3	R^4	R^5	Products[a]	Relative rate[b]	f_p
H	H	H	Me	Me	3'	1.2×10^{-3}	
H	H	Me	H	H	4'	4.7×10^{-5}	28×10^{-5}
H	H	H	Me	H	c	1.2×10^{-5}	
H	H	H	H	H	4'	5.9×10^{-6} }	35×10^{-6}
H	Me	Me	H	H	4'	5.9×10^{-6} }	
H	Me	H	H	H	4'	3.5×10^{-7}	21×10^{-7}
Me	H	H	H	H	4'	2.4×10^{-8}	14×10^{-8}

(a) The position of nitration in the phenyl ring is indicated. Yields were quantitative.
(b) With respect to benzene.
(c) Two products were formed; the major one was either 1-(2-methyl-4-nitrophenyl)-or 1-(2-methyl-5-nitrophenyl)-pyrazole, the minor one was not identified.

N-nitration has been mentioned[217]. It is interesting to note that 4-ethyl-1-phenylpyrazole is nitrated at C-3 by acetyl nitrate[491].

Nitration of 3-phenylpyrazole in acetic anhydride (table A.5) gives 1-nitro-5-phenylpyrazole[489b,499a], and 4-phenylpyrazole is also *N*-nitrated[489b]. 1-Nitro-5-phenylpyrazole is converted into 3-(*p*-nitrophenyl)pyrazole in concentrated sulphuric acid[499a].

Much less is known about the nitration of arylazoles other than the arylpyrazoles, especially as regards any possible changes in orientation which might arise from change of nitrating conditions (but see 1-phenylimidazole, table A.5). If we consider the data on nitration with nitric acid or mixed acids it appears to be the case that nitration of the phenyl group is less completely *para*-oriented in the imidazoles than in the pyrazoles. Among the 2-phenylimidazoles orientation is mainly *para* in the phenyl group until 2-phenylimidazole-4,5-dicarboxylic acid is reached, when it becomes mainly *meta*. In this series, except in the cases of 2-phenylimidazole-4-carboxylic acid and -4,5-dicarboxylic acid, total yields of products were not much greater than 50%, and the behaviour of imidazole itself (p. 62) suggests that it is hazardous to try to draw conclusions from these results in the absence of mechanistic studies. The case of 1-phenylimidazole was mentioned above; the product of nitration decomposes in sulphuric acid[217]. With 5-phenyltetrazole we seem to reach a case of major *meta*-orientation without the presence of extra substituents in the heterocyclic nucleus.

The historical importance of the nitration of arylazoles has been mentioned (pp. 1 *et seq.*). Concerning one of these reactions some confusion exists. From 5-methyl-1-phenylpyrazole-3-carboxylic acid Knorr[1c] (p. 1) obtained by nitration with fuming nitric acid or with mixed acid a nitrophenyl compound, m.p. 122–4 °C. This is recorded in Beilstein's *Handbuch* [500] as 5-methyl-1-(*p*-nitrophenyl)pyrazole-3-carboxylic acid. Later repetition of the nitration gave an uncharacterised product, m.p. 153–4 °C, whilst two ring-synthetic methods gave what must be regarded as authentic 5-methyl-1-(*p*-nitrophenyl)pyrazole-3-carboxylic acid, m.p. 216–7 °C[501]. It seems possible that the nitrations gave mixtures.

In view of the behaviour of aminopyridines upon nitration (Schofield, pp. 172–3) some interest attaches to the cases of the amino-azoles, but not much is known about them. 3-Acetamidopyrazole is converted by mixed acids into 3-acetamido-4-nitropyrazole, but attempts to nitrate 5-amino-1-isopropyl-3-methylpyrazole or its acetyl derivative gave only tars[323c].

4-Ethoxycarbonylamino-imidazole was successfully nitrated at C-5 [374d].

Experiments with amino-1,2,4-triazoles reveal behaviour similar to that of aminopyridines[514a], though in all but the particular case of 3-amino-1,2,4-triazole itself and of 5-amino-1-methyl-1,2,4-triazole[514d], C-nitration is not possible, and this is also the case with 5-aminotetrazoles. In all of the cases recorded below, the nitramines were obtained by dissolving the nitrates of the amines in concentrated sulphuric acid.

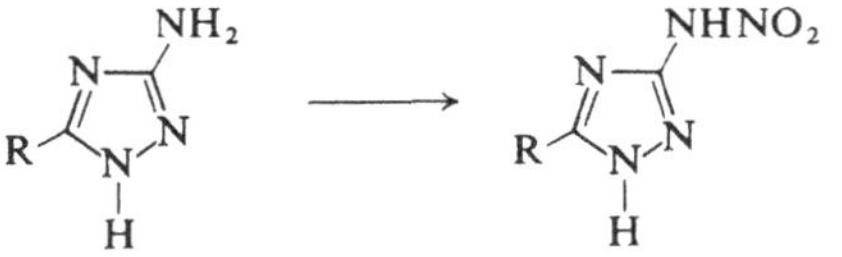

R = H, Me, Et [514a]

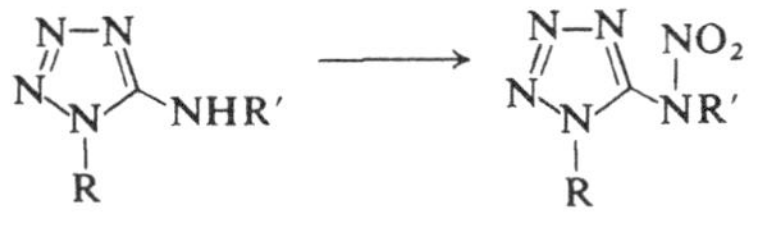

R = H, R' = H [515b]

R = Me, R' = H
    Et   H
    H   Me } [515d]
    H   Et

A few simple pyrazolones have been nitrated at C-4 with nitric acid. These include 3-methylpyrazol-5-one, 1-ethoxycarbonyl-3-methylpyrazol-5-one[516b], and 1-benzyl-2,3-dimethylpyrazol-5-one[428d]. 3,4-Dimethylpyrazol-5-one does not react with nitric acid, and 1-ethoxycarbonyl-3,5-dimethylpyrazol-5-one merely loses the ethoxycarbonyl group[516b]. 1,5-Dimethylpyrazol-3-one is nitrated at C-4[517]. Among the imidazolones only the hydantoins appear to have been nitrated[487e,518]; since 5,5-dimethylhydantoin also gave a nitro derivative, whilst 1-methyl-hydantoin did not, that from hydantoin is thought to be 1-nitrohydantoin, a conclusion supported by ring-opening experiments, and by ultraviolet spectroscopic data[519]. 1,2,4-Triazol-3-one is nitrated at C-5[430c], and 1-methyl-1*H*-1,2,4-triazol-5-ones at C-3 (the isomeric 3-ones were not nitrated) with fuming nitric acid[461].

A number of nitropyrazolones have resulted from the oxidation of the corresponding nitroso compounds with nitric acid (see p. 71). It should be noticed that where nitropyrazolones appear to be formed directly by nitration of pyrazolones with nitric acid the process of nitration by nitrosation followed by oxidation, familiar with such reactive derivatives of benzene as phenol, may in fact be operating. For neither the simple pyrazolones mentioned above, nor the arylpyrazolones discussed below is there any evidence on this point. The same is true for the nitrations with nitric acid of the other azol-ones.

Few arylazoles containing amino groups in the heterocyclic ring have been nitrated. 5-Amino-3-methyl-1-phenylpyrazole gives the 1-*p*-nitrophenyl compound when nitrated with mixed acid[330b] (the *para*-orientation is not proved but most probable). Like halogenation (p. 47), nitration of 3-anilino-5-phenylpyrazole causes tri-substitution, giving 4-nitro-3-(2',4'-dinitroanilino)-5-phenylpyrazole[437a,b,c]. 3-Picrylaminopyrazoles are nitrated at C-4[490b].

3-Amino-5-aryl-1,2,4-triazoles, in contrast to their alkyl analogues (see p. 66), do not give nitramines, but are said, without exemplification, to be nitrated in the aryl ring[514a]. Obviously, *N*-nitration followed by rearrangement is a possibility with these cases.

Some 1-arylpyrazol-5-[330g,392e†] and -3-ones[330c,e] have been 4-nitrated with nitric acid. 1-(*m*-Bromophenyl)-3-methylpyrazol-5-one gives the 4,*p*-dinitro compound[520], and 3-methyl-1-phenylpyrazol-5-one readily gives the corresponding product, picrolonic acid[392i,521]. Picrolonic acid itself is said to give 3-methyl-4,4-dinitro-1-(*p*-nitrophenyl)-pyrazol-5-one with nitric acid[522].

In contrast, nitration of 3-methyl-1-phenylpyrazol-5-one with mixed acid gives 3-methyl-1-(*p*-nitrophenyl)pyrazol-5-one[324i,523]. The apparently different behaviour of nitric acid and mixed acid in this case could be due to nitration with the former proceeding by nitrosation. However, the difference requires experimental verification.

Dilute nitric acid nitrates 3-hydroxy-1-phenylpyrazole at C-4[131].

A number of antipyrines have been nitrated[330h,392b,k,442a], giving the 4-nitro compounds, and 3-antipyrines behave similarly[330e,524].

† 3-Methyl-4-nitro-1-phenylpyrazol-5-one was in fact obtained by treating 3-methyl-1-phenylpyrazolid-5-one with nitric acid, and the example is relevant only if the pyrazolidone is assumed to be first oxidised under these conditions to the pyrazolone. This is not unlikely[392e].

Dinitrogen tetroxide is said to add to the 3,4-double bond in antipyrine, giving by loss of nitrous and nitric acid, 4-nitro- and 4-nitroso-antipyrine [525].

1-Methyl-3-phenyl-1,2,4-triazol-5-one is nitrated at the *para*-position of the phenyl ring by mixed acids[516a], and 1-aryl-5-ethoxy-1*H*-1,2,4-triazoles are also nitrated in the aryl group[526].

Compared with the pyrroles (Schofield, p. 80), the azoles have yielded fewer examples of the replacement by nitration of groups other than hydrogen. With mixed acid, 4-bromo-3-methylpyrazole gives 3-methyl-4-nitropyrazole[426c]. Whilst 4,5-dibromoimidazole could not be nitrated [374a], the following are interesting examples of nitro-deiodination[122].

	R	R'	R''	R'''		R	R'	R''	R'''
	H	I	H	I	$\xrightarrow{\text{HNO}_3/\text{H}_2\text{SO}_4}$	H	I	H	NO$_2$
	H	Me	I	I	⟶	H	Me	NO$_2$	I
	Me	Me	I	I	⟶	Me	Me	NO$_2$	I (7 parts)
						Me	Me	I	NO$_2$ (3 parts)

In the pyrazolone series phenylazo groups have been displaced in nitration; an example is the conversion of 3-methyl-4-phenylazopyrazol-5-one by nitric acid in acetic acid into 3-methyl-4-nitropyrazol-5-one (94%) [527b].

Very little is known about electrophilic substitutions into azole *N*-oxides. However, with mixed acid 1-methylpyrazole 2-oxide gives high yields of 1-methyl-5-nitropyrazole 2-oxide, and, under slightly different conditions, of a dinitro-compound which is possibly 1-methyl-3,5-dinitro-pyrazole 2-oxide. The dependency of the rate of mono-nitration on the acidity shows that reaction occurs via the free base[528]. In similar circumstances 1,4,5-trimethylimidazole 3-oxide is nitrated, as the free base at C-2[529]. The point is not yet established, but it is possible that these free base reactions occur at the encounter rate. The two points of interest are the probable activation of the azole nuclei, and the unusual orientation of substitution by comparison with reactions involving azole cations (and possibly, but less certainly, by comparison with the behaviour of azole free bases) (see p. 62). The cases are interesting for the parallels which they provide with the behaviour of pyridine 1-oxide as compared with that of pyridine (Schofield, pp. 174, 280).

3.10   Nitrosation

Unlike pyrrole, which can be *C*-nitrosated (Schofield, p. 80), the parent azoles have been neither *C*- nor *N*-nitrosated[2b,c]. Nitrosopyrazoles are, however, intermediates in the direct conversion of pyrazoles into their diazonium or diazo derivatives, as are nitrosopyrroles in the corresponding transformation of pyrroles (Schofield, p. 80). Thus, 3,5-dimethylpyrazole in buffered nitrous acid gives the 4-diazonium cation and thence the 4-diazo compound ($\sim$ 70%). 5-Ethoxy-3-methylpyrazole readily gives the diazo compound[486].

Nitroso-de-trimethylsilylation is useful in the pyrazole series. Thus, in the presence of trifluoroacetic acid both 4-trimethylsilyl- and 3,4-bis-trimethylsilyl-pyrazole give 4-nitrosopyrazole. With isoamyl nitrite alone 3-trimethylsilylpyrazole gives 3- and 4-nitrosopyrazole, 4-trimethylsilyl-pyrazole gives 4-nitrosopyrazole, and 3,4-bis-trimethylsilylpyrazole gives 4-trimethylsilyl-3-nitroso- and 4-nitroso-pyrazole[530].

In the arylazoles which have been examined, nitrosation occurs in the heterocyclic ring (table 12). The use of amyl nitrate produces the nitro as well as the nitroso compound.

The process of directly introducing a diazo group gave only a poor yield with 3,5-diphenylpyrazole, and failed with 3-phenylpyrazole.

Table 12. The nitrosation of arylazoles

Pyrazoles	Reagent	Product	Ref.
1,3,5-Ph$_3$	HNO$_2$-H$_2$SO$_4$	No action	392d
	'Gaseous HNO$_2$'	Green compound;	
	in ether	4-NO-1,3,5-Ph$_3$?	392d
Imidazoles			
5-Me-2-Ph	AmNO$_2$-NaOEt	No action	532d
2-Me-5-Ph	AmNO$_2$-NaOEt	2-Me-4-NO-5-Ph	532d
4-Ph[a][b]	AmNO$_2$-NaOEt	5-NO-4-Ph	532c
2,5-Ph$_2$[a][b]	AmNO$_2$-NaOEt	4-NO-2,5-Ph$_2$	532c
2,5-Ph$_2$[c]	AmNO$_3$-Et$_2$O	4-NO-2,5-Ph$_2$	146c
4,5-Ph$_2$[a][b]	AmNO$_2$-NaOEt	No action	532c
2,4,5-Ph$_3$[a][b]	AmNO$_2$-NaOEt	No action	532c

(a) No reaction with NaNO$_2$-Me.CO$_2$H. (b) No reaction with AmNO$_2$-Et$_2$O.
(c) As well as the nitroso compound, the nitro compound and products of ring-opening were formed.

Indirectly, success was achieved by mercurating 3-phenylpyrazole at C-4 (p. 60); the mercuri compound reacted with nitrosylsulphuric acid to give the diazo compound in poor yield. 3,4-Diphenylpyrazole reacted with nitrosylsulphuric acid in the presence of mercuric ions, but substitution occurred at the *para*-position of the 4-phenyl group[486a].

Attempts to nitrosate some 1-alkyl-5-amino-3-methylpyrazoles gave only tars, and sulphonyl and acyl derivatives did not react[323c]. However, 5-amino-1-methylpyrazole gave 5-amino-1-methyl-4-nitroso-pyrazole[531].

3-Methylamino- (like its 5-substituted analogues) and other 3-alkyl-amino-1,2,4-triazoles give the *N*-nitroso-alkylamines, rather than *C*-nitrosated compounds[268d,g,533]. The behaviour of primary amino derivatives of the azoles is discussed in §9.4.

5-Acylamino-3-phenylpyrazoles can be nitrosated at C-4[534b].

Attempts to nitrosate 5-amino-1-phenylpyrazole gave only tars, whilst 3-amino-1-phenylpyrazole gave 1,1′-diphenyl-3,3′-azopyrazole[323c]. In contrast, 5-amino-3-methyl-1-phenylpyrazole with nitrous acid in dilute hydrochloric acid gives a high yield of the 4-nitroso compound; only a minor proportion of diazotisation occurs[323c,537]. The use of strongly acidic media is stated to cause diazotisation[111c]. With amyl nitrite in ether, 5-(*p*-aminobenzenesulphonylamino)-3-methyl-1-phenylpyrazole gives its 4-nitroso derivative[417a], whilst sodium nitrite in acetic acid converts 5-amino-1,3-diphenylpyrazole into its 4-nitroso derivative and the related diazoamino compound[532a]. 3-Methyl-1-phenyl-4-aminopyrazole behaves differently from its 5-amino analogue, being smoothly diazotised[330k]. There is obvious similarity between the behaviour of 5-aminopyrazoles and 2-aminopyridines (containing electron-releasing substituents) on the one hand, and 4-aminopyrazoles and 3-aminopyridines on the other (Schofield, p. 176).

Several pyrazolones have been nitrosated. The examples illustrate compounds unsubstituted at C-4 which are nitrosated at that position. When C-4 is already substituted the results are not so clear. 3-Methyl-4-phenylazopyrazol-5-one is said to give a *N*-nitroso compound (ring nitrogen)[324i,523]. 1-Aminocarbonyl-3,4-dimethylpyrazol-5-one undergoes reaction with loss of the 1-aminocarbonyl group, forming an unstable *N*-nitroso compound[516b]. It has not been shown whether this is the same as the compound of unknown structure obtained from 3,4-dimethyl-pyrazol-5-one[74g,421c], and originally regarded[421c] as 3,4-dimethyl-4-nitrosopyrazol-5-one.

	R	R′	Reference
	H	H	421b
	H	Me	428c, 516b, 535–6
	PhCH$_2$	Me	428d
	H	.CO$_2$Me	421b
	.CONH$_2$	Me	465
	.CSNH$_2$	Me	465

1-Benzyl-2,3-dimethylpyrazol-5-one is nitrosated at C-4[428d,517].

Nitrosation of 1,2,3-triazolone[428b,429d] and 1,2,3-triazol-5-one-1-acetic acid[428a] gives deep violet-red and blue solutions respectively. In the former case sodium and potassium salts of the nitroso compound were isolated and gave on acidification yellow needles which exploded on heating[429d].

3-Methyl-1-phenylpyrazol-5-one [330c,392k] and several other 1-aryl-
pyrazol-5-ones [330g,392c,538] have been nitrosated at C-4. 1-Substituted
3-phenylpyrazol-5-ones [330d,428d], 3-phenylpyrazol-5-one itself [536],
and 5-methyl-1-phenylpyrazol-3-one [321f,330c] behave similarly.

Antipyrine [324h,392b,487a,539,540c] and a number of its analogues
[330e,h,442a,541] and also iso-antipyrine [330d] are readily nitrosated at
C-4. In contrast, 3-antipyrine is said to be difficult to nitrosate [330a,
524], and it is not clear that the nitroso derivative has been obtained.
As mentioned already, both 4-nitro and 4-nitroso compounds can be
obtained from the reaction of antipyrine with dinitrogen tetroxide (see
p. 68).

Sulphur analogues of pyrazolones have been nitrosated [330e,f].

1-Phenylpyrazol-3,5-dione is nitrosated at C-4 [321a], and nitrosation
of 3-chloro-2-methyl-1-phenylpyrazol-5-one gives 2-methyl-4-nitroso-1-
phenylpyrazol-3,5-dione [321k].

1-Phenyl-1,2,3-triazol-5-one and its 4-carboxylic acid both give 4-nitroso-
1-phenyl-1,2,3-triazol-5-one [19d].

The replacement of sulpho by nitroso groups is observed in the nitrosation of 3-methyl-1-phenylpyrazol-5-one-4-sulphonic acid [427].

In a number of cases the use of an excess of nitrous acid leads to the
formation of a nitro compound by oxidation of the initially formed
nitroso compound. Examples are common in the pyrazolone [330c,392k,
428d] and antipyrine [330d,e] series, and include one in which the
reagent was amyl nitrite [538a].

3.11 Sulphonation and other S-substitutions

Hot fuming sulphuric acid sulphonates pyrazole [392g]; the product is
almost certainly pyrazole-4-sulphonic acid, for it differs from a ring-
synthetic compound which must be pyrazole-3-sulphonic acid [542].
3-Methyl-[1c] and 3,5-dimethyl-pyrazole [437c] behave similarly.

With hot 50–60% oleum imidazole gives imidazole-4-sulphonic acid
[412a,543], and 2-methyl-[453a], 4-methyl-[543], 4-bromo-[374a], and
4-bromo-2-methyl-imidazole [11] are all similarly sulphonated. Under
severe conditions 2-methylimidazole gives some 2-methylimidazole-4,5-
disulphonic acid, and there are slight indications that imidazole can be
disulphonated [453a]. Sulphonation of imidazole has also been effected
with sulphur trioxide and air [324o]. In contrast to 4-bromo-[544],
4-chloro-1-methyl-, and 5-chloro-1-methyl-[545], which have been
successfully chlorosulphonated, 4-acetamido-imidazole gave only decomposition products [544].

1-Phenylpyrazole is readily sulphonated at the *para*-position of the
phenyl ring. Further reaction of the product occurs at C-4 in the
pyrazole ring. From 1-phenylpyrazole, chlorosulphonic acid produces a

mixture containing some of the 4,4'-disulphonyl chloride; in contrast, chlorosulphonic acid in chloroform gives 1-phenylpyrazole-4-sulphonyl chloride. 4-Chloro-1-phenylpyrazole under severe reaction conditions reacts at the *para*-position[546]. It is surprising that with oleum, 3-methyl-1,5-diphenylpyrazole gives 3-methyl-1,5-diphenylpyrazole-4-sulphonic acid and 3-methyl-1-phenyl-5-(*p*-sulphophenyl)pyrazole, no substitution occurring in the ring attached to nitrogen[446]. Chlorosulphonation of some 1-aryl-3,5-dimethylpyrazoles occurs at C-4[540j].

4-Phenyl- and 4-(β-naphthyl)-imidazole are said to give *N*-sulphonic acids by reaction with pyridine–sulphur trioxide[452b].

Chlorosulphonation of 2-phenyl-2*H*-1,2,3-triazole almost certainly occurred at the *para*-position of the phenyl group[510a].

Little is known about the sulphonation of azolones lacking aryl groups. 3-Methylpyrazol-5-one is said to give the 4-sulphonic acid[324*l*].

Several workers[324*l*,350d,427,462,540e,f] have sulphonated 3-methyl-1-phenylpyrazol-5-one. Generally the product has been described as the 4-sulphonic acid, but Möllenhoff[462] obtained the *p*-sulphophenyl compound. In fact, it appears that the product depends on the method used, and Möllenhoff obtained the *p*-sulphophenyl compound because his method of hydrolysing the reaction solution hydrolysed the sulphonic acid group from C-4. Sulphonation in oleum appears to give the disulphonic acid by both possible routes; sulphonation at C-4 and at the *para*-position of the phenyl group, followed by disulphonation. The 4-sulphonic acid is formed predominantly and is itself fairly rapidly sulphonated even at ordinary temperatures. When sulphuric acid is used the primary product is still mainly the 4-sulphonic acid, but since both it and the disulphonic acid are more readily hydrolysed than the *para*-sulphonic acid the latter accumulates and is isolated[427].

Other 1-arylpyrazol-5-ones have given products described as the 4-sulphonic acids[324*l*], and antipyrines have been sulphonated[350d, 540e] and chlorosulphonated[547] at C-4 (in the case of antipyrine an early worker[462] represented the product as the *para*-acid).

1,3,5-Trimethylpyrazole reacts vigorously with sulphur monochloride, giving a resin and 25–35% of di-4-(1,3,5-trimethylpyrazolyl) sulphide. Other 1-alkyl and 1-aryl compounds behave similarly. An excess of sulphur monochloride gives some 4-(1,3,5-trimethylpyrazolyl) chlorosulphide with 1,3,5-trimethylpyrazole[126c]. Surprisingly, the sulphide is also formed from trimethylpyrazole and thionyl chloride; even in the presence of aluminium chloride no sulphoxide is produced. A product from 1-phenylpyrazole and thionyl chloride, first described as di-4-(1-phenylpyrazolyl) sulphoxide[59f], is evidently the sulphide[126c].

5-Acetamido- and 5-amino-3-methyl-1-phenylpyrazole react at C-4 with *p*-nitrobenzenesulphenyl chloride[548].

In an unusual reaction 3-methyl-1-phenylpyrazol-5-one reacts with thiophenols in dimethyl sulphoxide to give the 4-arylsulphides. An ylide intermediate is said to be involved[549].

3.12 Thiocyanation

Direct thiocyanation of 1-phenylpyrazole could not be effected, but 4-acetoxymercuri- and 4-chloromercuri-1-phenylpyrazole gave 1-phenyl-4-thiocyanatopyrazole with thiocyanogen[59a].

Antipyrine has been directly thiocyanated[350a,b], a reaction which is the subject of several patents[324f,487b]. 1,2-Diphenylpyrazol-3,5-dione gives both the 4-thiocyanato- and the 4,4-dithiocyanato compounds [550].

3.13 Reactions at the nitrogen atoms

The presence of two or more nitrogen atoms in the azoles multiplies the number of ways in which an *N*-unsubstituted compound can be converted into an *N*-substituted one. The possibilities, recalling those existing with pyridine derivatives containing tautomerisable substituents (Schofield, p. 180), are illustrated here. The first process is that of quaternisation of the pyridinic nitrogen atoms followed by, or accompanying (S_E2') proton loss. With compounds in which the imino-hydrogen atom is already replaced, say by an alkyl group, quaternisation will result. The second process is the S_E2 reaction, and should be related to the energy needed to localise two electrons on the nitrogen atom being attacked. The third process is that of electrophilic attack on the conjugate base of the parent heterocycle (S_E2cB)[141].

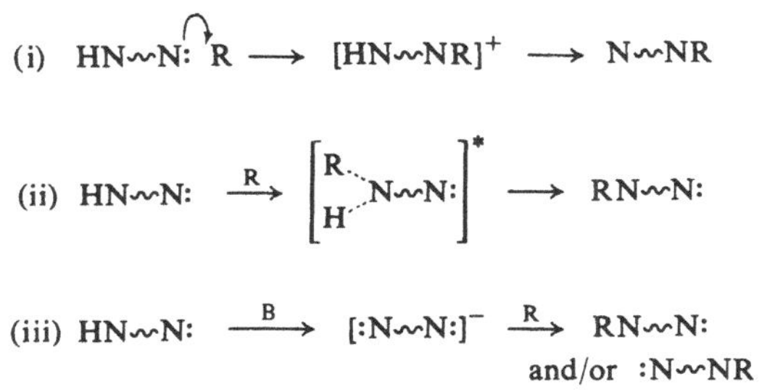

(i) $HN\!\sim\!N\!:\ R \longrightarrow [HN\!\sim\!NR]^+ \longrightarrow N\!\sim\!NR$

(ii) $HN\!\sim\!N\!: \xrightarrow{\ R\ } \left[\begin{matrix} R\cdots \\ H \end{matrix} N\!\sim\!N\!: \right]^* \longrightarrow RN\!\sim\!N\!:$

(iii) $HN\!\sim\!N\!: \xrightarrow{\ B\ } [:N\!\sim\!N:]^- \xrightarrow{\ R\ } RN\!\sim\!N\!:$
and/or $:N\!\sim\!NR$

Whilst these possibilities are clearly present, *a priori*, there is, in fact, except in the case of the alkylation of imidazoles, very little information about the mechanisms concerned in most of the *N*-substitutions to be mentioned. Sometimes the experimental conditions permit reasonable guesses; thus, reactions carried out with metal salts of azoles seem likely to be of the S_E2cB type.

3.13.1 *The attachment of alkyl and substituted alkyl groups: substitutive alkylation*

Pyrazoles have been *N*-alkylated by being heated with alkyl halides[63a,b, 66a-g,100b,g,i], or substituted alkyl halides such as methyl bromoacetate [233] or ethylene chlorhydrin[488]. The mechanisms of these reactions are not known and might be S_E2 or S_E2'. The latter seems probable in the light of knowledge about imidazoles (see p. 80). In these reactions there are formed salts from the halogen hydracids liberated, and also small proportions of quaternary salts from the *N*-alkylpyrazoles and the alkyl halides. An unsymmetrical pyrazole usually gives both possible *N*-alkylated isomers[66b,e,f,233]. Thus, 3-phenylpyrazole with methyl bromide at 100°C gives a mixture of 1-methyl-3-phenyl- and 1-methyl-5-phenyl-pyrazole[66b]. On the other hand, 3-chloro-5-phenylpyrazole with ethyl iodide gives only 5-chloro-1-ethyl-3-phenylpyrazole[100b]. 3-Methyl-pyrazole with ethyl bromide or methyl iodide gives comparable amounts of the possible isomers[66f], whilst 3-phenylpyrazole and ethyl bromide provide mainly 1-ethyl-3-phenyl-, and only a little 1-ethyl-5-phenylpyrazole [66g]. Methyl 3-methylpyrazole-5-carboxylate gives methyl 1,5-dimethyl-pyrazole-3-carboxylate with methyl iodide[100i]. This last case, and that of 3-chloro-5-phenylpyrazole are interesting. It might be argued that in the ester the tautomer (74) will be the dominant form, whilst with the chloro-compound the base-weakening influence of the halogen upon the adjacent nitrogen atom will render (75) the major form. In S_E2' reactions (74) and (75) would provide the isomers actually isolated.

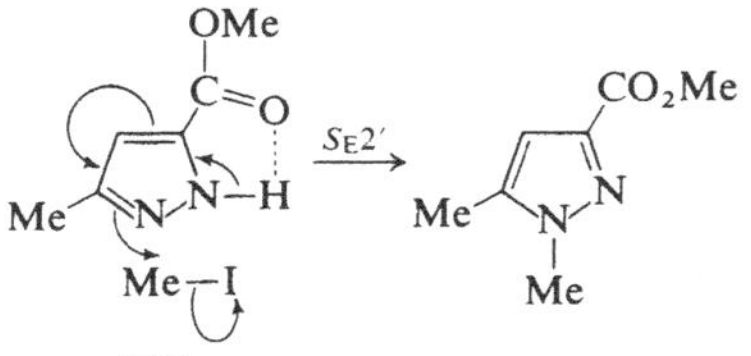

(74)

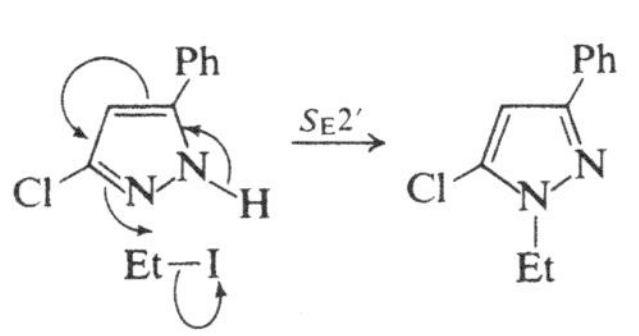

(75)

As suggested above, it is likely that alkylations in which pyrazole salts are used are of the S_E2cB type. Silver pyrazole and methyl iodide at 120°C give 1-methylpyrazole[2f,84a,392g], and the salt reacts similarly with β-chloroalanine[28]. Pyrazole[363b], 3-methyl-, and 3,5-dimethyl-pyrazole have been *N*-alkylated with alkyl halides and alcoholic sodium alkoxides[66a].

An unsymmetrical pyrazole can give two isomers when *N*-alkylated by these methods, but the proportion of each formed depends on the conditions and the nature of the substituents present[62b,c,63a,66b,d,e,f,g,j, 100a,b,d,i]. It is most probable that in a number of cases where the formation of one product is reported modern techniques of analysis might reveal the presence of the second isomer, and where two isomers are reported to be formed might not confirm the reported proportions of each one. A study of this problem is desirable, and in the meantime the following report can only be accepted with caution. Thus, sodium 3-chloro-5-methylpyrazole with methyl iodide in absolute ether is said to give 3-chloro-1,5-dimethylpyrazole, whilst in moist ether a little of the isomer is also formed. Silver 3-chloro-5-methylpyrazole with methyl iodide gives equal amounts of the two isomers, and the use of dimethyl sulphate and alkali produces both 3-chloro-1,5- and 5-chloro-1,3-dimethyl-pyrazole, the former predominating[62b,100a,i]. 3-Methylpyrazole with methyl iodide and sodium methoxide gives about equal parts of 1,3- and 1,5-dimethylpyrazole; ethyl bromide under the same conditions gives a similar result, but from benzyl chloride the main product is 1-benzyl-3-methylpyrazole[66f]. 3-Nitropyrazole with aqueous ethanolic potash and methyl iodide gives 1-methyl-3- (74%) and 1-methyl-5-nitro-pyrazole (26%)[246].

3-Chloro-5-methylpyrazole with ethanolic sodium ethoxide and ω-halogeno-alkanoic esters gives products formulated as the 1-substituted compounds, but the orientations are not established[551e]. With the same base and benzyl chloride, ethyl pyrazole-3-carboxylate is also believed to give the 1-benzyl compound, as is 3-phenylpyrazole with ethyl bromoacetate, with which pyrazole also reacts[363d]. Under these conditions ethyl pyrazole-3,4-dicarboxylate and methyl iodide give both possible isomers[553]. Sodamide has also been used as the base in the *N*-alkylation of pyrazoles[554b].

The use of aqueous alkali and dimethyl sulphate has already been instanced above, and these reagents are probably the most used for the purpose of *N*-methylating pyrazoles. They have been used with 4-nitro-[488] and 3,5-dimethyl-4-nitro-pyrazole[74b]. With 4-iodo-3-methyl-pyrazole they give 4-iodo-1,3-dimethylpyrazole[473].

Several pyrazoles have been *N*-alkylated by being heated with an alkyl halide and potassium carbonate[323d]. In this reaction with ethyl bromo-acetate, toluene or 2-ethoxyethanol have been used as solvents[473].

Potassium pyrazole with chloroform in benzene gives tri-1-pyrazolyl-methane[52a]. The sodium salt of 3,5-dimethylpyrazole (prepared with sodium in toluene) has been normally *N*-alkylated[555].

3,5-Dimethylpyrazole is substituted on nitrogen by the $PhCO.(CH_2)_2$ group by being heated with $PhCO.(CH_2)_2NMe_2$[555].

Diazomethane in ether has sometimes been used to *N*-alkylate

pyrazoles, e.g. 3-chloro-5-methylpyrazole, 3-methoxycarbonyl-4-phenyl-
and 3-methoxycarbonyl-5-phenyl-pyrazole. The main products from the
last two are the isomers *N*-methylated adjacent to the ester grouping[66j],
but the proportions of isomers, e.g. from 3-methoxycarbonylpyrazole vary
with the conditions[556a,b]. The product from 3-chloro-5-methyl-4-nitro-
pyrazole is probably a mixture[426a].

Methylations with diazomethane, like those with some other reagents,
might when they produce mixtures of isomers be taken to indicate the
occurrence of tautomerism in *N*-unsubstituted pyrazoles. Whether they do
so or not depends on the mechanisms of the reactions. An older theory
of the action of diazomethane suggested that this reagent placed a

methyl group upon the atom, in this case nitrogen, which held the proton. Another possibility is that the diazomethane abstracts the proton and the resulting methyldiazonium cation then reacts with the mesomeric pyrazolyl anion. In this case the formation of two products merely reveals this mesomerism, and even if the older mechanism applies the result demonstrates the existence of tautomerism without providing any evidence about the position of the equilibrium involved[208d,557]. For the pyrazoles the mechanism is not known.

Some of the differing consequences of using various reagents or conditions from among the examples quoted above are illustrated above. As has been stressed, these reports must be accepted with caution.

Reaction of an imidazole with one equivalent of an alkyl halide[6b, 9b,16d,e,54d,e,66h,84a,90b,c,408-9,558] or sulphate[16d,453b,496-8, 558-60] effects *N*-alkylation. The former reaction has been carried out neat, or in ether, alcohol, or benzene, the latter without a solvent or in water. The use of ether as solvent[54e] is disadvantageous[561], and in all of these reactions a degree of quaternisation can occur. With mustard gas in dilute aqueous solution at pH = 8.5 imidazole gives (76) and a quaternary picrylsulphonate supposedly of the cation (77)[560]. The structure (77) is not established and is highly improbable (see p. 96). As with the pyrazoles, an unsymmetrical imidazole generally provides both possible *N*-alkyl derivatives[16d,453b,497-8,559]. This is not always the case; 4-nitroimidazole with methoxymethyl chloride gave 1-methoxymethyl-4-nitroimidazole and other halides gave the analogous products, but yields were not high (21-54%). These products were orientated by n.m.r. spectroscopy[496]. 1,2,4-Trimethyl-5-nitroimidazole was obtained (64%) from 2,4-dimethyl-5-nitroimidazole and dimethyl sulphate at 100°C [496], and in the same way 4-nitro-5-(*p*-acetamidophenyl)imidazole gave 1-methyl-5-nitro-4-(*p*-acetamidophenyl)imidazole[508b]. In quantitative studies imidazole was *N*-ethylated by reaction with ethyl methanesulphonate (see p. 81).

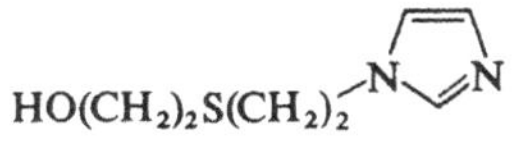

$$HO(CH_2)_2S(CH_2)_2 \quad\quad\quad\quad S$$

(76) 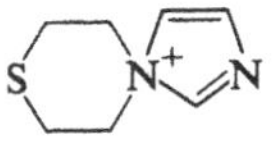   (77)

An important reaction of the present kind is that in which fusion of an acylated sugar with an imidazole in the presence of a small proportion of chloroacetic acid gives an imidazole nucleoside. The method has been applied to the reaction of tetra-*O*-acetyl-β-D-ribofuranose with 2-nitroimidazole and 4-bromo-5-nitroimidazole. The latter gave only the 1-ribofuranosyl-5-bromo-4-nitroimidazole. These reactions are believed to be S_E2', and to occur through a carbonium or acetoxonium ion[562a,b].

Imidazoles have also been *N*-alkylated by reaction of their silver salts with alkyl halides. Examples are the reaction of silver imidazole with acetyl bromide in xylene[563], and of silver 4-nitroimidazole with methyl iodide in benzene which gives only 1-methyl-4-nitroimidazole[564]. Two other examples are illustrated here[564-5].

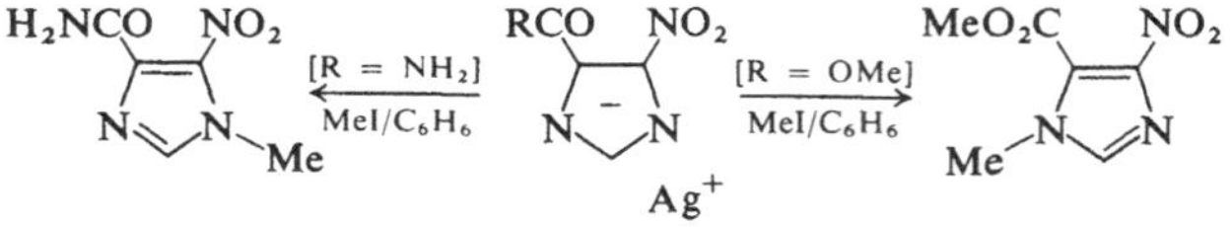

Alkylation of silver salts is also used in the formation of imidazole nucleosides from acylglucosyl halides, the first example being the preparation of 1-glucosopyranosyl-5-methylimidazole (perhaps with some of the 3-isomer) from silver 4-methylimidazole and α-acetobromoglucose[566]. The reaction has been used frequently[562a,565-8], and proceeds with Walden inversion. Sometimes both possible isomers are formed. The chloromercuric salts are also useful, sometimes reacting when the silver salts fail[562a].

The reactions of the silver salts of several imidazoles with triphenylmethyl chloride have been reported. Silver imidazole and 4,5-diphenylimidazole give the *N*-triphenylmethyl compounds, and that from the diphenyl compound rearranges to 4,5-diphenyl-2-triphenylmethylimidazole when heated[569a]. Other 4,5-diarylimidazoles behave similarly, the *N*-triphenylmethyl compounds giving 2-triphenylmethyl compounds when heated, though in some cases the original imidazole and decomposition products are formed[569b]. Silver 2,4,5-trisubstituted imidazoles do not react with triphenylmethyl chloride, but silver 4-phenyl- and 2,4-diphenylimidazole give *N*-triphenylmethyl compounds. Silver 2-phenylimidazole gave 2-phenyl-1-triphenylmethylimidazole, which isomerised to 2-phenyl-4-triphenylmethylimidazole when melted. Silver 2-t-butylimidazole gave 2-t-butyl-4-triphenylmethylimidazole[569c].

Potassium imidazole gives 1-alkylimidazole by reaction with alkyl halides in various solvents at their boiling points or in sealed tubes[375].

Imidazoles have been methylated with methyl iodide in sodium methoxide[570] or ethoxide[508a] solutions, and with sodium methyl sulphate and sodium ethoxide[571]. Since 4-(3,4-dichlorophenyl)imidazole gave 1-methyl-4-(3,4-dichlorophenyl)imidazole with methyl iodide in ethanolic sodium ethoxide the products so obtained from several 4-aryl-imidazoles have been assumed to possess this orientation[508a]. The use of aqueous methanolic sodium hydroxide with methyl iodide or butyl bromide gave good yields of 1-alkylimidazoles[561].

Dimethyl sulphate and caustic soda have been used to alkylate imidazoles[16a,b,d,453b], and also methyl iodide or other alkyl halides with potassium carbonate and acetone. In contrast to methyl sulphate alone (see p. 77) the latter reagents convert 4-aryl-5-nitroimidazoles

Table 13. Isomer ratios in the N-methylation of imidazoles[a]

Imidazole[b]	MeI	Me2SO4	CH2N2	Me2SO4/OH-	Ag salt with MeI
4-Br[497]		1:34			
4-CN[16b,d]	2:1			2.9:1	
4-CHO[559,573]		5-CHO-1-Me only			1,4-isomer only
4-CO2Me[559]		1,5-isomer only			
4-Me[16d]					
4-NO2[453b,497,564]		1:350	1:45	2.2:1	
4-Ph[453b,497]	Both isomers[c]	5:1	1:2	3:1	
4-Ar-5-NO2[508b]		4-Ar-1-Me-5-NO2			
4-Br-5-Me[453b]		5-Br-1,4-Me2 only			
4-Br-5-NO2[497]		4-Br-1-Me-5-NO2 only		2:3	
4-Br-5-Ph[453b]		5-Br-1-Me-4-Ph only			
4-CONH2-5-NO2[564]					4-CONH2-1-Me-5-NO2 only
4-CHO-5-Me[559]		5-CHO-1,4-Me2 only			
4-CO2Me-5-NO2[564]					5-CO2Me-1-Me-4-NO2 only
2-Me-4-NO2[498]		1:50			
4-Me-5-NO2[497,564]		233:1d			1,5-Me2-4-NO2 only
2-Me-4-Ph[572]				2.7:1e	
4-NO2-5-C6H4NO2(p)[497]		1-Me-5-NO2-4-C6H4NO2(p) only			
4-NO2-5-CH:CHPh[564]					1-Me-5-NO2-4-CH:CHPh only
2,4-Br2-5-Me[453b]		1:45f	1:10f	1:1	
2-Br-4-Me-5-NO2[497]		2-Br-1,4-Me2-5-NO2 only			

(a) For di-substituted products the ratio given is that of 1,4-isomer : 1,5-isomer. (b) In naming substituents the numbering used carries no implications for the tautomeric composition of the compound. (c) MeI-K2CO3-acetone. (d) In favour of 1,4-Me2-5-NO2. (e) In favour of 1,2-Me2-4-Ph. (f) In favour of 2,5-Br2-1,4-Me2.

into a mixture of both possible N-alkyl compounds[508b], and with 2,4-dimethyl-5-nitroimidazole they give 1,2,5-trimethyl-4-nitroimidazole [496].

Finally, diazomethane has been used to N-methylate imidazoles[453b, 570]. The dominant formation of 5-substituted 1-methylimidazoles when the eventual 5-substituent is a group with a high electron density has been interpreted to indicate initial formation of an ion pair $[\text{Im}^-\text{CH}_3\text{N}_2^+]$ in which the cation is situated near to the nitrogen atom next to the substituent[208d].

Some details of the results of methylating imidazoles with various reagents are collected in table 13. It should be noticed that yields from these reactions were generally not quantitative.

The work of Ridd[141] has provided a sound basis for the interpretation of these results. The reaction of 4(5)-nitroimidazole with dimethyl sulphate in dilute aqueous sodium hydroxide containing 10% of ethanol is homogeneous, and proceeds at a convenient rate at 25 °C. In these circumstances the nitroimidazole is present almost completely as the anion and the mechanism should be S_E2cB. From the observed kinetic form,

$$\text{Rate} = k_2[\text{Me}_2\text{SO}_4][\text{Im}],$$

and isomer ratio (about 11% of the product is 1-methyl-5-nitroimidazole, and the rest is the isomer; this ratio for homogeneous methylation differs appreciably from that reported earlier (table 13)) the rate coefficient for each nitrogen atom was calculated. Anhydrous formic acid containing sodium formate was a convenient medium for studying the reaction in acidic circumstances, and the change in rate with sodium formate concentration led to the conclusion that

$$\text{Rate} = k_2'[\text{Imidazole}]_{\text{molecular}}[\text{Me}_2\text{SO}_4],$$

i.e. the mechanism was S_E2'. At least 86% of 1-methyl-5-nitroimidazole was formed, and no trace of the isomer was obtained. The kinetic results are expressed in the diagrams. In both cases the ratios of the nucleophilic

Rate coefficients ($l\ mol^{-1}\ s^{-1}$) for methylation of 4(5)-nitroimidazole with methyl sulphate

S_E2cB (aqueous solution, 25 °C) S_E2' (Formic acid, 50 °C)

activities of the two nitrogen atoms are much smaller than the ratio of their basicities (p. 25). If a linear free-energy relationship exists so that $k_{rate} \propto K^{-a}_{equilib.}$, then the data are consistent with $a \sim 0.3$. Such a relationship may be general. In imidazoles with [$-I$] substituents (R) at C-4(5), the predominant tautomer should be 4-R-imidazole, and methylation in an S_E2cB process should give mainly 1-methyl-4-R-imidazole. With [$+I$] derivatives the situation should be reversed. With 4(5)-nitro-imidazole the conjugate base reacts about 10^3 times faster than the neutral molecules. Accordingly, the transition from S_E2cB to S_E2' should occur at about pH = pK_{acid} – 3, that is, at about pH 6–11 for negatively substituted imidazoles. When $0 < a < 1$ the change from S_E2cB to S_E2' should change the main product of methylation.

Examination of the data summarised in table 13 in the light of these considerations, shows that the preparative (heterogeneous) conditions give qualitatively the results to be expected. The discrepancies observed with halogeno-compounds in alkali and with 4(5)-phenylimidazole may be due, respectively, to the occurrence of S_E2' substitution of the imidazole dissolved in methyl sulphate and to steric factors.

The rates of reaction of some imidazoles with ethyl methanesulphonate in water at 35 °C have been reported[574]. Imidazole, 2-methyl-, 4-methyl-, and 2,4,5-trimethyl-imidazole give second order rate constants linearly related to pK_a: evidently steric effects are almost absent in these reactions. Although 'a constant concentration of sodium hydroxide' was added 'to maintain the amine in the reactive form', these were evidently S_E2' reactions.

1,2,3-Triazole has been alkylated by reaction with propyl bromide, allyl bromide, ethyl bromoacetate, 2-chloropropionitrile, and β-phthalimido-ethyl bromide in the presence of sodium ethoxide. In each case the 1-alkyl compound predominated, ratios of 1-alkyl-1H-1,2,3-triazole to 2-alkyl-2H-1,2,3-triazole varying between 4 : 1 and 3 : 2. Use of the silver salt did not much change the ratios, whilst reaction with excess triazole alone gave almost exclusively the 1-alkyl compounds. The products were recognised by n.m.r. spectroscopy[235]. Similarly, methylation of 1,2,3-triazole with caustic soda and dimethyl sulphate gave both isomers, but 1-methyl-1H-1,2,3-triazole predominated[77c]. Under these conditions 4,5-dibromo-1,2,3-triazole gave equal proportions of both isomers, but yields were not quantitative[575].

Diazomethane produces roughly equal proportions of 1-methyl-1H- and 2-methyl-2H-1,2,3-triazole from 1,2,3-triazole[276b], but 3-benzoyl-oxy-4-methyl- and 3-benzoyloxy-1,2,3-triazole give the 2-methyl compounds[576e].

Most alkylations of 1,2,4-triazoles have been effected with a sodium alkoxide as the base. In this way methylation, ethylation, and allylation give the 1-substituted 1,2,4-triazoles. These and other results are collected in table 14. It will be seen that 4-alkylation has been postulated to occur to a very small degree in one case, and that in two instances where it has been held to produce the sole product the structures of these products have not been proved.

The alkylation of tetrazoles commonly produces both possible products (table 15). The data are inadequate to reveal any general trends.

The situation existing in the pyridine series when alkylation of compounds containing tautomerisable substituents is considered (Schofield, p. 180), is complicated in the azoles by the possibility in the latter of substitutive alkylation occurring at a nuclear nitrogen atom and also by the greater likelihood of *C*-alkylation. The case of quaternising alkylation (see p. 95) is more properly compared with the reactions of the pyridines.

Little or nothing is known about the substitutive alkylation of amino-pyrazoles. The cases of some aminopyrazolones are discussed below.

With dimethyl sulphate and alkali 4-amino-5-aminocarbonyl-1,2,3-triazole gave equal parts of both isomers, whilst the formyl derivative of this amine gave 4-aminocarbonyl-5-formamido-2-methyl-2*H*-1,2,3-triazole [149c].

Benzyl chloride and caustic soda, and dimethyl sulphate and caustic soda, alkylate 3-phenyl-5-ureido-1,2,4-triazole at a nuclear nitrogen atom, but the reasons for preferring the structures (78), R = Me or PhCH₂, over the other possibilities are not convincing[318c,586].

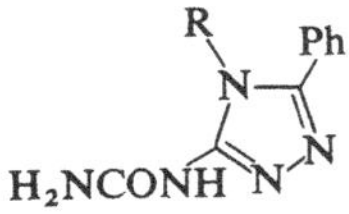

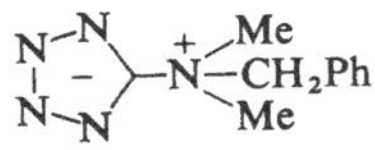

The alkylation of 5-aminotetrazole has been studied in some detail. With dimethyl sulphate in water the sodium salt gives 5-amino-1-methyl-1*H*-1,2,3,4- and 5-amino-2-methyl-2*H*-1,2,3,4-tetrazole, the former predominating[83e,583,587]. The total yield of these compounds was high, and the products of further methylation (1- and 2-methyl-5-methylamino tetrazole and 5-imino-1,3- and -1,4-dimethyltetrazole, see p. 100) were isolated in very low yields (≯ 1%). Methyl and ethyl iodide, allyl bromide, benzyl chloride, chlorohydrin, and ethyl sulphate similarly gave mixtures of 1- and 2-alkyl-5-aminotetrazoles in which the former predominated[123,581c], as did diazomethane[583]. The particular case of benzylation and substituted-benzylation has been examined several times. The 'α-monobenzyl' compound formed with benzyl chloride in the presence of caustic soda[25c] was probably 5-amino-1-benzyl-1*H*-1,2,3,4-tetrazole[515a,588a]. Experiments with benzyl chloride and bromide and

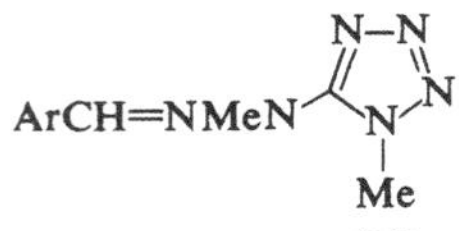

Table 14. *N*-Alkylation of 1,2,4-triazoles

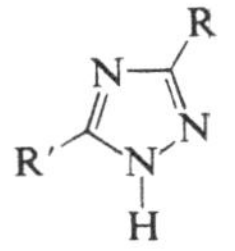

R	R'	Conditions	Products	Ref.
H	H	MeI–NaOMe–MeOH	1-Me	22f, 577a
H	H	EtBr–NaOEt–EtOH	1-Et	22f
H	H	C_3H_5Br–NaOEt–EtOH	$1\text{-}C_3H_5$	22f
H	H	PhCH$_2$Cl–NaOEt–EtOH	Probably 1-CH$_2$Ph	363c
H	H	BrCH$_2$CO$_2$Et–NaOEt–EtOH	Probably 1-CH$_2$CO$_2$Et	366
H	H	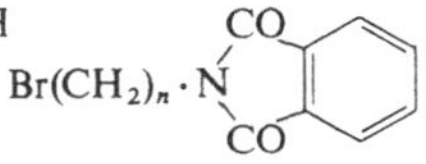 Br(CH$_2$)$_n$·N(phthaloyl) –NaOEt–EtOH	Probably 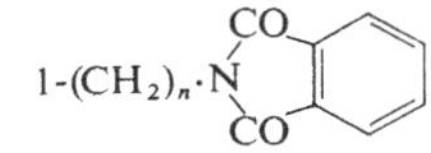 1-(CH$_2$)$_n$·N(phthaloyl)	366
H	H	Na salt–benzene–Ph$_3$CCl	Probably 1-Ph$_3$C	175a
	phthaloyl·N·(CH$_2$)$_2$			
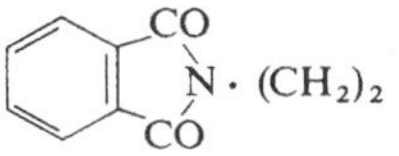	H	MeI–NaOEt–EtOH	1-Me-3- and 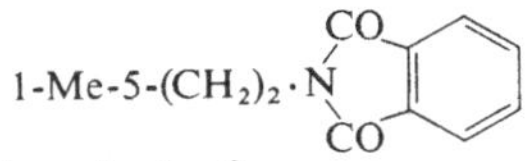 1-Me-5-(CH$_2$)$_2$·N(phthaloyl) in ratio 1 : 2	366
Me	H	Na salt–benzene–Ph$_3$CCl	Probably 1-Ph$_3$C	175a
Me	Me	MeI–NaOMe–MeOH	1,3,5-Me$_3$	577a
Me	Me	CH$_2$N$_2$–MeOH	1,3,5-Me$_3$	577a
Me	Me	EtI–NaOEt–EtOH	1-Et-3,5-Me$_2$ (58%) + 4-Et-3,5-Me (~1%?)	577a
Me	Me	CH$_3$CHN$_2$	1-Et-3,5-Me$_2$	577a
Me	Ph	MeI–NaOMe–MeOH	1,5-Me$_2$-3-Ph	577b
Me	Ph	CH$_2$N$_2$	1,3-Me$_2$-5-Ph and 1,5-Me$_2$-3-Ph in ratio 3.7 : 1	577b
Ph	H	MeI–NaOMe–MeOH	1-Me-5-Ph and 1-Me-3-Ph in ratio 1 : 2	577b
Ph	H	CH$_2$N$_2$	1-Me-5-Ph and 1-Me-3-Ph in ratio 1.6 : 1	577b
Ph	Ph	MeI–NaOMe–MeOH	1-Me-3,5-Ph$_2$	577a

some *p*-substituted compounds showed reaction to occur at the amino group to the extent of about 10% of the proportion of nuclear attack. The proportions of the 1- and 2-substituted compounds formed were very similar and did not vary much [590d].

As well as the 1- and 2-benzyl compounds this method also produces the ylide (79) [589] from 5-dimethylaminotetrazole.

5-Aminotetrazole gives both the 1- and 2-ethoxycarbonylmethyl

Table 15. *N*-Alkylation of tetrazoles

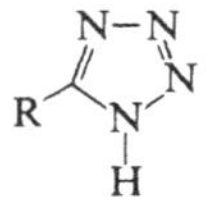

R	Conditions	Products	Ref.
H	Me_2SO_4	1-Me + 2-Me	94
	Ag salt–EtI–C_6H_6	1-Et + 2-Et	80c
Me	MeI or Me_2SO_4–NaOH	1,5-Me_2	95
	CH_2N_2–EtOH–Et_2O	1,5-Me_2 (9.3%) + 2,5-Me_2 (47.7%)	95
	$BrCH_2CO_2Et$–Et_3N–C_3H_6O	1- + 2- isomer (> 2 : 1)	584
CF_3	Na salt–$C_8H_{17}Br$–MeCN	1-C_8H_{17} + 2-C_8H_{17} in ratio 1 : 18	578
MeO.$(CH_2)_2$	Me_2SO_4–NaOH–H_2O	1-Me (47.5%) + 2-Me (42%)	579b
$C_6H_{11}.(CH_2)_2$	p-$O_2NC_6H_4CH_2Br$–K_2CO_3–H_2O–EtOH	2-p-$O_2NC_6H_4CH_2$ (probably)	93
CO_2H	Ag salt–EtI	2-Et	580
CO_2Et	CH_2N_2	2-Me	94
CH_2CO_2Me	Et_3N salt + RX	Both isomers	584
Ph	MeI–NaOH–H_2O–acetone	2-Me (56%)	154
	MeI–NaOH–H_2O–acetone	1-Me (20%) + 2-Me (80%)	581b
	MeI–NaOH–H_2O–acetone	1-Me (16%) + 2-Me (55%)	585
	Me_2SO_4–NaOH–H_2O	1-Me (43%) + 2-Me (29%)	153b
	Me_2SO_4–Na_2CO_3–80% acetone	1-Me (16%) + 2-Me (54%)	153b
	CH_2N_2–tetrahydrofuran	1-Me + 2-Me	153b
	EtI–EtOK–EtOH	Mixture?	512a
	$PhCH_2Br$–K_2CO_3–H_2O–EtOH	2-$PhCH_2$	93
	p-$O_2NC_6H_4CH_2Br$–K_2CO_3–H_2O–EtOH	2-p-$O_2NC_6H_4CH_2$ (probably)	93
β-$C_{10}H_7$	MeI–NaOEt–EtOH	?	90a,e
	EtI–NaOEt–EtOH	?	90a,e
Various aryl groups	Dialkylaminoalkyl halides–NaOH–H_2O–acetone	One isomer in each case (assumed 2-isomer)	513b, 582
	MeI–NaOH–H_2O–acetone	Both isomers (mainly 2-Me)	585
CN	Ag salt–EtI–C_6H_6 or Et_2O	2-Et	80c,e
Cl	$BrCH_2CO_2Et$–Et_3N–C_3H_6O	1- + 2- isomer (1 : >7)	584
Br	CH_2N_2	1-Me + 2-Me	583

compounds with ethyl bromoacetate in the presence of triethylamine [584].

Acting in the absence of alkali, alkylating reagents effect both substitutive and quaternising alkylation. These reactions are considered below, with the quaternisation of alkyl-5-aminotetrazoles.

In the related methylation of arylhydrazones of 5-hydrazinotetrazole, the main product is the 1-methyl compound, with some of the dimethyl compound (80), and a small amount of the compound monomethylated in the side chain. Increasing the proportion of alkali present increases the proportion of dimethylation, and with some hydrazones the hydrazones of 5-hydrazino-2-methyl-2H-1,2,3,4-tetrazole were also formed[591a,b]. N.m.r. spectroscopy can be used to orientate the methyl derivatives of 5-aminotetrazole, but coincidence of signals prevents this with the methyltetrazol-5-ylhydrazones[591c].

The alkylation of pyrazolones has received considerable attention, particularly in connection with the preparation of antipyrine and its analogues. For convenience these reactions will be divided into two sections, namely those starting from *N*-unsubstituted compounds, and those starting from *N*-substituted compounds.

The first reported alkylation of 3-methylpyrazol-5-one[76e] was unusual. By analogy with hydrazones it was expected that this compound would be reduced by being heated with sodium methoxide. In fact, ring opening to the extent of 41%, giving nitrogen and butyric acid, occurred, together with *C*-alkylation to produce 3,4-dimethylpyrazol-5-one (39%). 4-Ethylation and -propylation were similarly effected. These reactions recall the similar alkylations of pyrroles and phenols (Schofield, p. 68). Other alkylations of 3-methylpyrazol-5-one[62b,100a] are summarised in the diagram. The general formation of 1-alkyl-3-methylpyrazol-5-ones by

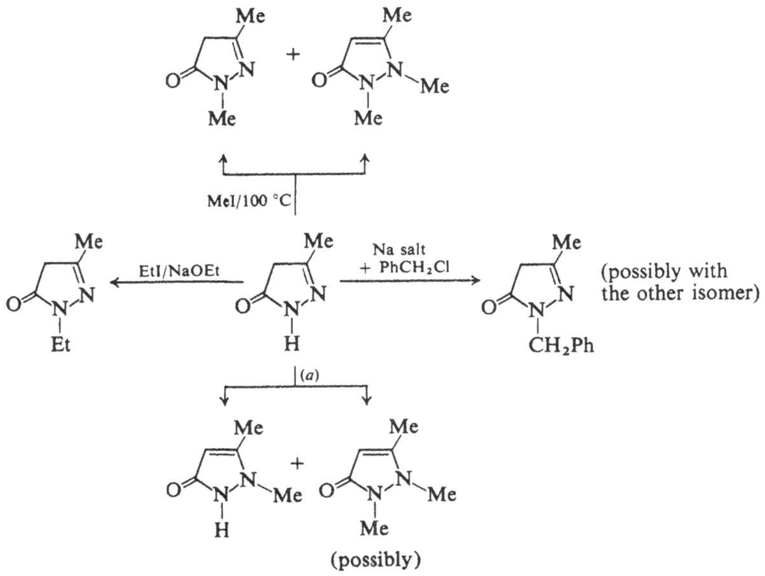

(a) NaOMe/MeOH/MeI *or* $C_7H_7SO_3Me$

(a) The text of ref. 100b says NaOMe/MeOH/MeI, the
experimental section NaOMe/MeOH/$C_7H_7SO_3Me$

reaction with alkyl halides alone has been confirmed[554a]. From these
reactions in the absence of base, the products are, of course, hydrogen
halide salts, from which the alkyl derivatives are obtained by neutralisation. A slightly different situation is found with 3-phenylpyrazol-5-one
[100b], as shown. 3-Methyl-4-phenylpyrazol-5-one heated with methyl
iodide gives 1,2,3-trimethyl-4-phenylpyrazol-5-one[554a].

3-Pyrazolone reacts with xanthydrol, but it is not clear whether the
product is a 1,2- or a 1,4-disubstituted compound[592].

In 1884 Knorr discovered antipyrine (81), R = Ph, R' = R'' = Me,
R''' = H, which he prepared by heating 3-methyl-1-phenylpyrazol-5-one
with methyl iodide in methanol in a sealed tube at 100°C[392a,b,540a].
He later[392k] prepared antipyrine by ring-synthesis, and several
analogues by similarly alkylating other 3-substituted 1-arylpyrazol-5-ones
[392a,c,k,l]. Since Knorr's original observations, numerous analogues of
antipyrine have been prepared by such alkylations[324g,467b,540d,541b,
554a,b,588b,593a], and derivatives of 3-antipyrine (82) have been
similarly prepared[330c].

(81)   (82)

Many variations of reagent and conditions for preparing antipyrine
itself by methylating 3-methyl-1-phenylpyrazol-5-one have also been
reported[324h,539a,c,540c,595a-c].

The ethylation of 3-methyl-1-phenylpyrazol-5-one was early described

[392*l*], but attempts to prepare higher homologues of antipyrine gave a variety of results[596], shown below. *C*-Alkylations occurring under other conditions are also shown here.

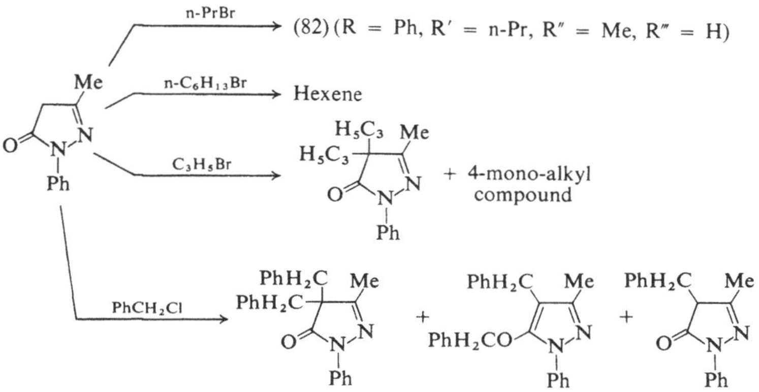

$$\text{n-PrBr} \longrightarrow (82)\,(R = Ph, R' = n\text{-Pr}, R'' = Me, R''' = H)$$

$$\text{n-C}_6\text{H}_{13}\text{Br} \longrightarrow \text{Hexene}$$

$$\text{C}_3\text{H}_5\text{Br} \longrightarrow \quad + \text{ 4-mono-alkyl compound}$$

$$\text{PhCH}_2\text{Cl} \longrightarrow$$

Isoantipyrine (81), R = R' = Me, R'' = Ph, R''' = H, is formed by heating 1-methyl-3-phenylpyrazol-5-one with methyl iodide[330d]. In general the 2-methylation of 1-alkylpyrazol-5-ones[324g,554a] is more difficult than that of 1-arylpyrazol-5-ones, and exchange of alkyl groups can occur[554a]. Thus, when 3-methyl-1,4-di-isopropylpyrazol-5-one is heated with methyl iodide it seems that some 1,2,3-trimethyl-4-isopropyl-pyrazol-5-one is formed, and 3-methyl-4-phenyl-1-isopropylpyrazol-5-one gives the 1,2,3-trimethyl compound, though 1-ethyl-3-methyl-4-isopropyl-pyrazol-5-one gives the 1-ethyl-2,3-dimethyl compound. The exchanges have been represented as shown in the diagram.

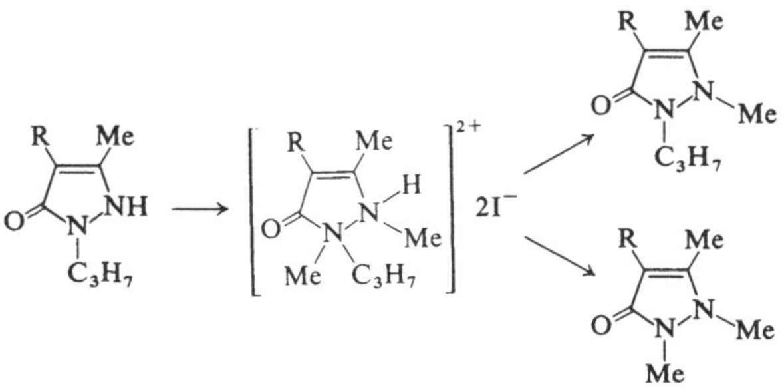

Methylation of 3-methyl-1-phenylpyrazol-5-one, other than by heating with an alkyl halide, was first described by Knorr; the use of methyl iodide with sodium methoxide was said to effect the successive introduction of two methyl groups at C-4. 3,4-Dimethyl-1-phenylpyrazol-5-one similarly gave 3,4,4-trimethyl-1-phenylpyrazol-5-one[392k]. Later, the six products shown were reported to arise from the methylation of 3-methyl-1-phenylpyrazol-5-one under these conditions (experimental details were not given)[392f].

Other workers[597], using sodium methoxide, methanol, and dimethyl sulphate, prepared 5-methoxy-3-methyl-1-phenylpyrazole, and from the pyrazolone in boiling sodium hydroxide solution by adding dimethyl sulphate, a high yield of antipyrine. *C*- and *O*-benzylations, using sodium ethoxide and benzyl chloride, have been described[596], and *p*-nitrobenzyl chloride effects 4-substitution[388b]. Ethyl chloroacetate and sodium ethoxide give *O*-substitution[329e].

3-Methyl-1-phenylpyrazol-5-one is substituted at C-4 by reaction with chloroacetone and sodium hydroxide, and this[388c] and other[388b] 4-substituted compounds give analogues of antipyrine when methylated in presence of alkali. Similarly 3-methyl-1-phenyl-4-isopropylpyrazol-5-one with alkyl halides, sodamide, and dioxan, is said to give analogues of antipyrine[554].

1-Phenylpyrazol-3-one is *O*-methylated with dimethyl sulphate and aqueous alkali[131] and gives the *O* ether with allyl bromide and potassium carbonate in acetone[131], but with aqueous alkali a 35% yield of 2-allyl-1-phenylpyrazol-3-one[598].

In contrast to these base-catalysed alkylations is the 4-alkylation of 3-methyl-1-phenylpyrazol-5-one with triarylcarbinols or their ethers in acetic acid containing hydrochloric acid[357a,b].

Diazomethane has not been much used with pyrazolones, but some 3-methyl-1-thiazolylpyrazol-5-ones are reported to give analogues of antipyrine with this reagent[594]. In contrast, 3-methyl-1-phenylpyrazol-5-one gives mainly the *O*-methyl ether, with only a trace of antipyrine[17b].

The behaviour of 4-amino-5-methyl-1-phenylpyrazol-3-one on methylation is shown below[330c]. With alkyl halides, 4-aminoantipyrine gives
the 4-alkylamino compounds[324a,b,e,g,392m,540b,593b,595d]. Such
reactions were and remain important because of the value of the
products as drugs; 4-dimethylaminoantipyrine ('Aminopyrine', 'Pyramidone') was first prepared by Stolz[324a,b,392m,540b], by reaction of
aminoantipyrine with methyl iodide in methanolic potash.

The acylaminopyrazolone (83), R = H, gives (83), R = Me, with
diazomethane[599].

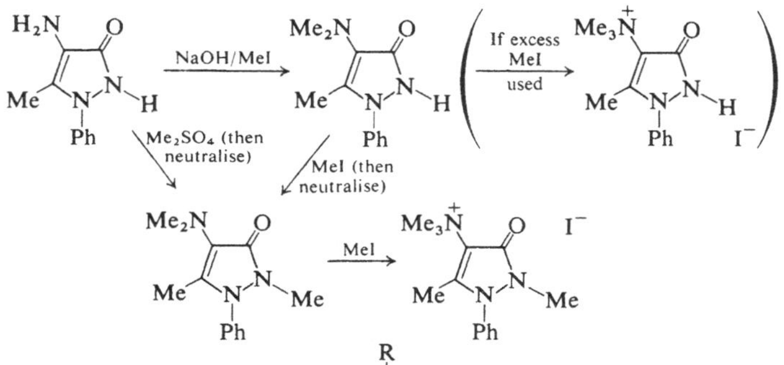

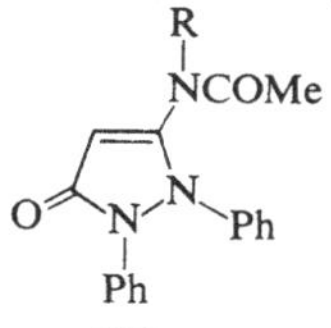

(83)

In the alkylation of 1-phenylpyrazol-3,5-diones both carbon and
oxygen compete successfully against nitrogen[551a,600b], as shown, and
alkylation at C-4 is generally observed[339f]. 4-Butyl-1,2-diphenylpyrazol-
3,5-dione ('Phenylbutazone') was introduced in 1946 for use against
rheumatoid arthritis; it and its homologues are readily prepared by
reaction of 1,2-diphenylpyrazol-3,5-dione with alkyl halides and alkali
[359,601a-c], as already indicated (p. 35). 4,4-Dialkylation can also be
effected[358].

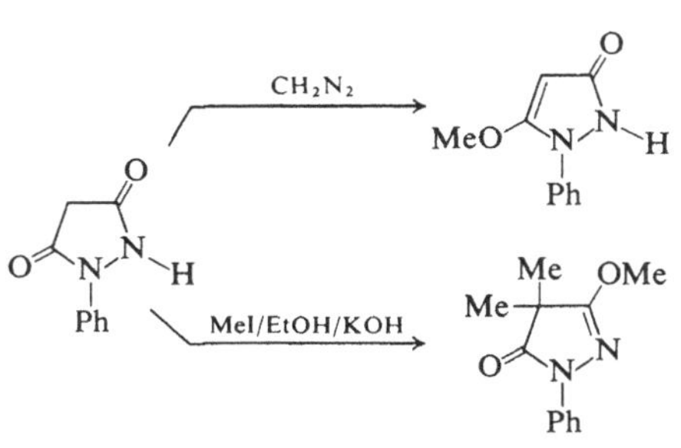

Table 16. The alkylation of hydroxy-1,2,3-triazoles[a]

Starting Material		Products				

(structures: HO-1,2,3-triazole with R', R; and product structures as drawn)

R	R'					
H	H	38%	2%		43%	5%
H	Ph	+	+		+[b]	+[b]
H	CO$_2$Et	27%	–		+[b]	+[b]
Me	H	+	+		–	–
CH$_2$Ph	H	+	+	+	–	
CH$_2$Ph	H	100%[c]	–	–	–	
Ph	Me[d]	62%	12%	25%	–	
Me	Ph	+	42%	23%[e]	–	
Me	CO$_2$Me	45%	54%	–	–	
Ph	Ph[f]	30%	48%	12%	–	
Ph	CO$_2$Et	–	[OEt compound][g]	–	–	

(structure: HO-triazole, N-Me) +(R=Me, R'=H)   –   –   +(R'=H)

(structure: HO-triazole, N-Ph (h)) (structure with Et)   –   –   (structure: EtO-triazole, N-Ph)

(structure: HO-triazole N-N-Me) (structure: MeO-triazole N-N-Me) + One isomer [576e]

(structure: Ph HO-triazole N-N-Me) (structure: Ph MeO-triazole N-N-Me) + (structure: O-triazolone Ph, N-Me, N-Me) [576e]

(a) Unless otherwise stated CH$_2$N$_2$–Et$_2$O–MeOH was used[576]. In this table + indicates that product was present, – that it was absent. (b) Two products were

Probably because the products lack the practical value of the pyrazolones, the alkylation of imidazolones has been very little investigated.
With alkali and dimethyl sulphate, imidazol-2-one-4-carboxylic acid is
methylated at both nitrogen atoms[602]. With methanolic potash and
methyl iodide, hydantoin gives 3-methylhydantoin[487e,518], and diazomethane methylates 1-ethylhydantoin at N-3[603].

Recently the methylation of hydroxy-1,2,3-triazoles with diazomethane
has been reported. Other reagents have been much less used. The results
(table 16) show that attempts to monomethylate a compound initially
un-substituted on nitrogen fail, dimethyl compounds always being formed.
Commonly, whether the starting compound was substituted on nitrogen
or not, all of the possible O,N-dimethyl compounds were formed. The
invariable formation in appropriate cases of the meso-ionic derivative (84)
as an important proportion of the total product is interesting.

(84)   (85)

According to the conditions, methylation in the presence of base of
5-nitro-1,2,4-triazol-3-one can give either the 1- or the 4-methyl compound
[606]. Alkylation in the presence of bases of some 1-substituted 5-
hydroxy-1H-1,2,4-triazoles[61a,607e] and some 4-substituted 5-hydroxy-
4H-1,2,4-triazoles[268e,487d,608b] has been represented as giving
products of the type (85). This orientation has been proved for the
compound (85), R = Me, R′ = H, R″ = NH$_2$, derived from (85), R =
R′ = H, R″ = NH$_2$, and is probably correct in analogous cases[268e],
but with other compounds it is not clear that isomeric structures can be
excluded.

Some results obtained by alkylating urazole and its derivatives are
given in table 17. The reactions with diazomethane have been used in a
discussion of the enolisation of urazoles[600a].

The reactions of 5-hydroxytetrazole and its derivatives with diazomethane are illustrated[583]. All of the possible dimethyl compounds
except one appear to be formed; the structure of the product represented

---

obtained but definite assignments to these structures were not possible. (c) Using
MeI–NaOH–MeOH[576d]. (d) Esterifying conditions failed[429a]. (e) Assumed
structure. (f) Dimroth[429a], by treating 5-hydroxy-1,4-diphenyl-1H-1,2,3-triazole
with MeI–NaOMe–MeOH or Me$_2$SO$_4$–NaOH, obtained a compound, m.p. 126°C,
which he regarded as 5-methoxy-1,4-diphenyl-1H-1,2,3-triazole. The true methoxy
compound has m.p. 86–7°C[576c] and Dimroth's product remains unidentified.
See also ref. 604. (g) Using EtI–silver salt. (h) Using Et$_2$SO$_4$–NaOH–EtOH[576d,
605].

Table 17. The alkylation of urazole and its derivatives

Starting compound		Reagent	Products														Ref.
R	R'		R	R'	R''	R	R'	R''	R	R'	R''	R	R'	R'	R	R'	
H	H	Deficiency of CH₂N₂							H	Me	Me						600a
H	H	Excess CH₂N₂							Me	Me	Me				Me	Me	600a
Ph	H	K salt + alkyl halide	Ph	Alkyl (a little)								Ph	Alkyl				609d
Ph	H	K salt + MeI							Ph	Me	Me						600a
Ph	H	Ag salt + EtI	Ph	Et		Ph	Et	Et									609a
Ph	H	Ag₂ salt + EtI				Ph	Et	Et									609a
Ph	H	Deficiency of CH₂N₂	Ph	Me								Ph	Me	(5%)			609a, d
Ph	H	Excess CH₂N₂	Ph	Me (quantitative)					Ph	Me	Me						609a, d
Ph	H	CH₂N₂							Ph	Me	Me				Me	Me	600a
Ph	Me	CH₂N₂				Ph	Et	Et	Ph	Et	Me	Ph	Me		Ph	Et	609a, d
Ph	Et	Ag salt + EtI				Ph	Et	Et									609a
Ph	Et	Na salt + MeI or CH₂N₂															609a
Me	Et	CH₂N₂													Me	Me	609d
Ph		CH₂N₂													Ph	Et	600a
Ph		K salt + EtI													Ph	Me	609d
Ph		CH₂N₂															600a

Table 17. Continued.

Starting compound			Reagent	Products												Ref.
R'	R	R'		R'	R	R'	R''	R	R'	R'	R''	R	R'	R	R'	
	H	Me	CH_2N_2									H	Me	Me	Ph	600a
	Me	Me	CH_2N_2									Me	Me			600a
	H	Ph	CH_2N_2													600a

Starting compound:

Me–N, HO, R'=OH, N–Ph

Products:

Me, OEt, N, Ph (first product)

Me–N, Ph, N–Et (second product)

	Reagent	Product 1	Product 2	Ref.
	EtI–40% EtOH	10%	90%	609c
	Ba salt–EtI–40% EtOH	6.5%	93.5%	609c
	Na salt–EtI–40% EtOH	58.7%	41.3%	609c
	K salt–EtI–40% EtOH for 1h		90% of a total of 30.35% of ethylated product	609c
	K salt–EtBr–40% EtOH for 1h		87.7% of a total of 22.5% of ethylated product	609c
	Ag salt–EtI	65%	35%	609c

as (86) (a trace of which is also formed from 5-hydroxy-1-methyl-tetrazole) is probable but not completely established. With ethyl bromo-acetate in the presence of triethylamine, 5-hydroxytetrazole gave only the 1,4-disubstituted tetrazolone[584].

A number of observations have been made concerning the alkylation of azol-thiones. Various 1-aryl-3-methylpyrazol-5-thiones react at the sulphur atom with methyl iodide[330f]. Similarly a number of *N*-unsub-stituted imidazol-2-thiones have been alkylated on the sulphur atom under conditions which include heating with an alkyl halide[425b], or with alcoholic hydrogen chloride[374c], and heating the alkylating agent with the sodium salt of the thione in water[610] or liquid ammonia [143b]. 4-Phenylimidazol-2-thione with dimethyl sulphate and alkali gave the *S*-methyl compound and both possible *N*,*S*-dimethyl compounds[611]. 1-Methyl- and 1-phenyl-imidazol-2-thione give the methylthio compounds when treated with methyl iodide in chloroform, or with dimethyl sulphate and aqueous potassium carbonate[12,612].

Similarly, *N*-unsubstituted 1,2,4-triazol-5-thiones give methylthio

compounds when heated with methyl iodide[86a], whilst with *N*-unsub-
stituted compounds methyl iodide[86b] or an alkylating agent and a
base[612,613a-c] have been used. 5-Hydroxy-1-phenyl-1,2,4-triazol-3-
thione ('1-phenyl-3-thiourazole') has been alkylated as shown above[609a].

Alkylation of the sulphur atom rather than a nitrogen atom in
1-substituted tetrazol-5-thiones is also generally observed. Reagents used
have been alkyl halides in aqueous alcohol[614c], alkyl halides with
sodium ethoxide[24a,83d] or sodium hydroxide[614b], dimethyl sulphate
and potassium carbonate[612], the silver salt with methyl iodide[24a],
and diazomethane[83d]. In the Mannich reaction 1-aryltetrazol-5-thiones
give what are probably the 4-dialkylaminomethyl compounds[614a].

3.13.2 The attachment of alkyl and substituted alkyl groups: quaternising alkylation

The incidental formation of quaternary salts during the substitutive
alkylation of pyrazoles has been mentioned; it happened in the reaction
of pyrazole with methyl iodide[2f,84a]. Knorr and Kohler[392j] who
prepared 1-methylpyrazole methiodide by heating pyrazole with methyl
iodide in methanol, proved it to be 1,2-dimethylpyrazolium iodide by
degrading it to 1,2-dimethylhydrazine. Commonly quaternary pyrazolium
salts have been prepared by heating at about 100°C a mixture of a
N-substituted pyrazole and an alkyl halide. The earliest examples seem
to have been the quaternisation with methyl iodide of 1-phenyl- and
4-methyl-1-phenyl-pyrazole[60d]. 5-Chloro-3-methyl-1-(*m*-nitrophenyl)-
pyrazole was quaternised with dimethyl sulphate[330h]. The structures
of the salts from 1-phenylpyrazole and methyl and ethyl iodide as well
as of 1,2-diethyl- and 1-ethyl-2-methylpyrazolium iodides, were proved
by degradation to hydrazines[615b]. 3,4,5-Trimethyl-1-phenylpyrazole
[69d] and 1-alkyl- or -aralkyl-pyrazoles containing alkyl groups[63a,d,e,
66f,100c], chloro and methyl groups[100a,d], and methoxycarbonyl and
methyl groups[66f] have been quaternised without difficulty. The
examples illustrated, like the degradations already mentioned, show that
a 1-substituted pyrazole is quaternised at N-2. Other examples are found
among the 3,4-tetramethylene-pyrazoles (tetrahydroindazoles)[63a,d,e].

Reactions in which *N*-alkylpyrazoles are converted into different
N-alkylpyrazoles by being heated with an alkyl halide go through
quaternary salts (see §9.13).

As in the pyrazole series so with the imidazoles, quaternisation to
some degree has often accompanied substitutive alkylation. The corresponding quaternary salts were obtained by boiling imidazole with ethyl
bromide or benzyl chloride[6b], or with methyl iodide, chloracetic acid
or ethyl chloracetate[408].

Numerous examples have been described of the quaternisation of
1-alkylimidazoles with alkyl, alkenyl, or aralkyl halides[16a,54c,d,e,66h,

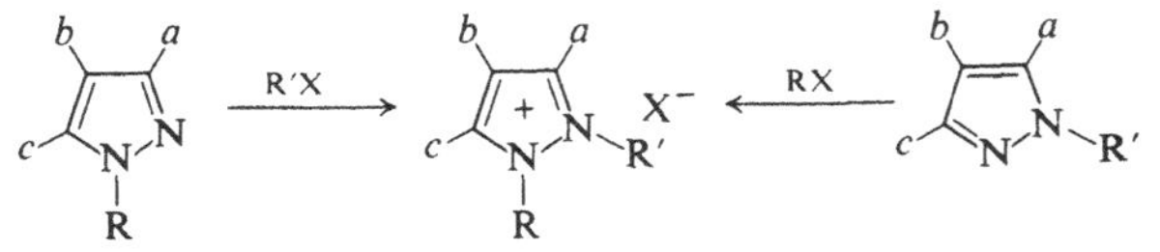

R = R' = a = Me; c = Cl; b = H [100a]
R = CH$_2$Ph; R' = a = Me; b = H; c = Cl [100a]
R = a = Me; b = H; c = Cl; R' = CH$_2$Ph [100a]
R = c = Me; a = b = H; R' = CH$_2$Ph [66f]
R = CH$_2$Ph; a = b = H; c = R' = Me [66f]
R = c = R' = Me; a = CO$_2$Me; b = H [66f]
R = c = R' = Me; a = b = H [66f]
R = a = b = R' = Me; c = Cl [100d]
R = a = R' = Me; b = Cl; c = H [100d]
R = a = R' = Me; b = NO$_2$; c = H [100d]

90c,373a,408,616,617b], ethyl chloracetate[408], or phenacyl bromide [618]. *N*-Alkylimidazoles containing aryl[497], halogen[373a,c,374b], and nitro groups[497] have also been quaternised. The reactions occur readily; 1-methyl- or 1-ethyl-imidazole and methyl iodide react together most vigorously[53,54d]. *N*-Arylimidazoles have been quaternised by fusing with methyl toluene-*p*-sulphonate[411b].

The first evidence to demonstrate that quaternisation of a 1-alkyl-imidazole proceeds at the unsubstituted nitrogen atom was obtained by Pinner and Schwarz[90c], who showed that the quaternary salt from 1-methylimidazole and amyl bromide was decomposed by alkali to give both methylamine and amylamine. Other such examples were described. Later authors observed the formation of the same quaternary salt from different alkylimidazoles: see diagram.

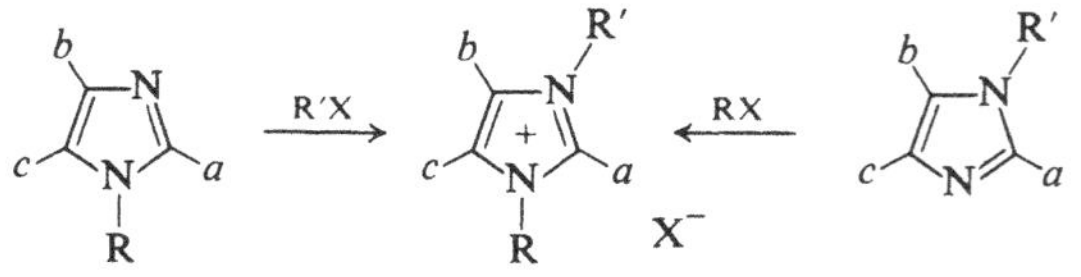

R = b = R' = Me; a = c = H [16]
R = R' = Me; b = Cl; a = c = H [373a]
R = R' = Me; b = Br; a = c = H [374b]
R = Me; R' = n–Pr; a = b = c = H [66h]
R = Et; R' = n–Pr; a = b = c = H [66h]
R = R' = Me; b = NO$_2$; a = c = H [497]
R = R' = Me; b = Ph; a = c = H [497]

The rates of reaction of some 1-substituted imidazoles with ethyl iodide in ethanol[619] and in acetone have been measured[620]. The expected sequence of reactivities, 1-Me > 1-PhCH$_2$ > 1-Ph is observed. Only qualitative observations have been recorded regarding substituent effects; thus, 1-methyl-4- and -5-chloroimidazole react readily enough with methyl iodide but less easily than does 1-methylimidazole[373a].

Quaternisations of imidazoles in which the initial *N*-substituent is other than an alkyl, aralkyl, or aryl group are known. Thus 1-benzoyl-4-phenylimidazole is quaternised with triethyloxonium tetrafluoroborate. Methanolysis of the product gives 1-ethyl-5-phenylimidazole [621]. The derivative of 2-methylimidazole with the group -PS(Ph).NEt$_2$ attached to nitrogen is quaternised on the ring with methyl iodide [622].

Wolff [76c] prepared the first quaternary salt of the 1,2,3-triazole series by heating 1,5-dimethyl-1*H*-1,2,3-triazole with methyl iodide at 100°C. Much later [623b] the first reactions we show were regarded as demonstrating that quaternisation of 1,2,3-triazoles proceeded at N-3.

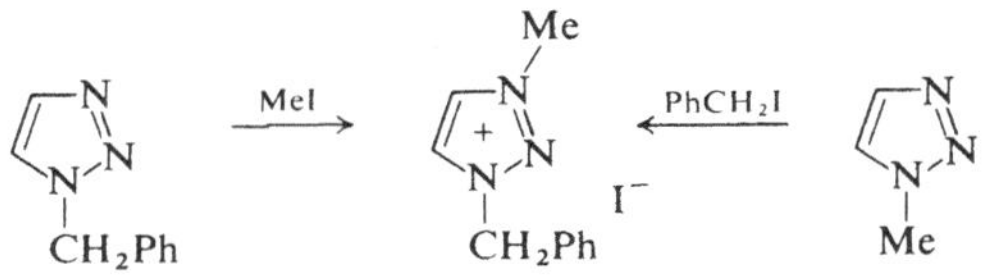

The quaternary salt was said to be homogeneous, but if the experiments excluded a 1,2-substituted structure for the cation, they did not strictly exclude the 1,1-disubstituted structure. This unlikely possibility was removed by the reactions shown next [208a].

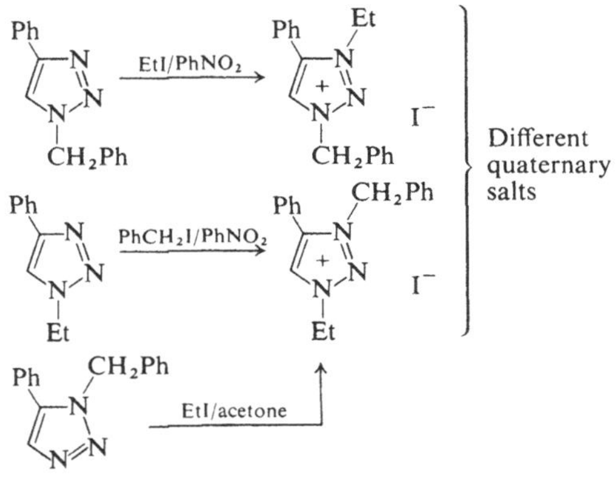

1-Phenyl-1*H*-1,2,3-triazole reacted with methyl iodide in acetone–ether during 2-5 weeks at room temperature [542]. It is never clear from these various reports whether the frequently longer times or more severe conditions of reaction used (compared with those used for diazoles) are necessary for successful quaternisation of 1,2,3-triazoles.

1-Substituted 1*H*-1,2,3-triazoles can be quaternised with methyl toluene-*p*-sulphonate, but the 2-substituted compounds do not react with the reagent, with methyl iodide or dimethyl sulphate, or merely give very low yields of products. In contrast, 2-methyl- and 2-phenyl-2*H*-1,2,3-triazole are both efficiently quaternised with methyl fluorosulphonate (n.m.r. spectroscopy shows the two groups attached to nitrogen to be on different nitrogen atoms)[576k].

Early workers quaternised several 1-aryl-1H-1,2,4-triazoles, containing C-alkyl[20a,624-5] or C-aryl groups[91], with methyl or ethyl iodide, without being able to assign structures to the products. Correct proof of the structure of a quaternary salt in this series was obtained as illustrated.

$$MeNH_2 + MeNHNH_2$$

The salt was degraded to methylamine and methylhydrazine, and similar evidence proved the product from 3,5-dimethyl-1-phenyl-1H-1,2,4-triazole and methyl iodide to be 3,4,5-trimethyl-1-phenyl-1,2,4-triazolium iodide [613a]. The same general mode of quaternisation is in other cases attested by evidence of a different kind. Thus, the salts formed from methyl toluene-p-sulphonate and both (87) and (88) with the appropriate second components give the same cyanines, proving the salt from (87) to be (89)[613a]; a C-methyl group in such compounds is only reactive when situated between the two substituted nitrogen atoms (see p. 242). [540g,613a]. The quaternary salts (90) from 3-methyl-1-phenyl-1H-1,2,4-triazole do not show reactivity of the C-methyl group. This circumstance is further illustrated by the properties of the compound obtained by treating the sodium salt of 3-methyl-1,2,4-triazole with triphenylmethyl chloride (table 14); with methyl iodide this gives a quaternary salt which does not react with p-nitrosodimethylaniline[175a] and is therefore probably (91).

It can then generally be assumed that 1-substituted 1*H*-1,2,4-triazoles
are quaternised at N-4 (thus, the product from 1-dodecyl-1*H*-1,2,4-triazole
and ethyl iodide is probably 1-dodecyl-4-ethyl-1,2,4-triazolium iodide
[282]), and that 4-substituted 4*H*-1,2,4-triazoles are quaternised at N-1
or N-2 (thus, 4-aryl-3,5-dimethyl-4*H*-1,2,4-triazoles give 4-aryl-1,3,5-
trimethyl-1,2,3-triazolium iodides [613a,617c]), but obviously triazoles of
the type (92) could still give two quaternary salts. If either *a* or *b* is a
methyl group its properties in the quaternary salt give evidence for the
structure of the latter, as illustrated by the examples containing sulphur
substituents which are quoted below.

The quaternary 1,2,4-triazolium salts so far mentioned were prepared
using alkyl halides or toluene-*p*-sulphonates as quaternising agents. Triethyloxonium tetrafluoroborate in ethylene chloride has been used to
quaternise 5-chloro-3-methyl-1-phenyl-1*H*-1,2,4-triazole [626]. This last
reagent, in nitromethane, converts 1,2,4-triazole itself into a mixture of
both possible *N*-methyl compounds and the 1,4-dimethyl quaternary salt
[621]. It also converts 1-acetyl-1*H*-1,2,4-triazole into 1-acetyl-4-methyl-
1,2,4-triazolium tetrafluoroborate which, on methanolysis, gives 4-methyl-
4*H*-1,2,4-triazole [621].

A number of 1,2,4-triazolium salts have been prepared from 3-arylazo-
1,2,4-triazoles or the corresponding 5-carboxylic acids by combined
substitutive and quaternising methylation, e.g. by reaction with dimethyl
sulphate in *o*-dichlorobenzene or dimethylformamide, sometimes in the
presence of magnesium oxide [540i,601d,608a,c,617,e,h,i,j,627].

An early attempt to quaternise 2,5-diphenyl-2*H*-1,2,3,4-tetrazole by
heating it at 100°C for 3 hours with methyl iodide failed [23a]. In
similar conditions 5-methyl-1-(3,4-dimethylphenyl)tetrazole and related
compounds were quaternised with methyl iodide, neat or in boiling
isopropanol or with methyl benzenesulphonate on the steam-bath [628-9].

The structures of the quaternary salts from 5-methyl-1-phenyl- and
5-methyl-1-(3,4-dimethylphenyl)tetrazole with methyl iodide were proved
to be (93*a*) and (93*b*) respectively by alkaline degradation to methylamine

Me
 \
  N—N
  ‖ + ⟍
Me   N   N
  \ ‖
   N   I⁻
   |
   Ar

(93) *a*, Ar = Ph; *b*, C$_6$H$_3$Me$_2$(3,4)

and an aryl azide [613b]. Several other related quaternary salts, as well as
1,4,5-trimethyltetrazolium iodide, have been described [617d]. Ring-opening
reactions prove the salt from 1-ethyltetrazole and ethyl toluene-*p*-
sulphonate to be 1,4-diethyltetrazolium toluene-*p*-sulphonate [406b], and
that from 2-methyltetrazole and methyl benzenesulphonate to be 1,3-

dimethyltetrazolium benzenesulphonate[407a]. The general presumption from these results is that 1- and 2-substituted tetrazoles are both quaternised at N-4. However, this is not completely true; 1-methyl-5-phenyl-1*H*-1,2,3,4-tetrazole kept with methyl iodide for 90 days at room temperature gives starting material (55%), 2-methyl-5-phenyl-2*H*-1,2,3,4-tetrazole (6%), and a mixture of 1,3-dimethyl- and 1,4-dimethyl-5-phenyl-1,2,3,4-tetrazolium salts (37%). The relative stabilities of the two quaternary salts lead to the production from the reaction components of 2-methyl-5-phenyl-2*H*-1,2,3,4-tetrazole at higher temperature (see §9.13).

We shall now consider the quaternisation of azoles containing substituents which can compete with the ring nitrogen atom for the alkylating agent.

5-Amino-1-arylpyrazoles are quaternised at N-2 by alkyl halides[330b, 552c,d]. As in the pyridine series (Schofield, p. 181), the anhydronium or conjugate bases of the quaternary cations are quaternised at the exocyclic nitrogen atom, as in the example shown[330h].

In contrast to the 5-amino compounds, 4-amino-5-chloro-3-methyl-1-phenylpyrazole reacts with methyl iodide to give 5-chloro-4-dimethylamino-3-methyl-1-phenylpyrazole hydriodide, and the free base from this salt is quaternised by methyl iodide at the exocyclic nitrogen atom[551f, 552e].

Little is known about the quaternisation of amino-imidazoles or aminotriazoles. 5-Amino-4-aminocarbonyl-1-benzyl-1*H*-1,2,3-triazole is quaternised at N-3 by methyl toluene-*p*-sulphonate[149b], and 4-amino-4*H*-1,2,4-triazoles are quaternised at N-1[630a]. 5-Amino-1-methyl-1*H*-1,2,4-triazole is quaternised at N-4[514d].

As indicated above (p. 82), alkylating agents reacting with 5-aminotetrazole in the absence of alkali effect both substitutive and quaternising alkylation. The reaction of the amine with alkyl halides was early studied [25c], though correct identification of the products took a long time [515e]. At one stage the products formed when various 1-alkyl-5-amino- and 5-amino-1-aryl-1*H*-1,2,3,4-tetrazoles were heated with alkylating agents (methyl benzenesulphonate, dimethyl sulphate, diethyl sulphate, and alkyl halides) were believed to be 1-substituted 5-alkylaminotetrazoles [515a]. The differences in properties of these products and compounds unambiguously possessing such structures[579a] showed this view to be wrong, and a 1-alkyl-5-aminotetrazole heated with an alkylating reagent gives, in fact, a 1,4-dialkyl-5-aminotetrazolium salt. Usually the reaction

mixture is basified and the product isolated as the anhydronium base, a
1,4-dialkyl-5-iminotetrazole[515c]. These formulations are supported by
sequences such as those illustrated[155,515c]. In fact, ring alkylation
occurs, generally at both possible sites, as in the next cases shown[123].
The structure of (94) is adopted from the proved[587] structure of (95).
The formation of 1,2-dialkyl-5-iminotetrazoles has not been
observed.

As illustrated in one instance above, further quaternisation of 1,4-
dialkyl-5-iminotetrazoles proceeds, as would be expected, at the exocyclic
nitrogen atom. In another example 1-benzyl-4-methyl-5-iminotetrazole
gave with methyl benzenesulphonate the quaternary salt from which
basification produced 1-benzyl-4-methyl-5-methyliminotetrazole[123].

Numerous 1-alkyl-5-aminotetrazoles have been quaternised by heating
with alkylating agents, and the products identified by their preparation
from two different 1-alkyl compounds reacting with complementary
alkylating agents, or by hydrogenolysis of benzyl groups as illustrated
above[515e,f,g].

When we turn to derivatives of hydroxyazoles we find some complicated results in the pyrazolone series, involving the quaternisation of
antipyrines. In the simplest circumstances a quaternary salt arises from
the reaction of antipyrine (or an analogue) with an alkyl halide at 60°C,

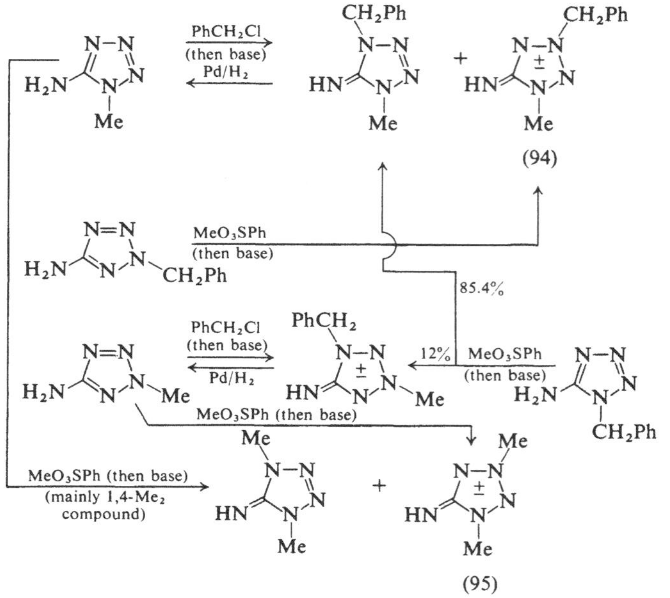

(94)

(95)

or more slowly at room temperature. The structures illustrate the situation. Experiments at higher temperatures produce more complicated results; antipyrine with methyl iodide at 80–200 °C, or (96), R = R' = Me, treated similarly, gives 4-methylantipyrine and 3,4,4-trimethyl-1-phenylpyrazol-5-one. The proportion of the trimethyl compound increases with temperature; it does not arise by the direct isomerisation of 4-methylantipyrine, the presence of methyl iodide being necessary. Next shown are reactions related to these [615c]. In the case where (R = H) later workers detected also the formation of some 4-benzyl-3-methyl-1-phenylpyrazol-5-one [596]. The *C*-alkylations have been attributed to the formation of quaternary salts by attack of the alkylating agent at N-2 with subsequent migration to C-4; loss of the *N*-benzyl group from such a quaternary salt generates an antipyrine with methyl in place of benzyl [596,615c].

Acyloxy- and alkoxy-pyrazoles are quaternised at nitrogen [329e].

As illustrated here, meso-ionic sulphur derivatives of imidazole form quaternary salts with methyl iodide [572].

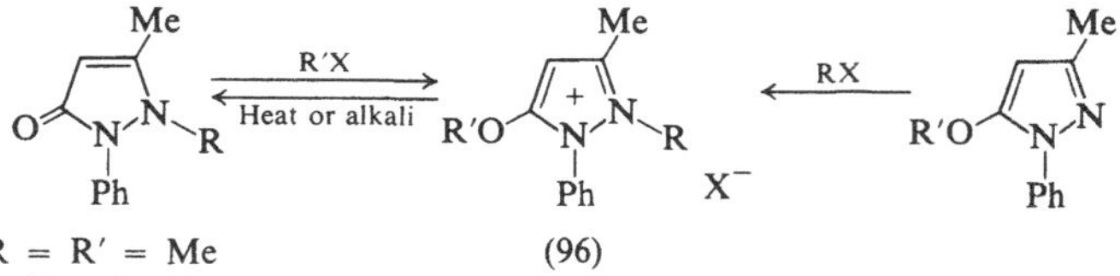

R = R' = Me   (96)
or R = Me, R' = Et

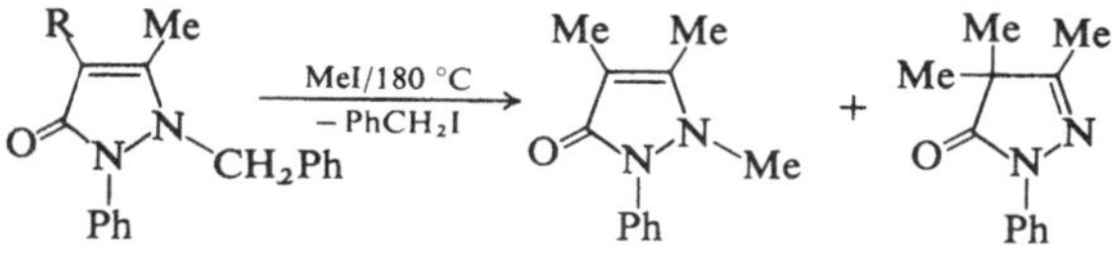

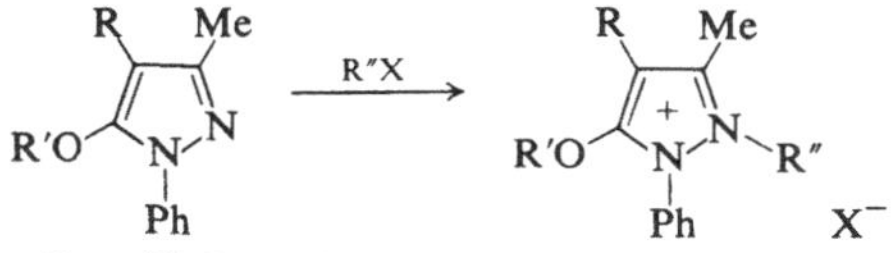

$$R = H, R' = PhCO, R''X = MeI$$
$$R = Me, R' = Et, R''X = BrCH_2CO_2Et$$

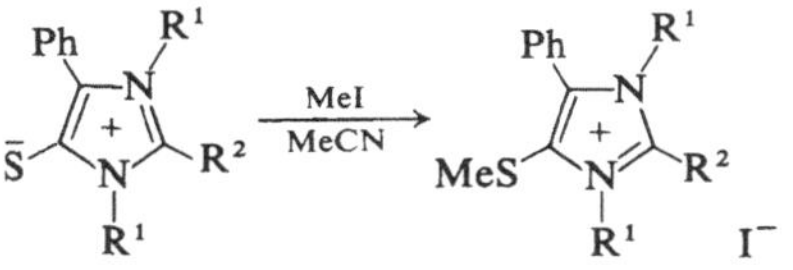

$$R^1 = R^2 = Me; \quad R^1 = C_4H_9, \quad R^2 = CHMe_2$$

In the 1,2,3-triazole series, quaternary salts (97), R = Me, R' = H, R'' = Me or Et, have been obtained from meso-ionic compounds of the type (84) by reaction with methyl or ethyl iodide[576b], but attempts to obtain the quaternary salts from several other meso-ionic compounds of this type failed[576c]. This is not surprising in view of the reactions which occur between 1-substituted 5-methoxy-1,2,3-triazoles and alkyl halides (table 18). Commonly, meso-ionic compounds of the type (84) are formed, no intermediate quaternary salt being detectable or isolable. In a few cases, however, the quaternary intermediates have been either detected or isolated (table 18).

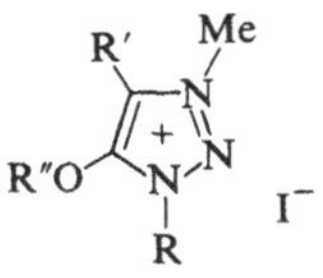

(97)

Numerous examples of the formation of quaternary salts from the sulphur analogues of antipyrine have been described. As would be expected, these reactions, some of which are given in table 19, occur at the sulphur atom, whilst the methylthio compounds react at nitrogen, as shown here[330f]. The related selenium compounds behave similarly [330a,d].

Table 18. Reactions of 1-substituted 4- and 5-acyloxy- and 5-alkoxy-1,2,3-triazoles with alkyl iodides

Quaternary salt not detected	R	R'	R''	R'''X	Ref.
	Me	Ph	Me	MeI	576c
	Me	EtO$_2$C	Me	MeI	576c
	PhCH$_2$	H	Me	MeI	576c
	Ph	Me	Me	MeI	576c
Quaternary salt detected	Ph	H	Me	MeI	576c
	Ph	H	Me	EtI	576c
Quaternary salt isolated	Me	H	Me	MeI	576a,b[a]
	PhCH$_2$	H	PhCO	MeI	576d

| | | | | MeI | 576a,b[a] |
| | ditto | | | PhCH$_2$I | 576c |

(a) The quaternary salt very readily decomposed in solution (CHCl$_3$) to give the meso-ionic product.

Table 19. Some quaternary salts formed from sulphur analogues of antipyrine

a	b	d	e	Ref.
Me	Me	Ph	H	330d
Me	Ph	Me	H	330a
Me	Ph	Me	Me	330c
Ph	Me	Me	H	552b
Ph	Me	Me	PhCO	321h

Sulphur derivatives of 1,2,4-triazole show the same pattern of behaviour, but in the cases where reaction occurs at nitrogen the problem of orientation arises. This can be solved by noting the effect of the quaternisation upon the reactivity of *C*-attached substituents (table 20). Two related reactions in the tetrazole series are shown[612,631].

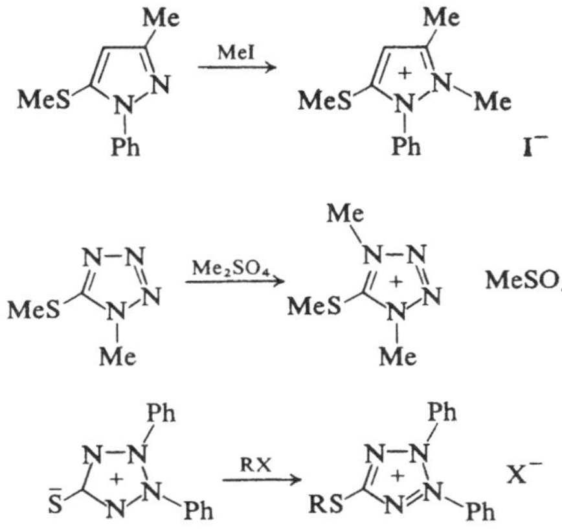

RX = MeI and ClCH₂CO₂H

RX = MeI and ClCH$_2$CO$_2$H

3.13.3   The attachment of aryl groups: substitutive arylation

Azoles have been successfully *N*-arylated by reaction with halogeno-benzenes, activated towards nucleophilic attack, under a variety of conditions, most commonly in the absence of strong bases (table A.6).

Included in table A.6 are several examples of the *N*-arylation of imidazole by unactivated halogenobenzenes in the presence of cuprous bromide; these reactions may differ mechanistically from the others.

In the pyrazole series the less basic the compound the greater was the reaction time, and with substituted pyrazoles the least hindered product appeared to be formed preferentially[114b]. 3,5-Di-t-butyl-, 3,5-di-t-butyl-4-methyl- and 4-t-butyl-3,5-dimethyl-pyrazole did not react with 2,4-dinitrofluorobenzene either in boiling ethanol or boiling xylene[114h]

N-Picrylpyrazoles have been prepared in good yield by heating together picryl chloride and *N*-acylpyrazoles[636]. 1-Acetyl-3-methyl-pyrazole gives 3-methyl-1-picrylpyrazole.

Not much work has been reported on the arylation of azoles contain-ing tautomerisable substituents. 3-Amino-1,2,4-triazoles react first at the amino group with picryl chloride, as does 4-amino-4*H*-1,2,4-triazole[490a] see diagram.

With 2-bromopyridine and related heterocycles, the sodium salt of 2,3-dimethylpyrazol-5-one gives 3-aryloxy-1,5-dimethylpyrazoles rather than *N*-arylated products[517].

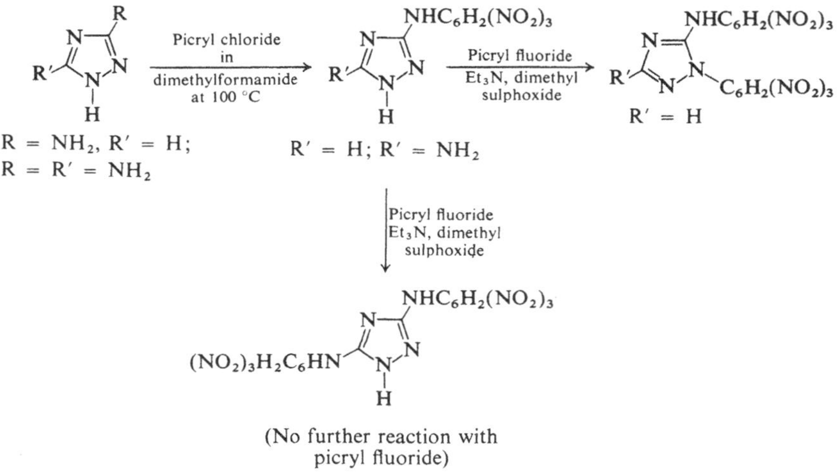

Picryl chloride in dimethylformamide at 100 °C

R = NH₂, R' = H;
R = R' = NH₂

Picryl fluoride Et₃N, dimethyl sulphoxide

R' = H; R' = NH₂

Picryl fluoride Et₃N, dimethyl sulphoxide

R' = H

Picryl fluoride Et₃N, dimethyl sulphoxide

(No further reaction with picryl fluoride)

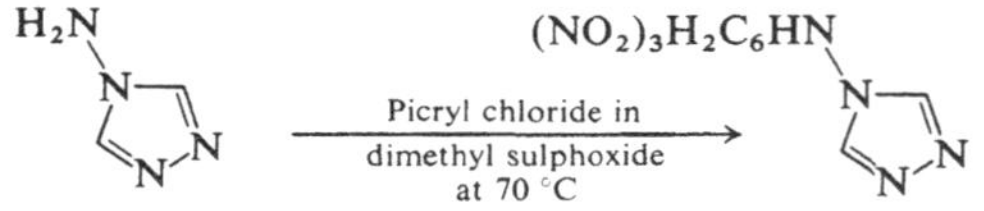

Picryl chloride in dimethyl sulphoxide at 70 °C

Table 20. Some quaternary salts formed from sulphur derivatives of 1,2,4-triazole[a]

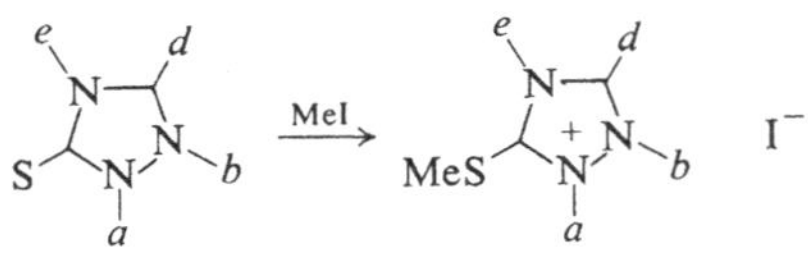

a	b	d	e	Ref.
Ph		Me	Ph	223c
Ph		Me	Me	613a
Me	Ph	Me		613a
	Me	Me	Me	223c, d
	Ph	Me	Me	613a

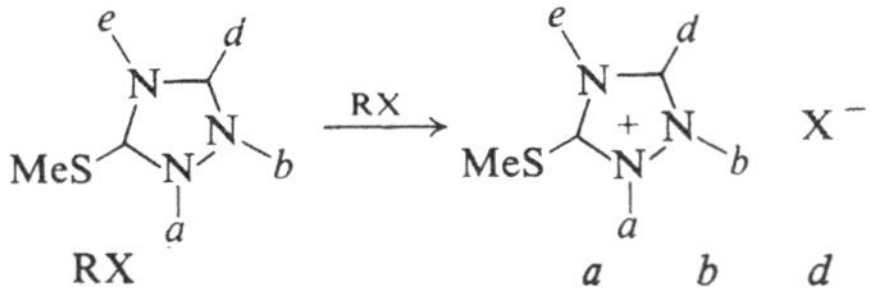

a	b	d	e	RX	a	b	d	e	Ref.
Ph	Me			MeO₃S.C₇H₇	Ph		Me	Me	613a[b]
Ph	MeS			Could not be quaternised					613a
	Ph	Me		MeI or MeO₃S.C₇H₇		Ph	Me	Me	613a[c]
	Ph	Me		Me₂SO₄	Me		Ph	Me	612[d]
	Me	Me		MeX		Me	Me	Me	613a, 617c[e]
	Me	Et		MeX		Me	Me	Et	613a, 617c[e]
	Me	Ph		MeX		Me	Me	Ph	613a, 617c[e]

(a) Skeletal structures are used to facilitate tabulation. (b) The *C*-methyl group was unreactive. (c) The *C*-methyl group was reactive but the MeS group was not. (d) The MeS group was reactive. (e) The *C*-methyl group was reactive.

3.13.4 Formation of N-substituted-alkyl azoles by addition reactions

Azoles unsubstituted at nitrogen react readily with olefinic compounds activated to nucleophilic attack (table A.7). Acidic and basic catalysts have been used, but many of the reactions proceed satisfactorily in their absence. The reactions are of the type illustrated, Z being an activating group. Of the same kind is the addition of 1,2,3-triazole to aziridinium tetrafluoroborate to give 1-(1H-1,2,3-triazolyl)ethylamine[235], and of 3,5-dimethylpyrazole to benzoquinone to give (98)[126a].

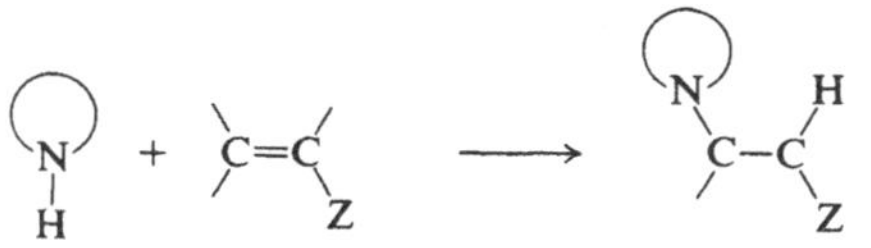

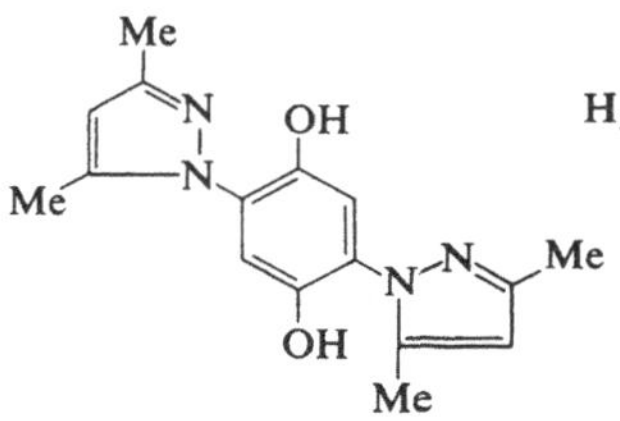

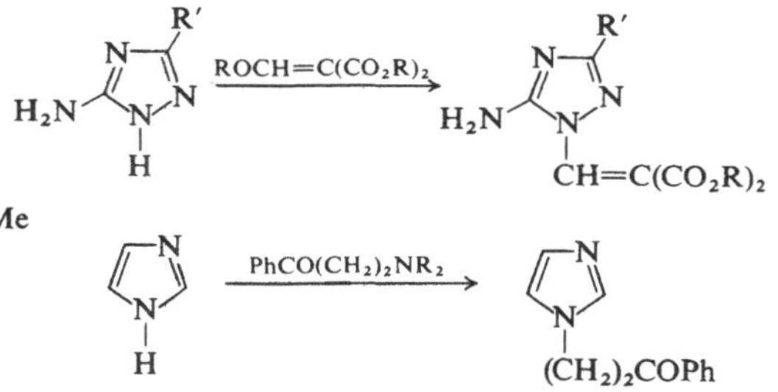

4,5-Dibromo-1,2,3-triazole adds to mesityl oxide in the expected way, giving the 2-substituted triazole, and the same product is formed from the salt of the triazole with dimethylamine and acetone. The dimethylamine salt of tetrazole similarly reacted with acetone, but that of 2,4,5-tribromo-imidazole did not. The corresponding triethylamine salts did not react with acetone[575].

The reactions of 5-amino-1,2,4-triazoles with alkoxymethylene malonic esters[540h], and of imidazole with Mannich bases[637], illustrated above, may be mechanistically similar, though this has not been demonstrated.

Azoles also react with acetylenic compounds. The conditions under which N-vinylimidazoles arise from imidazoles and acetylene[324k] make it unlikely that these are simple electrophilic additions.

Diels and Alder first studied the reaction of pyrazole with dimethyl acetylenedicarboxylate[356]: with ether as solvent they obtained two isomeric products, m.p. 158°C and 139°C, each formed from one molecule of the ester and two of pyrazole. The lower melting compound was obtained on only one occasion. 3,5-Dimethylpyrazole gave a similarly constituted product, and also a 1 : 1 addition product. Subsequent work left the situation confused. For the product from pyrazole Reimlinger[432a] claimed to prove the structure (99) without saying what was the melting point of the product to which this structure was attributed. Later, Reimlinger[432b] obtained a 2 : 1 addition product from dimethyl pyrazole-3,4-dicarboxylate and dimethyl acetylenedicarboxylate, which he formulated as arising from reaction at nitrogen. Pyrazole itself with dimethyl acetylenedicarboxylate in ether containing a little sodium methoxide gave a product of the same melting point (139 °C) as that obtained on only one occasion by Diels and Alder.

Acheson and Poulter[638a] re-examined the problem, and obtained from pyrazole, 3,5-dimethylpyrazole, and 3,4,5-trimethylpyrazole the products (100), R = R' = R'' = H; R = R'' = Me, R' = H; R = R' = R'' = Me, that from pyrazole having the m.p. 158 °C. From 3,4,5-trimethylpyrazole (101) was also obtained. They could not reproduce Diels and Alder's isolation from 3,5-dimethylpyrazole of the product corresponding to (101) (but see the statement in ref. 638e). The various products clearly arose by reaction at nitrogen and not at carbon.

(99)   (100)

(101)   (102)   (103)

Reimlinger and his co-workers appear to have resolved the confusion. Pyrazole reacts with dimethyl acetylenedicarboxylate in carbon tetrachloride to give mainly the 1 : 1 product of *cis* addition[432d,e] (102), R = CO_2Me. This product reacted with dimethyl acetylenedicarboxylate in ether containing a little sodium methoxide to give, stereospecifically, a product of the type (100), m.p. 142 °C, and in ether without the basic catalyst to give both this product and an isomer, m.p. 167 °C [432e]. The compound m.p. 142 °C presumably corresponds to that of Diels and Alder, m.p. 139 °C, and Diels and Alder's compound, m.p. 158 °C, must have been an impure form of the product, m.p. 167 °C. *N*-Substituted pyrazoles gave only tars with dimethyl acetylenedicarboxylate[638a].

In neutral conditions pyrazole reacted with methyl propiolate to give (102), R = H, but in basic conditions a second molecule of pyrazole was added. Similarly, pyrazole and methyl tetrolate under neutral conditions gave the 1 : 1 product of (probably) *cis* addition[432e].

Diels and Alder[356] also examined the reactions of imidazoles with dimethyl acetylenedicarboxylate. The 1 : 1 product obtained from 4-methylimidazole was wrongly formulated as being formed by reaction at C-2; it is, in fact, analogous to (101), but of unproved stereo-chemistry[638c]. 2-Methylimidazole reacts similarly, but imidazole did not give a crystalline product, and when 1-methylimidazole was used only hexamethyl mellitate could be isolated[638c]. In contrast, 1,2-dimethyl- and 2-ethyl-1-methyl-imidazole gave the compounds (103), R = Me and Et, respectively[356,638b,639]. This behaviour is in marked contrast to that of *N*-substituted pyrroles (Schofield, p. 82), which behave as dienes in these reactions.

1,2,3-Triazole with dimethyl acetylenedicarboxylate gives the analogue of (100), reaction occurring at N-1. The same is true of 1,2,4-triazole and of 3-methyl-1,2,4-triazole, but only in the former case does n.m.r. spectroscopy demonstrate reaction to have occurred at N-1[638d].

The reactions producing compounds (103) are certainly reactions which begin at nitrogen (*cf.* the reactions involving pyridine, Schofield, p. 193), and the same is probably true of other reactions which end in modification of the azole ring.

Not much is known of reactions of the kind under discussion occurring with azoles containing tautomerisable substituents. One, involving 5-amino-1,2,4-triazole, was mentioned above, and three others are summarised in the diagram[614b,640].

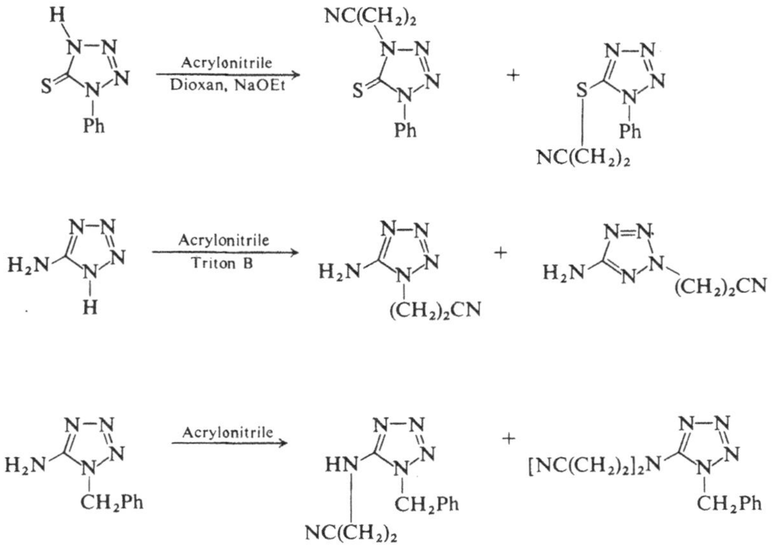

In the pyridine series addition to the nitrogen atom necessarily causes quaternisation (Schofield, p. 191). Corresponding reactions with azoles are rare. One is illustrated here[542].

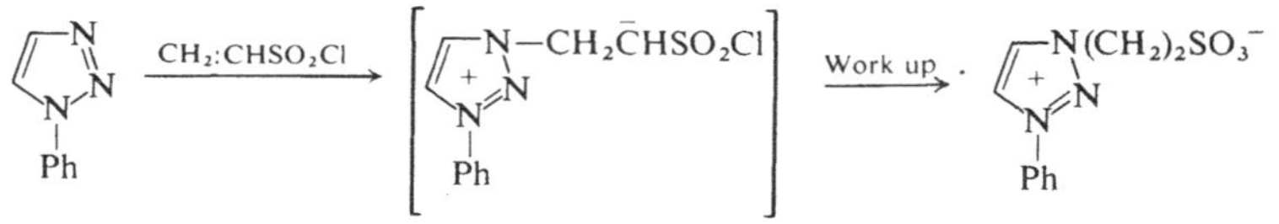

3.13.5 The attachment of acyl groups

Pyrazole and its *N*-unsubstituted derivatives have been *N*-acylated in various ways. Acyl chlorides have been used alone[52a,63a,66c,e,74d, 100b,c,f,h,126a,b,e,150h,189a,392g,431c,555,642b,644-5,648] or together with pyridine[100b,h,189a,218,646]. Special cases are the use of chloroformic esters[52a,63c,66i,100f] and phosgene. With the latter reagent in tetrahydrofuran, 3,5-dimethylpyrazole gave the *N,N*'-carbonyl-bispyrazole[150h] whilst from 3-methylpyrazole the *N*-chlorocarbonyl compound was obtained[100c]. Thiocarbonyl compounds have been similarly prepared from thiophosgene[642c], and from thiobenzoyl chloride[649b].

N-Acylation has also been effected with carboxylic acid anhydrides [52a,74d,100c,f,431c,556a] and with mixed anhydrides of the type RCO.O.COEt[189a,642a,b]. The stereoisomeric methyl β-(acetylthio)-acrylates both gave methyl 1-acetylpyrazole-3-carboxylate with methyl pyrazole-3-carboxylate[556a].

Pyrazoles react with butyl isocyanate[555] or phenyl isocyanate[63a, 66i,150h] giving *N*-aminocarbonyl derivatives, and with phenyl cyanate giving compounds of the type (104)[647].

The sodium or silver salts of pyrazoles have been *N*-acylated with acid chlorides[100b], and with carbamoyl chlorides[100c,555].

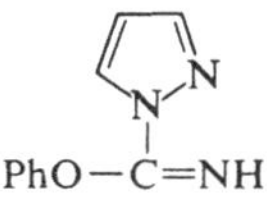

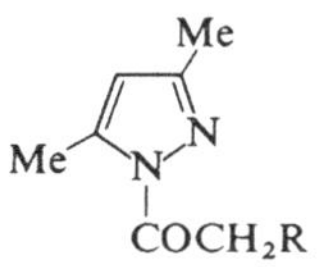

The formation of *N*-acylpyrazoles from pyrazole Grignard reagents is discussed later (§9.10). An acylation which is presumably not an electrophilic substitution produces compounds (105) from 3,5-dimethyl-pyrazole and diazoketones $RCOCHN_2$ by photochemical or thermal decomposition of the latter[642b].

Formally, two isomeric *N*-acyl derivatives might be expected to be formed from suitably substituted pyrazoles, but most often only one, the more stable of the isomers results. The less stable isomer can often be prepared by ring-synthetic methods, and is usually converted, by heating, into the stable form. This situation is illustrated by the examples[100h]. The structures of these and other pairs of isomeric acyl derivatives have usually been attributed on the basis of synthetic

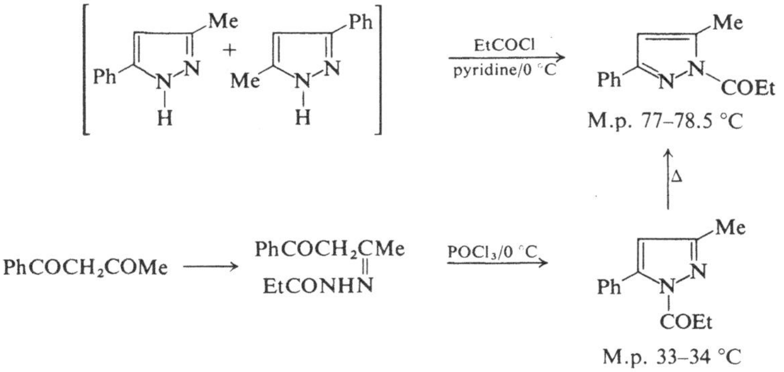

$$
\left[\text{Ph–[pyrazole: Me, N–N–H]} + \text{[pyrazole: Ph, Me, N–N–H]} \right] \xrightarrow[\text{pyridine/0 °C}]{\text{EtCOCl}} \text{Ph–[pyrazole: Me, N–N–COEt]}
$$

M.p. 77–78.5 °C

$$ \Delta \uparrow $$

PhCOCH₂COMe → $\underset{\text{EtCONHN}}{\overset{\text{PhCOCH}_2\text{CMe}}{\|}}$ $\xrightarrow{\text{POCl}_3/0\,°\text{C}}$ Ph–[pyrazole: Me, N–N–COEt]

M.p. 33–34 °C

evidence, combined in some instances with refractivity data[63g,100f,h], though these are not always helpful. Sometimes the acylation reaction and the ring-synthesis from a diketone give the same product, and further evidence is then necessary, as in the case shown[100h].

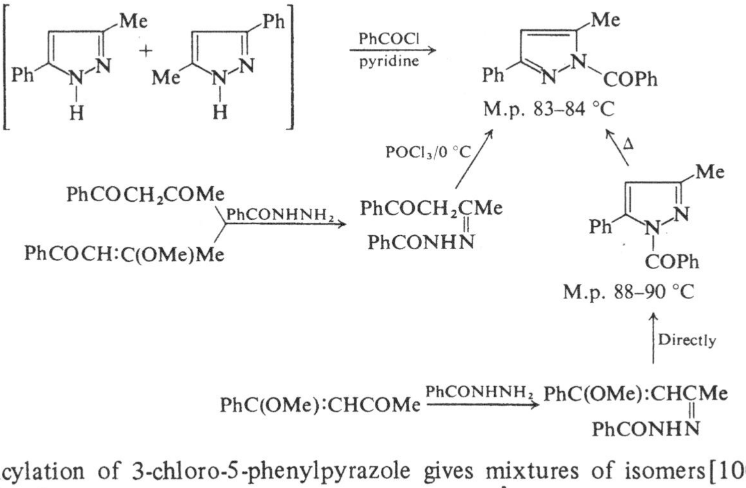

$$
\left[\text{Ph–[pyrazole: Me, N–N–H]} + \text{[pyrazole: Ph, Me, N–N–H]} \right] \xrightarrow[\text{pyridine}]{\text{PhCOCl}} \text{Ph–[pyrazole: Me, N–N–COPh]}
$$

M.p. 83–84 °C

$$ \text{POCl}_3/0\,°\text{C} \nearrow \qquad \nwarrow \Delta $$

PhCOCH₂COMe

PhCOCH:C(OMe)Me $\xrightarrow{\text{PhCONHNH}_2}$ $\underset{\text{PhCONHN}}{\overset{\text{PhCOCH}_2\text{CMe}}{\|}}$ → Ph–[pyrazole: Me, N–N–COPh]

M.p. 88–90 °C

$$ \uparrow \text{Directly} $$

PhC(OMe):CHCOMe $\xrightarrow{\text{PhCONHNH}_2}$ $\underset{\text{PhCONHN}}{\overset{\text{PhC(OMe):CHCMe}}{\|}}$

Acylation of 3-chloro-5-phenylpyrazole gives mixtures of isomers[100b].

The formation of the more stable isomers in the cases mentioned, and in others[100d,f,i], raises interesting questions. There are, for example, reasons to believe[63g] that of the two tautomeric forms, 3-phenyl-1*H*-pyrazole (106) predominates over 5-phenyl-1*H*-pyrazole (107). In a S_E2 acylation (106) would give 1-acyl-3-phenylpyrazole (108), and from an S_E2' reaction 1-acetyl-5-phenylpyrazole (109) would result. The alternative possibilities are shown below. The product obtained is in fact (108)[100h]. If, then, (106) is the strongly dominant tautomer, and if

(106) and (107) react at comparable speeds, (108) must result from an S_E2 reaction, or from rearrangement of the initially formed isomer. Rearrangement under some of the conditions of acylation would not be surprising, but acylation has frequently been carried out in pyridine solution at 0 °C. In the latter circumstances rearrangement does not seem to have been observed with complete unambiguity starting from a less stable isomer. It is obviously essential, in any particular case (such as that of the *N*-benzoyl-3-methyl-5-phenylpyrazoles above) to prove that a possible case of isomerisation relates to the cyclised compound, and not to the hydrazone precursor.

In the circumstances, where uncertainty exists concerning the structures of pairs of isomeric *N*-acylpyrazoles, more particular discussion of mechanism is, of course, pointless. Table 21 lists the melting points of a number of pairs of isomers, with the structures at present attributed to them. The ease of rearrangement of less stable to more stable isomers depends on the acyl group, and increases roughly in the order o-nitrobenzoyl$<$benzoyl$<$acetyl$<CONH_2$ [100h].

Table 21. [100h]

| | | Less stable | | More stable | |
R	R'	Ac	Less stable (m.p./°C)	More stable (m.p./°C)
Ph	Me	MeCO	45.5–46.5	41
		EtCO	33–34	77–78.5
		PrCO	34–35.5	72–72.5
		PhCH$_2$CO	104.5–105.5	58–59.5
		CO$_2$Et	65–66	73.5–74.5
		PhCOa	88–90	83–84
		o-MeC$_6$H$_4$CO	63–65	36.5–37.5
		m-MeC$_6$H$_4$CO	79–80	56–57
		p-MeC$_6$H$_4$COb	83–85	68–70
		o-NO$_2$C$_6$H$_4$CO	157–157.5	107–108
		H$_2$NCOc		154–156
Me	H	H$_2$NCO	117–119	127–128
		o-NO$_2$C$_6$H$_4$CO	130	120
Ph	H	H$_2$NCO	133–134	142–143
		CO$_2$Et	58–59	(B.p. 190°/13 mm)
		o-NO$_2$C$_6$H$_4$CO	151–152	107.5–108.5

(a) See text. (b) Heat does not isomerise the less stable form but merely de-acylates it. (c) The absence of a less stable isomer is not surprising in view of the ease of migration of ·CONH$_2$ [66c,100c].

The N-acylpyrazoles need further examination both as regards their structures and the mechanisms of their formation. N.m.r. spectroscopy should prove valuable in this connection. The study of the single product, probably 1-acetyl-3-t-butylpyrazole, obtained by acetylating 3-t-butyl-pyrazole, and of the unambiguous structure (110) has already been mentioned (§2.2.3). More recently the technique was applied to methyl 1-acetyl-3-carboxylate as mentioned above, but seems not to have been generally applied so far.

(110)

Wyss[6b] first attempted to *N*-acylate imidazole with acetyl and benzoyl chloride and acetic anhydride. Perhaps the ease of hydrolysis of the products (p. 184) was the cause of his failure. Although benzoylation by the Schotten–Baumann method proceeds normally with 2,4,5-tribromo-imidazole[650] and with 4-phenylimidazole and its derivatives[507,621], the method is not generally suitable because of ring-opening which succeeds *N*-benzoylation (p. 169).

Knorr[392g] obtained *N*-acyl- and -aroyl-pyrazoles by treating the base with half a molecular equivalent of an acid chloride in ether. Gerngross[381c] used the method, with benzene as solvent, in the first successful preparation of 1-benzoylimidazole and some of its homologues. 1-Acetylimidazole was prepared similarly[138], and subsequently the method has been frequently applied[150a,b,c,651-5]. Tetrahydrofuran is a useful solvent in these reactions, and has been used for the reaction of imidazole with ethyl chloroformate[150g]. Both tetrahydrofuran and benzene have been used as solvents in the preparation of 1,1'-carbonyldi-imidazole from imidazole and phosgene[150f,656]. Thiophosgene reacts similarly to phosgene[150j,642c], and thiobenzoyl chloride thiobenzoyl-ates imidazole[649b]. Imidazole reacts with diphenylcarbamoyl chloride in alcohol or acetone[657].

Some imidazoles have been *N*-acetylated by being heated with acetic anhydride[651]. Imidazole, 2-methyl-, 4-methyl-, and 4,5-dimethyl-imidazole have been so acetylated. From the reaction with 2-methyl-imidazole the stable 1-acetyl-2-methylimidazolium acetate was isolated, and n.m.r. spectroscopy indicated the product from 4-methylimidazole to be 1-acetyl-4-methylimidazole[658a].

Imidazole has been *N*-acetylated with isopropenyl acetate[458a,b,659]. With this reagent 2-benzoyl-4-phenylimidazole is believed to give 1-acetyl-2-benzoyl-4-phenylimidazole, which is also obtained by the use of acetic anhydride–sulphuric acid[458a,b].

N-Acylation by reaction of acyl chlorides with silver[138] or potassium imidazole[375] has been described.

N-Acylimidazoles have been prepared from acetyl or benzoyl chloride, ethyl chloroformate, phosgene[190b,324n], or thiophosgene[660] and 1-trimethylsilylimidazole.

What kind of product is formed from an imidazole and an isocyanate [581a] depends upon the conditions. In tetrahydrofuran imidazole and organic isocyanates give the expected *N*-acyl compounds[150g,i]. With 4,5-disubstituted imidazoles containing at least one aryl group, *N*-acylation, *C*-acylation, and further reactions can occur (p. 32). With *m*-chloro-phenylcyanate, imidazole gives the compound corresponding to (104)[647].

An important reaction in which *N*-acylimidazoles are formed is that of transacylation between a carboxylic acid and an *N*-acylimidazole (p. 187). Imidazole Grignard reagents can also be *N*-acylated (p. 229).

1-Acetylimidazole can be formed enzymically, probably by the following sequence[661]:

Acetyl phosphate + coenzyme A $\rightleftharpoons$ Acetyl coenzyme A + phosphate,

Acetyl coenzyme A + imidazole $\rightleftharpoons$ 1-Acetylimidazole + coenzyme A.

Because the imidazolyl group of a histidine residue may be involved in the mechanism of reaction of such hydrolytic enzymes as trypsin and chymotrypsin, imidazoles have been much studied as catalysts of ester hydrolysis. Imidazole is an effective catalyst, and it owes this property to the intermediate formation of 1-acylimidazoles. In the hydrolysis of esters with good leaving groups, such as *p*-nitrophenyl acetate, the free base is the effective catalyst, and the more basic the imidazole derivative the more effective it is. In the hydrolysis of *p*-nitrophenyl acetate catalysed by imidazole the formation and hydrolysis of 1-acetylimidazole can be followed by the change in optical density at 245 nm[281a,b, 662]. The presence of 1-acetylimidazole as an intermediate in ester hydrolyses has also been shown by trapping experiments[663].

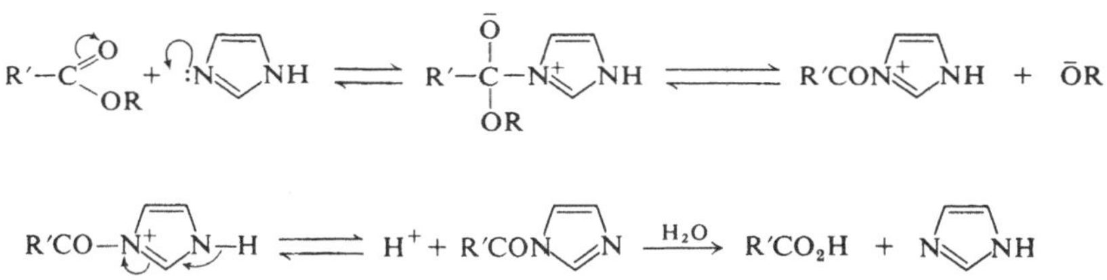

The catalysis of the hydrolysis of esters with poorer leaving groups, such as *p*-cresol acetate, involves the imidazole anion. For imidazoles with $pK_a \leqslant 4$ the catalysis by the anion is the chief reaction. 2-Substituents in the imidazole nucleus cause a large steric effect, whereas at C-4(5) they are important only because of their effect upon basicity.

In general the rates of these hydrolyses can be written as follows:

$$\text{Rate} = [\text{ester}] \, \{k_n[\text{Im}] + k'_{gb}[\text{Im}] + k_{gb}[\text{Im}]^2 + k_A[\text{Im}^-]\}$$

where k_n relates to the type of nucleophilic catalysis proceeding through the *N*-acylimidazole, k'_{gb} relates to general-base catalysis (111), k_{gb} to general-base catalysis (112), and k_A to catalysis by the anion of the imidazole. k_{gb} increases in importance with the concentration of imidazole, k_A predominates at high pH values, and k'_{gb} and k_{gb} gain in relative importance in the hydrolysis of esters whose leaving groups are stronger bases than the imidazoles[281c].

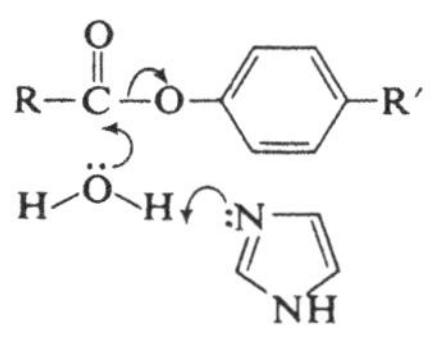

(111) 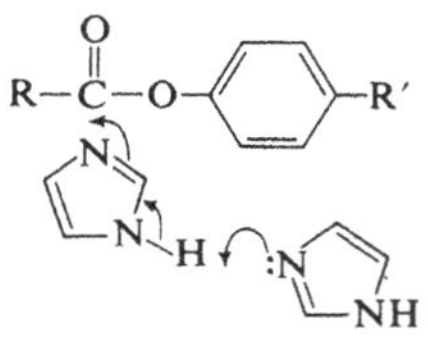   (112)

Two particular examples of the catalysis by imidazoles of ester hydrolysis will be mentioned. In the first, shown here, intramolecular catalysis occurs by internal transacylation[448a]. The second concerns the hydrolysis of vinyl acetate in aqueous solution catalysed by imidazole. 1-Acetylimidazole is formed very rapidly and reaches its steady-state concentration quickly. Acetaldehyde, produced in the reaction, reacts with imidazole to give a product, which is acetylated by 1-acetylimidazole, giving (113)[658b].

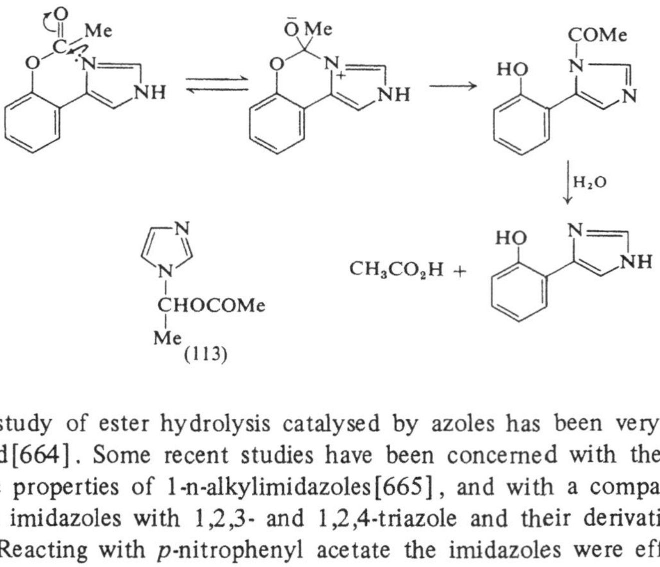

(113)

The study of ester hydrolysis catalysed by azoles has been very widely extended[664]. Some recent studies have been concerned with the catalytic properties of 1-n-alkylimidazoles[665], and with a comparison of some imidazoles with 1,2,3- and 1,2,4-triazole and their derivatives [666]. Reacting with *p*-nitrophenyl acetate the imidazoles were effective as the neutral molecules and the triazoles as their anions.

The mechanism of ester hydrolyses inhibited by imidazole and its derivatives is not fully understood[667].

Unsymmetrical imidazoles should, of course, be capable of providing two isomeric *N*-acyl derivatives. Such isomers have not been described. The potential value of n.m.r. spectroscopy in the determination of the structures of *N*-acylimidazoles is illustrated by the case of 1-acetyl-4-methylimidazole already mentioned (p. 114).

Baltzer and v. Pechmann[275] reported 1,2,3-triazole and 1,2,3-triazole-4-carboxylic acid to be acetylated by reaction with acetic

anhydride. That they could not purify the products is not surprising in
view of the expected properties of such compounds (p. 185). *N*-Benzoyl-
1,2,3-triazole was prepared by the Schotten–Baumann method, or from
silver triazole and benzoyl chloride, but its ready hydrolysis led to
discrepancies in the reported melting points[18h,19a,b,75a,275]. Since
the early work, with appropriate means of isolation, 1,2,3-triazole and
its 4-methyl, 4,5-dimethyl, and 4,5-dibromo derivatives have been
successfully *N*-acetylated with acetyl chloride or acetic anhydride[74d].
With benzoyl chloride in pyridine, 4,5-diphenyl-1,2,3-triazole gives
1-benzoyl-4,5-diphenyl-1*H*-1,2,3-triazole[153c].

1,2,3-Triazole is *N*-acylated by organic isocyanates[150i].

Until recently, *N*-acyl-1,2,3-triazoles were recorded without regard to
the position of the acyl group. Methods now available have made
possible the orientation of some *N*-acyl-1,2,3-triazoles. Thus, 4,5-dimethyl-
2-trimethylsilyl-2*H*-1,2,3-triazole (p. 127) reacts with acetyl chloride in
benzene to give 1-acetyl-4,5-dimethyl-1*H*-1,2,3-triazole, as is shown by
n.m.r. spectroscopy. Heated at 100 °C this compound changes into
2-acetyl-4,5-dimethyl-2*H*-1,2,3-triazole (a change strongly catalysed by
acid), a compound previously obtained from the reaction of 4,5-dimethyl-
1,2,3-triazole with acetyl chloride[74d,190d]. 1,2,3-Triazole itself reacts

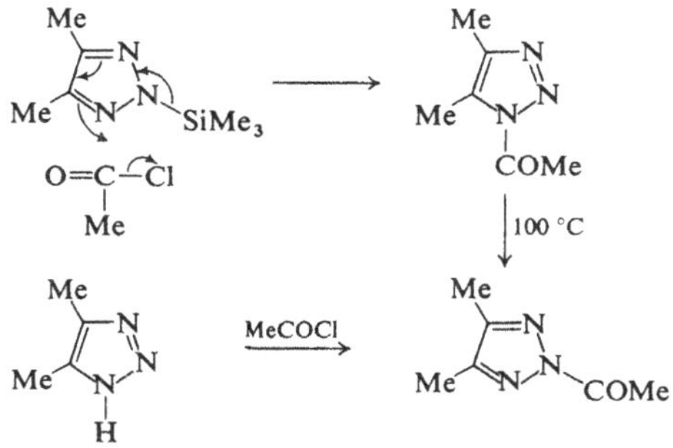

with acetyl chloride in the cold to give 1-acetyl-1,2,3-triazole with a
trace of the 2-isomer. The 1- is converted into the 2-isomer by heating
at 150 °C[190d]. This approach to the isomerism of *N*-acyl-1,2,3-triazoles
has been generalised[190f]. A number of 2-trimethylsilyl compounds
have been made by ring-synthetic methods, or by reaction of a triazole
with trimethylsilyl chloride (see p. 127). Their structures are confirmed
by infrared and n.m.r. spectroscopy, and by measurement of their dipole
moments. They react with acetyl chloride in benzene at room temperature as shown here. As will be seen, in the unsymmetrical cases the
1-acetyl-5-substituted isomers were the minor products. At equilibrium
(apparently at about 100 °C) the mixtures of isomers contained rather
more than 10% of the 1-acyl, and about 90% of the 2-acyl compound,

R—N N—SiMe₃ (R') ⟶ R—N N (R', COMe) + R—N—COMe N (R') + Me₃SiCl

R	R'		
n-C₃H₇	H	Major products	15%
n-C₄H₉	H	Major products	10%
Ph	H	Major products	
Me	Me	One product in high yield	
H	H	One product in high yield	
Me	H	Major product	25%

except in the cases of the derivatives of 1,2,3-triazole itself where the proportion of 1-acetyl-1*H*-1,2,3-triazole at 100°C was 22%, and of 4,5-dimethyl-1,2,3-triazole of which the N-2 acetyl derivative occurred solely. The low dipole moment of triazole itself suggests that the dominant tautomer is 2*H*-1,2,3-triazole (p. 9). In the light of this fact, the reactions of triazole with acetyl chloride in benzene at 0°C to give chiefly 1-acetyl-1*H*-1,2,3-triazole (with ⊅ 10% of the isomer), and of 4,5-dimethyl-triazole to give chiefly 1-acetyl-4,5-dimethyl-1*H*-1,2,3-triazole (with ⊅ 5% of the isomer), are readily understood, as are the acylations discussed above, as S_N2' reactions.

N-Acyl-1,2,4-triazoles have been prepared by the method, familiar from other examples discussed above, of treating the triazole with half an equivalent of an acyl chloride in benzene or tetrahydrofuran[150a,c]; among other chlorides, ethyl chloroformate[150g] and phosgene[150f] have been used. Boiling acetic anhydride, or acetyl chloride with the sodium or potassium salt of a triazole in benzene have been used[577a]. An old report[75a] that benzoyl chloride converted 1,2,4-triazole into dibenzoylhydrazine has been contradicted[577c]. *N*-Acyl-1,2,4-triazoles have been formed by transacylation (p. 187).

N-Acetyl-1,2,4-triazole and carbonyl-di-1,2,4-triazole are shown by n.m.r. spectroscopy to be the 1-acyl compounds[150m,223a].

1,2,4-Triazole reacts normally with phenyl isocyanate[581a], and the reaction of 3,5-dimethyl-1,2,4-triazole with potassium cyanate has been described[668]. With *o*-anisyl cyanate, 1,2,4-triazole gives the analogue of (104)[647].

N-Trimethylsilyl-1,2,4-triazole reacts with acetyl chloride, ethyl chloroformate, and phosgene to give the 1-acyl triazoles[190b].

The very reactive 2-acetyltetrazole can be prepared from acetyl chloride and silver tetrazole[150a]. Benzoyl chloride converts tetrazole into dibenzoyl urea, whilst in pyridine dibenzoylhydrazine is formed[75a]. Tetrazole and various 5-substituted tetrazoles give with acetyl chloride the 2-acetyl derivatives, which when heated undergo ring modification (see p. 165). 2-Aroyltetrazoles have not been isolated from reactions using aroyl chlorides because ring modification occurs so readily [153a,e]. 2-Acyl- and 2-aroyl-tetrazoles, and related compounds, are the un-isolated intermediates in numerous ring-modifying reactions (p. 165).

Many examples of the reactions of acylating agents with amino-pyrazoles have been reported. 1-Substituted 4-amino [330k,552c] and 5-amino-pyrazoles [323e,330b,j,417a] give the expected acylamino derivatives with formic acid, acetic anhydride, acid chlorides, cyanates, and carbon disulphide. 4,5-Diamino-1-arylpyrazoles give 4,5-diacylamino derivatives [417a,552c] and 4-amino-1-aryl-5-phenylaminopyrazoles react at the 4-amino group [330k,552d]. Surprisingly, 5-benzylamino-3-methyl-1-phenylpyrazole could not be acylated [330b], and 1-substituted 5-phenyl-aminopyrazoles also do not react [552d]. 5-Iminopyrazoles give quaternary salts with acid chlorides, as shown below, and subsequent pyrolysis of these salts (p. 244) is a route to acylaminopyrazoles not directly obtain-able, as in the examples just mentioned.

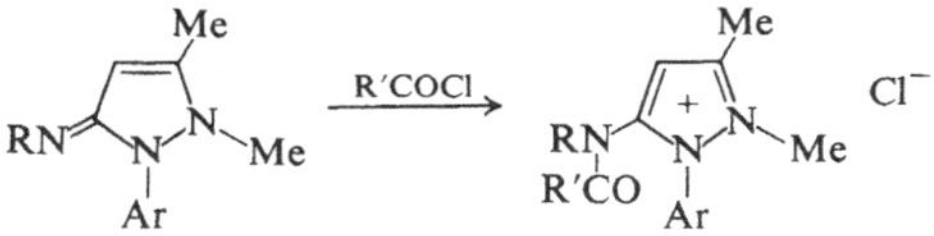

4-Aminopyrazoles not substituted at the ring nitrogen atoms give 1-acetyl-4-acetylaminopyrazoles with acetic anhydride [146a,473]; the ring acetyl group is easily removed by hydrolysis [473]. 3-Aminopyrazoles also give diacetyl derivatives readily [534a,669a,b,c], by reaction at both the amino group and at a nuclear nitrogen atom; the nuclear acyl group is very readily hydrolysed [534a,669b,670b].

3-Acylaminopyrazoles have been obtained by heating 3-aminopyrazoles with esters; subsequent acetylation then occurs at a ring nitrogen atom [534b]. The di-acyl derivatives have usually been prepared from 3-amino-pyrazoles or 3-acylaminopyrazoles by reaction with acetic anhydride [534a, b], or with an acid chloride in presence of caustic soda or pyridine [534b, 669c], but a second acyl group was introduced into 3-acetylamino-4-bromopyrazole by reaction of its sodium salt, formed in tetrahydrofuran, with dimethylaminocarbamoyl chloride [607a]. 3-Aminopyrazoles give ureas with cyanates [669c]. With acetic anhydride, 5-amino-1-aminothio-carbonyl-3-phenylpyrazole gave 1-acetyl-5-acetylamino-3-phenylpyrazole [672].

2-Aminoimidazole [274c] and some of its derivatives [274c,673] give on acetylation or benzoylation what are almost certainly 2-acylamino-imidazoles. The crude product from the benzoylation of 2-aminoimidazole by the Schotten–Baumann method may have contained di- or tri-benzoyl derivatives [274c]. Similarly, 4-aminoimidazoles give 4-acylamino compounds or sometimes 4-acylamino compounds in which a ring nitrogen atom is also acylated [418a,b,452a,674b]. Catalytic reduction of 4-nitro-imidazole in acetic anhydride–acetic acid gives such a diacetyl compound in which the nuclear acetyl and the acetylamino group are readily identified by their carbonyl stretching frequencies (1720 and 1680 cm^{-1}, respectively); boiling water removes the nuclear acetyl group giving 4-acetylaminoimidazole [544] (which has been prepared by ring-synthesis [174]). 4-Aminoimidazoles react with cyanates at the amino group to give ureas [418a,b,452a].

Acetylation of 3-amino-1,2,4-triazole with acetyl chloride in tetra-hydrofuran at 20 °C gives a product once formulated as 1-acetyl-3-amino-1H-1,2,4-triazole, whilst acetic anhydride at 100 °C gives what was regarded as 1-acetyl-3-acetylamino-1H-1,2,4-triazole. Hydrolysis of the latter gave 3-acetylamino-1,2,4-triazole, also formed from the supposed 1-acetyl-3-amino-1H-1,2,4-triazole by thermal rearrangement at 150 °C [150d,190a,675b]. Benzoyl chloride behaves similarly to acetyl chloride [150d] and the monobenzoyl product similarly isomerises on heating, and the process has been generalised with several 5-substituted 3-amino-1,2,4-triazoles [514b,d]. Russian workers concluded from spectroscopic evidence that the products originally formulated as 1-acyl-3-amino-1H-1,2,4-triazoles were in fact 4-acyl-3-amino-4H-1,2,4-triazoles [514c,e], so the original examples of acetylation mentioned above should be formulated as below (with some doubt remaining about the orientation of the diacetyl compound). The acylation of 3-anilino- [676d] and 3-ureido-1,2,4-triazole [318c,586] falls into this pattern. With acetic anhydride, 3-picrylamino-1,2,4-triazole gives what is either 1-acetyl-3-picrylamino-1H-1,2,4- or 4-acetyl-3-picrylamino-4H-1,2,4-triazole [490a]. Acetylation of

(114)

3-amino-1,2,4-triazole with ethyl acetate in the presence of sodium ethoxide gives 3-acetylamino-1,2,4-triazole[677d] A patent[607g] describes the reaction of 3-amino-1,2,4-triazole with dimethylamino-carbonyl chloride in tetrahydrofuran containing sodium hydride as giving 3-amino-1- and 5-amino-1-dimethylaminocarbonyl-1*H*-1,2,4-triazole, but its formulation of the acetylation of 3-amino-1,2,4-triazole does not appear to agree with what has been reported above.

With acetic anhydride 5-amino-1-methyl- and 5-amino-1,3-dimethyl-1*H*-1,2,4-triazole give monoacetyl derivatives[268d]. The product from 5-amino-1-methyl-3-phenyl-1*H*-1,2,4-triazole and benzoyl chloride–pyridine has been formulated as (114).

The acylation of 4-amino-4*H*-1,2,4-triazoles proceeds normally at the amino group[223b,678].

The consequences of acetylating guanazole have been formulated as in the scheme shown; the orientations of the nuclear acetyl groups are not established[675b]. With ethyl formate or formamide, 1-phenylguanazole is said to give (115)[679].

(115)

5-Aminotetrazoles are mono-acylated under mild conditions, probably at N-1, but heating of the products or acylation under vigorous conditions causes ring modification (p. 165).

As mentioned already (p. 33), pyrazolones can undergo *O*-, *N*- and *C*-acylation. We are concerned at this point only with instances where the outcome is *O*- or *N*-acylation, or both.

1-Phenylpyrazol-3-one with pyridine-acetic anhydride gives 3-acetoxy-1-phenylpyrazole[131], and 1-aryl-5-methylpyrazol-3-ones are probably

O-benzoylated[330a,e]. 3-Phenylpyrazol-5-one was supposed by early workers[421c,428c] to be converted by acetic anhydride into 1-acetyl-3-phenylpyrazol-5-one, and a di-acetyl derivative was also obtained. 3-α-Furylpyrazol-5-one reacted similarly with acetic anhydride[680]. Phenyl isocyanate was also said to react at N-1 in the latter case[680]. Later workers[681] re-examined the acylation of 3-phenylpyrazol-5-one, and their conclusions are tabulated here. Structural assignments were made on the basis of chemical tests and cannot all be regarded as conclusive. Spectroscopic evidence has been adduced in support of structures (116) and (117)[158].

The results of acetylating 3-methylpyrazol-5-one are summarised below; the structures of the products are based on spectroscopic evidence[158]. Similar evidence[158] supports the formulation of the product obtained from 3-methylpyrazol-5-one by the Schotten–Baumann method[62a] as 1-benzoyl-5-benzoyloxy-3-methylpyrazole.

4-Hydroxypyrazoles have been *O*-mono- and *O*,*N*-di-benzoylated under Schotten–Baumann conditions[76a].

M.p. 170–2 °C

M.p. 38 °C

Ac₂O steam-bath

Vigorous acylation of 1-substituted pyrazol-5-ones can cause *C*-acylation (p. 33), but it seems that under Schotten–Baumann conditions benzoyl chloride first effects *O*-benzoylation. Thus 3-methyl-1-phenylpyrazol-5-one gives (118) (as shown by the subsequent quaternisation of this product, p. 102)[329e]. Acetic anhydride probably reacts in the same sense with 3-methyl-1-phenyl-4-triphenylmethylpyrazol-5-one[357b]. 3-Methyl-1-phenylpyrazol-5-thione is *S*-benzoylated under Schotten–Baumann conditions[330f], whilst with benzoyl chloride thiopyrine gives (119)[552a]. 1-Aryl-5-methyl-4-hydroxypyrazoles[440a] have been represented as undergoing *O*-benzoylation.

(118)   (119)

The acetyl derivatives of imidazol-2-ones have usually been written as *N*-mono- or *N,N'*-di-acetyl compounds[353b,c,602,682], or not assigned structures at all[354,683], whereas derivatives of imidazol-2-thiones have been represented as *S*-acetyl or *N,S*-diacetyl compounds[143,674a]. The structures of the acetyl derivatives of imidazol-5-ones are also uncertain [684], but that from 4-methyl-2-phenylimidazol-5-one is formulated as 4-acetoxy-5-methyl-2-phenylimidazole because it is basic[325]. Derivatives of 1-substituted imidazol-5-ones are also presumably *O*-acetyl compounds [575].

1,2,3-Triazol-4-one gave a dibenzoyl derivative by the Schotten–Baumann method[429d], or with pyridine–benzoyl chloride[576e]; the product is 2-benzoyl-4-benzoyloxy-2*H*-1,2,3-triazole[576e], which is easily hydrolysed to 4-benzoyloxy-1,2,3-triazole. Failure to acylate some *N*-substituted triazolones by this method has been reported[429a]. The use of the sodium salt of 4-methoxycarbonyl-1-phenyl-1,2,3-triazol-5-one with benzoyl chloride[429a], and of several *N*-substituted triazolones with acetic anhydride–pyridine[605] or with benzoyl chloride–pyridine [429a,526,576d,686] was successful. Spectroscopic evidence suggests that *O*-acyl derivatives are thus formed from 1-substituted 1,2,3-triazol-4-[605] and -5-ones[576d].

Meso-ionic compounds of the type (84) react with benzoyl chloride to give the 4-benzoyloxy quaternary triazolium salts, which can be de-quaternised (p. 246)[687].

Acetyl derivatives have been prepared from several 1-phenyl-1,2,4-triazol-5-ones[61a,b]. Neither the structures of these derivatives, nor those of several 4-phenyl-1,2,4-triazol-5-ones (which were represented as *N*-acyl compounds)[676c] have been clarified.

A considerable amount of work on the acylation of 1-phenylurazole has been reported but the results are chaotic. Boiling with acetic anhydride appears to produce an unstable diacetyl derivative. This is easily hydrolysed to a monoacetyl compound, m.p. 173 °C, which has also been prepared under other conditions (e.g., by treating silver 1-phenylurazole with acetyl chloride) and represented as 2-acetyl-1-phenylurazole. Neither this structure nor those of other compounds in this series can be regarded as established[609a,b,d,688a,689].

An acetyl derivative of 1-phenyltetrazol-5-one has been described[83d].

3.13.6 The attachment of sulphonyl groups

Pyrazole reacts with benzenesulphonyl chloride in pyridine to give 1-benzenesulphonylpyrazole[59d], and various sulphonyl derivatives of pyrazoles have been similarly prepared[189a,318i,690-1]. The product from 3-methylpyrazole and toluene-*p*-sulphonyl chloride is shown by n.m.r. spectroscopy to be 3-methyl-1-toluene-*p*-sulphonylpyrazole[691].

Various sulphonyl derivatives of imidazoles have been prepared by reaction of a sulphonyl halide with the heterocycle in the presence of triethylamine or aqueous caustic soda[448c] or from potassium imidazole and sulphonyl chlorides[375]. An equally common practice has been to treat two molecular equivalents of the imidazole with one of the sulphonyl halide in ether or tetrahydrofuran[150c,*l*,448c]. Imidazole reacts with sulphuryl chloride to give 1,1′-sulphonyl-di-imidazole and thionyl chloride behaves similarly[150m] (thionyl chloride produces the same product from 1-trimethylsilylimidazole[190g]). Sulphonyl derivatives have also been prepared from imidazolides[150k]. The toluene-*p*-sulphonyl derivative prepared from 4-methylimidazole is proved by n.m.r. spectroscopy to be 4-methyl-1-toluene-*p*-sulphonylimidazole, and histidine and related compounds also give products of this orientation[448c]. With arenesulphonyl chlorides 1-methylimidazole gives the quaternary sulphonyl derivatives[692].

Sulphonyl derivatives of 1,2,3-triazole (p. 354) appear to have been made only by ring-synthesis. Various arenesulphonyl derivatives of 1,2,4-triazole have been prepared by the use of an arenesulphonyl chloride in the presence of triethylamine, but their orientations do not appear to be certain[693j].

Toluene-*p*-sulphonyl chloride in pyridine effects ring-opening with some 5-substituted tetrazoles, but the intermediate *N*-toluene-*p*-sulphonyl derivatives were not isolated (p. 166).

The reactions of 3-aminopyrazole with toluene-*p*-sulphonyl chloride (TsCl) and its analogues (R.SO_2Cl; R = Ph, .C_6H_4.Cl-*p*, .C_6H_4.NHAc-*p*) in pyridine are illustrated. The structures of the products are supported by alternative syntheses and a study of their physical properties[172c].

(120)

Acetylsulphanilyl derivatives of some 5-substituted 3-aminopyrazoles have been described[307a,323e,f,669c]. With acetylsulphanilyl chloride in pyridine, 4-aminopyrazole gives a bis-sulphonyl derivative, hydrolysis of which gives what is probably 4-(*p*-aminobenzenesulphonylamino)-pyrazole[307a]; the corresponding derivative of 4-amino-3-β-pyridyl-pyrazole has been described[694].

Various sulphonyl derivatives, especially sulphanilyl derivatives of 1-substituted 3-[323g], 4-[330k], and 5-aminopyrazoles[323e,f,330j,548, 617g] have been prepared. bis-Sulphonyl derivatives (120), R = R', R ≠ R', of 5-amino-1-methylpyrazole arise from reaction with sulphonyl chlorides in pyridine[172a,b].

Arenesulphonyl derivatives of 2-aminoimidazoles have been prepared using sulphonyl chlorides in pyridine[617a,695].

Sulphonyl derivatives of 3-amino-1,2,4-triazoles, again especially sulphanilyl derivatives, have been described, and have usually been represented as sulphonylaminotriazoles[324m,693e,695-6]. However, 3-amino-5-methyl-1,2,4-triazole was said to react at a nuclear nitrogen atom with acetylsulphanilyl chloride[697a], and 3-amino-5-phenyl-1,2,4-triazole gives two products, formulated as 5-amino-3-phenyl-1-toluene-*p*-sulphonyl- and 3-phenyl-5-toluene-*p*-sulphonylamino-1,2,4-triazole, when it is toluene-*p*-sulphonylated under Schotten–Baumann conditions[514b]. 4-Amino-4*H*-1,2,4-triazole is, of course, sulphonylated at the amino group[696].

5-Aminotetrazole reacts with arenesulphonyl chlorides in pyridine or aqueous sodium carbonate to give products which early workers regarded as sulphonylaminotetrazoles. These are in fact sulphonylguanyl azides, formed by opening of the tetrazole ring (p. 167). These azides re-cyclise to sulphonylaminotetrazoles in the presence of sodium hydroxide. 5-Dimethylaminotetrazole evidently reacts with arenesulphonyl chlorides in pyridine to give 2-arenesulphonyl-5-dimethylamino-2*H*-1,2,3,4-tetrazoles, but ring-opening readily follows[697c]. 5-Amino-1-methyl- and 1-methyl-5-methylamino-tetrazole were said not to react with toluene-*p*-sulphonyl chloride in pyridine[697b], but more recently the 1-methyl compound has been found to give products mono- and di-toluene-*p*-sulphonylated at the amino group[195c], and similar derivatives have been obtained from 5-amino-2-methyl-2*H*-1,2,3,4-tetrazole[195e].

Arenesulphonyl derivatives have been prepared from several 1-substituted pyrazol-3-[330a,c,e] and pyrazol-5-ones[617f,693b,d], and have usually been represented as *O*-derivatives. However, the benzenesulphonyl derivative of 3-methyl-1-phenylpyrazol-5-one has been written as the *O*-[330f] and the *N*-derivative[698].

N-Toluene-*p*-sulphonyl derivatives of 1,2,4-triazol-5-ones have been prepared by ring-synthetic methods (p. 355).

3.13.7   *N-Oxidation*

In the azole series *N*-oxides or, in the case of structures in which there are no other *N*-substituents, the tautomeric *N*-hydroxy compounds have been prepared almost solely by ring syntheses. The process of *N*-oxidation, which is of such great importance in pyridine chemistry (Schofield, p. 197) is known only in two instances.

For unknown reasons simple *N*-methylpyrazoles generally resist oxidation by the usual reagents appropriate to this kind of reaction[246, 699]. However, with hydrogen peroxide in acetic acid 1-methylpyrazole gives 1-methylpyrazole 2-oxide in moderate yield[245], and with peroxytrifluoroacetic acid 5-amino-1-methylpyrazole gives 1-methyl-5-nitropyrazole 2-oxide[700].

With hydrogen peroxide, imidazole gives oxamide, whilst with other reagents imidazole derivatives give urea. Perbenzoic acid destroys the imidazole ring, producing ammonia and urea. With the use of a 2 : 1

ratio of perbenzoic acid to imidazole it was possible to isolate an intermediate in the oxidation, which was formulated as (121) but which needs re-examining. Imidazole-4,5-dicarboxylic acid resisted oxidation[701].

4,5-Hexamethylene-1,2,3-triazole is easily autoxidised and the product has been represented as 2-hydroxy-4,5-hexamethylene-2H-1,2,3-triazole [702].

(121) (122) (123) (124)

3.13.8 N-Trimethylsilylation and N-trimethylstannylation

With hexamethyldisilazan, pyrazole, 3,5-dimethylpyrazole, imidazole, and 1,2,4-triazole give N-trimethylsilyl derivatives[190b,324n]. 5-Amino-tetrazole reacts at N-1 and the amino group[190c,h], but the reagent causes ring-opening with tetrazole[190c]. Tetrazole has been successfully trimethylsilylated (at N-1) with trimethylsilyl chloride and trimethylamine, and this method works with 5-phenyltetrazole (at N-2[190c]) (see p. 253). 1,2,3-Triazole and 4-methyl-1,2,3-triazole react similarly at N-2[190f]. Imidazole and 1,2,4-triazole have been trialkyl- and triaryl-stannylated by reaction with the corresponding stannyl oxides, or by treating their sodium salts with stannyl chlorides[703].

3.13.9 Miscellaneous reactions

When trimethylamine borane reacts with pyrazole, (122), R = H, is formed, and substituted analogues have been prepared. Alkali metal borohydrides give di- (123), tri- and tetra-(1-pyrazolyl)borates, which form chelates with first-row elements such as copper(II). 1,2,4-Triazole behaves similarly. The chemistry of these compounds has been studied[704a]

Pyrazole, imidazole, and 1,2,4-triazole react with penta- or tetra-carbonyliron, mercury tetracarbonylferrate(II), ferrocene, or dicarbonyl-cyclopentadienyliron to give, respectively, iron(II) complexes of the form $Fe(N_2C_3H_3)_2$, $Fe(N_2C_3H_3)_20.5(N_2C_3H_4)$, and $Fe(N_2C_3H_3)_2$. Iron(III) complexes have also been described[705].

With thionyl chloride, imidazole gives 1,1'-thionyldi-imidazole, and sulphuryl chloride[150k] and sulphur dichloride[649a] behave similarly. 1-Trimethylsilylimidazole and sulphur dichloride react in the same way [190b].

With sulphur trioxide, imidazole gives the salt (124), whilst 1-methyl-imidazole and chlorosulphonic acid produce a zwitterion analogous to the product from pyridine (Schofield, p. 161)[706].

3,5-Dimethylpyrazole reacts with amidodichlorophosphates in the presence of triethylamine giving the phosphoric triamides containing two *N*-linked pyrazole nuclei. In contrast, 1-substituted pyrazol-5-ones react at their oxygen atoms, giving amidophosphates[707].

With phosphoryl chloride and alkali, imidazole gives 'diphosphoryl-imidazole' (125)[708-9]. 1-Methylimidazole gives an analogous compound [710]. The catalysis by imidazoles of the solvolysis of tetrabenzyl pyrophosphate in 1-propanol involves intermediates containing P-N bonds; one such (126) is formed when silver imidazole reacts with dibenzyl-phosphoryl chloride[711]. Other compounds containing P-N bonds are known[622].

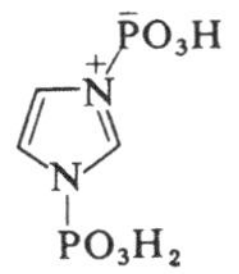

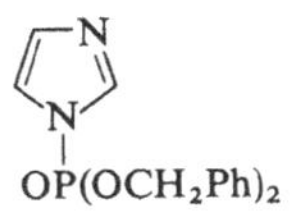

The zwitterion (127) is formed from 1-methylimidazole and tetra-cyanoethylene oxide[618].

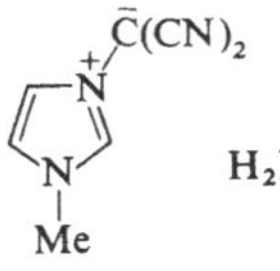

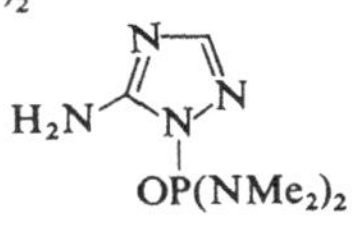

3-Amino-1,2,4-triazoles give compounds of the type (128) with bis-dimethylaminophosphoryl chloride[675a,c].

Sodium 3,5-dimethylpyrazole, sodium 3,4,5-trimethylpyrazole, and sodium 4-iodopyrazole all react with cyanogen bromide to give the corresponding *N*-cyano compounds. However, from sodium pyrazole itself the 'trimeric' product, 2,4,6-tri-(1-pyrazolyl)-1,3,5-triazine results, and silver pyrazole gives 1-aminocarbonylpyrazole[712a].

5-Amino- and 5-dimethylamino-tetrazole react at N-1 with cyanogen bromide, but the products undergo ring-opening and have not been isolated (p. 163). 5-Methylaminotetrazole suffers ring-opening in the same way, but also gives the exocyclic cyano-derivative, as does 5-imino-1,4-dimethyltetrazole[407b,713].

3.14 Intramolecular electrophilic substitutions

The reactions to be discussed are mainly based on amino-azoles, and have been quite extensively studied because of interest in the properties of the products.

Application of the Combes, Conrad–Limpach, or Knorr reactions to 1-substituted 5-aminopyrazoles leads necessarily to cyclisation at carbon. Thus, from 5-amino-3-methyl-1-phenylpyrazole and acetylacetone Bülow [604a] obtained (129a), whilst ethyl acetoacetate gave (129b). 5-Amino-1,3-diphenylpyrazole similarly gave (129c) or its isomer[534a], and it is obvious that in general with unsymmetrical β-keto-esters two directions of cyclisation are possible. Several other examples of reactions of these types

(129)

(130)

(a) R = Ph, R' = R" = R‴ = Me
(b) R = Ph, R' = R‴ = Me, R" = OH
(c) R = R' = Ph, R" = OH, R‴ = Me
(d) R = Ph, R' = R‴ = Me, R" = OH

have been reported[125a,534c,670b]. The reaction of 5-amino-3-methyl-
1-phenylpyrazole with ethyl acetoacetate was thoroughly investigated by
Russian workers[125a]. In acetic acid (129d) was formed, whilst the
crotonate from the two reactants gave, when heated in Dowtherm,
Bülow's original product (129b). Similar reactions occurred with 5-amino-
1-phenylpyrazole, and both possible isomers were obtained from 4-amino-
1-phenylpyrazole. The crotonates from 5-amino-1-methyl- and 5-amino-1-
benzyl-pyrazole have also been cyclised in the sense of the Conrad–
Limpach reaction[172d,714].

An activating ring-substituent can impose cyclisation at carbon even
when the ring-nitrogen atoms of a 5-aminopyrazole are not substituted.
Thus, 3-aminopyrazol-5-one, and 1-methyl- and 1-phenyl-3-aminopyrazol-
5-one, all react with some β-dicarbonyl compounds to give compounds
of the type (130), R = H, Me, or Ph, (tautomerism being neglected, as
in previous cases)[192b]. In other cases cyclisations involving 3-amino-
pyrazol-5-one occur both at carbon and nitrogen, and the direction
observed depends on the conditions[335f,642d,715a,716].

Many examples of reactions of these types with 3-aminopyrazoles
unsubstituted at ring nitrogen, leading to pyrazolo[1,5-*a*]pyrimidines,
have been recorded[172d,534a,b,c,670b,c,671b,d,714,715c,717]. 3-Amino-
pyrazole gives derivatives of the same ring system with esters of propiolic
acids, but in acetic acid cyclisation occurs at C-4. Tetrolic ester gives
both kinds of cyclisation[712b].

Cyclisation in polyphosphoric acid of the crotonate from acetoacetic
ester and 5-amino-1-(*o*-toluenesulphonyl)pyrazole occurs at nitrogen with
loss of the sulphonyl group[172e].

With 1-substituted 3-aminopyrazoles cyclisation both at carbon and
nitrogen have been reported, as illustrated[125a,671c].

The Skraup reaction has not been extensively applied to aminopyrazoles. With 3-amino-1-phenylpyrazole cyclisation occurs at C-4[323i],
whilst 4-amino-1-phenyl- and 4-amino-3-methyl-1-phenyl-pyrazole cyclise
at C-5[59e]. The reaction failed with 4-amino-5-methyl-1-phenylpyrazole
[59e]. 1-Substituted 5-aminopyrazoles react normally[323i]. In all of
these cases yields were only moderately good.

Fischer cyclisations of the derivatives formed from ketones and 'antipyrine hydrazone' (131) are accompanied by demethylation, as in the formation of (132) from cyclohexanone[323j]. 3-Aminopyrazole in the Japp–Klingemann reaction, or 3-hydrazinopyrazoles reacting with *a*-diketones, give products such as that (133) from the hydrazine and benzil[419a].

(131)

(132)

(133)

(134)

(135)

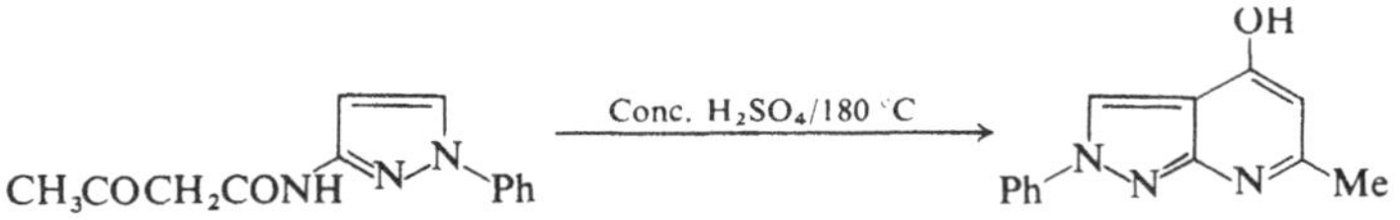

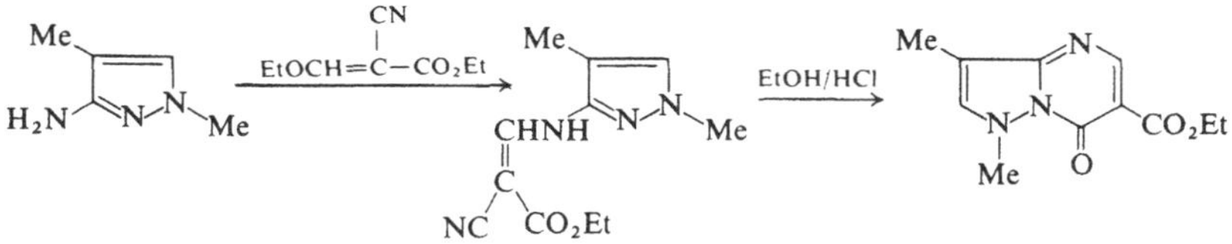

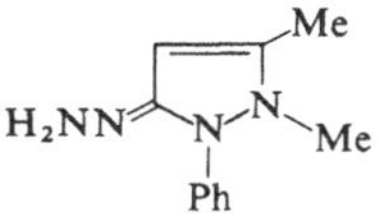

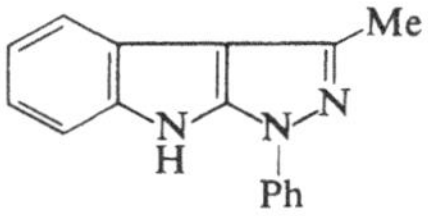

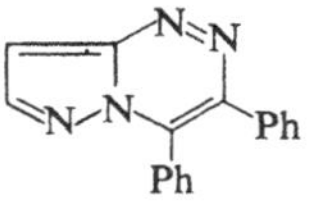

Dibromides of the type (134) are readily cyclised to the quaternary compounds (135)[704b,718].

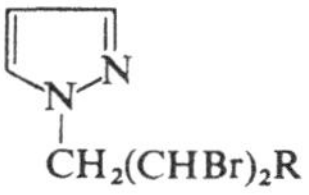

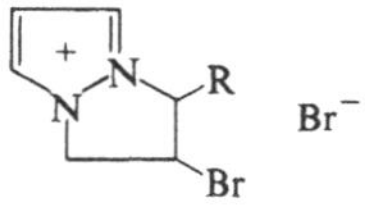

In reactions of the kinds discussed above 2-aminoimidazoles yield derivatives of the imidazo[1,2-*a*]pyrimidine system. Thus, 2-aminoimidazole itself with ethoxymethylenemalonic ester gives (136), R = H, R′ = CO_2Et, R″ = OH, and with acetoacetic ester (136), R = Me, R′ = H, R″ = OH (tautomerism being neglected)[719a], and other examples are known, some with further-substituted 2-aminoimidazoles[390b,719a, 720-1]. *β*-Diketones have also been used[390b].

With acetylacetone, 4-amino-5-methylimidazole gives the corresponding derivative of imidazo[1,5-*a*]pyrimidine[320b].

(136)

(137)

(138)

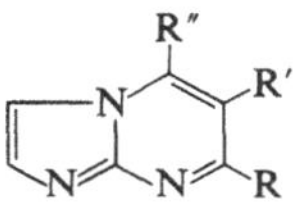

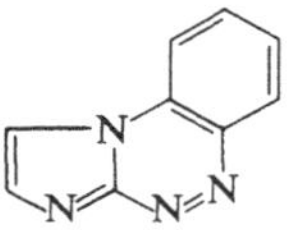

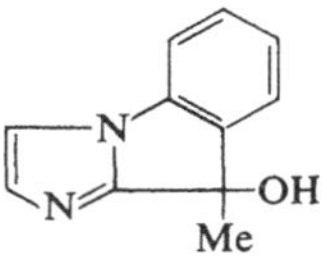

Internal diazo-coupling occurs to give (137) when 1-(*o*-aminophenyl)-imidazole is diazotised. If the 5-position is occupied, coupling occurs at C-2 but with greater difficulty[411d]. The reactions are interesting in view of the inertness of 1-phenylimidazole to diazo-coupling.

In an unusual reaction 1-(*o*-acetylphenyl)imidazole gives (138) when heated with cupric oxide, potassium carbonate, and pyridine[251b].

Not many cyclisations in the 1,2,3-triazole series have been reported. Cyclisation at carbon occurs when 1-amino-1*H*-1,2,3-triazole reacts with acetoacetic ester[722,723a]. The same reagent[722,723a] or acetylacetone [723c] necessarily cause cyclisation at nitrogen with 5-substituted 4-amino-1,2,3-triazoles. A quite different type of reaction produces (139) when 1-(*o*-azidophenyl)-1,2,3-triazole is heated, and triethyl phosphite converts dimethyl 1-(*o*-nitrophenyl)-1,2,3-triazole-4,5-dicarboxylate into the related product[177b].

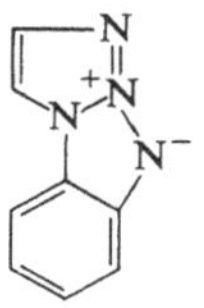

(139)

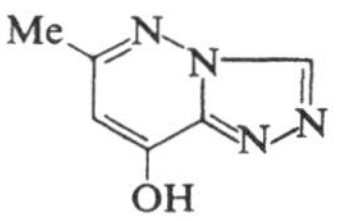

(140)

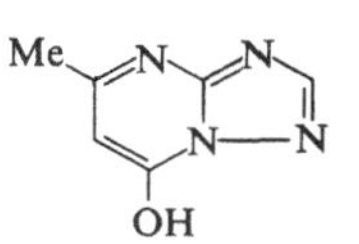

(141)

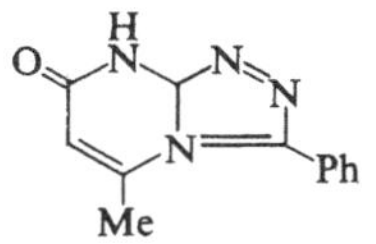

(142)

In contrast to this situation, cyclisations of 1,2,4-triazoles have been widely studied. Bülow[724a,b] first prepared derivatives of *s*-triazolo-[4,3-b]pyridazine from 4-amino-4*H*-1,2,4-triazole and β-diketones or β-ketoesters. The product from ethyl acetoacetate was written as (140). Later work supports this, and several examples of these reactions have been reported[111a,630b,693f,725-6].

Bülow was also the first to carry out the reaction between β-diketones or β-ketoesters and 3-amino-1,2,4-triazole. He formulated the products as derivatives of *s*-triazolo[2,3-*a*]pyrimidine, that from acetoacetic ester being written as (141)[527a]. Except for the problem of tautomerism this structure has since been verified[719a,b,725]. Many examples of such cyclisations involving 3-amino-1,2,4-triazoles and β-difunctional reagents of varied character have been described, and cyclisation in the sense of (141) generally assumed[593c,d,607c,d,617k,*l*,670a,d,e,677a-e, 693a,i,719c,d,720,727]. In contrast, 3-amino-5-phenyl-1,2,4-triazole is said to react with acetoacetic ester to give the *s*-triazolo[4,3-*a*]pyrimidine (142)[719e], and other cyclisations have been represented as going in

this direction[728]. Generally, it seems likely that cyclisations of these
types occur at N-2, giving s-triazolo[2,3-a]pyrimidines, but there may be
exceptions, and often definitive evidence for structure is lacking.
According to the conditions 3-amino-1,2,4-triazole and methyl propiolate
can give either of the two possible triazolopyrimidines[729b,c].

Oxidative cyclisation of 3-benzylidenehydrazino-1,2,4-triazoles with
lead tetra-acetate gives 3-phenyl-5H-1,2,4-triazolo[5,1-c]-1,2,4-triazoles,
and with cyanogen bromide 3-hydrazino-1,2,4-triazole gives the amino
derivative of this ring system[730].

Like the similar imidazole (see above), 1-(o-acetylphenyl)-1H-1,2,4-
triazole is cyclised when heated with cupric oxide, potassium carbonate,
and pyridine, giving the triazole analogous to (138)[251b].

(143)   (144)   (145)

5-Aminotetrazole gives derivatives of tetrazolo[1,5-a]pyrimidine with
β-difunctional reagents[719a,d,724c,729a,731]. Pyruvic acid and 5-hydra-
zinotetrazole give (143)[715b]. Araldehyde 1-methyl-5-tetrazolylhydrazones
give derivatives of (144) when oxidised with lead tetra-acetate[590c], and
the bromo-derivatives (145) of such hydrazones, and the corresponding
2-methyl compounds cyclise readily to compounds of the same type
[591d,e].

4 Behaviour in substitution reactions: nucleophilic substitution

4.1 Alkylation and substituted alkylation

The range of nucleophilic alkylations by organometallic compounds which can be effected in the pyridine series (Schofield, p. 200) is unknown in azole chemistry.

Without considerable activation additional to that provided by the imidazole ring itself, nucleophilic replacement of halogen by a substituted alkyl residue does not occur [374b,f]. 5-Chloro-1-methyl-4-nitroimidazole does react with sodiomalonic ester, the chlorine atom being replaced [732].

4.2 Amination and substituted amination

The chemistry of azoles shows nothing equivalent to the Tschitschibabin reaction (Schofield, p. 206). Sodamide can cause ring-opening of pyrazoles (p. 168), or removal of an N-aryl group (p. 219). 1-Alkyl-4,5-diphenyl-imidazoles did not react with sodamide [411a]. Amination with hydroxyl-amine in the presence of alkali has not been much used, but 1,2-dimethyl-4- and 1,2-dimethyl-5-nitroimidazole give respectively the related 5-amino and 4-amino compounds. Replacement of the hydrogen at C-4 appears to be easier than of the hydrogen at C-5 [733].

In contrast to this situation, amination by replacement of halogen is well known. Examples in the pyrazole series involve compounds substituted on one or both of the nitrogen atoms. The first example in the first category was the conversion of 4-benzoyl-5-chloro-3-methyl-1-phenyl-pyrazole into the 5-amino compound by heating with aqueous ammonia; aniline reacted similarly [551d]. 5-Iodo-3-methyl-1-phenyl-4-phenylazo-pyrazole, but not the corresponding chloro compound, reacted with ammonia, though other amines reacted with the chloro compound [552c]. 5-Chloro-3-methyl-4-nitro-1-p-nitrophenylpyrazole reacted with ammonia, aniline, or phenylhydrazine in the expected way [62d,321c,734].

5-Chloropyrazoles with carbonyl substituents at C-4 can react with hydrazine beyond the stage of halogen replacement[62a,321*l*,551d], as in the following example:

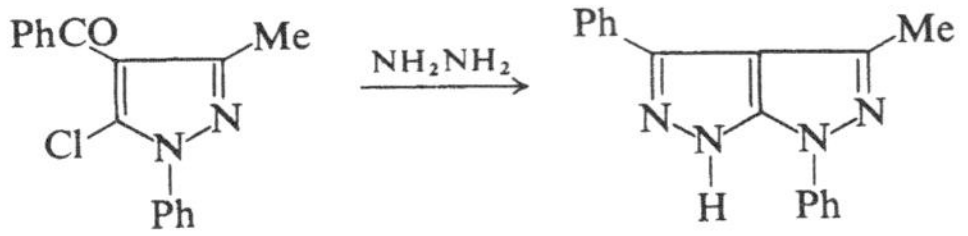

With an activated aniline derivative ring closure occurs[317]:

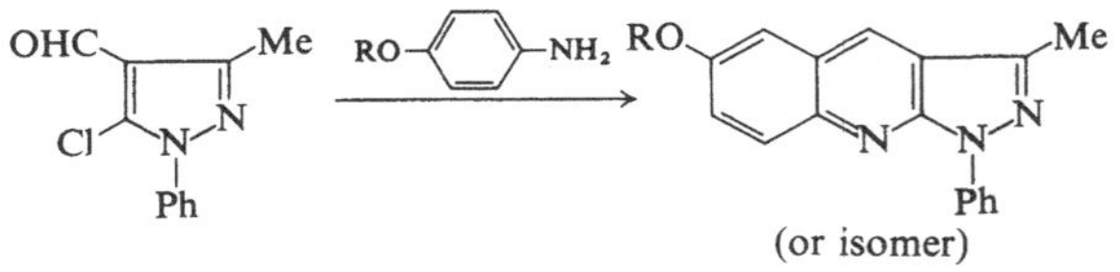

(or isomer)

A range of 5-chloropyrazoles revealed complicated behaviour in reaction with hydrazine (and with guanidine), each giving one or more products of the types shown[734].

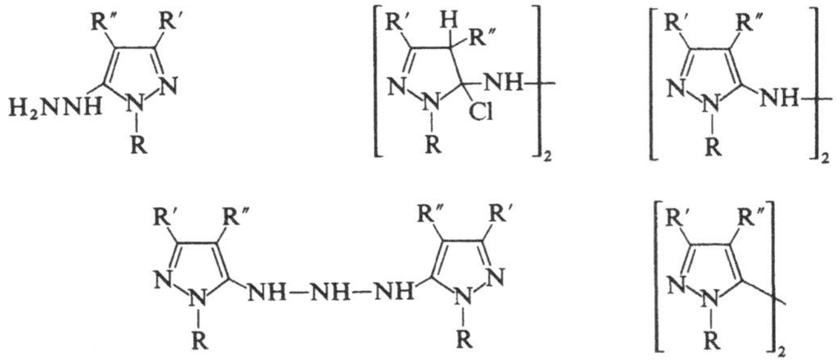

The reactions of 1,2-di-*N*-substituted pyrazoles relate mostly to the case of antipyrine chloride (146). With phosphoryl chloride and primary aromatic amines, antipyrine is converted via (146) into the products of amination, usually isolated as the imines (147)[551b].

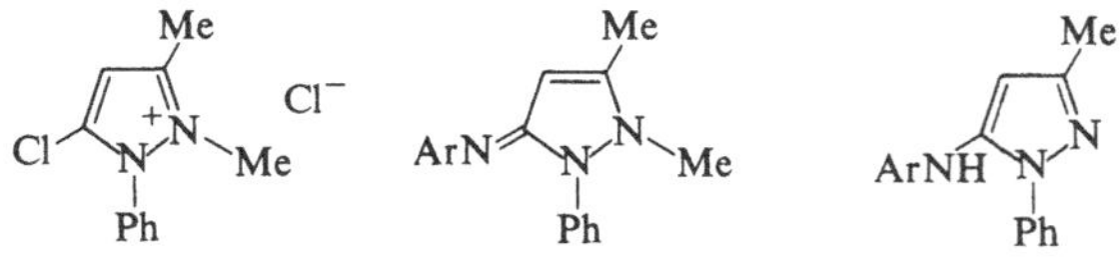

(146)   (147)   (148)

Antipyrine chloride can be prepared separately and used in the reaction[551c], which has also been applied to analogues of the chloride [330h] in reaction with amines or with hydrazines[323j,612,735g,736]. When higher reaction temperatures are used the quaternary salts which are the parents of the imines (147) are dequaternised so that the products of reaction are now of the type (148)[552d,735d]. Antipyrine chloride and its analogues can similarly react with ammonia, ammonium carbonate, or aliphatic amines to give imines or amines, analogous to (147) or (148) respectively, according to the conditions[329d,330b,551c, 735d].

The conversion of 5-methyl-1-phenylpyrazol-3-one into the corresponding 3-anil by reaction with aniline hydrochloride and phosphoryl chloride presumably goes via 3-chloro-5-methyl-1-phenylpyrazole[346b].

In the imidazole series some failures to replace halogen in amination reactions were early reported[374d,f]. 1-Alkyl-5-halogeno-4-nitro- and 1-alkyl-4-halogeno-5-nitro-imidazole do, however, react with ammonia [374d,494b] and with amines[139,733,737]. 1-Alkyl-2-bromo-4,5-diphenylimidazole reacted with ammonia, alkylamines, and arylamines in an autoclave or in dimethylformamide[738], and 2-bromo-1-methyl-5-nitroimidazole reacted with boiling piperidine[139]. Some quantitative studies are mentioned below.

The halogen atoms of 5-chloro-1-methyl-, 5-chloro-1-phenyl-, and 5-bromo-1-methyl-1*H*-1,2,3-triazole have been successfully replaced in reactions with ammonia and amines[139,148c,276b,429c,739]. Under appropriate conditions the product obtained can be that resulting from Dimroth rearrangement, which was first observed in this series (p. 159). In contrast, 4-chloro-1-methyl-1*H*-1,2,3-triazole failed to react with aniline [77c] and 4-bromo-2-methyl-2*H*-1,2,3-triazole to react with ammonia [276b].

Although amination by replacement of halogen is not important in the 1,2,4-triazole series since amino-1,2,4-triazoles can be prepared by ring-synthesis, and indeed are used as sources of halogeno-1,2,4-triazoles (a situation also found with the tetrazoles) (p. 139), the reaction has received some attention, initially without success[86a]. 3-Chloro- and 3,5-dichloro-1,2,4-triazole have been used successfully with ammonia and various amines[617m], as have 5-chloro- and 5-bromo-1-methyl-1*H*-1,2,4-triazole[139,269d].

1-Aryl-5-halogenotetrazoles react as expected with hydrazine[83e,608d].

Quantitative studies of the reactivities of halogeno-azoles towards amines are scarce. 4-Substituents influenced the rates of reaction of 1-aryl-5-chloro-3-methylpyrazoles with hexamethylene imine in the order $.NO_2 \gg PhN_2. > Ph.CO. \gg Br \gg H$ [740]. More important are Barlin's results [139] which, however, did not include a pyrazole derivative among the compounds studied. Rates and activation parameters were reported for the reactions of 5-bromo-1-methyl-1*H*- and 5-bromo-2-methyl-2*H*-1,2,3,4-tetrazole, 5-bromo-1-methyl-1*H*-1,2,4-triazole, 2-bromo-1-methyl-5- and 5-bromo-1-methyl-4-nitroimidazole, and 2-bromopyridine with piperidine, and less complete results were obtained for some other compounds. The wide range of reactivities, the complicated substitution patterns, and, in some cases, anomalous kinetic behaviour, limit the comparisons which can be made. 5-Bromo-1-methyltetrazole was more reactive than 5-bromo-

1-methyl-1,2,4-triazole or its 4-methyl isomer, or 4-bromo-1-methyl-1,2,3-triazole. 5-Bromo-2-methyltetrazole was more reactive than 4-bromo-2-methyl-2H-1,2,3-triazole (which failed to react with piperidine at 167.9 °C). 5-Bromo-1-methyl- was more reactive than 5-bromo-2-methyl-tetrazole; for comparable reaction rates for these two compounds there was a temperature difference of about 120 °C. 2-Bromo-1-methylimidazole did not react with piperidine at 200 °C (but see below). Generally large positional effects were observed, and two or three pyridinic nitrogen atoms were necessary to overcome the deactivating effect of one pyrrolic nitrogen atom towards nucleophilic attack. 2-Bromo-1-methyl-5-nitro- and 5-bromo-1-methyl-4-nitro-imidazole were both considerably more reactive than the corresponding 1,2,4-triazoles. Thus, a nitro-group is considerably more activating than a pyridinic nitrogen atom in these circumstances (*cf.* Schofield, p. 215).

The observation that 5-halogeno-1-methylimidazoles gave, with lithium piperidide and piperidine in ether, a mixture of the 4- and 5-piperidino compounds, coupled with competition experiments using pyrrolidine and piperidine, led to the suggestion that a hetaryne mechanism was operating[741]. These results have, however, been challenged[742], in particular the formation of the 4-piperidino compound could not be observed. From the reactions of 5-bromo- and 5-chloro-1-methylimidazole with lithium piperidide and piperidine in boiling ether the products were as follows.

From 5-chloro-1-methylimidazole	*From 5-bromo-1-methylimidazole*
5-chloro-1-methylimidazole, 70%	5-bromo-1-methylimidazole, 49%
4-chloro-1-methylimidazole, trace	4-bromo-1-methylimidazole, 3%
1-methyl-5-piperidino-imidazole, 15–20%	1-methyl-5-piperidino-imidazole, 25%
1-methyl-2-piperidino-imidazole, 3–5%	1-methyl-2-piperidino-imidazole, 16%

4-Chloro- and 4-bromo-1-methylimidazole were very unreactive and gave no trace of the 2- or 5-piperidino compounds, and 2-bromo-1-methyl-imidazole gave 1-methyl-2-piperidino-imidazole by reaction with piperidine at 200 °C for 60 h. The formation from the 5-halogeno compounds of their 4-halogeno isomers is not surprising since this isomerisation has been shown to occur in the presence of a strong base (p. 145). 5-Piperidino compounds presumably arise from 5-halogeno compounds, and it is suggested that the 2-piperidino compounds are formed as shown here.

A few examples are known where substituted-amination by replacement of an oxygen or sulphur function has been achieved. Thus, 4-hydroxymethyl-antipyrine when heated with phenyl isocyanate gives the anil (147), Ar = Ph, PhNH.CO.OCH$_2$. at C-4[743].

Attempts to replace methylmercapto or methoxyl groups in 1,2,4-

triazoles[86a,c] or tetrazoles[590a,b] by reaction with ammonia, amines, or hydrazine have generally failed, but 1-alkyl-5-methylmercaptotetrazoles are said to react with hydrazine[590a]. Quaternary salts are more reactive, and methylmercapto groups have been replaced from the 2-position in imidazolium, the 5-position in 1,2,4-triazolium, and the 5-position in 1,2,3,4-tetrazolium salts by reaction with acylhydrazines [612], triethylamine being used as a catalyst.

1-Methyl-3,5-dinitro-1H-1,2,4-triazole gives with hydrazine a mixture of 5-amino- and 5-hydrazino-1-methyl-3-nitro-1H-1,2,4-triazole[269c].

4.3   Arylation

It has already been remarked (p. 133) that nucleophilic alkylation by organometallic compounds is unknown in azole chemistry; the position regarding arylation is little different.

Methyl 1-phenylpyrazole-4-carboxylate reacted with phenyl magnesium bromide to give the expected carbinol, and also 4-benzoyl-1,5-diphenyl-pyrazole. The same product of phenylation was also formed in small yield from 4-benzoyl-1-phenylpyrazole[59j].

Although 3-methyl-1-phenylpyrazol-5-one did not react with phenyl magnesium bromide, the Grignard reaction did convert 4-arylazo-3-methyl-1-phenylpyrazol-5-one into the corresponding 5-aryl-4-arylazo-3-methyl-1-phenylpyrazoles[744].

4.4   Azidation

Most commonly azido-azoles have been prepared from diazonium salts by reaction with azide ions. In this way 3-aminopyrazole gave 3-azido-pyrazole[432h], 4-amino-3,5-dimethylpyrazole gave 4-azido-3,5-dimethyl-pyrazole[431a], and several 5-amino-1-arylpyrazoles gave the 5-azido compounds[745c]. The same reaction has been used with 5-amino-N-aryl-1,2,3-triazoles[693h,745a].

1-Aryl-5-chloro-1H-1,2,3,4-tetrazoles react readily with sodium azide in acetone[177c,608d], as does 5-chloro-4-ethyl-3-methyl-1-phenyl-1,2,4-triazolium tetrafluoroborate with sodium azide in methanol[626].

4.5 Cyanide formation

Cyanide formation by replacement of hydrogen has rarely been observed in the azole series (see the reaction of an imidazole N-oxide below). Interesting cases are the following, which are suggested to proceed by the S_N' mechanism. From the 1-methyl compound, 2-cyanomethyl-1-methylimidazole is also formed[746]. With potassium cyanide in ethanol 1,2-dimethyl-5-nitro-imidazole gives (149), and 1,2-dimethyl-4-nitro-imidazole behaves analogously[733]. The conversion of nitro into azoxy compounds by reaction with cyanide is not, of course, peculiar to the imidazole series.

(R = PhCH$_2$ or Me)

(149)

Equally rare is the conversion of a diazonium salt into a nitrile. 4-Amino-1-phenylpyrazole gave a diazonium salt which with potassium cuprocyanide gave 4-cyano-1-phenylpyrazole[747]. An earlier example, using 4-amino-1,3,5-trimethylpyrazole probably succeeded, but the product was not purified[392g]. The attempt to convert 5-amino-3-methyl-4-nitropyrazole into the cyano compound by diazotisation in hydrochloric acid and addition to potassium cuprocyanide gave the chloro compound [426a].

Cyanides have more frequently been prepared from halogen compounds. The reaction of 1-benzyl-4-bromopyrazole with cuprous cyanide required careful control of conditions[363b]. With 5-chloro-1,3-dimethyl-4-nitropyrazole, potassium cyanide in dimethylformamide has been used[693g].

N-Alkyl-halogeno-nitro-imidazoles evidently react with potassium cyanide readily only when the halogen atom and alkyl group are adjacent. Examples are found among the 1-alkyl-5-chloro- and 1-alkyl-5-bromo-4-nitro-imidazoles[373c,495a,748-9,750]. In contrast, 4-chloro-1-methyl-5-nitro-imidazole did not react with potassium cyanide in ethanol, but after 20 h at 120–130°C with potassium cyanide and potassium iodide in dimethylformamide gave 5-cyano-1-methyl-4-nitro-imidazole[565, 751]. This surprising transformation does not occur in the absence of iodide. It is suggested that transmethylation occurs giving 5-cyano-1,3-dimethyl-4-nitro-imidazolium, which decomposes to give 5-cyano-1-methyl-

4-nitro-imidazole preferentially. The dimethylformamide does not appear to be essential to the reaction [565,733].

Both 4-chloro-1,2,3-[745b] and 3-chloro-1,2,4-triazoles [617m] have been converted into cyanides by reaction with potassium cyanide in dipolar solvents.

Similar conditions effect the replacement of the methanesulphonyl group from 1-alkyl-2-methanesulphonyl-5-nitro-imidazoles [607b].

The sequence of reactions shown, paralleling that of great value in the pyridine series (Schofield, p. 225), has recently been observed with imidazole *N*-oxides [246].

(150) R = H, Me

4.6 Halogenation

Nucleophilic replacement of hydrogen by halogen is evidently unknown. However, with imidazole oxides 2-chlorination is effected by phosphoryl chloride, as in the examples illustrated above [246] (*cf.* Schofield, p. 228).

The standard transformation of an amino compound into the halogen derivative, via a diazonium salt, has been used frequently in the pyrazole series as a source of 3-halogeno-[70c,426a,432c,752] and 4-halogeno-pyrazoles [392g,431c,615a,747], and 4-halogeno-pyrazolones [330c]. 3-Diazopyrazole gives 3-iodopyrazole when boiled with hydriodic acid [432c].

1-Methyl-1*H*-1,2,3-triazole-4-diazonium cation gives 4-bromo-1-methyl-1*H*-1,2,3-triazole with hydrobromic acid and copper [276b]. Examples of the reaction are more numerous in the 1,2,4-triazole series, and have been long known [139,268f,365,431b,688b].

3-Halogeno-1,2,4-triazol-5-ones have been prepared in this way [268f, 676e]. Guanazole (3,5-diamino-1,2,4-triazole) has been converted into 3,5-dihalogeno-1,2,4-triazoles via tetrazotisation [83f,268f]. It should be noted that in the reaction of some 3-amino-1,2,4-triazoles with nitrous acid nitrosamines appear to be produced (p. 213); they have been converted into 3-halogeno-1,2,4-triazoles [430b,c].

5-Aminotetrazole has been diazotised and the diazonium salt, in the presence severally of cupric chloride, cupric bromide, and potassium iodide, converted into the 5-halogenotetrazoles[83h]. 5-Amino-1-aryl-tetrazoles have also given 5-halogeno compounds by diazotisation in presence of copper powder[83e].

Nucleophilic halogenation of azoles by exchange of one halogen atom for another has been observed fairly often. Thus, 1-aryl-4-arylazo-5-chloropyrazoles react with potassium iodide in boiling aqueous ethanol, sometimes with a little added hydrochloric acid, the chlorine atom being replaced by iodine[735h]. Exchange in pyrazolium quaternary salts occurs readily; the commonest form of the reaction is shown here[63f, 321b,c,d,e,323j,735c]. Halogen exchange occurs in similar circumstances

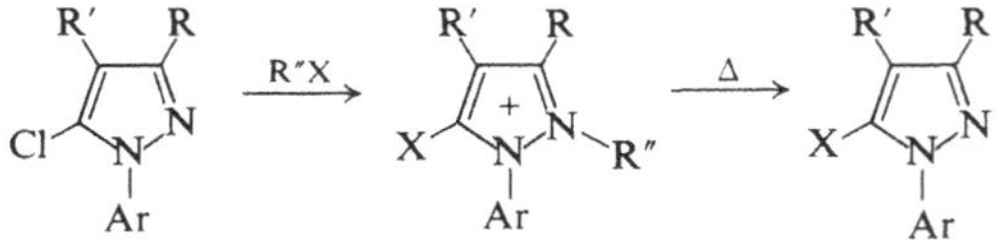

X = Br or I

in the pyridinium series (Schofield, p. 180). If quaternisation is carried out at lower temperatures than those needed for the second step, it can occur without halogen exchange[321e]. 4-Chloro-3,5-dimethyl-1-phenyl-pyrazole reacts at 100°C with methyl iodide to give 4-iodo-1,3,5-trimethyl-2-phenylpyrazolium iodide[330c], and 3- or 5-chloro-1-methyl-pyrazoles give the halogen-exchanged quaternary salts with methyl iodide [100b].

It was reported [753] that 2,5-dibromoimidazole-4-carboxylic acid *p*-bromoanilide gave the 2,5-dichloro compound with concentrated hydrochloric acid at 150°C. Attempts to repeat this reaction gave a mixture of halogen compounds, but led to other useful examples[434]. Thus, in boiling hydrochloric acid 2,4,5-tribromo- and 2-bromo-4,5-dichloro-imidazole gave 2,4,5-trichloro-imidazole, and 2,4-dibromo-5-nitro- gave 2,4-dichloro-5-nitro-imidazole. With potassium iodide in dimethylformamide 1-alkyl-4(5)-halogeno-5(4)-nitro-imidazoles exchanged bromine or chlorine for iodine[737].

With boiling hydrochloric acid, 3-bromo-1,2,4-triazol-5-one gave 3-chloro-1,2,4-triazol-5-one[268f], and in this, and related replacements of a nitro group, the probable importance of protonation in activating the heterocyclic nucleus was noted (Schofield, pp. 215, 230).

Chloro- and bromo-pyrazoles have in a large number of cases been prepared from hydroxypyrazoles or pyrazolones by reaction with phosphoryl halides. By far the commonest form of this reaction involves 1-arylpyrazol-5-ones reacting with phosphoryl chloride[321b,c,d,e,330e,g, 551d,552d,735a,b,c,e,f,754], phosphoryl chloride and pyridine[126b], or phosphoryl bromide[69a,321c]. Similarly, 1-aryl-3-chloropyrazoles have frequently been prepared from 1-arylpyrazol-3-ones[321f,330a,c,e, 735h,755]. 1-Aryl-3-hydroxypyrazol-5-ones can give the products of either mono- or di-substitution[321i,551a,735h]. 4-Hydroxy-1-phenylpyrazole gave 4-chloro-1-phenylpyrazole in very poor yield[76a], the lower reactivity as compared with that of the types of compound already mentioned recalling the similar behaviour of 3-hydroxypyridine (Schofield, p. 231). 1-Methylpyrazol-3-[100b], and 1-methylpyrazol-5-ones[98,100a,330d], and *N*-unsubstituted pyrazol-3-ones[100a,b,114a, 321j,330d,693c,756-8] all give the corresponding chloro or bromo[114a, 444] derivatives, more complicated products sometimes resulting on prolonged reaction[758].

Antipyrine and its analogues give with phosphoryl chloride the corresponding chloropyrazolium quaternary salts (as 146)[100a,321b, 441c,551f,735b,759], which can lose an alkyl chloride to give the chloropyrazole[321b,441c,760] (p. 243). 3-Antipyrine[330a] and iso-antipyrine[330d] behave similarly. With thionyl chloride, antipyrine-4-carboxylic acid gives the carboxylic acid chloride and also 5-chloro-3-methyl-1-phenylpyrazole-4-carboxylic acid[383c].

Reactions of the present kind are much rarer in the imidazole series. The formation of 2-chloro-4,5-diphenylimidazole from 4,5-diphenyl-imidazol-2-one by reaction with phosphoryl chloride was reported in 1907[353a], the corresponding preparation of 2-chloroimidazole itself in 1967[212]!

Similarly, few applications of the reaction to triazolones and tetrazo-lones have been reported; examples include 1-phenyltriazol-3- and -5-ones [89,626], and 1-phenyltetrazol-5-one[83e].

Less familiar than the reactions in which halogen replaces an oxygen function are some rare instances of the displacement of a nitro group. Thus, 3-nitro-1,2,4-triazol-5-one is converted by boiling hydrobromic acid into 3-bromo-1,2,4-triazol-5-one[268f], and 5-nitro-1,2,4-triazole and its 1-methyl derivative, and also the related 3,5-dinitro compounds react similarly with hydrobromic or hydrochloric acid[269d].

The imidazole *N*-oxides (150), p. 139, are converted by phosphoryl chloride into the 2-chloroimidazoles[246] (*cf.* Schofield, p. 228).

4.7 Hydroxylation and substituted hydroxylation

Quaternary azolium salts are attacked by hydroxyl ion, but ring-opening occurs (p. 167). Straightforward hydroxylation has only rarely been observed. 1,2-Dimethyl-4- and -5-nitroimidazole show signs of reacting with hydroxyl ion, and prolonged treatment causes loss of ammonia[733]. Fusion of 1-methyl-4,5-diphenylimidazole with potash gives 1-methyl-4,5-diphenylimidazol-2-one[761].

In the pyridine series the reactions of the 1-oxides with acetic anhydride are important means of preparing 2-pyridones (Schofield,

p. 234). As already noted, azole *N*-oxide chemistry is relatively undeveloped, but it has been reported that with acetic anhydride, 1-methylpyrazole 2-oxide gives 3-acetoxy-1-methylpyrazole and 1-acetyl-2-methylpyrazol-5-one[245].

1-Benzyl-4,5-dimethylimidazole 3-oxide (150), R = Me, reacts with acetic anhydride to give 1-benzyl-4,5-dimethylimidazol-2-one[246]. In the reactions illustrated completion of the substitution is prevented by the substituents present[762].

In the following examples[114e] nucleophilic hydroxylation follows initial electrophilic attack, but aromatisation is not completed (p. 52).

R' = Me, Br

Replacement of an amino by a hydroxyl group in an azole has rarely been reported. Whilst 5-amino-1,3-diphenylpyrazole is converted by concentrated hydrochloric acid into 1,3-diphenylpyrazol-5-one[532b], 3-amino-5-phenylpyrazole is unaffected[763]. 1,4-Disubstituted 5-iminotetrazoles are readily converted into tetrazol-5-ones[25c,515e].

The formation of hydroxy compounds from diazonium salts seems to be lacking from pyrazole chemistry, perhaps because of the surprising stability of the pyrazole diazonium salts (p. 214). 4-Methyl-2-phenyl-2*H*-1,2,3-triazole-5-diazonium cation has been hydrolysed to the hydroxy compound[764]. The conversion of 5-aminotetrazole into the hydroxy compound by diazotisation and hydrolysis has long been known[25,83h, 583]; carried out in sulphuric acid in the presence of copper sulphate it

is a useful preparation[148g]. 5-Amino-1-methyltetrazole gave the hydroxy
compound when diazotised in dilute nitric acid but 5-amino-2-methyl-
tetrazole behaved differently (p. 147)[583].

Halogen atoms in pyrazoles are not readily replaced by hydroxyl.
However, substitution has been observed with 5-chloro-1-phenylpyrazoles
containing a benzoyl[551d] or a phenylazo[735h] group at C-4.
Similarly, 5-chloro-3-methyl-4-nitropyrazole gives indications of reacting
with hydroxyl anions[426a], and the chlorine atom of 5-chloro-3-methyl-
4-nitro-1-*p*-nitrophenylpyrazole can be replaced by alkoxide groups[62d].

The replacement of a halogen atom by an alkoxy, aryloxy, or hydroxy
group in imidazoles which also contain an activating nitro group has
frequently been carried out[122,318j,737,765]. In the replacement of
bromine from 4-bromo-1,2-dimethyl-5-nitroimidazole by alkoxide, blue-
green colours appear, and are attributed to the intermediate complex
(*cf.* Schofield, pp. 244-5). In contrast, 5-bromo-1,2-dimethyl-4-nitro-
imidazole did not react under the conditions of these experiments[733].

Halogen atoms in *N*-substituted 1,2,3-triazoles have been replaced by
alkoxyl without assistance from a nitro group[429c,576c]. Similar
replacements have been effected with *N*-substituted 1,2,4-triazoles, with
[766] or without[269d] a nitro group being present. 3-Chloro-1,2,4-
triazole gives the triazolone with sodium hydroxide in glycol[617m].

5-Chloro- and 5-bromo-tetrazole are not attacked by aqueous alkali,
but prolonged treatment converts 5-iodotetrazole into tetrazolone whilst
the use of sodium ethoxide gives, surprisingly, tetrazole[83h]. 1-Substi-
tuted 5-halogenotetrazoles readily give the hydroxy[83e,583,608d],
alkoxy[583,598,608d], or aryloxy[767-8] derivatives.

As would be expected, chlorine atoms in pyrazolium quaternary salts
are more readily replaced, and antipyrine chloride (5-chloro-2,3-dimethyl-
1-phenylpyrazolium chloride) and its analogues are hydrolysed to antipyrines[321b,330h]. 5-Chloro-1,2-dimethylpyrazolium salts give 1,2-
dimethylpyrazol-5-ones in the same way[100a,321j].

Interesting combinations of nucleophilic hydroxylation with electrophilic bromination have been observed in the reactions of azolium salts
with alkali. The 1,2,3-triazolium compounds have been most studied.
When 4-bromo-1,3-dimethyl-1,2,3-triazolium tosylate is treated with 1M-
sodium hydroxide solution at 100 °C the equilibrium shown is quickly
set up. Evidently (151) is converted into a ylide which is further
brominated by (151), the details of the bromination mechanism being
uncertain. Subsequently, (151) and the dibromo compound undergo very
much slower nucleophilic substitution giving the 1,3-dimethyl-1,2,3-
triazolo-4-oxides, (152), R = H, and (152), R = Br, respectively. The
ability of triazolium salts such as (151) or the unbrominated analogue to
form ylides was shown by their undergoing base-catalysed nuclear
deuteration. 1,3·Dimethyl-1,2,3-triazolium reacted slowly in alkali with

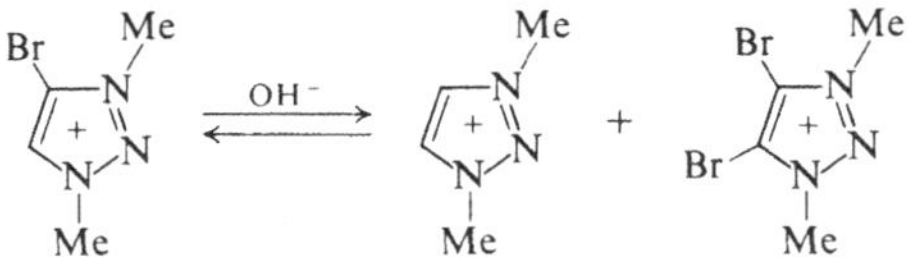

(151)

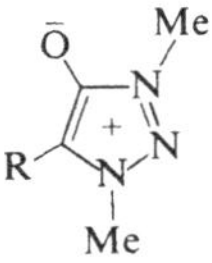

(152)

N-bromoacetamide giving (152), R = Br; evidently bromination and di-bromination occur by the ylide mechanism, and hydrolysis of the dibromo compound gives the anhydronium base (152), R = Br[576f].

Other triazolium salts behave similarly. Thus, in aqueous alkali the illustrated equilibria exist. In this case the dibromo compound is slowly decomposed by alkali to give an unidentified product, and no oxides analogous to (152) were recognised. This is also true of the reaction of 1,3-dibenzyl-1,2,3-triazolium with *N*-bromoacetamide in alkali, but 1-methyl-3-phenyl-, 1,4-dimethyl-3-phenyl-, and 1,5-dimethyl-3-phenyl-1 2,3-triazolium gave oxides exactly as described above for the 1,3-dimethyl compound. Rates of base-catalysed hydrogen–deuterium exchange for the triazolium salts were measured and related to both their ease of acceptance and abstraction of Br$^+$ in these reactions[576h].

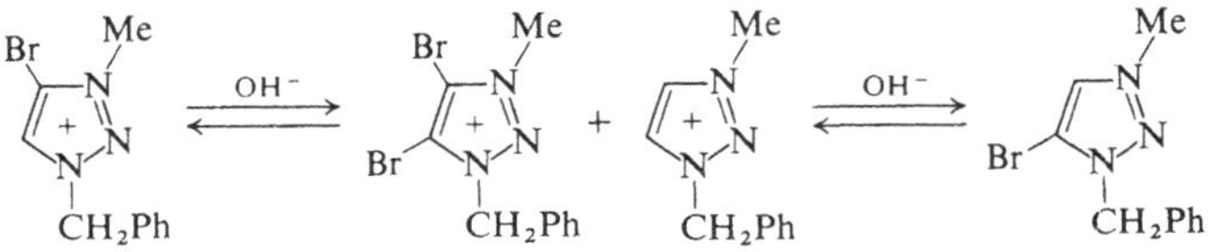

Methoxide ion converts a mixture of 1,3-dimethyl-1,2,3-triazolium and the corresponding dibromo compound into (152), R = H, in high yield. Clearly the dibromo compound brominates 1,3-dimethyl-1,2,3-triazolium by the ylide mechanism, subsequent nucleophilic substitution gives 4-methoxy-1,3-dimethyl-1,2,3-triazolium, and the latter is demethylated to the anhydronium base[576i]. Other examples are known[576j].

In similar processes 1,2-dimethyl- and 1-methyl-2-phenyl-1,2,3-triazolium cations react, by the ylide mechanism, with *N*-bromoacetamide and alkali to give the 5-bromo compounds; subsequent reaction with alkali produces the triazolones[576k].

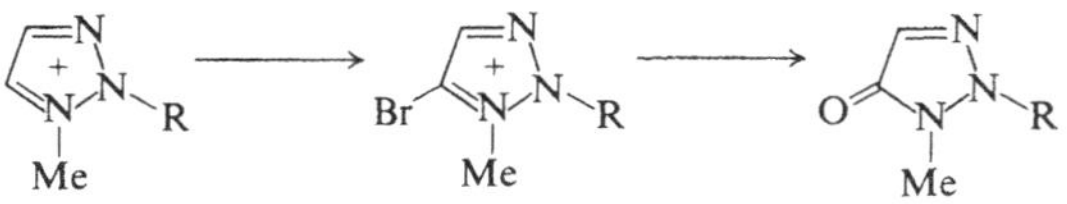

R = H or Me

With alkali 4-bromopyrazolium salts give cine substitution, as in the following examples (only the major products are shown). 4-Chloro-1;2-dimethylpyrazolium behaves similarly. In contrast 4-bromo-1,2,3,5-tetramethylpyrazolium, with no nuclear hydrogen atom, merely undergoes some degree of de-quaternisation, whilst 3-bromo compounds such as 3-bromo-1,2,5-trimethylpyrazolium give the corresponding pyrazolones by nucleophilic substitution. A combination of nucleophilic substitution and self-bromination via an ylide is seen in the reactions shown. Base-

catalysed hydrogen–deuterium exchange rates show the bromine atom strongly to acidify nuclear hydrogen atoms in pyrazolium salts with consequences such as the following: 1,2-dimethylpyrazolium is unaffected by N-bromoacetamide and alkali in 14 days, whilst 4-bromo-1,2-dimethylpyrazolium gives 3,4-dibromo-1,2-dimethylpyrazol-5-one, presumably via the di- and tri-bromopyrazolium cations and nucleophilic substitution.

Mechanistic studies of the cine substitutions of 4-halogenopyrazolium salts by hydroxide (and methoxide ion) illustrated above, whilst not conclusive, suggest that they proceed to a minor extent by an inter-halogenation mechanism (producing a dibromo compound which is then hydrolysed; 4-bromo-1,2,3-trimethylpyrazolium gives with methoxide some 4-bromo-1,2,3-trimethyl-pyrazol-5-one), and to a major extent by a process of the following kind. However, in place of this last an abnormal addition–elimination reaction might be involved, but not hetaryne formation [576g].

Ylide formation and transbromination of the kind mentioned above is clearly related to the phenomena mentioned above in the reactions of metal amides with halogeno-imidazoles (p. 136) and the migrations which occur when 5-bromo-1-methylimidazole reacts with lithium piperidide (p. 136).

Hydroxyazoles have frequently been prepared by the hydrolysis of alkoxyazoles. As in the pyridine series (Schofield, p. 247), there is no evidence for the mechanism of these reactions, which may or may not be nucleophilic substitutions. A few examples are conveniently noted at this point. In the pyrazole series the de-alkylation was effected using hydrochloric or sulphuric acid, with heating in a sealed tube[8b,76b,329a]. Quaternisation[329e] or the presence of electron-attracting substituents [76b] facilitate the reaction. In the 1,2,4-triazole series the reaction is useful because of the existence of ring-synthetic methods for preparing alkoxy compounds, and many examples have been reported[86b,268f, 676a]. Hydrochloric acid has usually been used to hydrolyse 5-alkoxy-tetrazoles[24,83g,583], but with 5-aryloxytetrazoles alkali was used[769].

Hydroxyazoles have also been obtained from azoles containing sulphur functions. As in the pyridine series (Schofield, p. 249), 5-methylthio quaternary pyrazolium salts are hydrolysed by alkali to pyrazol-5-ones [552a,b], and the same reaction has been used with 1,4-dimethyl-5-methylthiotetrazolium[612]. 5-Methylsulphonyl-1-phenyltetrazole[83d] and tetrazole-5-sulphonic acid[24c], and some of its 1-substituted derivatives[83d] also give the related tetrazolones with alkali. In the latter reaction[24a] and with 1-phenyltetrazole-5-sulphonamide[770b] hydrochloric acid has also been used, although this reagent left 1-methyltetrazole-5-sulphonamide unchanged.

4.8   Nitration

The reactions to be described in this section are those in which a diazonium group is replaced by a nitro group. Such reactions are not common.

The diazonium fluoborate prepared from 3-amino-4-phenylpyrazole gave 22% of 3-nitro-4-phenylpyrazole when added to sodium nitrite and copper in water[70c]. Similarly, the addition of hydrochloric acid to sodium nitrite and 5-amino-1-methylpyrazole-4-carboxylic acid amide at 100°C gave 1-methyl-5-nitropyrazole-4-carboxylic acid; reaction at 0°C produced only the nitrosamine, whilst reverse addition gave the pyrazolo-triazine by internal coupling[771].

The reaction in the imidazole series has attracted attention because of the natural occurrence of 1-methyl-2-nitroimidazole ('azomycin'). 2-Amino-imidazole gives the nitro compound by diazotisation in the presence of copper sulphate or of cuprous ions and sodium nitrocobaltate, and the 1-methyl and other analogues have been prepared similarly[607f,772-3].

3-Nitro-1,2,4-triazole-5-carboxylic acid[269b] and 5-nitrotetrazole[318a] have both been prepared by this method. 5-Amino-2-methyl-2H-1,2,3,4-tetrazole behaves in a variety of ways when diazotised; in acetic acid it gives 2-methyl-5-nitro-2H-1,2,3,4-tetrazole, but in nitric acid the diazo-amino compound or the ditetrazolyl amine is produced, depending on the acid concentration[583].

4.9 Sulphur bond formation

The commonest type of nucleophilic substitution reaction leading to the formation of a sulphur bond proceeds by displacement of halogen. 5-Chloro-1-phenylpyrazoles containing electron-attracting groups such as phenylazo[735h] or benzoyl[330f] at C-4 react readily with sulphides or mercaptans to give the related thiones or alkylthio derivatives. 4-Bromo-1-phenylpyrazol-5-ones react similarly with sodium thiophen-oxides[774]. Numerous examples have been reported of the conversion of antipyrine chloride (5-chloro-2,3-dimethyl-1-phenylpyrazolium chloride) and analogous pyrazolium compounds into thiones by reaction with sodium or potassium hydrogen sulphide[321h,j,k,330a,c,e,551b,f,552a,b, 759]. Quaternary pyrazolium salts also react readily with sodium or potassium sulphite, halogen being replaced to give compounds exemplified by (153) and (154)[320a,d,321j,552b].

(153)   (154)

5-Halogeno-4-nitroimidazoles with or without a 1-methyl group react with ammonium sulphite or mercaptans to give the corresponding sulphur derivatives[494d,498,544]. Several of these reactions, and also that of 4-chloro-1-methyl-5-nitroimidazole, have been effected with ammonium or sodium sulphide, or thiourea[494c]. Conversion of halogenoimidazoles into sulphonic acids by reaction with sodium sulphite succeeds only when activation additional to that provided by the imidazole ring nitrogen atoms is present; otherwise sulphite merely reduces the halogen compounds (p. 222). Thus, the major reaction of 2,4,5-tribromoimidazole is reduction although some 4-bromoimidazole-5-sulphonic acid is produced [374a]. Generally a nitro group is able to introduce sufficient activation to ensure that substitution occurs but this is not always the case; whilst 2-bromo-1,4-dimethyl-5-nitroimidazole gives the 2-sulphonic acid, no sulphonic acid was obtained from 2-bromo-1,5-dimethyl-4-nitroimidazole [412b]. 4-Bromo-5-nitro-, 5-bromo-1-methyl-4-nitro-, and 5-chloro-1-methyl-4-nitro-imidazole readily give the sulphonic acids[374b].

3-Chloro-1,2,4-triazole is readily converted into 3-alkylthio- and 3-arylthio-1,2,4-triazoles[617m,775]. Similarly, by reaction with potassium xanthate, 5-chloro-1-p-chlorophenyltetrazole gives the thione[608d].

Examples of the formation of thiones from the related oxygen compounds are rarer. 3-Methyl-1-phenylpyrazol-5-one with 'phosphorus pentasulphide' in xylene gives the thione[69c,330f]. Isoantipyrine (1,2-dimethyl-3-phenylpyrazol-5-one) [330d] and 5-chloro-2,3-dimethyl-1-(p-nitrophenyl)pyrazolium[330h] react similarly with potassium hydrogen sulphide.

5 Behaviour in substitution reactions: radical substitution

Radical substitution of azoles has not been much studied. The reactions about which something is known are alkylations and arylations.

5.1 Alkylation

Methylations have been carried out with methyl radicals generated by the decarboxylation of acetic acid with ammonium peroxydisulphate, catalysed by silver. Pyrazole gave about 10% of 3-methylpyrazole and imidazole gave about 50% of 2-methylimidazole. Several 1-alkylimidazoles were also methylated at C-2, but 1,2-dimethylimidazole did not react. The method might be useful for obtaining some 2-alkylimidazoles, but separation of product from starting material would be a problem on the large scale [776].

5.2 Arylation

The first phenylation of 1-methylpyrazole was carried out with thermally decomposing benzoyl peroxide or N-nitrosoacetanilide; 1-methylimidazole was also phenylated using N-nitrosoacetanilide or diazoaminobenzene. In both cases conversions of 10–20% were achieved, 94% of 5-, 5% of 3-, and 1% of 4-phenyl substitution occurring with 1-methylpyrazole, and 67% of 2-, 33% of 5-, and <2% of 4-phenylation with 1-methylimidazole. Competition reactions gave rates relative to that of benzene of 0.62 and 1.2 for 1-methylpyrazole and 1-methylimidazole, respectively [777b]. The derived partial rate factors are [778d] for 1-methylpyrazole, $f_3 = 0.18$, $f_4 = 0.03$, and $f_5 = 3.4$, and for 1-methylimidazole, $f_2 = 4.8$, $f_4 = 0.14$, and $f_5 = 2.2$. Later use of benzenediazonium tetrafluoroborate with 1-methylimidazole produced phenylated products in isomeric proportions similar to those already mentioned, but which depended on the concentration of the salt; parallel studies with pyridine and nitrobenzene confirmed the radical character of the substitutions and suggested that

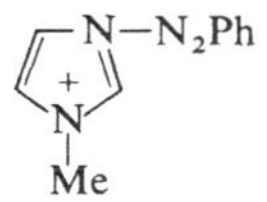

(155)

some substitutions were occurring in the cation (155), which was more reactive than the free base[777c].

In later work phenylations (by decomposition of benzoyl peroxide at 118 °C) in which the heterocycle was used as the solvent were compared with those in which acetic acid was the solvent. The results are summarised in table 22[778a,b]. Clearly, the proportion of substitution occurring adjacent to the 'pyridinic' nitrogen atom is increased by protonation,

Table 22. Phenylation of 1-methylpyrazole and 1-methylimidazole[778a,b]

| | Products (%) | | Relative rates[a] | |
| | | | In acetic | In excess |
Substrate	In acetic acid	In excess heterocycle	acid	heterocycle
1-Methylpyrazole	5-Phenyl (59.5)	5-Phenyl (92)	1.43	0.57
	3-Phenyl (40.5)	(3+4)-Phenyl (8)		
1-Methylimidazole	2-Phenyl (100)	2-Phenyl (58)	9.0	5.7
		5-Phenyl (42)		

(a) With respect to benzene: determined by comparison with nitrobenzene (relative rate = 3), the performance of which was not affected by change of medium.

without appreciable change in the overall yield of phenylated products. The relative rate of reaction observed in this work for 1-methylimidazole differs appreciably from the figure quoted above.†

When benzoyl peroxide was decomposed at 80 °C in 1-phenylpyrazole products of phenylation and benzoyloxylation were obtained. The distribution of products was similar to that observed with biphenyl, but with a somewhat higher proportion of benzoyloxylation. Phenylation gave roughly 30% of substitution at C-3 of the pyrazole ring, 54% at the *ortho-*, 6% at the *meta-*, and 10% at the *para-*position of the phenyl ring. Noteworthy are the high proportion of *ortho-*substitution and the unusually selective attack at C-3. Qualitatively the pattern of substitution for benzoyloxylation resembled that for phenylation[777a].

† It has been suggested that 1-methylimidazole may be present as a quaternary species when benzoyl peroxide is used (e.g. by conversion into 1-benzoyl-3-methylimidazolium) and that this species would be responsible for the high relative rate. For this reason the result obtained with diazo-aminobenzene or *N*-nitrosoacetanilide (relative rate 1.2) is preferred (Professor B. M. Lynch, private communication).

5.3 Miscellaneous reactions

To what extent the process of benzoyloxylation mentioned above for 1-phenylpyrazole involved the pyrazole nucleus is not known. 2,4,5-Triphenylimidazole with benzoyl peroxide gave 4-benzoyloxy-2,4,5-triphenyl-4*H*-imidazole [779].

As in the pyridine series (Schofield, p. 256) this section of azole chemistry contains a variety of reactions which are not easy to classify, often because the mechanism operating is not understood. Other reactions are recognisable as belonging to one of the classes already discussed, but which for various reasons, such as those of traditional usage, have been included here.

6.1   Additions

Pyrazole is not reduced with sodium and amyl alcohol [2f] (sodium and ethanol at 130°C and 20 atmospheres are said to convert it into pyrazoline [324j]), and these reagents generally do not attack N-unsubstituted pyrazoles [111c]. As reported at other points, acidic reducing agents can be used to reduce or remove substituents without affecting the pyrazole nucleus. In contrast, 1-substituted pyrazoles are reduced with sodium and alcohol; in particular, many 1-arylpyrazoles have been converted into pyrazolines [60a,321d,392d,f,k,506a,735c,780-1]. Further reduction produces in some cases 1-amino-3-arylaminopropanes [60a,c,65b] by ring-opening. Whilst 5-ethyl-4-methyl-1-phenylpyrazole gave the open-chain diamine, 3,5-dimethyl-1-phenylpyrazole gave the pyrazolidine, benzene, and 3,5-dimethylpyrazole [65b]. The pyrazolidine was also formed when 1,5-dimethoxycarbonyl-3,4-diphenylpyrazole was reduced with zinc and acetic acid [782]. The reduction with sodium and alcohol to give a pyrazoline, followed by the production of a red colour when the pyrazoline was oxidised was known as the 'Pyrazoline reaction' and was used as a test for the pyrazole ring.

Pyrazole and 1-phenylpyrazoles are easily hydrogenated over palladium to give the pyrazolines or, under more severe conditions the pyrazolidines [525]. 1-Phenylpyrazol-5-one gives the pyrazolidone by catalytic reduction, whilst antipyrine gives a dihydro-derivative or, under more severe conditions, 1-phenyl-2,3-dimethylpyrazolidine [525].

Hydride reduction of antipyrine and isoantipyrine, and of some quaternary pyrazolium salts, has been carefully studied. The products from antipyrine[783] are shown. Di-isobutylaluminium gave the same result. A hindering group, such as t-butyl, at C-3 prevented reduction

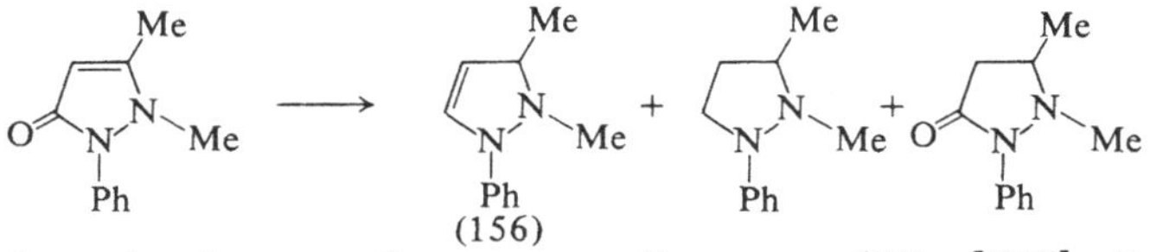

(156)

[784a]. Isoantipyrine gave the corresponding pyrazolidine[785]. Enamines such as (156) are not reduced by lithium aluminium hydride[785] and the pyrazolidine must be formed from the pyrazolidone[783]. A detailed mechanism has been proposed for the hydride reduction of pyrazolones [784b].

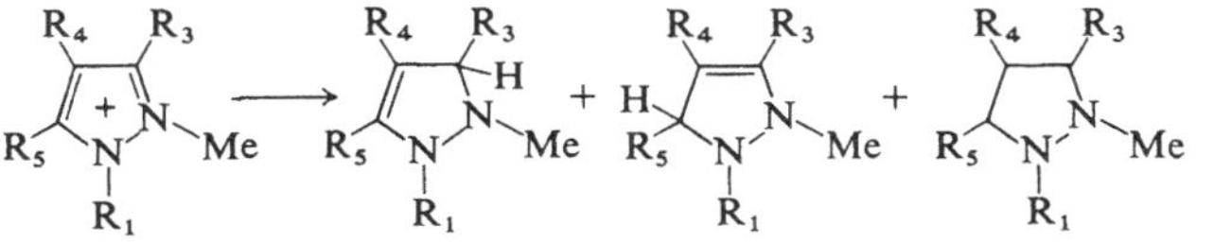

R_1	R_3	R_4	R_5			
Me	Ph	H	H		100%	
Me	Ph	H	Me	15%	85%	
Me	Ph	Me	H		100%	
Me	Me	Ph	H	6%	94%	
Ph	Me	H	Me	70%	<5%	30%
						(*cis* + *trans*, 5 : 1)

From pyrazolium iodides, lithium aluminium hydride produced predominantly 3-pyrazolines, but pyrazolidines and alkylhydrazines (by ring-opening) can also appear. Some results are given here[114i,784a]. The following is an example of ring-opening[114i].

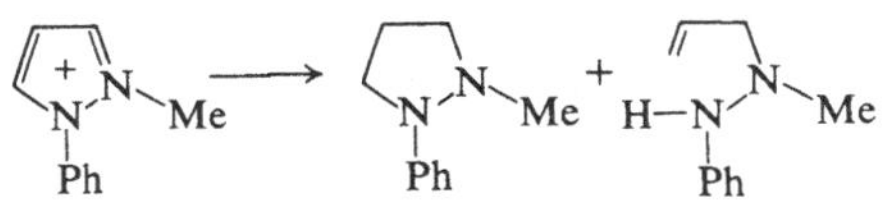

Electrochemical reduction of 1,5-diphenyl-3-styrylpyrazole produces the Δ^2-pyrazoline as one of the products[792].

Imidazole is not reduced by sodium and alcohol or by tin and hydrochloric acid[6c]. Imidazole and 2,4,5-trimethylimidazole resist catalytic hydrogenation[786], and the early report[786] that lophine could be hydrogenated to give the corresponding imidazoline was wrong (only the phenyl groups are reduced)[787]. The claim[318h] to have produced the imidazolidine by catalytic hydrogenation of 2-methyl-4,5-diphenyl-imidazole must be doubted.

With sodium borohydride in ethanol, 1-methyl-3-phenethylimidazolium iodide gives N,N-dimethyl-N'-phenethylethylenediamine (92%). The reaction in general gives both possible products, as shown[788].

(Major product)

The tetrazole nucleus can survive the reduction of substituents attached to it but examples are known where the ring is ruptured. Thus, 5-(p-nitrobenzenesulphonylamino)tetrazole is reduced by a variety of reagents to sulphanilylguanidine[770a]. In the fission illustrated there is no evidence for the formation of a stable intermediate[789].

(R_1 and R_2 = alkyl, aralkyl or $-(CH_2)_5-$)

The behaviour of tetrazolium salts on reduction is important. In studying the antibacterial properties of these compounds Kuhn and Jerchel[790] observed their reduction to formazans, with consequent staining of the nuclei of the bacterial cells. This reaction has been intensively studied and used, and a very large literature has accumulated about it[791]. In addition to the enzymatic reduction, tetrazolium salts can be converted into formazans by many of the common reducing agents such as ammonium polysulphide, sodium amalgam, ascorbic acid with dilute alkali, sodium dithionite, and so on. Our proper concern here is with the primary products of the reduction, but despite extensive studies by chemical and biochemical methods, and the use of polarographic and e.s.r. spectroscopic techniques, it is still not entirely clear what these are. The evidence indicates the occurrence of one-electron reduction of the tetrazolium salt (and one-electron oxidation of the formazan) to produce a radical, that from 2,3,5-triphenyltetrazolium chloride having the structure (157).

6.2   Oxidations

Some examples of the oxidation of aminophenylpyrazoles with potassium permanganate, causing rupture of the benzene ring, were mentioned in chapter 1. This reagent can also disrupt the pyrazole ring; some aryl-pyrazoles gave the arene-carboxylic acid, nitrogen and carbon dioxide[1d, 752,793-4].

With ozone in chloroform pyrazole gives glyoxal, hydrazine, and carbon dioxide, whilst 3-methylpyrazole gives methylglyoxal, glyoxal, hydrazine, and some pyruvic acid. 3,5-Dimethylpyrazole gave diacetyl, methylglyoxal, acetic acid, formic acid, nitrogen, and some hydrazine and nitric acid, whilst 3-methyl-1-*p*-nitrophenylpyrazole gave an insoluble ozonide [795].

Peracetic acid oxidised 3,5-diphenylpyrazole to benzoic acid, 1,2-dibenzoylhydrazine, and 2,5-diphenyl-1,3,4-oxadiazole [794].

With peroxytrifluoracetic acid some aminopyrazoles gave the nitro compounds (p. 211) or the *N*-oxide (p. 126), but 3-aminopyrazole was degraded [700]. Michaelis and his co-workers [330j] described the oxidation of 5-amino-1-aryl-3-methylpyrazoles to compounds which they called 'azipyrazoles'; thus, 5-amino-3-methyl-1-phenylpyrazole with nitrous acid, or hydrogen peroxide in aqueous acetic acid, gave 'azipyrazole' itself. More recently somewhat modified conditions (90% hydrogen peroxide in acetic acid) had to be used to produce azipyrazole, which was shown to be 3-phenylazocrotononitrile, m.p. 109 °C [796]. One scheme suggested for the production of this compound is shown here. Other workers [745c]

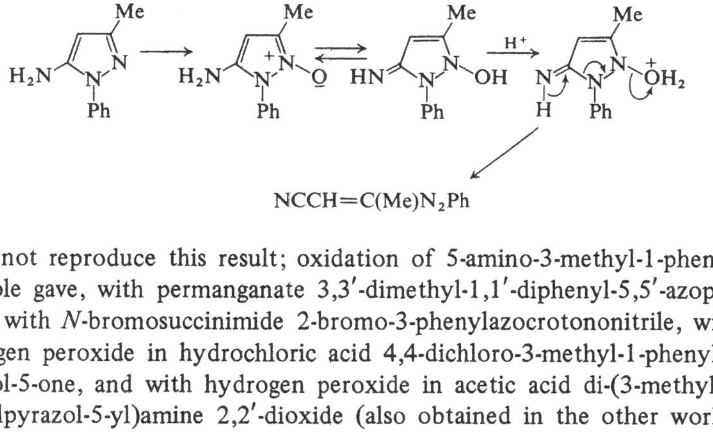

$$NCCH=C(Me)N_2Ph$$

could not reproduce this result; oxidation of 5-amino-3-methyl-1-phenylpyrazole gave, with permanganate 3,3'-dimethyl-1,1'-diphenyl-5,5'-azopyrazole, with *N*-bromosuccinimide 2-bromo-3-phenylazocrotononitrile, with hydrogen peroxide in hydrochloric acid 4,4-dichloro-3-methyl-1-phenyl-pyrazol-5-one, and with hydrogen peroxide in acetic acid di-(3-methyl-1-phenylpyrazol-5-yl)amine 2,2'-dioxide (also obtained in the other work [796]). Phenyl-iodoso-acetate produced a 3-phenylazocrotononitrile, but it differed (m.p. 61–2 °C) from that described in the other work. The mechanism proposed there for the oxidation neglects the important possibility that a nitrene intermediate is involved.

Electrochemical oxidation of 3-methyl-1-phenylpyrazole under alkaline conditions gave pyrazole-3-carboxylic acid, but under acid conditions benzoquinone was formed [797].

The 'quinonoid' character of the intermediates produced by oxidising pyrazolones with lead tetra-acetate [798-9] is shown by the nature of the products formed when they are trapped with dienes such as 2,3-dimethylbuta-1,3-diene [798].

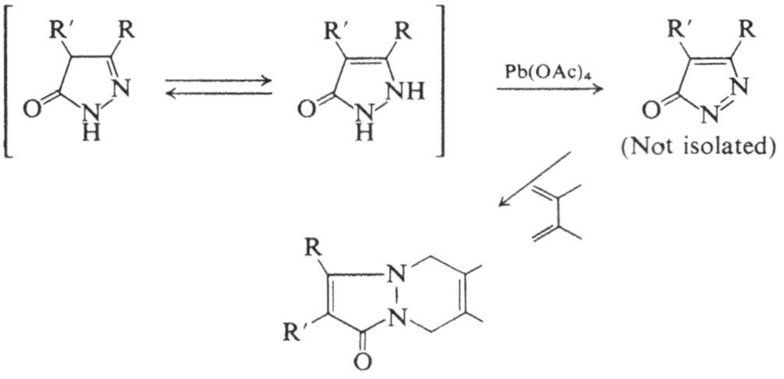

With Frémy's salt 1,4-dihydroxypyrazoles give 3,4-diazacyclopenta-dienone monoxides, but peracids take the reaction further, producing 1,3,4-oxadiazin-6-ones[800].

Ring-opened products are formed by treating antipyrines with nitrous acid under some conditions[539b], and antipyrine is degraded by electrochemical oxidation[797]. Several reagents oxidise pyrazolones to 4-(5'-oxopyrazolinyl)pyrazol-5-ones and 4-hydroxypyrazol-5-ones[801]. Some pyrazoles are converted by alkaline hypobromite into products which arise from ring-opening of an initially formed *N*-bromopyrazolone: the case of 3,4-dimethylpyrazole has been represented as shown[77e].

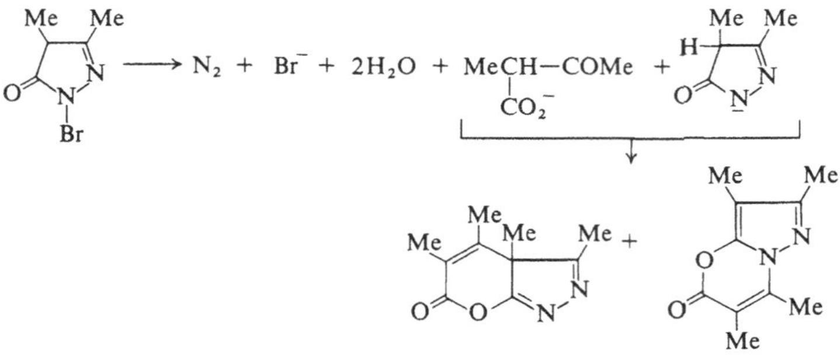

Imidazole can be degraded by permanganate to formic acid and carbon dioxide[6b]. Similar extensive degradation occurs with 1-ethyl-2-methyl-imidazole[54c], and pilocarpine[90c,802]. Similarly, selenium dioxide destroys 2-methylimidazole[803].

In several of the cases mentioned the resistance of the imidazoles to oxidation with chromic acid was noted. This is not always found; 4,5-diphenylimidazol-2-thione is converted by this reagent into dibenzoylurea. Nitric acid converts it into benzil[425c].

Imidazole and some of its homologues were oxidised with hydrogen peroxide to oxamide[9e]. In contrast, 4-ethyl-2-phenylimidazole has been shown more recently to be converted by alkaline hydrogen peroxide into carbonate, benzoate, propionate, and 5-ethyl-3-phenyl-1,2,4-oxadiazole.

The reaction seems to be general for 2,4-dialkyl-, 2,4-diaryl-, 2-alkyl-4-aryl- and 2-aryl-4-alkyl-imidazoles. In contrast, 4,5-diphenyl- and 2,4,5-triphenyl-imidazole were unaffected. 2- and 4-Phenylimidazole gave only benzoic acid, which may have been formed from the oxadiazoles[804].

The reaction of imidazole with perbenzoic acid has already been mentioned (p. 126).

Some degradations of imidazoles by electrochemical oxidation have been reported[805].

4,5-Diaryl-2-(*p*-dimethylaminophenyl)imidazoles are oxidised by bromine to give the bromide hydrobromide of the cation (157a); permanganate and perchloric acid give the perchlorate[806b]. In other circumstances bromine can degrade the imidazole nucleus, as is indicated by the formation of ammonium bromide during the bromination of imidazole[374a]. With bromine or *N*-bromosuccinimide in aprotic solvents imidazole gives tribromoimidazole (p. 50), but in aqueous media nuclear substitution is accompanied by ring degradation, producing glyoxal, ammonia, and possibly formamide. With *N*-bromosuccinimide in aqueous buffers or acids the nuclear substitution is suppressed (or overtaken) and formation of glyoxal enhanced by increasing acidity. Tribromoimidazole is itself degraded[448b].

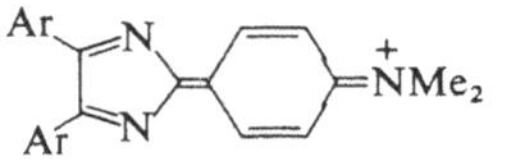

(157a)   (158)

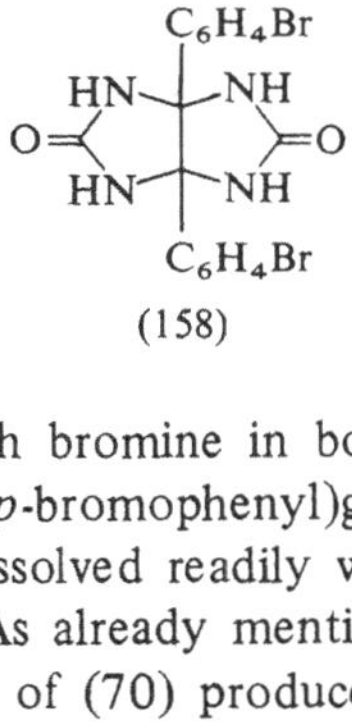

4,5-Diphenylimidazol-2-one with bromine in boiling acetic acid gave 4,4'-dibromobenzil and 3a,6a,-di-(*p*-bromophenyl)glycoluril (158). 4,5-Di-(*p*-bromophenyl)imidazol-2-one dissolved readily when bromine was added to its suspension in acetic acid. As already mentioned (p. 54) the product was (70). The hydrolysis of (70) produced 4,4'-dibromobenzil and urea, and it was the reaction of the urea with 4,5-diacetoxy- or 4,5-dihydroxy-4,5-di-(*p*-bromophenyl)-2-imidazolidinone from (70) which produced the glycoluril (158)[468].

Histidine is degraded to iodoform and oxalate during iodination. The reaction proceeds by mono- and di-iodination, and has been studied kinetically and quantitatively[475b].

In the present context lophine (2,4,5-triphenylimidazole) has been studied more than any other imidazole. With chromic acid it gives benzamide and benzanilide[8a], but greatest interest has centred on the involvement of lophine in the phenomenon of chemiluminescence. This was first observed by Radziszewski[807]; with alcoholic potash lophine gave ammonia and potassium benzoate but when air was present the decomposition was accompanied by the emission of light. Later workers by irradiation of lophine obtained a product regarded as a photo-oxide and *N,N'*-dibenzoylbenzamidine[808]. It now seems clear that the chemiluminescence produced by the action of alkali and air on lophine and its analogues involves a hydroperoxide as an intermediate. This probably decomposes intramolecularly to give the excited singlet state of a diaroylarylamidine salt which is the light emitter. The hydroperoxide anion could arise from the imidazole anion and oxygen or by removal of a proton from the hydroperoxide itself[809,810].

Lophine with ethanolic potash and aqueous ferricyanide gives a piezochromic dimer (159)[811] which in organic solvents gives an intensely violet solution of the radical (160). The solutions give the yellow dimer (161), and the radical with hydrogen peroxide gives the hydroperoxide with consequent chemiluminescence[810]. The radical can also be generated from lophine by oxidation in a non-polar solvent with lead dioxide, and it is in equilibrium in solution with a dimer which has been formulated in a way different from that mentioned above[806a].

(159)   (160)   (161)

Photo-oxidation of lophine or 1,2,4,5-tetraphenylimidazole in dilute methanolic solution in presence of methylene blue produces the corresponding dibenzoylbenzamidine; it is argued that in these circumstances hydroperoxides cannot be intermediates. Imidazole itself reacted slowly giving 4,5-dimethoxyimidazolidone. Mechanisms were proposed for these transformations[812].

Generally, the nuclei of 1,2,3- and 1,2,4-triazoles are stable to oxidation. However, 4-amino-4*H*-1,2,4-triazoles are oxidised by lead tetra-acetate, presumably via a nitrene, to give nitrogen and a nitrile[813a].

6.3   Thermolyses

Simple pyrazoles and imidazoles are generally stable to heat (as in dis-
tillation). The same is to some extent true of triazoles and tetrazoles,
but thermal instability amongst triazole and tetrazole derivatives is not
uncommon (p. 10, and below). Imidazole is said to decompose at
590 °C but the nature of the process is not known [814]. Under severe
conditions (650 °C and 0.05 mmHg) 1-cyano-3,5-dimethylpyrazole gives
dienes, whilst 1,3,5-trimethylpyrazole (800 °C and 0.05 mmHg) reacts in
a different way [815].

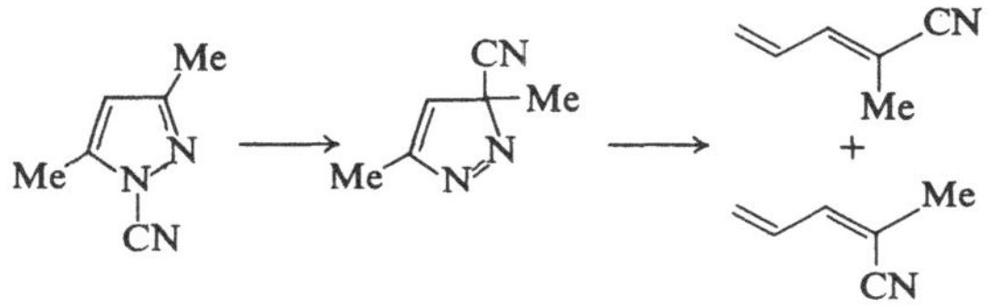

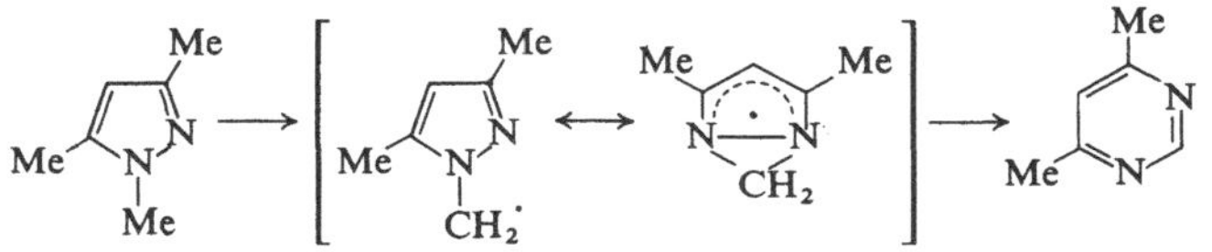

5-Azido-1,3,4-triphenylpyrazole when heated in ligroin at 40–50 °C
gave a deep red product formulated as below [745c]. Analogous products
behaved similarly (*cf.* the oxidation of 5-aminopyrazoles above, p. 155).

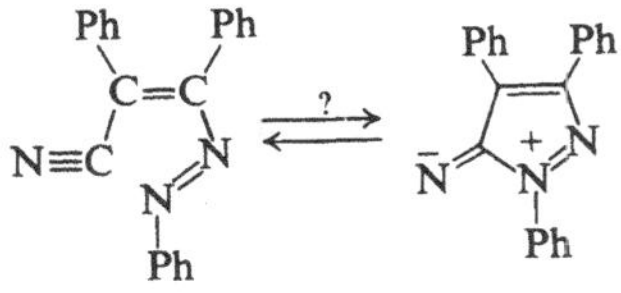

The observation that certain 1,2,3-triazoles are related tautomerically
to diazoalkanes was first made by Dimroth, initially in the case of
5-hydroxy-1,2,3-triazoles [429a,816a,c]. The results show that electron-
attracting ability in 'R' moves the equilibrium in favour of the diazo-
compound.

The conversion of 4-benzeneazo-5-hydroxy-1-phenyl-1*H*-1,2,3-triazole
in boiling acetic acid into 2-phenyl-2*H*-1,2,3,4-tetrazole-5-carboxylic acid
anilide has been represented as proceeding by the same kind of ring-
opening [276a]. The same kind of ring-opening is probably involved in
the reactions shown; the acid may catalyse the ring-opening as well as
providing the nucleophile important in subsequent steps [429a,817].

Closely analogous to these reactions of 5-hydroxy-1,2,3-triazoles is the
Dimroth rearrangement of the analogous 5-amino compounds. This
rearrangement, the nature of which was first clearly recognised by

R	R'	K [a]
H	.$CONH_2$	2.26
H	.CO_2Me	36
$PhCH_2$	.CO_2Me	118
p-Tolyl	.CO_2Et	120
Ph	.CO_2Me	300
p-Br.C_6H_4.	.CO_2Et	555
p-$O_2N.C_6H_4$.	.CO_2Me	Very large

(a) K = [Diazo-compound]/[Triazole] in ethanol at 25°C.

Dimroth for examples in the triazole series, occurs widely in hetero-aromatic chemistry (Schofield, p. 269)[818]. The following examples, which generally occur at temperatures higher than those required for the hydroxy-compounds, illustrate the rearrangement[816b,d,819]. Dimroth correctly postulated the general mechanism of these rearrangements. Later workers[148c] used Dimroth's titration data to estimate the equilibrium positions in these cases (see p. 161) and noticed that the shift of the equilibrium with change of solvent was in the same direction as that observed with tetrazoles (the more basic the solvent the more the equilibrium favoured the more acidic tautomer (with no substituent on ring nitrogen)). They examined several isomeric pairs and established equilibria from both sides in homogeneous melts at 184-5°C. Electron-attracting groups in 'R' moved the equilibrium in favour of the *N*-unsubstituted isomer, and the equilibrium constants were linearly correlated with σ-constants. The similarity of the situation in this respect to that in the tetrazole series was again noted, and kinetic measurements on the isomerisations strengthen the analogy[820b].

Recently isomerisations of several 5-amino-1-benzyl-1*H*-1,2,3-triazoles were reported. These were effected in hot aqueous potash, giving the equilibrium mixtures. The 5-benzylamino compounds were stable at 20°C for not less than 3 months, but were converted into the equilibrium mixtures in hot neutral solvents[149e]. 4-Substituted 5-amino-1-phenyl-

	R	R'	Conditions	% 5-RNH
(A)	Ph	.CO$_2$Et	Boiling pyridine	
			At m.p. (126 °C)	
			Boiling EtOH/EtONa	
			EtOH (150 °C, 3 h)	76.3
			C$_6$H$_6$ (150 °C, 3 h)	56.3
(B)	Ph	.CO$_2$Et	EtOH (150 °C, 3 h)	76.2
			C$_6$H$_6$ (150 °C, 3 h)	58.2
(A)	Ph	H	At m.p. (110 °C)	
(A)	Ph	Ph	Boiling pyridine	
			Melt (> 180 °C ?)	75
(B)	Ph	Ph	Melt (> 180 °C ?)	75
(A)	Ph	Me	Boiling water or pyridine	

1H-1,2,3-triazoles when heated in acetic anhydride gave the acetyl derivatives of the 5-anilino compounds; the latter give back the acetyl derivatives of the amino compounds on long treatment in boiling acetic anhydride[723b] (cf. the behaviour of N-acyltetrazoles below). The intramolecular cyclisation of some derivatives of 5-amino-1,2,3-triazole can give products formed by Dimroth rearrangement[723d], as shown.

The Dimroth rearrangement is subject to acid catalysis. In the case of 5-amino-4-ethoxycarbonyl-1-phenyl-1H-1,2,3-triazole this has been studied kinetically. Formation of an amidinium-like cation facilitates the rearrangement[821].

R = $-$C(Me)=CHCO$_2$Et *or* $-$COCH$_2$COMe

Recalling the Dimroth rearrangement is the observation that 1-cyano-1H-1,2,3-triazole is in equilibrium with a-diazo-N-cyanoethylidenimine. This system was prepared from cyanogen azide and acetylene, and similar systems were obtained using 1-propyne, 1-butyne, and 1-hexyne. However, ethoxyacetylene gave the diazo-compound with a temperature-independent n.m.r. spectrum[822].

A different kind of open-chain compound is formed by loss of nitrogen in the thermal ring-opening of 1-benzoyl-4,5-diphenyl-1H-1,2,3-

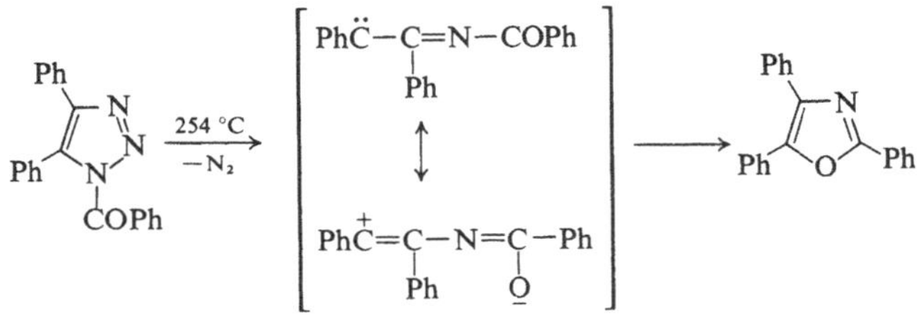

triazole. It cyclises to an oxazole[153c]. The intermediate is an imino-carbene. Its behaviour has been contrasted with that of the nitrile imine (in which all C-octets are filled) formed in the thermolysis of 2-benzoyl-5-phenyl-2H-1,2,3,4-tetrazole (see p. 165). Iminocarbenes are probably formed in the thermolysis of phthalimidotriazines, as illustrated[823].

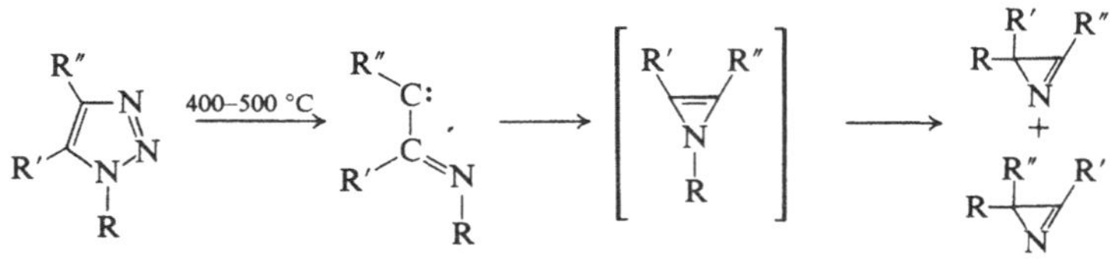

Related to these reactions are those occurring when some α-chloro-heteroaromatics, such as 2-chloropyridine, are heated with 1,2,3-triazole [635].

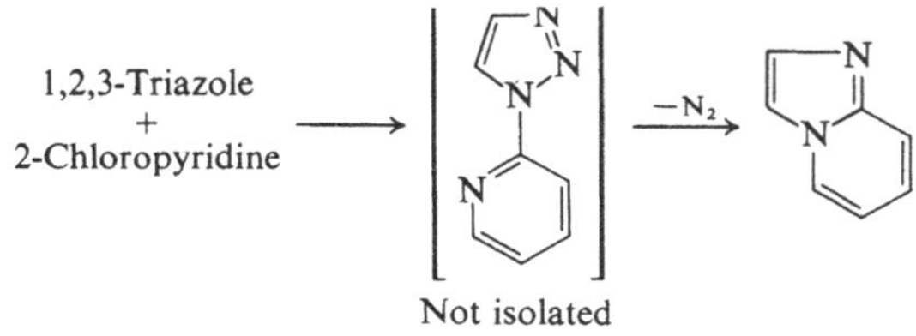

Not isolated

4-Phenyl- and 4,5-diphenyl-1,2,3-triazole both give nitrogen on thermolysis, but the only identifiable product of consequence was 2,3,5,6-tetraphenylpyrazine from the latter[824].

Lithiation of some 1-phenyl-1H-1,2,3-triazoles at C-5 causes ring-opening via an ylide. Thus, in the illustration below, when R = Ph lithiation in tetrahydrofuran at –20 to –40°C, followed by a rise to room temperature, causes ring-opening. In the cases R = H or Me the opening required higher temperatures[484].

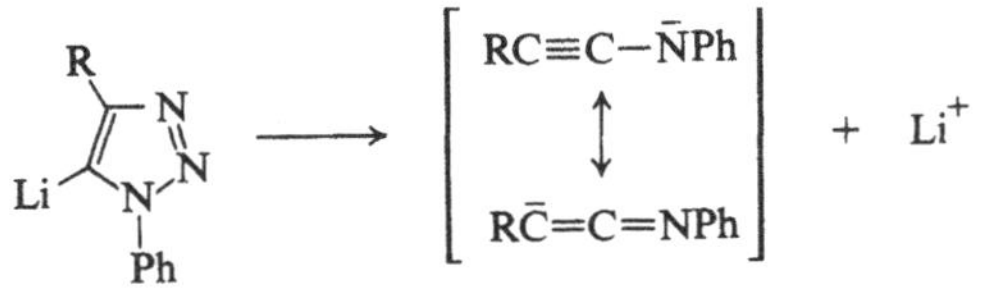

The exocyclic nitrene formed from 5-azido-1,4-diphenyl-1H-1,2,3-triazole on melting (70°C), or in inert solvents at temperatures as low as 50°C, is in equilibrium with an open-chain form[745a]. The related carbenes from 5-diazomethyltriazoles did not behave analogously[745b].

In the tetrazole series numerous examples of thermolytic ring-openings have been observed, many of them analogous to those already discussed for the triazoles.

1-(*o*-Nitrophenyl)-1*H*-1,2,3,4-tetrazole-5-thiol gave in boiling benzene 5-(*o*-nitrophenylamino)-1-thia-2,3,4-triazole. When the thiol was heated in aqueous alkali, nitrogen and sulphur were lost and *o*-nitrophenylcyanamide formed[177c].

The observation of a thermal Dimroth rearrangement in the tetrazole series was first reported in 1953, and the intermediacy of a *C*-azido-formamidine ('guanylazide') was recognised[579a]. The existence of the equilibrium was demonstrated in the case of 5-benzylamino-1-methyl-tetrazole. Other examples were reported[825], including the isomerisation of 5-benzylamino- into 5-amino-1-benzyl-tetrazole, and the conversion of 1-acetyl-5-benzylaminotetrazole into 5-acetylamino-1-benzyltetrazole. A more general demonstration of the existence of the equilibrium followed, and a measurement of equilibrium constants for reactions in ethylene glycol at 193°C showed the equilibrium to move in favour of the 5-substituted-amino compound as the substituent became more electron-attracting. The equilibrium was well correlated with pK_a, and in the case of aryl substituents with σ-constants[581d]. The kinetics of these processes have also been studied[581e]. Heating 5-amino-1-aryl- or 5-arylamino-tetrazoles with acetic anhydride brings about the isomerisation, with the production of 5-arylamino-1-acetyl- and 5-acetylamino-1-aryl-tetrazoles[826a]. Longer treatment with acetic anhydride brings about a different change, discussed below. The Dimroth rearrangement of tetrazoles is very general[590d].

The relationship between aminotetrazoles and *C*-azidoformamidines has long been known, salts of the latter giving the former on basification [25a,827]. Thermochemically the aminotetrazoles are more stable than the *C*-azidoformamidines[51a]. Treatment of 5-aminotetrazoles with cyanogen bromide gives *N*-cyano-*C*-azidoformamidines[407b,713]. *C*-Azidoformamidines formed in the ring-opening of 5-hydrazino- and

5-anilino-tetrazole in acid give products evidently arising by a Curtius reaction[820a]. When 5-amino-1-phenyltetrazole is heated at 98–100 °C with caustic soda it is rapidly decomposed to ammonia, aniline, carbonate, and azide[581e], which are the expected products of hydrolysis of phenyl-*C*-azidoformamidine.

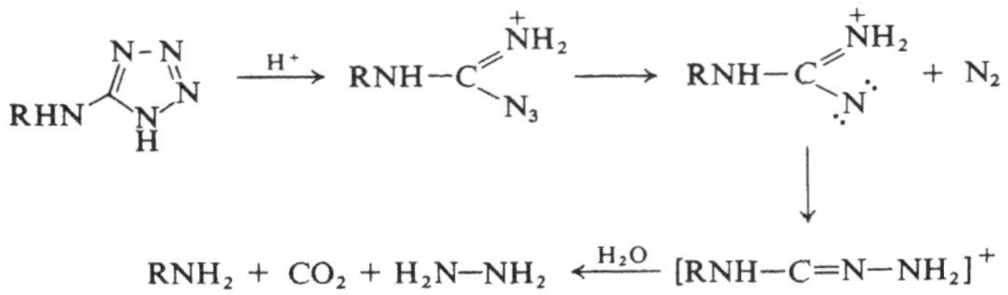

$$RNH_2 + CO_2 + H_2N-NH_2 \xleftarrow{H_2O} [RNH-C=N-NH_2]^+$$

The violent decomposition (with 'Feuererscheinung') of 5-phenyl-tetrazole above its melting point (218 °C) was early recorded[512a]. More careful heating gave nitrogen, 3,5-diphenyl-1,2,4-triazole, and 3,6-diphenyl-1,2,4,5-tetrazine[512b]. A later study of the reaction, in which decomposition was effected in solvents such as mesitylene, revealed, in addition to the products mentioned, 3,6-diphenyl-dihydro-1,2,4,5-triazine, 4-amino-3,5-diphenyl-4*H*-1,2,4-triazole, 2,4,6-triphenyl-1,3,5-triazine, and hydrazoic acid. These could be accounted for by postulating decomposition of 5-phenyltetrazole to the nitrilimine, Ph.C:N̄-N̄H and nitrogen, and also to benzonitrile and hydrazoic acid[828c]. The rates of thermolysis of 5-phenyltetrazole and some of its analogues in various solvents have been measured[829].

. The unusual decompositions of some 5-substituted tetrazoles shown here were supposed to proceed by initial halogenation of the 5-substituent. Accordingly, compounds of the halogenated type were prepared and thermolysed, with the results also shown[830b].

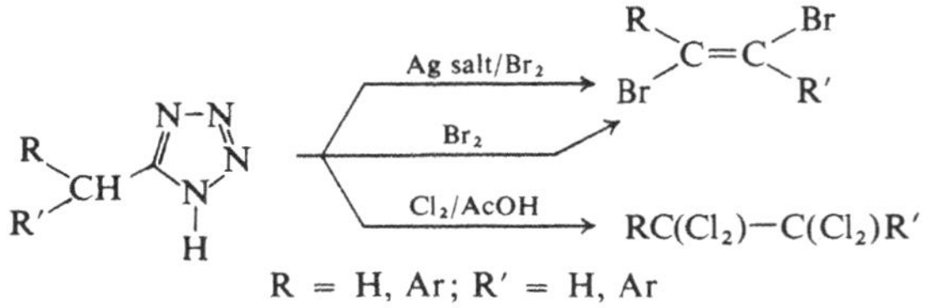

$R = H, Ar; R' = H, Ar$

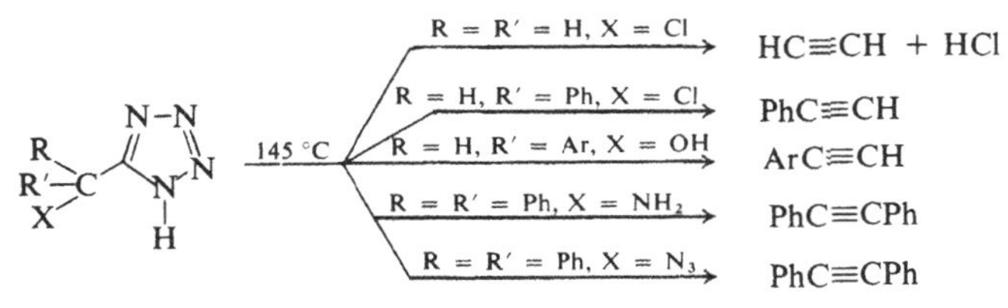

5-(1-Phenanthryl)-1-phenyltetrazole, when heated above its melting point, gives *N*-phenanthryl-*N*-phenylcarbodiimide, formed by a C→N migration of the phenanthryl group, with loss of nitrogen, and also the di-phenanthryl- and di-phenyl-carbodiimides formed by subsequent redistribution. This kind of change occurs more generally [830a,831]. Ring-closure of the product can give a 2-arylbenzimidazole. In a related reaction 1-trimethylsilyltetrazole gives *N,N*-di-trimethylsilylcarbodiimide [190c]. In these reactions the azidoformamidine isomeric with the tetrazole is presumably formed first and loses nitrogen (*cf.* the acid-catalysed case above).

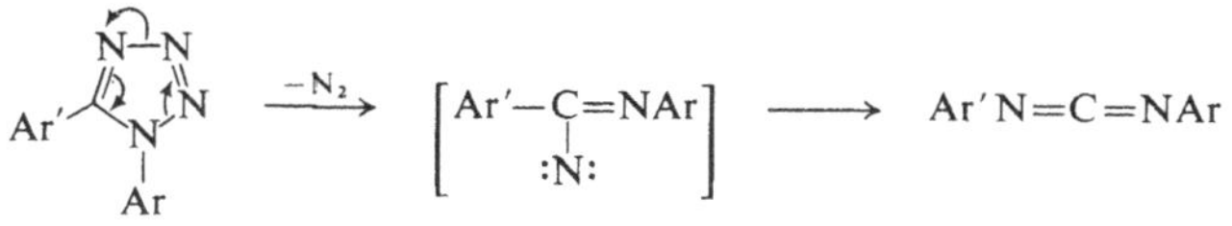

The tetrazole-azidoformamidine interconversion has been studied in several 1,5-ring-fused tetrazoles [832].

2,5-Disubstituted tetrazoles decompose on thermolysis to nitrogen and a nitrile imine. The latter can be captured; thus, 2,5-diphenyl-2*H*-1,2,3,4-tetrazole thermolysed in the presence of aniline gave Ph.C(NHPh):NNHPh, and in the presence of benzonitrile, 1,3,5-triphenyl-1*H*-1,2,4-triazole [153b, 828a,b]. These reactions do not occur so readily as those of the 2-acyl compounds (see below). The kinetics of decomposition of 2,5-diaryl-2*H*-1,2,3,4-tetrazoles in 1-chloronaphthalene at 165.8 °C have been studied, and substituent effects correlated with σ-values [833].

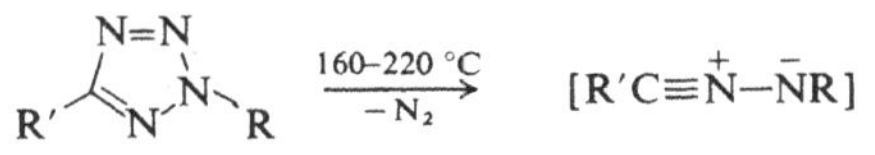

$$R = R' = Ph; \quad R = Ph, R' = Me; \quad R = Me, R' = Ph$$

In 1897 Pinner [834], by boiling 5-(*p*-tolyl)tetrazole with acetic anhydride, obtained a compound which he formulated as the *N*-acetyl derivative of *p*-toluamidine. Huisgen later corrected the structure to that of 2-methyl-5-*p*-tolyl-1,3,4-oxadiazole. Stollé observed the corresponding transformation of 5-aminotetrazole [83h]. Later, 5-amino-1-(*p*-nitrophenyl)-tetrazole or its acetyl derivative, and 5-(*p*-nitrophenylamino)tetrazole, were found to be converted by prolonged treatment in boiling acetic anhydride into 2-methyl-5-*p*-nitrophenyl-1,3,4-oxadiazole. 5-Acetylamino-1-phenyltetrazole did not react [826]. Huisgen and his co-workers [153a, 828f,835] have generalised this conversion of tetrazoles into oxadiazoles for many 5-aryl- and 5-alkyl-tetrazoles and 5-aminotetrazole with aroyl chlorides. From 5-substituted tetrazoles and acetyl chloride the *N*-acetyl-tetrazoles could be isolated, and on subsequent thermolysis in xylene or pyridine at 110–140 °C gave nitrogen and the oxadiazole. *N*-Acetyltetrazole itself lost nitrogen but the oxadiazole was not isolated. The reactions were represented as ring-openings of 2-acyl-2*H*-1,2,3,4-tetrazoles to give intermediates which, by loss of nitrogen, gave acyl nitrilimines. 2-Acetyl-5-phenyl-2H-1,2,3,4-tetrazole has been isolated and decomposed in this way

As well as acid chlorides, other acylating reagents can be used, giving a variety of heterocyclic products; thus, imidochlorides produce 1,2,4-triazoles[828d], isocyanates give 1-oxa-3,4-diazoles, thio-isocyanates or thio-acid chlorides give 1-thia-3,4-diazoles[828e]. 5-Phenyltetrazole with p-toluenesulphonyl chloride gave 3,6-diphenyl-1,4-di-p-toluenesulphonyl-1,3-dihydro-1,2,4,5-tetrazine and 3,6-diphenyl-1,2,4-triazine, these products arising from two molecules of the ring-opened intermediate[828e].

The kinetics of these reactions, using various combinations of 5-aryl-tetrazoles and aroyl chlorides, in pyridine at 70°C have been studied. The ring-opening is the rate-determining step, and substituent effects agree with expectations for the above processes; electron-releasing substituents in the 5-substituent, and electron-attracting ones in the acyl group facilitate reaction, though the effects are not large[828g]. In connection with 5-amino-1-p-nitrophenyl- and 5-p-nitrophenylamino-tetrazole being converted by boiling acetic anhydride into the oxadiazole an alternative possible mechanism was pointed out[826a]. It involved the 1-acetyl-tetrazole. Whereas in Huisgen's mechanism N-3 and N-4 are lost from a 2-acyl-2H-1,2,3,4-tetrazole, this proposal requires loss of N-2 and N-3 from a 1-acyl-1H-1,2,3,4-tetrazole. 5-Phenyltetrazole labelled at N-1(4) with ^{15}N reacted with benzoyl chloride to give the oxadiazole containing half of the label, in agreement with Huisgen's mechanism[826b].

Other examples of this rearrangement include that of 5-vinyltetrazole with acetic anhydride[836], and 5-perfluoroalkyltetrazoles with perfluoro-acyl chlorides[837].

The reaction of 5-phenyltetrazole with *p*-toluenesulphonyl chloride was mentioned above. 5-Aminotetrazole reacts with arenesulphonyl chlorides in pyridine to give arenesulphonylazidoformamidines ('arene-sulphonylguanyl azides', $ArSO_2.NH.C(N_3):NH$). These are cyclised by caustic soda to 5-arenesulphonamidotetrazoles. Thus, if caustic soda is used in the original sulphonylation the tetrazole is isolated[697b,838]. Earlier products described as arenesulphonylaminotetrazoles were the azides[694,697a,770a].

As already mentioned (p. 126), 5-dimethylaminotetrazole with arene-sulphonyl chlorides in pyridine reacts at N-2, and ring-opening follows ($\rightarrow N_2 + Me_2N.C:\overset{+}{N}-\bar{N}.SO_2Ar$), and the intermediate reacts with the pyridine ($\rightarrow \overset{+}{P}hN.C(NMe_2):NNHSO_2Ar$) [697c].

It should be recorded that early workers observed the conversion of tetrazole by benzoyl chloride in pyridine into dibenzoyl urea, and by benzoyl chloride into dibenzoylhydrazine[75a] and of 1-phenyltetrazole by benzoyl chloride into dibenzoylurea[83d]. More recently acyl chlorides or anhydrides were observed to convert 1-amino-1*H*-1,2,3,4-tetrazole into nitrogen and 2-substituted 5-acetylamino-1-oxa-3,4-diazoles[839].

Some ring-degradations of tetrazoles clearly proceed through the anions generated by removal of groups from C-5. Thus, 5-chloro-1-phenyl-tetrazole gives, with magnesium, sodium, or butyl lithium, nitrogen and phenylcyanamide[177c]. The 1-methyl- and 1-phenyl-5-tetrazolyl lithiums decompose in the same sense (*cf.* the triazole case above)[840]. Such reactions clearly do not differ fundamentally from those initiated in some systems by strong bases (see p. 170).

6.4   Attack of nucleophiles

Some reactions reviewed in this section involve nucleophiles acting in their roles of strong bases.

The ring-opening of pyrazolium quaternary salts by alkali, a useful method for proving the structure of the salts, has already been mentioned (p. 95).

C-(3,5-Dimethylpyrazol-1-yl)formamidine nitrate gives acetyl acetone 3,5-dimethylpyrazole, and aminoguanidine nitrate when heated with hydrazine[841]. Similarly, 3-methyl-1-phenyl-pyrazol-5-one gives 3-methyl-pyrazol-5-one and phenylhydrazine, and other examples are known[736]. Antipyrine similarly gives 3-methylpyrazol-5-one and 1-methyl-2-phenyl-hydrazine[736], and the latter is produced by decomposition of anti-pyrine with potash[842].

1-Phenyl- and 4-acetyl-5-methyl-1-phenylpyrazole are resistant to potassium t-butoxide in t-butanol. However, the introduction of more powerfully electron-attracting substituents (PhCO., O_2N., NC., *p*-Me.-$C_6H_4.SO_2$.) at C-4 leads to ring-opening with the production of β-anilinoacrylonitriles. The reaction is fairly general[843a,b] and can be initiated without the assistance of electron-attracting substituents by sodamide[843b,844]. It can be represented as follows. The key to the transformation is the generation of the anion, and this can also be achieved by heating some 4-substituted 1-phenylpyrazole-3-carboxylic acids with quinoline; the anion and carbon dioxide are produced[843a].

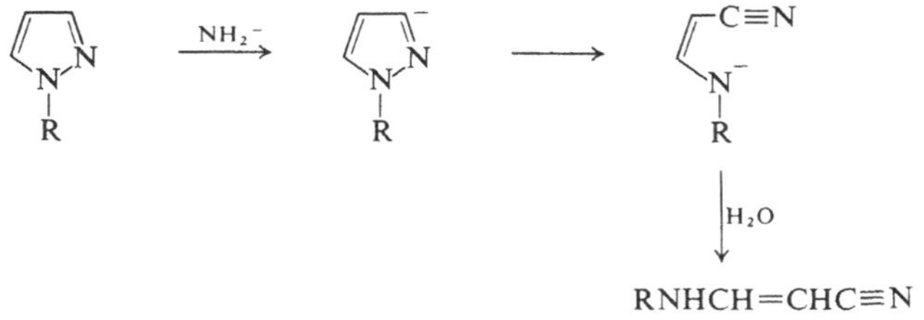

$$RNHCH{=}CHC{\equiv}N$$

In the imidazole series the hydrolytic cleavage of amino compounds was first observed by Fargher[274a]; reduction of 4-methyl-5-nitro-imidazole with stannous chloride and hydrochloric acid gave alanine and ammonia. The amino compound was supposed to be hydrolysed to the imidazolone which was further hydrolysed with ring-opening. Other examples, effected in the same way, include the conversion of 1-methyl-5-nitroimidazole into methylamine and glycine[497] and 1,4-dimethyl-5-nitroimidazole into the corresponding amine, alanine-*N*-methylamidine, and ammonia[16d]. Amino-imidazoles themselves have been degraded by heating with hydrochloric acid; thus, 5-amino-1-methyl-4-*p*-aminophenyl-imidazole gives ammonia, methylamine, and *p*-aminophenylalanine[497], and 4-amino-1-ethyl-2-propylimidazole-5-carboxylic amide gives ammonia, *N*-ethylglycine, and n-butyric acid (analogous examples were also reported)[495a].

Hydrochloric acid also cleaves hydrolytically 4-nitroso-imidazoles[146c, 532d,845a]; thus, 2-methyl-4-nitroso-5-phenylimidazole gives 3-benzoyl-5-methyl-1-oxa-2,4-diazole.

4-Nitroso-5-phenylimidazoles are similarly cleaved by phenylhydrazine hydrochloride[414b], hydrazine hydrochloride[414c], and semicarbazide hydrochloride[414d], giving the derivatives of these reagents with 3-benzoyl-1-oxa-2,4-diazoles. With hydroxylamine the result is different, 5-benzoylamino-4-phenyl-1-oxa-2,5-diazole being formed[845b].

The nature of the degradation of imidazole accompanying nitration in sulphuric acid (p. 62) is not understood, but oxidation is involved since nitrous acid appears.

Wallach first observed the decomposition of quaternary imidazolium salts to primary amines by the action of alkali[54c,d]. In the early reports quaternary salts with different groups on N-1 and N-3 were held to produce only one primary amine[408], but, as already mentioned (p. 96), later work showed both possible primary amines to be formed

[90c]; thus, 1-amyl-3-methylimidazolium salts gave methylamine, amylamine, formic acid, and methanol. The reaction is general[453d,632]. Intermediate products of the ring-opening can be observed[411b,633,846] if the quaternary salt is hydrolysed with ammonia or carbonate. For example, 1-methyl-4,5-diphenyl-3-(2,4-dinitrophenyl)imidazolium *p*-toluenesulphonate gives Ph.C(NMe.CHO):C(NHAr).Ph, where Ar = 2,4-dinitrophenyl[411b].

1,3,4,5-Tetraphenylimidazolium perchlorate treated with potassium t-butoxide in dimethyl sulphoxide in presence of air, gives benzil dianil. This reaction has been represented as proceeding via a 2-carbene which adds water at C-2 and after peroxidation is ring-opened[847].

The difficulties of using the Schotten–Baumann reaction to *N*-benzoylate imidazoles have been mentioned (p. 114). Bamberger and Berlé[848] first tried to benzoylate imidazole using Schotten–Baumann conditions. The unexpected product of the reaction was 1,2-dibenzamidoethylene (see also ref. 849). Similar ring-opening occurs with 4-methyl-[850a,c], 4-ethyl-[850e], 4,5-dimethyl-[368a], and 4-β-dimethylaminoethyl-imidazole [851]. Whilst 1-methylimidazole[90c], 4-nitro-, 4-carboxy-, 4-formyl-, and 4-carbethoxy-imidazole, and imidazole side-chain acids are not affected[850b], esters of the latter compounds suffer ring-opening[143a, 850b,c,d,852]. Pyridine–benzoyl chloride also causes ring-opening[853], and isovaleryl chloride acts similarly to benzoyl chloride[850d]. It is very likely[230] that ring-opening proceeds by nucleophilic attack on quaternary diacyl derivatives, as shown below.

Carbobenzoxy chloride and bicarbonate caused ring-opening of 1-(*p*-nitrophenyl)- and 1-(2,4-dinitrophenyl)-imidazole whilst leaving other 1-arylimidazoles unaffected. With two equivalents of carbobenzoxy chloride, imidazole gave $C_7H_7O.CO.N(CHO).CH:CH.NH.CO.OC_7H_7$, but with one equivalent 1-carbobenzoxy-imidazole could be isolated and subsequently ring-opened in the same sense with benzoyl chloride and bicarbonate[854]. These various ring-openings presumably proceed by nucleophilic attack on quaternary diacyl derivatives[381d].

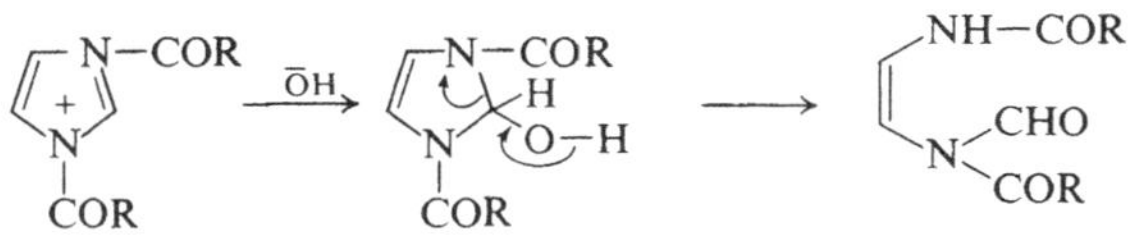

With benzoyl chloride and alkali 1,2,4-triazole is said to give dibenzoyl urea, whilst 1,2,3-triazole gives only its benzoyl derivative[75a].

1,2,4-Triazol-5-ones are ring-opened by acid hydrolysis; thus 3-methyl-1,2,4-triazol-5-one gives hydrazine, ammonia, carbon dioxide, and acetic acid. A study of the rates of 28 of these reactions occurring in 51.6% sulphuric acid at 130°C showed them to be increased by electron-attracting substituents and reduced by electron-releasing substituents.

The acidity dependence of rate recalled the behaviour of amides, and it was concluded that the hydrolyses involved attack of water on the carbonyl group [676e].

1,2-Diaryl-1,2,3-triazolium salts turn yellow with alkali, apparently with ring cleavage [855]. As already related, the alkaline degradation of 1,2,4-triazolium salts is valuable in structure determinations (p. 98).

The ring-opening by alkali of tetrazolium salts has also been mentioned (p. 99). 1-Aryl-4-methyltetrazolium salts give an aryl azide and methylamine [613b]. 1,4-Diethyltetrazolium gave, in contrast, nitrogen and diethylcarbodiimide, and the reaction is general [406b]. 1,3-Dimethyltetrazolium with 50% aqueous potash gave diazomethane and potassium methylcyanamide, a reaction which was represented as proceeding via the ylide [407a]. With primary and secondary amines, 1,4-dimethyltetrazolium picrate gave nitrogen and substituted guanidinium picrates [407a].

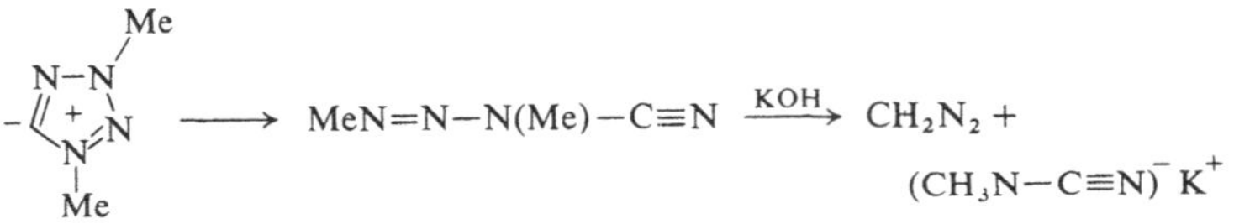

$$-\underset{\substack{| \\ Me}}{\overset{\substack{Me \\ |}}{\langle}} \overset{N-N}{\underset{N}{N}} \longrightarrow MeN{=}N{-}N(Me){-}C{\equiv}N \xrightarrow{KOH} CH_2N_2 +$$

$$(CH_3N{-}C{\equiv}N)^- K^+$$

6.5   Miscellaneous reactions

In reactions reminiscent of the Beckmann rearrangement, 3,5-disubstituted 4-nitrosopyrazoles react with phosphorus pentachloride to give ring-opened products R.C(Cl):N.N:C(CN)R' (*cf.* the reaction in the pyrrole series, Schofield, p. 89), which with ammonia give 5-amino-1,2,4-triazines [856]. 3-Methyl-4-nitroso-1,5-diphenylpyrazole is ring-opened by triethyl phosphite [857].

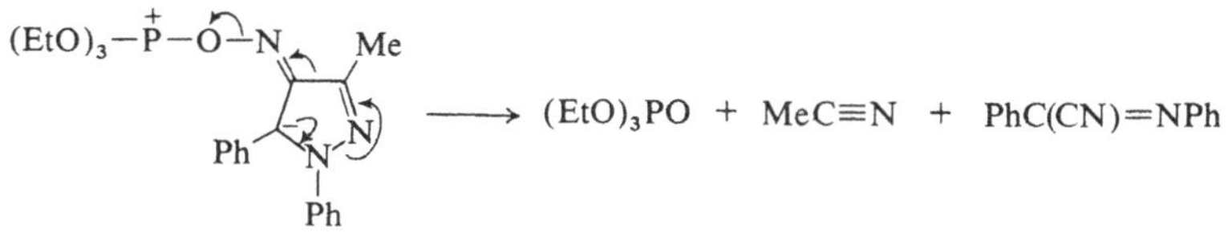

$$(EtO)_3{-}\overset{+}{P}{-}O{-}N \underset{\substack{Ph \\ | \\ Ph}}{\overset{Me}{N}} \longrightarrow (EtO)_3PO + MeC{\equiv}N + PhC(CN){=}NPh$$

6.6   Mass spectrometry

Although the mass spectrum of pyrazole was first mentioned by Jenning and Boggs [858], the first detailed examination of the mass spectra of pyrazoles was made by Nishiwaki [859a], who reported cleavage of the N-N bond and loss of HCN or MeCN from pyrazole or the N-unsubstituted methylpyrazoles. Loss of HCN from the molecular ion of N-alkylpyrazoles was negligible.

The formation of pyrimidinium and pyridazinium ion fragments from mono-, di-, tri-, and tetra-methylpyrazoles has been postulated[860a], and their presence in the mass spectra of benzylmethylpyrazoles confirmed[861b]. Ionisation has been correlated with the number of atoms in pyrazole systems, and used in identification[860b].

An extended study of pyrazoles and deuteropyrazoles confirmed the kind of fragmentation illustrated above[862]. Loss of HCN occurs from the 3(5)-position, but decompositions of metastable ions with loss of HCN from the $(M-H)^+$ ion show that the identity of the hydrogen atoms in this fragment is largely lost[862b]. A second important process is the loss of N_2 after initial loss of H· or a substituent, giving $C_3H_2R^+$ (perhaps a cyclopropenyl ion). Substituents exert a strong influence, e.g. 1-methylpyrazole loses $[CH_2CN]$·[862a,b].

With mono-, di-, and tri-phenylpyrazoles the migration of phenyl groups occurs, and a fluorenyl ion is formed[861a,863].

The mass spectra of fluoropyrazoles[859b] and pyrazolones[864] have been examined. Pairs of *syn*- and *anti*-1-phenylpyrazol-4-yl oximes give the same mass spectrum; no Beckmann rearrangement occurs, but the oxime hydrogen atom is transferred to the pyrazole nucleus with fission of the side-chain[865].

3- and 5-Amino-1-methylpyrazole give noticeable *M*-1 fragments, as well as those from loss of HCN[246].

1-Methyl-nitropyrazoles lose O·, NO, NO_2· and $NO^{·+}$. The dinitropyrazoles show complicated fragmentation patterns, initially because of the different sequences in which these fragments can be lost. 1-Methyl-5-nitropyrazole can lose HO· from the molecular ion, a process not observed with the 3- and 4-nitro isomers. Further, 1-methyl-5-nitropyrazoles eliminates HCO, probably by interaction of the methyl and nitro groups for this is not observed with 1-methyl-3-, 1-methyl-4-, or 1,3,5-trimethyl-4-nitropyrazole. The spectrum of the latter after loss of ·NO_2 was similar to that of 1,3,5-trimethylpyrazole[859a]. Both spectra exhibited prominent peaks at 56 m.u., evidently a characteristic of 1,5-dimethylazoles[246].

1-Methylpyrazole 2-oxide shows mass spectral behaviour typical of aromatic *N*-oxides, O· and HO· being lost from the molecular ion. After this the spectrum is similar to that of 1-methylpyrazole, H·, HCN, and $[CH_2CN]$· being lost. The principal loss of HCO seen in the mass spectrum of 1-methyl-5-nitropyrazole 2-oxide evidently occurs in a sequence different from those giving HCO from 1-methyl-5-nitropyrazoles [246].

Study of the mass spectra of simple imidazoles began only recently [866-8]. The spectra show pronounced molecular ions and fragmentation patterns, but skeletal re-arrangements are rare. The azirine cation (m/e = 40) is one of the principal fragments produced. At energies above the range 17–26 eV the azirine cation loses a hydrogen molecule, giving C_2N^+ [867]. Deuterium labelling has shown the loss of HCN from the molecular ion to be non-specific, although the loss does not readily occur from the 3- and 4-positions.

Mass spectrometry is useful in determining the type and position of substituents [869]. The mass spectrum of 1-methyl-4-nitroimidazole-5-carboxylic acid amide shows the elimination of water by an electron-impact, as distinct from the normal thermal process [870].

1-Methoxy-4,5-dimethyl-2-phenylimidazole (loss of ·CH$_2$ and ·OMe, and of MeCN from the fragment m/e = 171) is readily distinguished from 1,4,5-trimethyl-2-phenylimidazole 2-oxide (loss of ·O and ·OH). The spectra of 1-benzylimidazoles and 1-benzylimidazole 3-oxides are dominated by the tropylium ion (m/e = 91) [246].

2-Formyl-1,4,5-trimethyl- and 2-formyl-1,5-dimethylimidazole 3-oxides give very characteristic fragmentation patterns; loss of CO occurs, as it does from 2-formyl-1-methylimidazole [866a]. There appear to be two principal fragmentation schemes, one involving the loss of the *N*-oxide group and then the formyl group, the other involving the alternative process. The loss of H· which is observed with 2-formyl-1-methylimidazole is not evident in case of the *N*-oxides, presumably because the hydrogen radical is lost in the ·OH fragment. From 2-formyl-1,5-dimethylimidazole 3-oxide the loss of HCN occurs principally from the fragment of 95 m.u., the molecular ion having first lost ·O and then CO; however the loss of HCN from 2-formyl-1,4,5-trimethylimidazole 3-oxide occurs from the fragment of 109 m.u., the molecular ion in this instance having first lost ·OH and then CO. In the latter case a hydrogen radical can be removed from the methyl group on the carbon atom adjacent to the *N*-oxide function as well as from the formyl group, whereas in the former case the hydrogen radical is obtained only from the formyl group.

1,4,5-Trimethylimidazole-2-carboxylic acid 3-oxide shows principally loss of CO_2. Subsequent loss of ·O, ·OH and HCN are also observed[246].

1,4,5-Trimethyl-2-nitroimidazole 3-oxide forms fragments arising from loss of ·O, ·OH, NO, ·NO_2; however, the spectrum is complicated by the loss of HCO, a previously unreported observation for nitroazoles.

The spectrum of 1,4,5-trimethylimidazole 3-oxide after initial loss of ·O and ·OH, is similar to that of 1,4,5-trimethylimidazole.

The spectra of 1,5-dimethylimidazoles show a major fragment of 56 m.u. which evidently arises from the 1- and 5-positions (1,2-dimethylimidazoles do not give this fragment[866a]). The $C_2NH_4^+$ fragment also provides a principal peak in these spectra[246].

The mass spectra of di- and tri-phenyl-1,2,3-triazoles reveal the elimination of nitrogen[871], as do those of 1,2,4-triazole[872], and of 1,2,4-triazole 4-oxides[873]. Two flat-topped metastable peaks are seen in the spectrum of 1,2,4-triazole at m/e = 24.4 ($69^+ \rightarrow 41^+$ + 28) and m/e = 18.7 ($42^+ \rightarrow 28^+$ + 14). In the latter case deuterium labelling shows loss of a nitrogen atom to be occurring from the fragment $CH_2N_2^+$, formed by loss of HCN from the molecular ion. Possible representations of this are shown. The next scheme shown is suggested

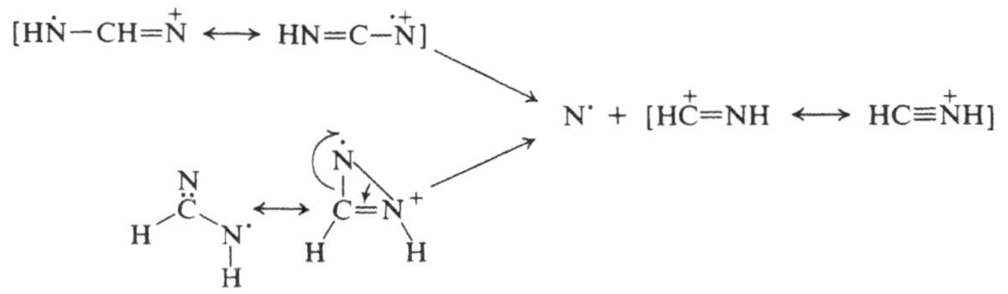

for loss of nitrogen from the molecular ion, the hydrogen transfer being supported by deuterium labelling[872].

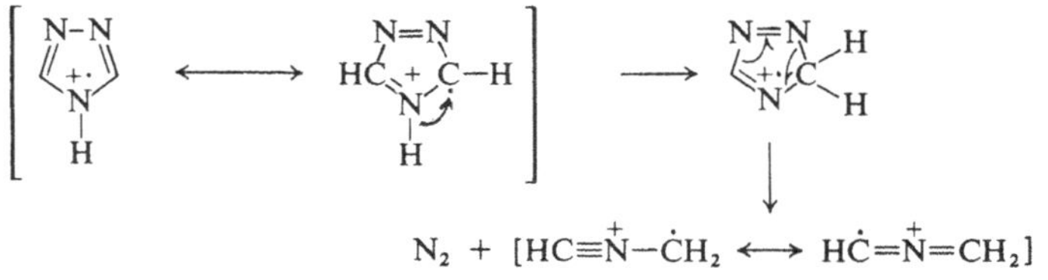

A number of 1,2,4-triazoles give mass spectra containing the ion m/e = 42 and the characteristic metastable peak at m/e = 18.7. Commonly the fragmentation can be represented as here[223e]. $[M-1]^+$ ions were not given by C-methyl compounds. Like 1,2,4-triazole itself, 3,5-diphenyltriazole showed appreciable loss of N_2, but other derivatives did not. 3-Alkyl(R) derivatives lose RCN dominantly from the molecular ion, and 3-chloro-1,2,3-triazole loses HCN and ClCN[223e].

5-Aminotetrazole also loses nitrogen, and in its mass spectrum flat-

The reaction schemes show fragmentation pathways producing metastable peaks:

$$-R^2CN \rightarrow \text{(intermediate)} \xrightarrow[\text{peak}]{\text{Metastable}} N^{\cdot} + R^3\overset{+}{C}=NR^1$$

$$-R^3CN \rightarrow \text{(intermediate)} \xrightarrow[\text{peak}]{\text{Metastable}} N^{\cdot} + R^2\overset{+}{C}=NR^1$$

topped metastable peaks are observed. The fragmentations $85^+ \rightarrow 43^+ \rightarrow 29^+ + 14$ represent the formation of HN_3^+ from the molecular ion, and the subsequent production of HN_2^+ and a nitrogen atom. Thermal decomposition could be a source of confusion and the origin of a fragment $m/e = 42$ is uncertain. This fragment gives CH_2N^+ and a nitrogen atom[874].

6.7   Ring expansions

Reaction of dichlorocarbene with pyrazole and with imidazole in the vapour phase is said to produce ring-expanded products, but details have not been given[875]. 3,4,5-Trimethylpyrazole adds dichlorocarbene generated under basic conditions to give 4-dichloromethyl-3,4,5-trimethyl-

pyrazolenine[876a]. With sodium ethoxide the latter gives a pyridazine [876b]. 3,4,5-Trimethylpyrazole with dichlorocarbene in neutral conditions gives four minor products: 4-chloro-3,5,6-trimethylpyridazine, 2-chloro-4,5,6-trimethylpyrimidine, 3,4,5-trimethyl-1-trichlorovinylpyrazole, and tris-(3,4,5-trimethylpyrazol-1-yl)methane[876a].

In either basic or neutral conditions 2,4,5-trimethylimidazole and dichlorocarbene give low yields of 5-chloro-2,4,6-trimethylpyrimidine. Reaction at nitrogen in the imidazole anion merely gives a product which is hydrolysed to starting material[876a].

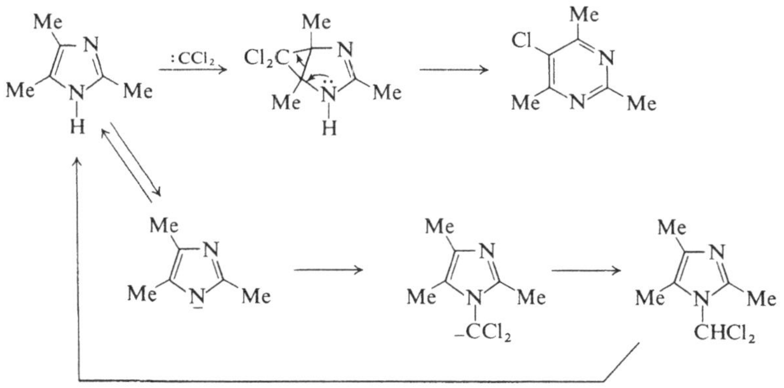

Lophine anion with tosyl azide gives 2,4,6-triphenyl-1,3,5-triazine, a change represented as attack at a carbon atom of the lophine nucleus followed either by production of the azide and expansion, or by concerted expansion[813b].

Dichlorocarbene does not ring-expand 3,5-dimethyl-1,2,4-triazole, but produces tris-(3,5-dimethyl-1,2,4-triazol-1-yl)methane[876a].

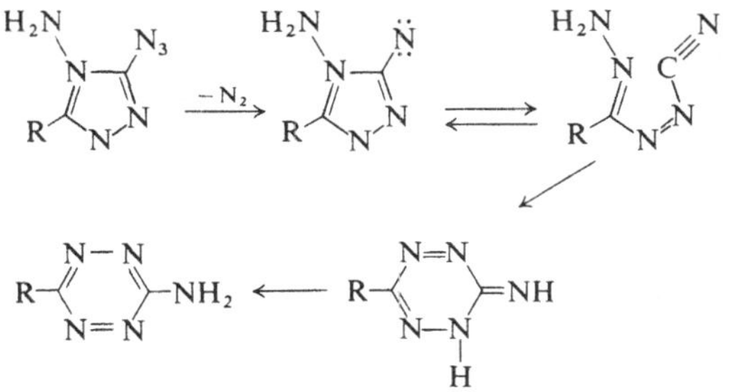

The generation of nitrenes has been postulated for some reactions discussed earlier in this chapter. The thermolysis of 4-amino-3-azido-4*H*-1,2,4-triazole gives 3-amino-1,2,4,5-tetrazines, and this reaction also is supposed to involve a nitrene[877a]. In contrast thermolysis of 3-azido-4-benzylideneamino-4*H*-1,2,4-triazoles gives nitrogen and *s*-triazolo-[3,2-*c*]-*s*-triazoles without modification of the original triazole ring[877b].

2,3,5-Triphenyltetrazolium cation reacts with diazomethane giving, by ring-expansion, a 'verdazyl'[878].

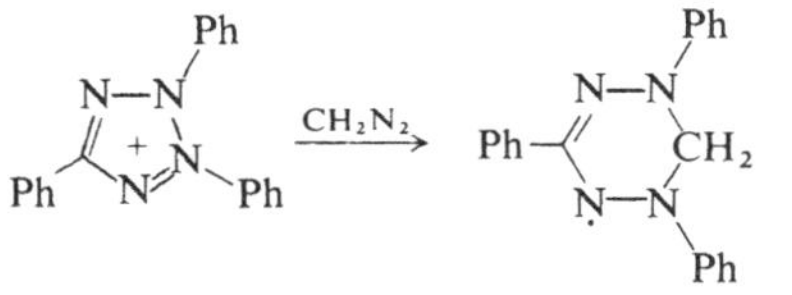

7 Photochemical reactions

The photochemical reactions to be considered in this chapter are those which involve the nuclei of azoles. They should be compared with the thermolyses discussed in Chapter 6.3.

The most important photochemical reaction of pyrazoles is that which transforms them into imidazoles. This has been observed in numerous cases including pyrazole itself, alkyl-, aryl-, and aralkyl-pyrazoles[879]. Some examples are tabulated. The reaction is conveniently carried out in 1,2-dimethoxyethane in presence of benzophenone, and the yields are in the range 20–32%. The sensitiser improves the yield but is not essential.

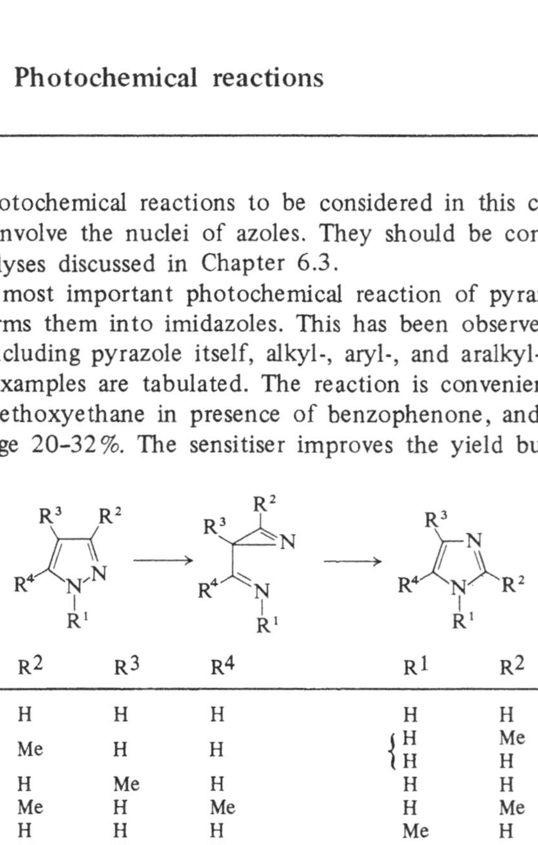

R^1	R^2	R^3	R^4	R^1	R^2	R^3	R^4
H	H	H	H	H	H	H	H
H	Me	H	H	{ H	Me	H	H
				{ H	H	H	Me
H	H	Me	H	H	H	Me	H
H	Me	H	Me	H	Me	Me	H
Me	H	H	H	Me	H	H	H
Me	Me	H	H	Me	Me	H	H
Me	H	Me	H	Me	H	Me	H
Ph	H	H	H	Ph	H	H	H
PhCH$_2$	H	H	H	PhCH$_2$	H	H	H

Electron-attracting groups (1-PhCO, 4-Cl, 4-NO$_2$) inhibit the reaction, resins being formed. 3-Amino-4-cyanopyrazole gives 4-amino-5-cyano-imidazole. These examples are all understandable in terms of the mechanism illustrated, *viz.* ring-opening to produce an azirine followed by ring-closure. The consequence is (allowing for tautomerism in the cases of *N*-unsubstituted pyrazoles) an interchange of N-2 and C-3 in the pyrazole. The case of 1,3,5-trimethylpyrazole does not fit this pattern [880a,b]. Irradiated in ethanol it gives very poor yields (7% and 2% respectively), and in cyclohexane better yields (26% and 10% respectively) of 1,2,4-trimethyl- and 1,2,5-trimethyl-imidazole. These two compounds are not interconverted under the conditions. The second product, but not the first, could be formed by the mechanism illustrated. It is suggested[880a,b] that 1,2,4-trimethylimidazole could arise via valence-isomerisation of the kind mentioned for imidazoles below, or by *two* sequences of the ring-opening and ring-closing type with 1,3,4-trimethylpyrazole as the intermediate. The significance of the latter suggestion is not clear; the first step could not be azirine formation, and the behaviour of 1,3,4-trimethylpyrazole on irradiation has not been examined.

Rearrangements of the kind described also occur with antipyrines. Irradiated in an alcohol antipyrine gives 1,5-dimethyl-3-phenylimidazol-2-one, and also the products of ring-opening. The results for reaction in methanol and in acetone are shown[881]. Irradiation of aminopyrine

(antipyrine with ·NMe$_2$ at C-4) in methanol gave 4-methoxy-1,5-dimethyl-3-phenylimidazol-2-one. This product did not arise from 4-methoxy-antipyrine by irradiation in methanol, but when additionally dimethyl-amine was present this reaction produced products of ring-opening[882].

Irradiation of 3,5-dimethyl-1-*o*-nitrophenylpyrazole in ethanol causes opening of the pyrazole ring and cyclisation through the nitro-group[883].

The conversion of imidazoles into pyrazoles evidently does not occur photochemically. However, alkylimidazoles can be isomerised to other alkylimidazoles[880a,b]. Thus, 1,4-dimethyl- and 1,2-dimethyl-imidazole are interconvertible in t-butanol, and 1,4,5-trimethylimidazole gives the 1,2,5-trimethyl isomer (40%) in ethanol, t-butanol, or cyclohexane. In cyclohexane 1,2,5-trimethyl- gives less than 5% of 1,4,5-trimethylimidazole, being largely destroyed. 1,2,4-Trimethylimidazole is hardly affected in t-butanol and whilst in cyclohexane it photoreacts it does not give the isomeric imidazoles. A possible mechanism for these interconversions involves a disrotatory valence-bond isomerisation, followed by a 1,3-sigmatropic shift and a second disrotatory isomerisation. Evidence for such a sequence is the photoconversion of 2-deutero-1,4-dimethyl-into 4-deutero-1,2-dimethyl-imidazole [880b].

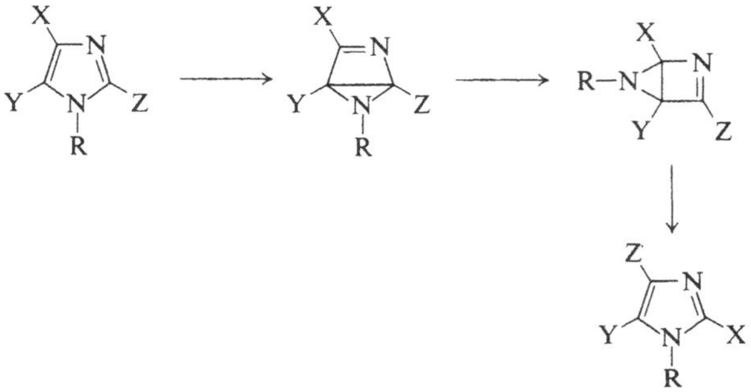

4-Phenyl-1,2,3-triazole gives nitrogen, phenylacetonitrile, benzonitrile, and methyl benzoate when irradiated in methanol. 5-Deutero-4-phenyl-1,2,3-triazole gives the same products, lacking deuterium. An intermediate such as Ph.C:C.NH$_2$ or Ph.CH:C:NH is suggested since either would lose deuterium, present initially, rapidly to the solvent. When irradiation is carried on in dichloromethane, benzonitrile and some 3,6-diphenyl-1,2,4,5-tetrazine are formed [458c,824].

1,5-Diphenyl-1H-1,2,3-triazole gives when irradiated in benzene benzalaniline, 2-phenylindole, and nitrogen. 1,4-Diphenyl- and 1,4,5-triphenyl-1H-1,2,3-triazole behave analogously, and the products are obtained in high yields. The reactions can be represented as shown [884].

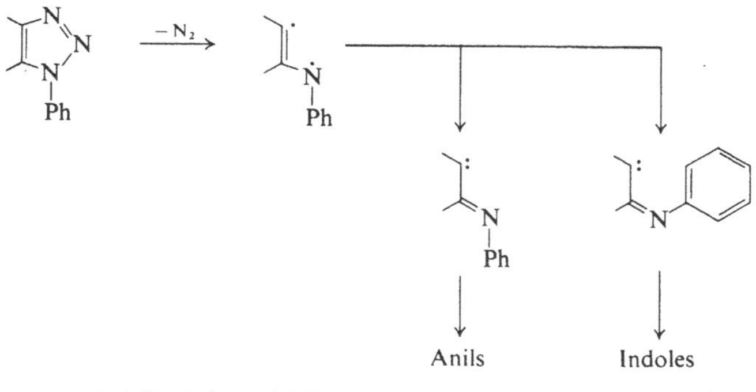

Anils   Indoles

In contrast, 2,4,5-triphenyl-2H-1,2,3-triazole is unaffected by irradiation. 2,4,5-Trimethyl-2H-1,2,3-triazole gives acetonitrile and a complex mixture; use of cyclopentene as a trap shows MeĊ:N.Ṅ.Me to be produced [885].

The lithium salts of several 4,5-disubstituted 1-*p*-toluenesulphonyl-amino-1*H*-1,2,3-triazoles decompose photochemically to disubstituted alkynes, nitrogen, and lithium-*p*-toluenesulphinate[886].

The photolysis of some meso-ionic 4-aryl-1-phenyl-1,2,4-triazol-3-ones gives aryl isocyanates, azo-compounds, and benzimidazoles (the latter presumably from Ph.N:C(R).Ṅ)[887].

Irradiation of 1,5-dimethoxycarbonyl-1*H*-1,2,3,4-tetrazole in benzene gave nitrogen and 5-methoxy-3-methoxycarbonyl-1-oxa-2,4-diazole (32%) [888a], whilst 1,5-diphenyl-1*H*-1,2,3,4-tetrazole gave 2-phenylbenzimidazole [888b]. These reactions could proceed via iminonitrenes. 2-Acetyl-5-phenyl-2*H*-1,2,3,4-tetrazole gives 2-methyl-5-phenyl-1-oxa-3,4-diazole[889].

Photolysis of 5-phenyltetrazole in dioxan[889] or tetrahydrofuran [890a] gives nitrogen quantitatively and 3,6-diphenyl-1,2-dihydro-1,2,4,5-tetrazine (which is converted into 3,5-diphenyl-1,2,4-triazole and benzo-nitrile as secondary products, and to 3,6-diphenyl-1,2,4,5-tetrazine by oxidation during work-up) in more than 70% yield. Ph.C:N.NH is inter-mediate in the reaction, and analogous decomposition of 2,5-diphenyl-2*H*-1,2,3,4-tetrazole produces Ph.C:N.NPh, as is shown by its capture with dimethyl fumarate[890a,891], and by the formation of Ph.C(N:Ph):C(N:NPh)Ph (ref. 891 reports benzil phenylosazone) (which gives aniline and 2,4,5-2*H*-1,2,3-triazole on irradiation)[890b,891]. For the decomposition of 5-phenyltetrazole ^{15}N-labelling and n.m.r. studies have shown that the nitrogen lost comes from N-3 and N-4[890b,c].

5-Phenyltetrazole anion (present in a solution of sodium methoxide in methanol) decomposes differently, giving nitrogen, benzyl methyl ether, and methoxide, presumably via phenylcarbene (the reactions in water and t-butanol are analogous)[890a]. The difference in behaviour between the anion, *N*-substituted tetrazoles, and 5-phenyltetrazole is attributed to the possibility of hydrogen bonding existing for the last compound[890b].

5-Phenoxy-1-phenyl-1*H*-1,2,3,4-tetrazole is converted by irradiation into biphenyl and 2-phenoxy-benzimidazole (which isomerises to 2-(*o*-hydroxyphenyl)benzimidazole)[892].

Perhaps even more than of the pyridines (Schofield, p. 270) it is true of the azoles that, despite the large amount of information available about their reactions, there is little that can readily be related to theoretical studies of aromatic reactivity. Such studies relate to the two-step mechanism of aromatic substitution (electrophilic, nucleophilic, or radical) either by evaluating localisation energies, using the Wheland intermediate as a model for the transition state, or by reference to electron distribution in the ground state. For the azoles, not only is the amount of quantitative information of the kind appropriate to such treatments very limited, but the actual incidence of this kind of substitution mechanism is far from universal. Ylide mechanisms, in particular, are being recognised increasingly. Where the two-step substitution mechanism does operate the first step is not necessarily rate-determining, so that comparison with theoretical calculations is invalidated. Additionally where the mechanism is of the appropriate kind there is the minor problem of deciding whether the results relate to azole molecules themselves, or to their anions or cations.

Amongst electrophilic substitutions acylations of N-substituted pyrazoles occur at C-4 but mechanisms have not been established. In some cases imidazoles react at C-4, but in others, such as that of 1-methylimidazole reacting with benzoyl chloride in the presence of triethylamine, at C-2, and some processes involve ylides. Carbonyl reactions yield no information of unambiguous significance. The situation in hydrogen–deuterium exchange is very complicated; ylide mechanisms can evidently be involved in the reactions of both pyrazole and imidazole, though with pyrazole the two-step mechanism appears to operate in some conditions either through the cation or the neutral molecule and in these circumstances C-4 is more reactive than C-3 and the free base is about 10^9 as reactive as the cation. For diazo-coupling the result with imidazole is clear; in an apparently normal two-step substitution the anion reacts initially at C-2. Unfortunately comparison with other azole anions cannot be made. Most chlorinations and brominations of pyrazoles occur at C-4, but mechanisms have mostly not been studied. Although the bromination of pyrazole, 1-methylpyrazole, and 3,5-dimethylpyrazole in water containing potassium bromide appears to be a two-step reaction involving bromine and the azole neutral molecules, producing substitution at C-4, there is a finite kinetic isotope effect. The mechanisms of the thoroughly-studied iodinations of imidazole and 1-alkylimidazoles exclude them from comparisons with the usual theoretical calculations. Nitration in sulphuric acid of both pyrazole and imidazole shows the cations to be reacting, substitution to occur at C-4, and both cations to be powerfully de-activated (partial rate factors: 2.1×10^{-10} and 3.0×10^{-9} for pyrazolium and imidazolium, respectively). With 1-methylpyrazole 2-oxide nitrations of the free base occurs at C-5, and the position is certainly activated. It seems likely, but has not been demonstrated, that sulphonation of pyrazole and imidazole, at C-4 in each case, involves the cations.

Thus, in one or two clear cut instances, pyrazole and imidazole cations react at C-4 in normal two-step processes, and in one case (diazo-coupling) the imidazole ion reacts first at C-2. Deuterium exchange also suggests C-4 to be the most reactive carbon atom in pyrazole itself.

It should be noted that in a number of electrophilic reactions with *N*-unsubstituted azoles reaction occurs first at nitrogen.

Nucleophilic substitution provides even less useful information. Where quantitative studies have been undertaken, as in the reactions of bromoazoles with piperidine, useful comparisons are difficult to make because of the presence of substituents. It appears that the activating influence of two or three pyridinic nitrogen atoms is required to counter the deactivating influence of one pyrrolic nitrogen atom.

Radical phenylation of 1-methylimidazole and 1-methylpyrazole shows the former to be slightly more reactive, and the latter to be slightly less reactive than benzene. In 1-methylimidazole the sequence of reactivities (partial rate factors in brackets) is C-2 (4.8) > C-5 (2.2) > C-4 (0.14), and in 1-methylpyrazole C-5 (3.4) > C-3 (0.18) > C-4 (0.03).

A number of molecular orbital studies of azoles relevant to the problem of reactivity have been reported. Several Hückel calculations [303-5,577b] have given π-electron densities, which some authors have supposed to be in agreement with orientations observed in electrophilic substitutions. Hückel and VESCF calculations on pyrazole and imidazole [306,893], and their anions and cations have been given; they match experiment to some extent as regards orientation, but are inadequately related to mechanisms.

The EHT-MO method has been used to calculate total electron densities for imidazole and pyrazole, and their cations and anions[308b, 894]. As regards imidazole, total electron densities and σ-complex energies alike suggest that in all three species C-4(5) will be the most reactive. The PPP–SCF method has been applied to pyrazole, imidazole, and the triazoles[313]. For pyrazole π-electron densities suggest C-4, and for imidazole C-5 (followed by C-2 and C-4) as the site for electrophilic attack.

Generally, experiment and theory make only the most tenuous of contacts in the field of aromatic reactivity amongst the azoles. The primary need is for more quantitative mechanistic studies.

9.1   C- and N-acyl, -carboxyl and -alkoxycarbonyl groups

9.1.1   N-acyl, N-carboxyl and related compounds

There has been considerable interest in the N-acyl derivatives of the
azoles (azolides) in recent years in view of the lability of the acyl
substituent in solvolysis and hydrogenolysis reactions. Some of the
properties of these compounds have already been discussed (infrared
(pp. 16-17) and n.m.r. (p. 22) spectra, synthesis p. 110 *et seq.*), melting
points (p. 113), rearrangement (p. 111), orientation of acylation (pp.
110, 117), and 1-acetylimidazole involvement in ester hydrolysis in
biological systems (p. 115 *et seq.*)). Table A.8 lists the melting points
of some N-acylazoles. In some cases the orientation of acylation is
uncertain as it was only recently that n.m.r. studies were brought to
bear on the structures of these compounds.

The compounds compare well with acyl halides and anhydrides as
regards their high degree of reactivity in nucleophilic reactions. Table 23
summarises the rates of hydrolysis of some simple azolides. Although
1-acetylpyrrole is not measurably hydrolysed, 2-acetyltetrazole decomposes
so rapidly into tetrazole and acetic acid on dissolving in water that the
rate of hydrolysis cannot be determined by usual methods[74d]. The
high degree of reactivity of the azolides is associated with the amide
nitrogen electron pair being involved with the π-system of the ring,
resulting in polarisation towards nitrogen in the N–(COR) bond. This
leads to weak acidity in unsubstituted compounds (p. 25) and ease of
nucleophilic reactions involving the carbonyl group of azolides regardless
of whether the mechanism is S_N1 or of the addition–elimination type.
The increase in reactivity with increase in number of ring nitrogen atoms
is explained by assuming that delocalisation of the lone electron pair of
the amide nitrogen (and hence also the electron attraction by the ring)
increases with the number of CH groups replaced by the more electronegative nitrogen atoms. Of the isomeric azolides, those with the greater

Table 23. Hydrolysis of azolides[a]

Azolide	$k' \times 10^5/s^{-1}$	$\tau_{1/2}/\mathrm{min}$
1-acetylpyrrole	→0	→∞
1-acetylpyrazole	1.27	908
1-acetylimidazole	28.1	41
2-acetyl-1,2,3-triazole	43.5	26.6
1-acetyl-1,2,4-triazole	180	6.4
2-acetyltetrazole	>2000	>0.5

(a) Values of k' and the half-lives ($\tau_{1/2}$) for the hydrolysis of N-acetylazoles in conductivity water at 25°C and pH 7.0 [150m].

number of adjacent nitrogen atoms show lower reactivity (e.g. pyrazolides are hydrolysed more slowly than imidazolides; 1,2,3-triazolides are hydrolysed more slowly than 1,2,4-triazolides) probably because two neighbouring nitrogen atoms cannot assume the same electron densities as those separated by a carbon atom.

Kinetic studies[74d,150a,b,261,655,897c,904a,h,916a,917-23] of the nucleophilic reactions of azolides have shown that their aminolyses and alcoholyses occur by a bimolecular, addition–elimination reaction mechanism, as does the neutral hydrolysis of the azolides of aromatic carboxylic acids[904a]. Aliphatic carboxylic acid azolides which are sterically hindered appear to be hydrolysed in pure water by a S_N1 process[150i]. The deacylation of azolides has been studied in detail by Staab[150a,b], Scott[916a], and Hüttel and Kratzer[895a]. Whereas Staab preferred a mechanism of the S_N1 type with an ionisation as the rate-determining step for the neutral hydrolysis of 1-acetylimidazole, 1-acetyl-1,2,4-triazole and 2-acetyltetrazole, this interpretation was criticised by Scott who

favoured a S_N2 type mechanism with a tetrahedral intermediate (though without very strong evidence). In their study of the neutral hydrolysis of a series of 1-acetyl-pyrazoles and -1,2,3-triazoles Hüttel and Kratzer observed a dependence of rates on solvent polarity, not in agreement with a S_N1 pathway, and also a retardation of the hydrolysis of 1-acetyl-pyrazole in D_2O by a factor of 3.4, thus favouring a two-step, bimolecular mechanism.

Aspects of the imidazole-catalysed hydrolysis of esters have already been discussed (p. 115). A kinetic study[655] of the hydrolysis, and imidazole-catalysed hydrolysis, of p-methyl, p-chloro-, and p-nitro-benzoylimidazole in aqueous solution showed that the rate of hydrolysis is first order in acylimidazole concentration. The isotope effect

$(k_1(H_2O)/k_1(D_2O))$ was 1.47, indicating that the imidazole catalysis appears to be pH (or pD) dependent. Imidazole is known to catalyse the hydrolysis of esters by a general base mechanism, the clearest example being imidazole-catalysed hydrolysis of 1-acylimidazoles[261], where nucleophilic attack by imidazole leads to no net reaction. The kinetics of hydrolysis of 1-alkyl-3-acetylimidazolium chloride have been measured[918a] at pH 5.5 showing a rate-dependence on the basicity of the leaving group, and a slower rate of hydrolysis than the corresponding 4-aminopyridinium compounds.

In the uncatalysed aminolysis of 1-acetylimidazole[919a,920] there is an unusually large sensitivity to the basicity of the attacking amine. Acyl functions with good leaving groups (e.g. phenyl acetates, acetylimidazolium ion[920]) readily undergo aminolysis without proton removal from the amine, as evidenced by similar reactivities of primary, secondary and tertiary amines[919a]. Thus, strong bases react with free 1-acetylimidazole in an uncatalysed, second order aminolysis reaction which is faster than a general base catalysis of hydrolysis; weak bases simply act as general base catalysts of hydrolysis. The limiting interpretation is that complete cleavage of the C–N bond has occurred in the transition state. That is to

$$RNH_2 + AcIm \underset{k_{-1}}{\overset{k_1}{\rightleftharpoons}} R\overset{+}{N}H_2Ac.Im^- \underset{k_{-2}}{\overset{k_2}{\rightleftharpoons}} R\overset{+}{N}H_2Ac + Im^-$$

(Ion pair)

$$K_A \updownarrow \pm H^+ \quad \pm H^+ \updownarrow K_I$$

RNHAc   ImH

say, the rate determining step is the dissociation of the ion pair $R\overset{+}{N}H_2Ac.Im^-$ to solvent-separated ions. If the dissociation step, k_2, is rate determining in the forward direction, encounter of the ions with rate constant k_{-2} is rate determining in the reverse direction. The reactions of 1-acetylimidazole with nucleophilic reagents reflect the fact that expulsion of an amine anion from an amide in aqueous solution is difficult or impossible without general acid–base catalysis. That these reactions occur at all with 1-acetylimidazole is largely a consequence of the much greater stability of the imidazole anion compared with anions of other amines. A substituent effect has been noted[897c] on the reaction rate of the aminolysis of 1-acyl-3,5-dimethylpyrazoles, which have been used as acylating agents[897g].

In the reaction of 4,5-disubstituted imidazoles, containing at least one aryl group, with phenylisocyanate the products are 4,5-disubstituted imidazole-2-carboxanilides and 2-phenylimidazo[1,2-c]hydantoins (p. 114). The latter compounds react with aniline to yield 2-carboxamidoimidazoles [208c], presumably via an aminolysis reaction. The dissociation constants for imidazole-1-carboxanilides are lowered by an increase in polarity of the medium, and by the introduction of electronegative substituents into the phenyl nucleus[925].

Most azolides react readily with alcohols, phenols and thiols. When 1-acetylimidazole reacts with weakly acidic thiols[917] there is a change in rate determining step with increasing imidazole buffer concentration.

$$RS^- + \overset{\overset{O}{\parallel}}{\underset{}{C}}\!\!{}_{Im} \underset{k_{-1}}{\overset{k_1}{\rightleftharpoons}} RS-\overset{O^-}{\underset{|}{\overset{|}{C}}}-Im \underset{K_1}{\overset{+H^+}{\rightleftharpoons}} RS-\overset{OH}{\underset{|}{\overset{|}{C}}}-Im$$

$$\downarrow k_2[ImH^+] \qquad\qquad \downarrow k_3[ImH^+]$$

$$-\overset{}{\underset{O}{\overset{|}{C}}}-SR + ImH$$

The imidazole-catalysed step is assigned to the breakdown of a tetra-hedral addition intermediate, and the other step to attack of thiol anion on free acetylimidazole. However, the concurrent uncatalysed reaction of acetylimidazolium ion with thiol anion does not undergo this change in rate determining step, nor does the phosphate catalysed reaction. Furthermore, phosphate can induce a rate increase under conditions in which imidazole has no further effect. In view of these observations it is suggested[917] that the results imply bimolecular reactions in aqueous solution proceeding by concurrent, independent paths which are not in equilibrium with each other with respect to transport processes. The pH-independent process may be formulated as the reaction of thiol anion with acetylimidazolium cation.

A study has also been made[921-2] of the kinetics of acetyl transfer from 1-acetylimidazolium[921] and 1-acetylimidazole[922] to phenols. The rate constants are in close agreement with those obtained for acetyl transfer to thiols. Furthermore, the acetyl transfer reactions of 1-acetyl-3-methylimidazolium chloride proceed with rate constants which, over a range of 10^6, do not vary significantly from those calculated[261] for the corresponding reactions of 1-acetylimidazolium ion. This demonstrates that the quaternary salt is a satisfactory non-dissociating model for 1-acetyl-imidazolium and provides evidence that many reactions of acetylimidazolium with reagents of type HY are in fact acid catalysed reactions of Y⁻.

As might be expected diazolides of carbonic acid, e.g. N,N'-carbonyl-diimidazole, exhibit very high reactivity towards nucleophilic reagents because of the electron attraction exerted from both sides of the carbonyl function. Although water hydrolyses N,N'-carbonyldiimidazole at room temperature within seconds with vigorous evolution of CO_2, the compound is crystalline and much more easily handled than phosgene which has similar reactivity. When equimolar proportions of N,N'-carbonyldiimidazole and carboxylic acids are mixed in inert solvents nearly quantitative yields of

$$\text{Im-CO-Im} + RCOOH \longrightarrow \left[\begin{array}{c} O \\ \parallel \\ C-R \\ O \\ \parallel \\ C-N \\ \parallel \\ O \end{array} \right]$$

$$RCO-Im + CO_2 + ImH$$

imidazolides are formed[904b]. Isotope studies on the reaction[904h] demonstrate a two-step mechanism in which the carboxylic acid reacts first with the N,N'-carbonyldiazole, which then yields, by subsequent intermolecular transacylation with the azole liberated in the first step, the corresponding azolide.

It should be mentioned at this stage that although this discussion places chief emphasis on imidazolides, other azolides react in a similar manner, albeit at different rates.

It is the transacylation reactions of imidazolides which give them their most useful synthetic applications, and lead to the formation of esters [904c,i,j], amides[904b,926], hydrazides[904b], hydroxamic acids[904b], peptides[656,927-8], anhydrides[904d,e], acyl halides[904k,n], C-acyltriphenylphosphine alkylenes[929], diacyl peroxides[150m,904f], and esters of peracids[904g,930]. Amides are formed by reaction of the free carboxylic acid with N,N'-carbonyldiimidazole in a 1 : 1 molar ratio, and then adding an equimolar quantity of amine after the evolution of CO_2 has subsided. Although the use of the diimidazolide in peptide synthesis gives favourable yields, and the mild reaction conditions normally

$RCO_2R' + CO_2 + 2ImH$ · $RCONR'R'' + CO_2 + 2ImH$

RCOOH
R'OH

(i) RCOOH
(ii) R'R''NH

$$\text{Im-CO-Im} \quad \begin{array}{c} \text{(i) RCOOH} \\ \overrightarrow{\text{(ii) NH}_2\text{NH}_2} \end{array} \quad RCONHNH_2 + CO_2 + 2ImH$$

(i) AcNHCH$_2$COOH
(ii) H$_2$NCHRCO$_2$Et

(i) RCOOH
(ii) NH$_2$OH

AcNHCH$_2$CONHCHRCO$_2$Et + CO$_2$ + 2ImH

$RCONHOH + CO_2 + 2ImH$

(Im = 1-imidazolyl)

minimise racemisation, considerable racemisation was found[928] to accompany the reaction of *N*-trifluoroacetylamino acids with *N,N'*-carbonyldiimidazole. The formation of anhydrides is synthetically useful only if the equilibrium is in favour of the anhydride. This can be achieved if the imidazolide reacts at room temperature with two moles

$$R'CO_2H \;+\; \overset{N}{\underset{}{\text{N}}}\!-\!COR \;\rightleftharpoons\; R'\!-\!CO\!-\!O\!-\!CO\!-\!R \;+\; ImH$$

$$R'CO_2H \;+\; R\!-\!CO\!-\!O\!-\!CO\!-\!R' \;\rightleftharpoons\; R'\!-\!CO\!-\!O\!-\!CO\!-\!R' \;+\; RCO_2H$$

$$ImH \;+\; RCO_2H \;\longrightarrow\; RCO_2^- \;\;\overset{N-H}{\underset{N}{\bigg\langle\!\!+\!\!\bigg\rangle}}$$

of carboxylic acid in such a way that an insoluble salt is formed with the imidazole which is produced. When *N*-trifluoroacetyl- or *N*-trichloro-acetyl-imidazoles are employed[999] it is possible to prepare symmetrical anhydrides even from carboxylic acids which do not form insoluble salts. Under these conditions there is irreversible formation of insoluble imidazole trifluoro- or trichloro-acetates.

Although normal transacylation of indole with 1-acetylimidazole leads to 1-acetylindole, when the reaction conditions are changed (e.g. slow addition of indole in acetic anhydride to a solution of 1-acetylimidazole in acetic anhydride at 125 °C) the major product is (165), formed by electrophilic attack at C-3 of indole by 1,3-diacetylimidazolium[926]. Hydrolysis of these compounds leads to aldehydes. The reaction is enhanced by the use of trifluoroacetyl derivatives[926].

(165)

When 1-acetylimidazole reacts with t-butanol in the presence of sodium t-butoxide the sodium enolate of 1-acetoacetylimidazole is formed[150m] in a reaction similar to a Claisen ester condensation. From this reaction it appears that in the case of tertiary alcohols, the alcoholysis is slow enough to allow a carbon–carbon condensation to occur when the α-carbon has an acidic hydrogen.

In addition to transacylations, imidazolides have a number of other synthetic applications. 1-Formylimidazole is a very effective formylating agent[150m], decomposing above its melting point (53-5 °C) practically quantitatively with evolution of carbon monoxide to form the unsubstituted heterocycle.

It had been noted[577a] that the attempted reduction with $LiAlH_4$ of 1-acetyl-3,5-dimethyl-1H-1,2,4-triazole to the ethyl compound was unsuccessful, resulting in deacetylation. This reaction, closely related to transacylation, results in the formation of aldehydes[642a,b,897b,904l]. The use of organomagnesium compounds yields ketones[904m]. Yields in both cases are high (50-95%) and the reactions involve nucleophilic attack at the carbonyl carbon atom.

N,N'-Carbonyldiimidazole has a number of reactions which are analogous to those of phosgene. In inert solvents ureas are formed with primary amines[150f], while secondary amines form imidazole-1-carboxamides[150g,323h]. When the reaction conditions are controlled, a 1 : 1 reaction between primary amine and imidazolide can result in the formation of imidazole-N-carboxamides which dissociate at low temperatures to form isocyanates.

With alcohols and phenols, N,N'-carbonyldiimidazole forms diesters of carbonic acid. The reaction is catalysed so effectively by catalytic amounts of sodium ethoxide or imidazolyl sodium that it occurs in most cases exothermically even at room temperature[150f].

Extremely unreactive imino derivatives are formed in the reaction of triphenylphosphine imides with N,N'-carbonyldiimidazole[932]. The reaction products dissolve in dilute mineral acids from which they are precipitated with ammonia; concentrated HCl cleaves them to imidazole, CO_2 and amine; they form stable salts with carboxylic acids, and stable 1 : 1 complexes with heavy metal salts.

Monoesters of phosphoric acid and H_3PO_4 react with N,N'-carbonyldiimidazole at room temperature to eliminate CO_2 and form imidazolium salts of imidazolides[933a,934b] ((ImH_2^+)($ROPO_2Im^-$)). Solutions of these salts are effective phosphorylating agents[935].

The reaction of N,N'-carbonyldiimidazole with water to give carbon dioxide has been recommended[936] for the microanalytical estimation of carbon and hydrogen in organic compounds. The reaction is reported to be quantitative, and an improvement over the use of PCl_5, and can be used with as little as 0.1 mg of compound.

The reactions of *N,N'*-thiocarbonyldiimidazole[150j,649a,897d,937-9] and the corresponding pyrazolides[897d,923] are very similar to those of the carbonyl compounds, although the thio compounds are less reactive. In a study[923] of the kinetics of ethanolysis of *N,N'*-thiocarbonyldipyrazoles the reaction was found to be first order with rate constants strongly dependent on substituents. A good linear correlation was observed between log k and Hammett σ-values with ρ = 4.8, indicating a S_N2 type of reaction mechanism. Use has been made of *N,N'*-thiocarbonyldiimidazole (166) in the one step synthesis of thiocarboxamides by reaction with dithiocarboxylic acids[649a,897d,923]. *N*-Thioacyl-1,2,4-triazoles are powerful thioacylating agents[939].

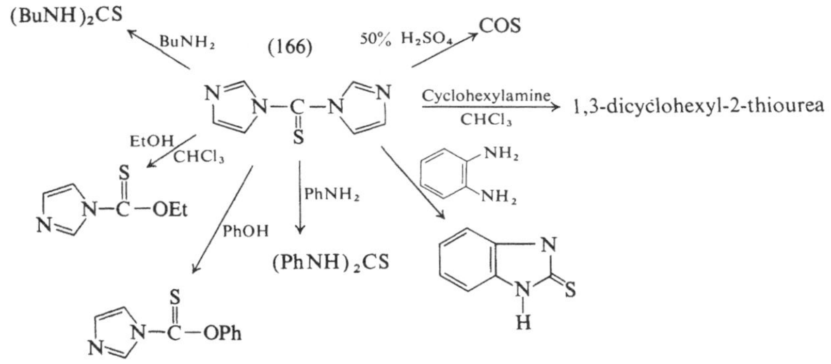

The use of 1-acetylimidazole as an effective terminating agent in solid phase peptide synthesis has been reported[940].

Deacylation of azolides is a process which may be accomplished very readily. The *N*-acyl group may be removed in weakly acid or basic media [675b], in a moist atmosphere[577a], with boiling water, or by heating [675b]. Electron-withdrawing substituents assist the hydrolysis process. Although reaction with ethyl iodide in a sealed tube failed to replace the acetyl group of 1-acetyl-3,5-dimethyl-1*H*-1,2,4-triazole by ethyl, the compound was readily hydrolysed by moisture and cleaved by picric acid. The 3,5-diphenyl compound proved to be more stable[577a].

1-Acyl-3,5-dimethylpyrazoles are amphoteric compounds in which the acidic properties predominate[898g]. They are weaker bases than the parent pyrazoles, but form picrates and perchlorates[307a]. The compounds exhibit a bathochromic shift in ultraviolet absorption in basic medium ascribed to the ionic form of the sodium salt formed at the hetero nitrogen. In acidic medium the absorption band shifts towards shorter wavelengths through formation of a cationic nucleus[898g].

1,1'-Carbonyldipyrazole reacts with cobalt(II) chloride in acetone to form a blue complex which hydrolyses to 2,2-(1'-pyrazolyl)propane[941].

Rearrangement reactions of *N*-acylazoles to the more stable *N*-acyl isomers have been discussed earlier for pyrazoles (p. 113) and 1,2,3-triazoles (p. 117), and mention has been made of the rearrangement of *N*-acyl-amino-1,2,4-triazoles to the amino-acyl compounds (p. 120). In the latter reaction a recent study[912a] suggests that the acetylation of 3-amino-1,2,4-triazole (167) to 1-acetyl-3-acetylamino-1*H*-1,2,4-triazole (169) involves intermediate formation of 1-acetyl-5-amino-1*H*-1,2,4-triazole (168), indicating that a migration of the *N*-acetyl group has occurred. Deuteration experiments point to a mechanism involving a rapid exchange of the acetyl group of (168) with CD_3COCl followed by a slow acetylation of the amino group[912a]. A base-influenced intermolecular process has been proposed to account for the rearrangement of 1-acyl-1*H*- to 2-acyl-2*H*-1,2,3-triazoles[907b].

The intermediacy of 1-benzoylpyrazoles in the oxidative debenzylation of 1-benzylpyrazoles has been noted[898i].

Unlike most 1-carbamoylimidazoles[150g,657] (which have the same reactions as isocyanates without the attendant health hazards), *N,N'*-diphenylcarbamoylimidazole is very stable, being unaffected by ethanol at 80°C or aniline at 100°C[657].

At 60-130°C, 2-acyl-5-substituted tetrazoles undergo ring fission with loss of N_2 and formation of 1,3,4-oxadiazoles (p. 165)[153a].

Pyrazole-1-carboxylic acids, which may be considered as *N*-substituted carbamic acids, are extremely unstable and spontaneously undergo decarboxylation. Only a few examples of these compounds are known as free acids[896b,942], although the derivatives (esters, amides, etc.) are quite stable and widely known.

9.1.2 *C-Acyl compounds (aldehydes and ketones)*

Table A.9 lists melting and boiling points for some compounds of this

Table 24. Infrared and ultraviolet[a] data for some imidazole aldehydes and ketones

Substituents	$\nu_{C=O}$/cm^{-1}	λ_{max}/nm (log ϵ) neutral species	λ_{max}/nm (log ϵ) anion	λ_{max}/nm (log ϵ) cation
2-CHO[483,803]	1700[803], 1680[483]	285 (4.10)[803]		
1-Me-2-CHO[375,803]	1680			
1-PhCH$_2$-2-CHO[375,803]	1675			
2-Ac[372]	1684, 1673			
4-Me-2-Ac[869,977b]	1678[977b], 1680[869]	290 (4.1) (in CH$_3$OH)[869]		
1-Me-2-COPh[981a]	1640	296 (4.15), 259 (3.89)		
1-Ph-2-COPh[981a]	1645	295 (4.04), 259 (4.06)		
1-PhCH$_2$-2-COPh[981a]	1640	297 (3.91), 259 (3.72)		
4-Me-2-COPh[869]	1620	255 (4.18), 335 (4.08) (in CH$_3$OH)		
4-CHO[983]		257 (4.08)	281 (4.23)	
1-Me-5-Ac[1003]	1670	255 (4.18)		238 (3.86)
1-Me-2-EtS-5-Ac[1003]	1660	290 (4.19)		235 (4.09)
				286 (4.06)

(a) Spectra of neutral species are for solutions in 95% ethanol except where indicated.

type and their derivatives. As a general rule throughout the *C*-acylazole series, substitution on nitrogen lowers melting points and increases solubility in solvents of low polarity.

The introduction of an acyl group at a pyrazole ring carbon atom gives rise to a bathochromic shift in the ultraviolet spectrum of the order of 25–40 nm[118]. Infrared studies of 4-acylpyrazoles indicate that the compounds exist in the ketone forms[1004]. 4-Acyl-3-methyl-1-phenylpyrazol-5-ones, though, appear to exist in both keto and enol forms[321h,1006]. The colourless keto form (170) crystallises from polar solvents, while the yellow enol form, (171) or (172), is obtained from non-polar solvents. It is believed[1006] that enolisation occurs in the side chain, so that structure (171) is preferred.

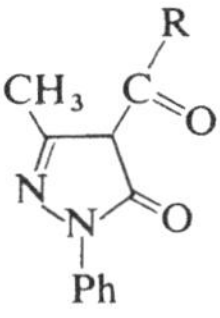

$$CH_3 \quad C{=}O \qquad CH_3 \quad C{-}OH \qquad CH_3 \quad C{=}O$$

(170) 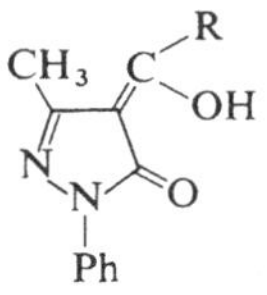   (171) 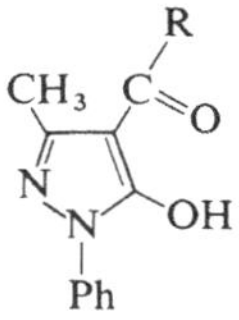   (172)

Acyl-substituted imidazoles also have distinctive infrared and ultraviolet spectra (see table 24), but no systematic study has been made of these compounds, probably because their synthesis is difficult. Ultraviolet spectroscopic evidence supports the aldehyde structure for imidazole-4-carboxaldehyde[124]. It is evident[372,977b] that an acyl group at C-2 of *N*-unsubstituted imidazoles makes the compounds more volatile and increases their solubility in non-polar solvents, perhaps[977a] because of intramolecular hydrogen bonding between the carbonyl oxygen and the NH proton competing with the normal intermolecular N–H‑‑N bonds which are formed in non-polar solvents, and in the liquid and crystalline states.

Dipole moments have been calculated for some imidazole-2-carboxaldehydes using dielectric data in benzene[108].

In meso-ionic compounds of type (173) the stability is attributed [1007] to effective delocalisation, by the 5-acyl group, of the exocyclic negative charge on the oxygen.

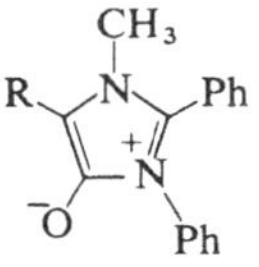

$$R = COCH_3;\ COC_2H_5$$

(173)

There are some interesting structural characteristics of the 1,2,4-triazole-3-carboxaldehydes. Although the *N*-substituted compounds have normal aldehyde properties, 5-aryl-1,2,4-triazole-3-carboxaldehydes (174)

dimerise in the solid state to carbonyl-free hemiaminals[995b,c] (175).
The free aldehydes exist in solution, but the dimers reform on removal
of the solvent. If the 5-substituent is not an aryl group, then the hemiaminal becomes less stable (*viz.* 1,2,4-triazole-3-carboxaldehyde exists
mainly as the aldehyde form in the solid state). Although the mass
spectra of the hemiaminals are consistent with normal aldehyde
properties, reducing properties are not demonstrated, and neither the
Cannizzaro nor benzoin reactions occur[995b,c]. Parallels may be drawn
with the behaviour of benzimidazole-2-carboxaldehydes[1008]. There is
considerable resemblance among the properties of 1,2,4-triazole-3-
carboxaldehyde and imidazole-4- and 2-carboxaldehyde. The observation
[970a] that the latter compound exists as a hydrate in acidic aqueous
solution might point to a similar property in the triazole. Recent studies
[1001] show that 1-substituted tetrazole-5-carboxaldehydes also form
stable hydrates and hemiaminals.

Although pyrazole aldehydes and ketones usually show typical
carbonyl reactions[59f,897f,943a,c,946,962c,975a] there is some evidence
that steric hindrance in the formation of functional derivatives may be
different from that observed in benzenoid compounds. While normal
carbonyl derivatives are formed by 4-acetyl-5-methyl-1,3-diphenylpyrazole
[962c], the bisulphite addition compounds of pyrazole-4-carboxaldehydes
could not be isolated[943a,c], even though 5-methyl-1-phenylpyrazole-3-
carboxaldehyde reacted normally with bisulphite. It is interesting to note
that 1,5-dimethylpyrazole-3-carboxaldehyde failed to produce a bisulphite
adduct[943a] and in general pyrazole-4-carboxaldehydes do not react
[943a,c]. In view of these results, steric factors can hardly be a problem.
In addition to forming acetals, hydrazones, oximes and carbonyl-
ammonias, pyrazole-3-carboxaldehydes react with hydrazine to form
azines[975a]. The base-catalysed condensations of pyrazole-3,4-dialdehydes
with ketones have been examined[944b,c]. The imidazole aldehydes
readily form acetals[483], oximes[375,983,1009] and other carbonyl
derivatives. The kinetics and mechanism of the reaction of imidazole-4-
carboxaldehyde with hydroxylamine have been studied[983]. With hydrazine hydrate, imidazole-4,5-dicarboxaldehydes give imidazo[4,5-*d*] pyrid-
azines (176)[969e]. With thio-, sulphinyl- and sulphonyl-diacetic acid
methyl esters benzimidazoles result[976a].

Both 1,2,3-[745b,895a,743d,984b] and 1,2,4-triazole aldehydes[175a,
995b,c,d,996] form normal carbonyl derivatives, although 1,2,3-triazole-

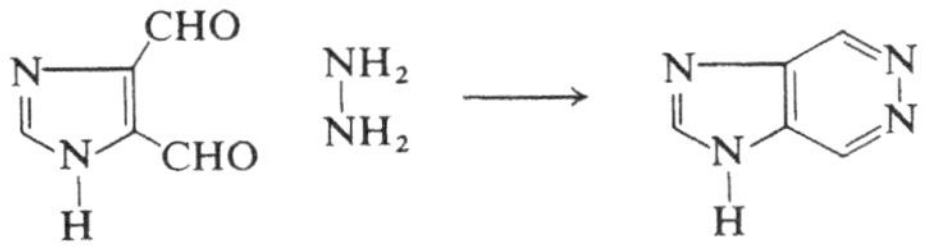

(176)

4-carboxaldehyde is reported[895a] to have more usual aldehyde properties (cf. pyrazole-3-carboxaldehyde) than 1,2,4-triazole-3-carbox-aldehyde. The oximes of 1,5-diaryl-1*H*-1,2,4-triazole-3-carboxaldehydes cannot be hydrolysed to the free aldehydes[175a]. Within the 1,2,4-triazole series there are considerable reactivity differences between the *N*-unsubstituted and *N*-substituted analogues. Whereas 3-benzoyl-1-phenyl-1*H*-1,2,4-triazole readily forms oximes[995e], analogues unsubstituted on nitrogen failed to form oximes even though dinitrophenylhydrazones formed normally. Only 3-benzoyl-5-phenyl-1,2,4-triazole appeared to form a thiosemicarbazone[995e]. Two oximes can be prepared from 3-benzoyl-1-phenyl-1*H*-1,2,4-triazole. Compounds of structures (177) (preferred over (178)) and (179) have been isolated and studied with regard to their behaviour in Beckmann rearrangement reactions[995e]. Whereas (177) reacted smoothly under a variety of Beckmann rearrangement conditions, (179) resisted rearrangement, perhaps because of extensive decomposition caused by the low migratory aptitude of the triazolyl group. The oximes are reduced with LiAlH$_4$ to amines[995e].

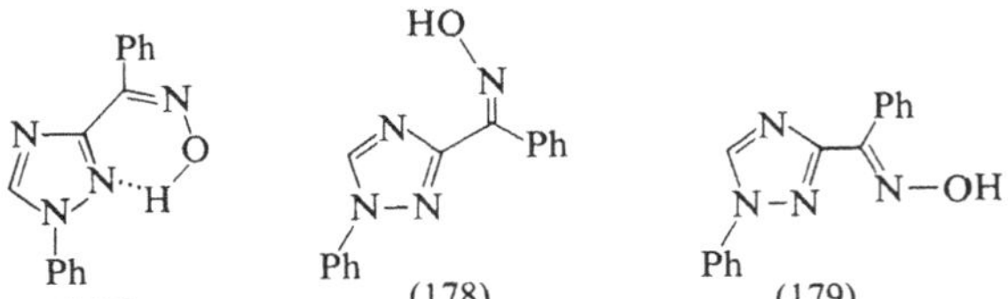

(177)   (178)   (179)

Normal carbonyl addition reactions occur with 5-acyl-tetrazoles[999-1001]. Although 3-methyltetrazole-5-carboxaldehyde is known as the semicarbazone[999], the unsubstituted aldehyde does not appear to have been reported. 5-Aroyltetrazoles react with hydroxylamine to form *syn*-oximes which undergo the Beckmann rearrangement[1000]. The Schmidt rearrangement also occurs[1000]. The *syn*- and *anti*-isomers of some 4-substituted pyrazolyl oximes have been prepared[945d] and their reactivity has been studied under a variety of Beckmann rearrangement conditions. In most cases involving ketoximes, amides are formed with phosphorus pentachloride, tosyl chloride, polyphosphoric acid, or concentrated sulphuric acid, although the last named was found to be a poor reagent. The *anti*-oxime (180) was an exception, reacting in polyphosphoric acid to yield 1-phenylpyrazole-4-carboxamide (181). Mixtures of

amides were formed in most cases, suggesting isomerisation of the *syn*-
and *anti*-oximes under the reaction conditions. Pyrazolyl aldoximes were
found to undergo elimination under Beckmann conditions[945d].

(180)   (181)

The effects of alkali on azole aldehydes have proved of considerable
interest. While 1-phenylpyrazole-4-carboxaldehyde undergoes a Cannizzaro
reaction after 65 hours in 50% potassium hydroxide[59a], pyrazole-3-
carboxaldehyde and its 5-carbethoxy derivative dissolve without alteration
in solutions of sodium hydroxide, from which they are precipitated on
neutralisation[77a,895a]. 3,3-Diphenylisopyrazole-5-carboxaldehyde is
insoluble in 2M sodium hydroxide but resinifies slowly (faster in 5M
sodium hydroxide). Even in 33% ethanolic sodium hydroxide no
Cannizzaro reaction is evident[77a].

Both imidazole-4-carboxaldehyde[20b] and the 2-carboxaldehyde[803]
resist the Cannizzaro reaction, but the 1-substituted compounds react
normally[803,972]. Similar considerations apply to the Perkin[20b] and
benzoin[803] reactions. 5-Nitro-1-substituted imidazole-2-aldehydes
condense readily with alkyl aryl ketones[1012]. In the 1,2,3-triazole
series, the 1-substituted-4-carboxaldehydes give a positive Cannizzaro
reaction[77a,895a]; 1-methyl-1*H*-1,2,3-triazole-4-carboxaldehyde is
converted into the carboxylic acid after only 4 hours in 2M sodium
hydroxide[77a]. These 1-substituted triazole aldehydes dissolve to some
extent in water to give neutral solutions, and are only sparingly soluble
in very dilute alkaline solution. On the other hand 1,2,3-triazole-4-
carboxaldehyde is acidic and may be recovered unchanged from alkaline
solution on neutralisation[77a,895a]. Presumably a sodium salt is formed
(*cf.* pyrazole-3-carboxaldehyde and imidazole-2- and -4-carboxaldehydes)
in which, on analogy with 3-acetylpyrazole[1013], the carbonyl function
is not involved in the salt formation. 1-Substituted 1*H*-1,2,4-triazole-3-
carboxaldehydes react readily in alkaline solution to yield Cannizzaro
products[175a], but *N*-unsubstituted-1,2,4-triazole-3-carboxaldehydes tend
to form stable hemiaminals which fail to undergo the dismutation
reaction[995b,c]. The formyl group of 1-substituted tetrazole-5-aldehydes
is removed by alkali treatment[1001]. Solubility in aqueous alkali is a
feature of the properties of 5-aroyltetrazoles[1000], which are also
degraded to benzoic acid by hot 60% sulphuric acid.

Pyrazole-4-carboxaldehydes condense normally with malonic acid[59a,
d,g] to yield acrylic acids. An examination has been made[59g] of the

influence of alkyl and aryl substituents on the reactivity of the aldehyde group in condensations with aniline, malonic acid, acetophenone, and *o*-phenylenediamine (formation of benzimidazoles). In the case of anil formation only the 3-methyl-1-phenyl- and 5-methyl-1-phenyl-pyrazole-4-carboxaldehydes reacted. With malononitrile, 1-methyl- and 1,3,5-trimethyl-pyrazole-4-carboxaldehydes condensed to form the 4-(2,2-dicyanovinyl)pyrazoles[946]. With 1-methyl-4-lithiopyrazole, 1-methyl-pyrazole-4-carboxaldehyde forms the secondary alcohol[74e].

Although *N*-unsubstituted imidazole aldehydes normally resist condensation reactions (see p. 197) nitroalkanes add to the carbonyl function [1009,1014-15]. Phenyllithium reacts in tetrahydrofuran with 2-acylimidazoles to form tertiary alcohols[1016b], and phosphorus oxychloride in dimethylaniline converts the ketones into diazafulvenes (182)[1016b].

(182) (183)

2-Phenyl-2*H*-1,2,3-triazole-4-carboxaldehyde undergoes an acyloin condensation to form the 'triazoin' (183)[1017], which is readily oxidised to the 'triazil'. This diketone forms a bisphenylhydrazone, and reacts with *o*-phenylenediamine to yield quinoxaline derivatives. 1,2,4-Triazole aldehydes substituted on nitrogen take part in Döbner and Perkin reactions[175a], but hemiaminal formation prevents similar reactivity in the *N*-unsubstituted analogues[995b,c]. With phenylmagnesium bromide in tetrahydrofuran, 5-aroyltetrazoles yield tertiary alcohols[1000].

Aldehyde groups on pyrazole are oxidised by silver oxide[77a,895b, 944a] or alkaline permanganate[59d,g,967b], while acetylpyrazoles are also smoothly oxidised to the corresponding carboxylic acids with sodium hypoiodite[955] or nitric acid[501,962d,e]. When 4-acetyl-3,5-dimethylpyrazole is treated with hypobromite oxidation does not occur; rather displacement of the acetyl function by bromine results[898g]. Although imidazole aldehydes and ketones are oxidised by usual reagents, 1-benzylimidazole-2-carboxaldehyde is reported to be fairly resistant to oxidation with selenium dioxide[483]. Silver oxide is successful in converting 1-phenyl-1*H*-1,2,3-triazole-4-carboxaldehyde into the corresponding carboxylic acid[895a].

A variety of reduction techniques has been employed with the *C*-acylazoles. Clemmensen reduction converted 4-acetylpyrazole into 4-ethylpyrazole[126d,898g], and 5-acetyl-1-methylimidazole could be reduced under acid conditions weaker than those generally required for Clemmensen reductions[979]. With zinc dust in acetic acid a mixture of the secondary alcohol and the ethyl compound was obtained, while borohydride reduction gave the alcohol only [979]. Wolff–Kischner reduction has been used for acylimidazoles [372] and 5-aroyltetrazoles [1000]. Catalytic hydrogenation of 3-aroyl-1,2,4-triazoles yields the secondary alcohols [995e]. Photochemical reduction of 4-aroyl-1,2,3-triazoles [1018] and 5-aroyltetrazoles [1000] gives quantitative yields of the pinacols.

Acyl groups at C-4 of pyrazole activate 5-halogens to such an extent that reaction with hydrazine results in the formation of pyrazolo-pyrazoles (p. 134) [62a]. Replacement of an acyl group by halogen has been noted in the reaction of 4-benzoyl-5-chloro-3-methyl-1-phenyl-pyrazole with phosphorus pentachloride [551d], and in the bromination of 1-phenyl-1*H*-1,2,3-triazole-4-carboxaldehyde [77a].

When 1,4-diphenyl-1*H*-1,2,3-triazole-5-carboxaldehyde is treated with sodium methoxide, quantitative nucleophilic deformylation occurs [745b]. This reaction is analogous to the well known reaction of chloral and also takes place in the presence of excess hydrazine. It is interesting to note that the isomeric 1,5-diphenyl-1*H*-1,2,3-triazole-4-carboxaldehyde was resistant to the decarbonylation.

9.1.3 *Carboxylic acids and derivatives*

Melting (and boiling) points for some azole-carboxylic acids and derivatives are listed in table A.10.

With the exception of tetrazole-5-carboxylic acid which decomposes spontaneously to tetrazole and is only known as its salts and acid derivatives [80d], the *C*-carboxylic acids are stable, crystalline compounds. Solubility characteristics are as expected. The water solubility increases with increasing numbers of carboxyl groups and decreases in the presence of aryl substituents, particularly when these are on nitrogen.

Salts are formed with a number of metals. For example, pyrazole-carboxylic acids form silver [447], lead and copper [1094], barium [1d, 1029b] and calcium [1029b] salts which are insoluble, or sparingly soluble in water. Pyrazole-tricarboxylic acids yield acidic salts with alkali metals, and even in the presence of mineral acids these are stable [962b].

There has been little systematic study of dissociation constants of azole-carboxylic acids (table A.10), although some have been examined in comparison with differently-substituted heterocycles [262,1095-6]. Although some of the properties of pyrazole-3-carboxylic acid are accounted for by an amphoteric structure such as (184), the pK_a value of 3.84 [1095] (*cf.* benzoic acid pK_a = 4.21) suggests that the acidic properties are far greater than those of α-amino acids. The presence of a nitro-substituent in the pyrazole ring greatly increases the acid character e.g. 3-methyl-4-nitropyrazole-5-carboxylic acid pK_a = 2.26. Basic pK values measured [262,1096] for a series of imidazole-carboxylic acids

(184)   (185)

indicate that the carboxyl functions exert a considerable base-weakening effect on the parent heterocycle (pK_a values: imidazole, 6.95; imidazole-4-carboxylic acid, 6.08; imidazole-4,5-dicarboxylic acid, 2.93; 2-methyl-imidazole-4,5-dicarboxylic acid, 4.25; 2-phenylimidazole-4,5-dicarboxylic acid, 3.00; ethylimidazole-4-carboxylate, 3.66). In view of this marked effect on basicity it is a little surprising that a 5-bromo-substituent in imidazole-4-carboxylic acids is not activated to nucleophilic displacement by KCN or Na_2SO_3[374f]. Zwitterionic structures of type (185) have been suggested for imidazole-4-mono- and -4,5-di-carboxylic acids[262]. Zwitterionic tautomers have also been implicated in decarboxylation mechanisms.

Azole-carboxylic acids form normal acid derivatives: pyrazole-carboxylic esters[70a,944a,967b], amides[426c,943a,962a,1029a,b,1046,1098], chlorides[943a,b,c,944a,959a,1043], hydrazides[432c,1097] and hydrox-amic acids[1059b]; imidazole carboxylic esters[274d,374f,453,570,907b, 915c,973a,e,1053a,1055-7], amides[375,570,1021b,1058b,c], chlorides [1100a] and hydrazides[374d,1101b]; 1,2,3-triazole carboxylic esters [915e,1063a,1064e,1066a], amides[943d,1055c,1063a,1066a,1081,1099], chlorides[943d,959b,1074,1081,1099]; 1,2,4-triazole carboxylic esters [175a,363c,910,1082,1091a], amides[175a 910,1087a], and hydrazides [175a]; tetrazole carboxylic esters[999,1016a,1064b], amides[80c,d,276a], chlorides[999], and hydrazides[1093f]. The reaction of 1-methyl-imidazole-4,5-dicarboxylic acid with phosphorus pentachloride is anomalous, giving 1-methyl-4-trichloromethylimidazole-5-carboxylic acid chloride [1102].

Decarboxylation of pyrazole-carboxylic acids occurs at elevated temperatures[62c,66g,77d,988h], the rate of reaction depending on the position of the carboxyl function. In general, pyrazole-4-carboxylic acids are decarboxylated least readily, and under certain conditions partial decarboxylation of polycarboxypyrazoles can be achieved. Pyrazole-5-carboxylic acids are reported[2a,f,1047b] to undergo decomposition at rather lower temperatures. Thus, when 1-phenylpyrazole-3,5-dicarboxylic acid is heated for a short period above its melting point the sole product is 1-phenylpyrazole-3-carboxylic acid[1029a]. This observation is not entirely in agreement with a recent study[988b] which suggests that the ease of decarboxylation parallels the acid strength (*viz.* pK_a values: 1-phenylpyrazole-5-carboxylic acid, 5.18; 1-phenylpyrazole-4-carboxylic acid, 5.93; 1-phenylpyrazole-3-carboxylic acid, 6.01).

Imidazole-carboxylic acids are also decarboxylated by heating them above their melting points[208c,284] often in the presence of copper-chromium oxide catalyst[274c,974a]. It is possible to remove one molecule of carbon dioxide from imidazole-4,5-dicarboxylic acid by heating the monoanilide[274c]. The kinetics and mechanism of decarboxylation of some 5-amino-imidazole-4-carboxylic acids have been studied[1103] and interpreted in terms of first order decarboxylation of both the acid and the zwitterion, whereas the anion is stable and is not decarboxylated. There is also evidence of hydrogen ion catalysis at low pH values, and of general base or nucleophilic catalysis. Transition metal ions decrease the rate of decarboxylation, presumably by complex formation. The generation of ylides by decarboxylation of imidazolium-2- and -5-carboxylates is considered to proceed through the zwitterionic tautomers[1050a,b]. Treatment with iodine in alkaline solution brings about the decarboxylation of imidazole-4-carboxylic acid with the production of 2,4,5-triiodo-imidazole[452b].

Triazole-[1066a] and tetrazole-carboxylic acids[80d,999,1104] are also decarboxylated by heating, the latter so readily that the silver salt of tetrazole-5-carboxylic acid is converted into tetrazole in boiling water [1104], while hydrolysis of the ethyl ester also gives tetrazole as the only product.

Carboxylic acid and ester functions in azoles are reduced by lithium aluminium hydride to alcohols[175a,999,1105] and by controlled potential reduction to aldehydes[970a]. In a study of the reduction of pyrazole-3,5-dicarboxylic ester, Greco and Pellegrini[944a] found that whereas diisobutylaluminium hydride in toluene formed the dialcohol, sodium or lithium aluminium hydrides in tetrahydrofuran gave methyl 4-formylpyrazole-3-carboxylate. The corresponding dithio-ester was reduced with lithium aluminium hydride or with Raney nickel to the dialcohol in low yield. The attempted reduction with lithium aluminium hydride of 4-cyano-2-methyl-2*H*-1,2,3-triazole-5-carboxamide resulted only in destruction of the heterocycle[1106].

Carboxylic acid functions do not appear to deactivate the 4-position in pyrazoles to electrophilic substitution, as bromination[66d,965b] and nitration[856] occur normally in 1-phenyl-pyrazole-3,5-dicarboxylic acid. Bromination also occurs at the 4-position of 3-methyl-1-phenylpyrazole-5-carboxylic acid[59h]. Halogenation of carboxyl-substituted imidazoles similarly does not appear to be inhibited[374a,f] (see Chapter 3).

Many of the reactions of the derivatives of the azole-carboxylic acids parallel those of the benzene series. Hydrazides are converted into acid azides[175a,432c], and then via urethanes into amines[432c]. The azides can take part in a Curtius rearrangement[1097]. Pyrazole hydroxamic acids undergo the Lossen rearrangement[1059b]. Methyl 1-phenyl-pyrazole-4-carboxylate reacts with phenylmagnesium bromide to form the tertiary alcohol along with some of the phenylated 4-benzoyl-product [59j] (p. 137). The acid chloride of 2-methyltetrazole-5-carboxylic acid takes part in a Friedel–Crafts reaction with benzene and aluminium chloride[999].

In the Hofmann reaction, pyrazole- and imidazole-carboxylic acid amides do not act normally. Although pyrazole-4-carboxamide reacts with potassium hypobromite to yield the amine, compounds having the 4-position unsubstituted are brominated by the reagent[426c]. In the cases of 4-bromo-3-methyl- and 3-methyl-4-nitro-pyrazole-5-carboxamides, a large excess of reagent led to the formation of the corresponding amines[426c]. Imidazole-4-carboxamides do not take part in the Hofmann reaction[374f].

The pyrolysis of esters of pyrazolecarboxylic acids takes place with cleavage of the carboalkoxy group and alkylation, although with large alkyl functions the alkylation is not observed[900g].

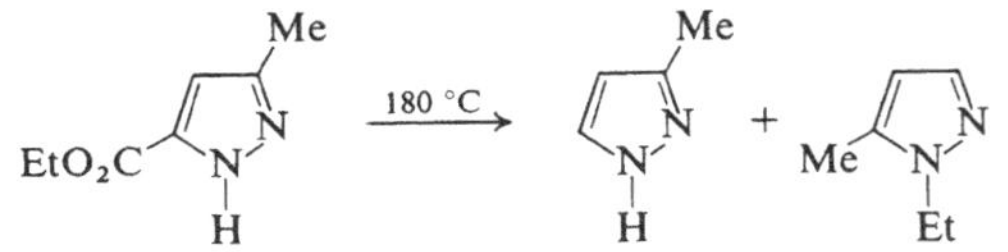

Pyrazole-4-[954] and -5-carboxylic acid[59h,954] silver salts take part in the Hunsdiecker reaction, but silver 3-carboxylates are either unaffected, or brominated at a vacant ring position[954].

1,2,4-Triazole-carboxylic acid hydrazides have been converted via the sulphonyl hydrazides, or by the Kalb–Gross method into the corresponding aldehydes[175a].

Although *o*-aminocarboxamides normally form 4-hydroxy-1,2,3-triazines when treated with nitrous acid there are some notable exceptions, namely 5-aminoimidazole-4-carboxamide[1107] and 5-aminopyrazole-4-carboxamide[1108] which yield fairly stable diazocarboxamides, which can be cyclised to triazines under acid, basic or neutral aqueous conditions[1108]. The action of formamide at 215°C cyclises 2-methyl-4-2*H*-1,2,3-triazole-5-carboxamide to 6-hydroxy-8-methyl-8-azapurine[149c].

9.2   *C*- and *N*-alkyl and substituted alkyl groups

9.2.1   *Substituents on ring nitrogen*

N-Alkylazoles usually have lower melting points (and decreased solubility in polar solvents) than the *C*-substituted isomers (e.g. 1-diphenylmethyl-imidazole, m.p. 85–6°C; 2-diphenylmethylimidazole, m.p. 237–8°C; 4-diphenylmethylimidazole, m.p. 179–80°C[111a]).

A number of methods are available for the removal of *N*-alkyl groups. When the substituent is benzyl, reductive debenzylation using sodium in liquid ammonia[77g,149c,363b,372], or catalytic hydrogenolysis[149b, 515c,e,g] are often effective. However, 1-benzyl-4-hydroxymethylpyrazole yields only 1-benzylpyrazole when treated with sodium–liquid ammonia [363b]. In a number of the cases where reductive debenzylation fails oxidative methods are successful. Bromo-substituted 1-benzylpyrazoles are smoothly oxidised by CrO_3 in 60% sulphuric acid to the benzoyl compounds with simultaneous hydrolysis to the pyrazole[898i]. The reaction appears to proceed only in the presence of electron-attracting substituents on the pyrazole nucleus. Thus, 1-benzyl-3,5-dimethylpyrazole is only partially oxidised even under drastic conditions[895b,d]. When tri(1-pyrazolyl)methane is heated with aqueous acetic acid the products are pyrazole (2 moles) and 1-formylpyrazole (1 mole)[52a], while 1-tri-phenylmethylpyrazoles are similarly hydrolysed[569c]. Substituted alkyl groups may also be removed under certain conditions. Imidazolyl-1-alkylhydroxamic acids are cleaved to imidazoles by polyphosphoric acid [1110]. When 1-cyanoethyl-3,5-dimethylpyrazole is boiled a high yield (82%) of 3,5-dimethylpyrazole results[898d]. In similar fashion, 1-hydroxy-methylpyrazoles decompose with loss of formaldehyde when heated above their melting points[74a]. 1-Aminomethylpyrazoles are decomposed by alkalies[74a].

Although methyl groups attached to ring nitrogen atoms are not normally regarded as 'reactive', they occasionally appear to exhibit this property. At high pD values (~ 13.5) 1,3,5-trimethyltetrazolium chlorides slowly exchange N-methyl protons for deuterium. No such exchange occurs in the 1,4,5-trimethyl isomers, for in both cases the N-methyl protons are relatively inert[1111a]. When C-5 of pyrazole is substituted it is possible for a 1-methyl substituent to be lithiated by butyllithium [1112], even though metalation normally occurs at a ring carbon. Similar observations have been made[484,840,981c] in the cases of 1-benzylimidazole and 1-phenyl-5-methyl-1H-1,2,3-triazole which may metalate on the exocyclic methylene function to some extent. A reaction between 1-methylpyrazole and diethyl oxalate in the presence of sodium ethoxide is reported[1113] to yield ethyl pyrazolyl-1-pyruvate.

There has been a number of reports of groups migrating from nitrogen to ring carbon atoms of azoles. The melting of 1-triphenylmethyl-imidazoles forms the 2-substituted isomers[569b], unless the 2-position is blocked, whereupon 4-substitution occurs[569c]. Methyl, ethyl, or isopropyl substituents at C-4 seem to inhibit the rearrangement[569c]. The observation[569a] that 1-triphenylmethyl-4,5-diphenylimidazole forms a blue melt is accounted for by the formation of radicals which are stabilised on rearrangement to the 2-position.

The vapour phase methylation of imidazoles with methanol in the presence of alumina[1114k] is accompanied by catalytic rearrangement of N-methyl groups to C-4 or C-5, followed by remethylation at the 1-position. It is likely that a free radical mechanism is also implicated in this instance. The thermal rearrangement of 1-alkylimidazoles to the 2-alkylisomers has been employed[54d,443,1115a] as a synthetic procedure, although a number of side products are obtained. Recently[981b] this reaction has been modified to produce high yields of 2-substituted imidazoles. The process is temperature dependent and appears to be an uncatalysed, concerted, intramolecular reaction which may follow a [1,5]-sigmatropic rearrangement sequence as proposed for similar migrations in 1-nitropyrazoles[1116a] and 1-alkylpyrroles[1117]. When the 2-position is blocked the N-alkyl substituent is lost. Some 'cracking' of propyl and butyl substituents is observed but the reaction has considerable versatility[981b].

9.2.2 Substituents on ring carbon
The effects of alkyl groups on pK_a values have been noted earlier (p. 24). There has also been discussion of the ultraviolet spectra of alkylazoles (p. 11). The mass spectra of alkyl-substituted 1,2,4-triazoles have been examined[1118a].

Alkyl groups, especially in the 3- and 5-positions of pyrazole[735c, 965c], are readily oxidised by permanganate to the corresponding carboxylic acids[60d,77g,363b,c,473,735c,898i], and sometimes to ketones[995e]. Although mild oxidation by air of antipyrine yields the aldehyde[1119], aldehydes are normally only approached by oxidation of the hydroxymethyl compounds with selenium dioxide[108,483], lead tetraacetate[976a,1009], or active manganese dioxide[924a,977a,1015]. Nitric acid oxidises 4-hydroxymethylimidazole to a mixture of the acid and aldehyde[20b]. Polyhydroxyalkyl substituents are cleaved to aldehydes by periodate[977c,1120a,1121]. Halomethyl groups are also readily oxidised, or converted via a Sommelet reaction, into carboxyl functions[59a].

As with other heterocyclic systems (pyridine, quinoline, thiazole) a methyl group joined to a carbon adjacent to a ring nitrogen atom (3- or 5-methylpyrazoles, 2-methylimidazole) is 'active'. Thus, the chlorination of 3,4,5-trimethylpyrazole results in polysubstitution up to .CCl_3 of the 3- and 5-methyl substituents, but the 4-methyl group is unaffected[77d]. A methyl group at C-2 of imidazole is reactive because of electron deficiency, but those in the 4- and 5-positions are not, even when the ring is cationic. 2-Methylimidazole condenses with benzaldehyde to form 2-styrylimidazole[1114i]. When a strongly electron withdrawing substituent (NO_2 or SO_3H) is present at C-4 or C-5, a 2-methyl substituent fails to condense with benzaldehyde[498,1011], but a 4- or 5-methyl group now becomes reactive[564,1115b,1122]. Protonation of the heterocyclic ring renders a 2-alkyl substituent particularly prone to proton loss, leading to the formation of stable anhydro bases such as the diazafulvenes (186)[1123].

(186)

The free radical bromination of 4,5-dimethyl-2-phenyl-2H-1,2,3-triazole results in substitution on an exocyclic methyl group[457]. In an unusual transformation, a 4-hydroxy-5-methyl-4H-1,2,4-triazole is acetoxylated at the methyl function by acetic anhydride[1124b]. Methyl groups at C-5 of tetrazole are active too, particularly in tetrazolium compounds. Deuterium exchange for C-methyl protons occurs readily in 1,4,5- and 1,3,5-trimethyltetrazolium iodides[1111a]. Metalation occurs readily at a 5-methyl group, as does condensation in basic medium with diethyl oxalate[1125a]. In the latter reaction the pyruvate ester initially formed is reconverted into the methyltetrazole by saponification with 10% aqueous sodium hydroxide.

The influence of alkyl substituents in pyrazole-4-aldehydes has been examined with regard to the reactivity of the formyl function in carbonyl condensations[59g] indicating that only the 3- and 5-methyl compounds form anils.

Halomethyl substituents undergo nucleophilic substitution reactions with the formation of the corresponding alcohols[457], amines[457] and cyanides[457,746]. Although 1-substituted-2-chloromethylimidazoles were reported[1126] to react with cyanide by an S_N1 reaction at C-5, a later study[746] maintains that the products of both nuclear and side-chain substitution (p. 138) are formed. Reaction with hydrazine at 200 °C converts the cyanomethyl substituents into methyl[1127a]. 3-Hydroxy-methyl-1,2,4-triazole may be transformed into the chloromethyl compound which condenses with *N*-formylaminomalonic ester to produce a histidine analogue[363c].

9.3   *C*- and *N*-alkenyl and -alkynyl groups

Table A.11 lists some representative compounds of this class. There has been a number of studies of alkenyl[898k,902b,1122,1148a,b,1152], and alkynyl[1133,1144a,1146a,1153] azoles in recent years, particularly with regard to the polymerisation of some of these compounds. Alkynyl compounds are as yet not well known but there have been spectroscopic studies[1142,1153] of some representative members. The ultraviolet spectra of 3-alkenylpyrazoles show a bathochromic shift of about 54 nm, whereas the 5-alkenyl isomers exhibit a hypsochromic shift of about 9 nm[902b]. In this respect, the compounds resemble phenylpyrazoles which are known to exhibit a steric effect in the 5-position. The use of n.m.r. spectroscopy has been of value in determining the regiochemistry of *cis* and *trans* isomers of alkenylazoles[902b,1148].

Oxidation of the unsaturated side-chain usually leads to carboxylic acids[564,1129,1130,1147], and sometimes to aldehydes[1145], while addition reactions are reported, in some cases, to occur normally. With bromine 3-methyl-1-(*p*-nitrophenyl)-5-propenylpyrazole undergoes substitution at C-4 and addition to the side chain[902b]. Normal addition also occurs in the bromination of 1-phenyl-4-vinylpyrazole[945c], and in the catalytic hydrogenation of 5-ethynylpyrazoles[1019g,1147]. On other occasions, however, addition reactions appear to be either hindered or prevented. 2-Vinylimidazole resisted epoxidation and hydroboration[981c]. Furthermore, 1-vinylimidazole does not react at low temperatures with hydrogen halides[1154a], but forms hydrochlorides when treated with hydrogen chloride in carbon tetrachloride at 25 °C. Only partial halogenation can be achieved, perhaps via initial complex formation, followed by addition at the vinylic double bond[1154a]. Catalytic hydrogenation with Raney nickel in alcohol can, however, be accomplished[1155]. This reactivity is paralleled in *N*-vinylbenzotriazoles which only form complexes with hydrogen halides, and undergo electrophilic addition with chlorine or bromine only with difficulty[1154b].

The steric effect of an alkenyl substituent at C-3 or C-5 of pyrazole results in the formation of mainly 1,3-disubstituted product on arylation with fluorodinitrobenzene[902b].

Many vinyl-pyrazoles[1128], -imidazoles[1137,1139], -1,2,3-triazoles [1055b], and -tetrazoles[1021i] are polymerised by free radical initiators.

Pyrazole-4-acrylophenones (187) condense with phenylhydrazine in acetic acid to form pyrazolylpyrazolines (188)[59g], while bipyrazolyls (190) result from the interaction of 3-alkynylpyrazole (189) and diazo-alkanes[1129].

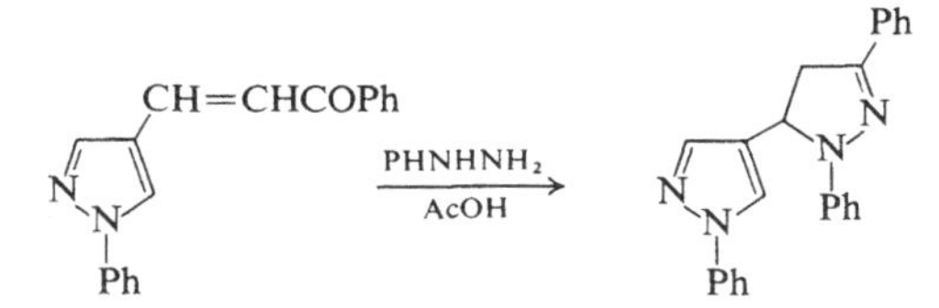

CH=CHCOPh

$\xrightarrow[\text{AcOH}]{\text{PHNHNH}_2}$

(187)

(188)

C≡CH

$\xrightarrow{\text{R'CHN}_2}$

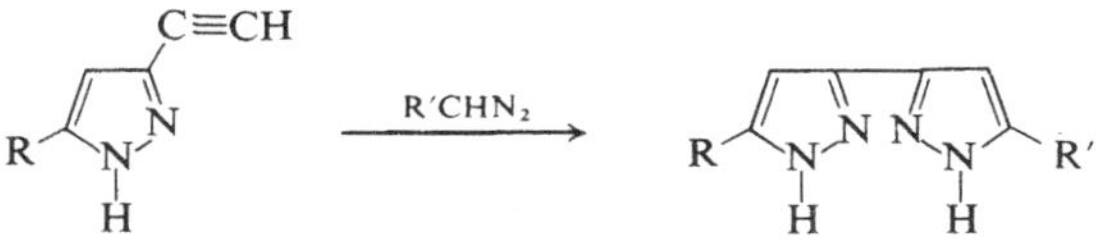

(189)

(190)

Azole-acrylic[945c,1152] and -acetylenic[1146a] acids are decarboxy-lated normally. Methyl-α-bromo-β-(1-phenyl-2-imidazolyl)acrylate is dehydrobrominated to the acetylenic acid[1146a]. Ethynyl alcohols of pyrazole are readily cleaved in the presence of alkali to ethynylpyrazoles and acetone[1133]. The decompositon of 5-nitro-2-(4-*trans*-styryl)-1-vinylimidazoles by sodium hydroxide has been studied in detail[1156].

The photofragmentation of spiropyrazoles has been studied[1157].

Thermal rearrangement of 1-vinylimidazole gives a 7 : 1 mixture of the 2- and 4-vinylisomers[981b]. Similar treatment of 1-allylimidazole yields equal quantities of 2- and 4-allylimidazoles[981b]. In the latter case, a Cope-type rearrangement may account for the high yield of 4-allylimidazole.

9.4   Amino, hydrazino, nitramino, nitrosamino, diazonium, diazo, azo and azoxy groups

The ionisation constants (p. 22), spectra (pp. 11, 16, 22) and tautomerism (pp. 11, 16, 22) of some of these derivatives have been discussed earlier. Some of the reactions of aminoazoles with acylating agents have already been noted (pp. 119-21), particularly with regard to the orientation of acylation in the nucleus or at the exocyclic nitrogen. There has been discussion of condensation and cyclisation reactions (p. 128) of the amino compounds. Nucleophilic displacement reactions of diazonium salts have been described (pp. 138-9, 142).

9.4.1   Amino groups

Physical data and references to some representative aminoazoles are listed in table A.12.

Coburn[912b] has suggested that some of the reactions of 2-amino-imidazole with picryl halides can be explained by its existence in dimethylformamide as the imino tautomer, but conclusions drawn from chemical reactions do not provide a sound basis for assignment of structure in a tautomeric system. In spite of some support for the imino structure[1202a-c], theoretical[157,315] and spectroscopic studies[590d, 1204-5] indicate that 5-aminotetrazole probably exists in the amino form under most conditions. Furthermore 3- (or 5-) aminopyrazoles have been shown by ultraviolet spectroscopy to exist in the amino form under the conditions of measurement[1206].

The 3-aminopyrazoles are only weakly basic compounds and not at all acidic. A variety of salts and double salts are formed by these compounds[330b]. An amino group in the 3- or 5-position renders C-4 very reactive to electrophilic substitution[1172]. 4-Aminopyrazoles react as normal aromatic amines.

Aminoimidazoles have been little studied because of synthetic difficulties in the case of the 2-amino compounds, and because 4-aminoimidazoles[58] (and 1-substituted-5-aminoimidazoles[1176a]) are unstable. 4-Aminoimidazole-5-carboxamides are important as biosynthetic precursors of the purines. Whereas 2-aminoimidazole is a monoacidic base which forms salts with acids, the 4-isomer is a diacidic base and forms dihydrochlorides and dipicrates. 4-Aminoimidazoles resemble true aromatic amines[274a]. However, even the salts of the 4-amino compounds are relatively unstable in aqueous solution at room temperature, and the base itself undergoes deamination and ring fission under the conditions of Van Slyke amino-nitrogen determination[418b].

Amino-1,2,3- and -1,2,4-triazoles, and aminotetrazoles react as normal aromatic amines. The first named are basic substances which form well-defined salts, but which retain, nevertheless, weakly acidic properties if the hydrogen atom attached to a ring nitrogen is unsubstituted. It has been demonstrated[1186] that 4-amino-1H-1,2,3-triazole is stable and is not involved in ring–chain tautomerism.

Numerous studies have been reported of the alkylation[902d,1005b, 1203a] and acylation[912a,1007,1184,1203b,1208-9] of aminoazoles. Much of this work has already been discussed (pp. 82, 100, 119). Some of Michaelis' early work[1005b] has recently been repeated[902d], showing that whereas methylation of 3-dimethylamino-1-phenylpyrazole gives the trimethylamino quaternary salt, methylation of 5-dimethylamino-1-phenylpyrazole occurs at N-2. Heating aminopyrazolium salts can cause migration of an alkyl group from nuclear to extranuclear nitrogen[1005b].

It is appropriate at this stage to consider the properties of alkyl-substituted aminotetrazoles. 1-Substituted 5-alkylaminotetrazoles (191) are very weakly basic solids which do not react readily with phenyliso-thiocyanate and which only form salts and picrates under anhydrous conditions[123]. The isomeric 2-substituted 5-alkylaminotetrazoles (192) are even weaker bases and only react sluggishly with phenylisothiocyanate [581c]. In contrast 5-dialkylaminotetrazoles (193) are solid acids[83d], and 5-monoalkylaminotetrazoles (194) are acidic compounds with few basic characteristics. The salts are usually unstable and sometimes explosive[1194a]. The 1,4-dialkyl-5-iminotetrazoles are hygroscopic, basic liquids (or low melting solids) which immediately form phenylthiourea derivatives, and also form mineral acid salts and picrates stable against hydrolysis[123].

Schiffs bases are formed normally from 3-, 4- and 5-aminopyrazoles [172f,330k,431c,552e,1158,1168,1210-1], 3-, 4- and 5-amino-1,2,4-triazoles[83f,268a,c,g,630a,913a,1212a,b,1213a-4], and 1- and 5-amino-tetrazoles[83e,591b,1215]. 2-Aminoimidazoles do not condense readily with aromatic aldehydes.

5-Aminopyrazoles condense with active methylene compounds to form addition products, which may cyclise if the pyrazole is not substituted on nitrogen (pp. 128-30). Condensation reactions involving carbonyl compounds also include the reactions of 3,5-diaminopyrazole with α-diketones[1171a], and the Paal–Knorr condensation of acetylacetone with 5-aminopyrazoles[1158]. In the latter reaction normal condensation produces 1-(pyrazol-5-yl)-2,5-dimethylpyrroles, but steric factors inhibit the reaction in some 1,4-disubstituted-5-aminopyrazoles. Acrylic esters [1019e] and acetoacetic esters[1211] condense with 3-aminopyrazoles

at the amino function. Although anils (195) form with acetoacetic esters in xylene at 20°C, at reflux temperatures with 20% excess ester the products are crotonates (196)[1211]. Michaelis and Schäfer[330j] have reported that 1-phenyl-3-methyl-5-aminopyrazole condenses with benz-aldehyde in the 4-position. Condensations with picryl halides occur to some extent at the exocyclic amino group in 3-methylpyrazole[490b]. In the case of 2-aminoimidazole, though, picryl fluoride is reported to react at a ring nitrogen atom in dimethylformamide, whereas picryl chloride reacts at the amino function[912b]. There has been little study of condensation reactions of aminoimidazoles because of the instability

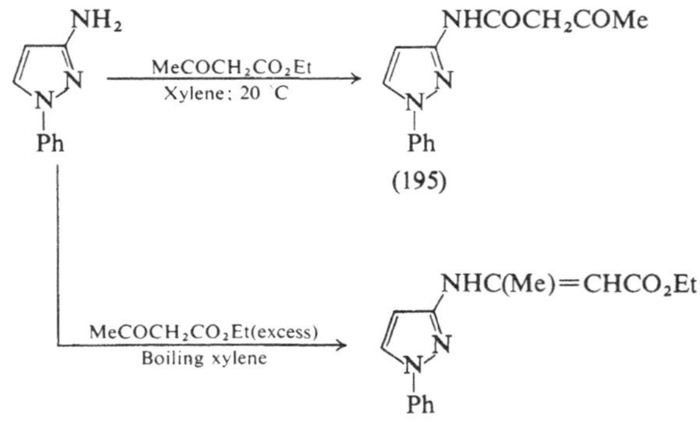

(195)

(196)

of the 4-amino compounds and the lack of aromatic amino character of the 2-isomers. In polyphosphoric acid 5-amino-1-imidazolyl-acetic and -propionic acids form internal lactams[1110]. 2,5-Diphenyl-4-aminoimidazole reacts with the corresponding 4-nitrosoimidazole to produce a dark blue product tentatively assigned the structure (197)[414a].

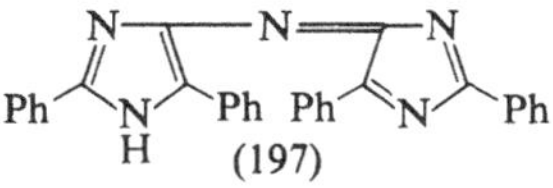

(197)

The cyclisations of 4-amino-1*H*-1,2,3-triazole-5-carboxamides to azapurines have been studied[149b,c,1106]. The 4,5-diamino-1*H*-1,2,3-triazoles show many of the common reactions of *ortho*-diamines[688a]. With nitrosobenzene in aqueous alkali, 2-phenyl-4,5-diamino-2*H*-1,2,3-triazole yields the corresponding 5-benzeneazo compound[1188].

Amino-1,2,4-triazoles condense readily with formaldehyde[630a], with isocyanates[911a], with methyl propiolate[729b], and by Michaelis' addition with acrylonitrile or ethyl acrylate[630a]. Benzoyl chloride induces the cyclisation of 3,4-diamino-4*H*-1,2,4-triazoles to triazolo-[3,2-*c*]-triazoles[676b].

Adducts form in the reactions of 5-aminotetrazole with acrylonitrile [640], and with α,β-unsaturated esters. While methyl acrylate only adds at a ring nitrogen, methyl propiolate yields a mixture of two isomeric 1 : 2 adducts[1019f].

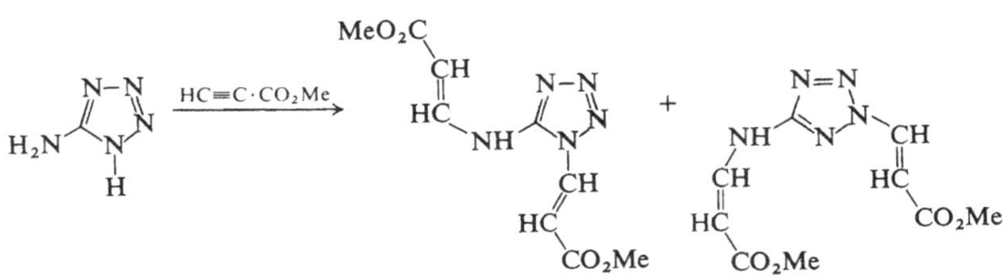

Little attention has been paid to the oxidation of aminoazoles (for ring-modifying oxidation see p. 155). 5-Amino-3-methyl-1-phenyl-pyrazole gives 3,3'-dimethyl-1,1'-diphenyl-5,5'-azopyrazole with dilute aqueous permanganate[745c]. Aminotetrazoles also form azo compounds on oxidation[83e]. When 4-amino-3,5-diphenyl-4H-1,2,4-triazole is subjected to lead tetraacetate treatment the major product is benzo-nitrile[1216]. Oxidation of 1-amino-imidazolium salts with aqueous bromine yields tetrazenes[1181].

The use of peroxides in acid medium with aminopyrazoles may lead to hydroxylamino products[330j,k], or to nitro compounds[700]. Although 3-aminopyrazole is degraded by peroxytrifluoroacetic acid, 3-amino-1-phenyl- and 5-amino-1-phenyl-pyrazoles form the corresponding nitro compounds. 5-Amino-1-methylpyrazole is anomalous in that the product is 1-methyl-5-nitropyrazole 2-oxide[700]. The same reagent converts 3-amino-1,2,4-triazole into the 3-nitro product[766].

The diazotisation of aminoazoles has received considerable attention. Aminopyrazoles yield typical diazonium salts with isoamyl nitrite[432c] or with nitrous acid[70c,83e,321l,330k,392m,419e,431c,552e,771,957b, 962e,1019b,1167,1193c,1217-18]. The diazotisation of 3-aminopyrazoles may sometimes be complicated by the concurrent formation of oximino compounds[1172], and if insufficient nitrous acid is used triazenes may result[330j].

Diazotisation of aminoimidazoles is rather more complicated. Under the conditions of the Van Slyke amino-nitrogen determination 4-amino-imidazole decomposes[418b]. However, 4-amino-5-methylimidazole differs from the parent compound in that it forms a stable diazonium salt[274a, 452a]. Normal diazotisation also occurs with 4-aminoimidazole-5-carboxamide[1176a,c,1219] and with 4-amino-2,5-diphenylimidazole[414a]. The difficulty of diazotising 2-aminoimidazole has been surmounted by carrying out the reaction in strongly acidic medium[607f,947g,1114m, 1220].

Amino-1,2,3-[1188] and -1,2,4-triazoles[431b,601d,608a,c,617i,688b, 766,995a,1193b,1222] form diazonium salts under standard conditions. Diazotisation in the presence of an oxyacid yields a more stable salt [431b]. Under mild diazotisation conditions nitrosamines may be isolated [430c,913b,1190,1193b]. On treatment with nitrous acid 4-amino-3,5-dihydroxy-4H-1,2,4-triazoles (urazines) lose the amino function[1199]. Although, as with the triazoles, mild diazotisation conditions may lead to the formation of nitrosamines[1093f], 5-aminotetrazoles form diazonium salts, in the manner of normal aromatic amines[83e,h,1047c, 1223]. Unlike the 1,5-isomer, 5-amino-2-phenyltetrazole fails to form a diazonium salt with nitrous acid[1093f].

Although nitrosation has not been reported for amino-imidazoles, and -1,2,3-triazoles, there have been a number of references to the formation of nitrosamino-1,2,4-triazoles[268g,430c,533,913b,1093a,1190,1193b, 1224a] and tetrazoles[83e,1093f,1226a]. Furthermore, 5-amino-1-methyl-pyrazole-4-carboxamide reacts with sodium nitrite and hydrochloric acid at 0 °C to form 1-methyl-5-nitrosaminopyrazole-4-carboxylic acid[771]. The nitrosamines obtained[913b] from 5-alkyl-3-amino-1-aryl-1*H*-1,2,4-triazoles appear to be more stable than those unsubstituted on ring nitrogen[1193b]. Stollé and Dietrich[83f] obtained both mono- and di-nitrosamines from guanazole when amyl nitrite in ethanol was used as the nitrosating agent. 5-Amino-1-phenyl-, -1-benzyl-, -2-benzyl- and 5-benzylamino-tetrazole are all reported[1224a,1226a] to form nitrosamines.

The action of concentrated sulphuric acid on 3-substituted -5-amino-1*H*-1,2,4-triazole nitrates produces nitramines[514a]. In the event that the 3-substituent is aryl, nitraminotriazoles are not formed – rather there is nitration of the aromatic ring.

In some cases deamination occurs with nitrous acid treatment[1181, 1227a]. Deamination has also been noted[515e] during attempts to acetylate 1,4-dibenzyl-5-iminotetrazoline by treatment with excess hot acetic anhydride. A product was obtained in which the =NH had apparently been replaced by =O, possibly by acetolysis. The product proved to be identical with that produced[25c] by warming the nitroso compound with acetic acid.

A number of rearrangements are found with aminoazoles. The thermal migration of alkyl groups in aminopyrazolium salts[1005b] has already been mentioned. Dimroth rearrangements of imidazoles[1007], 1,2,3-triazoles[1141d,1228], 1,2,4-triazoles[911b], and tetrazoles have already been discussed (pp. 159-63). 1-Benzamido-4-phenyl-1*H*-1,2,3-triazole rearranges in boiling concentrated hydrochloric acid to 5-amino-4-phenyl-1,2,3-triazole[988c].

In view of its importance as the biosynthetic precursor of the purines [1229], there has been considerable effort expended in studies of 4-amino-imidazole-5-carboxamide (and its corresponding nucleotides and nucleo-sides)[495b,1103a,b,1176c].

9.4.2 Hydrazino groups

Hydrazinoazoles are generally crystalline compounds which readily form crystalline salts with mineral acids. The compounds form hydrazones[83e, 269c 390b,430c,514a,591b,1191a,c,1212a,1213a,1226b] which may be reduced to the corresponding amino compounds[390b]. The kinetics of bromination of tetrazolyl hydrazones have been studied[1226b]. Oxidation with mercuric oxide of 5-arylhydrazinopyrazoles leads to the azo compounds[735g]. Similar oxidation behaviour has been noted in the imidazole series[1232]. When hydrazones of 1,2,4-triazoles are oxidised using lead tetraacetate, some acetoxylation occurs accompanied by oxidative cyclisation to triazolotriazoles[916b,1201].

When β-dicarbonyl compounds condense with 3-hydrazino-1,2,4-triazoles, pyrazolyl-substituted products are often obtained[268g,1191a,c].

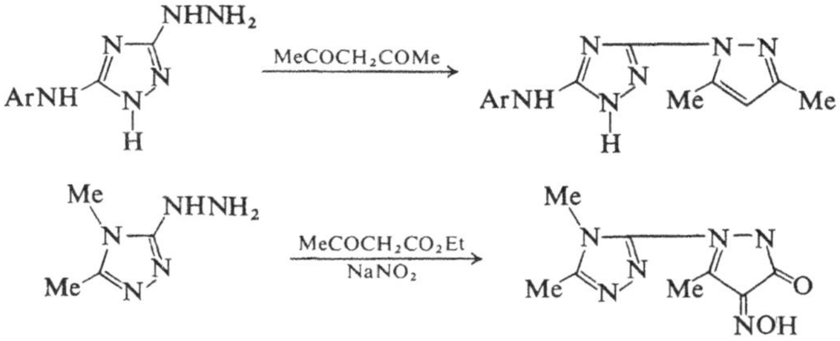

Reaction with nitrous acid converts a hydrazino function into an azido group[83e,f,269c,1213a]. Silver acetate removes the hydrazino function from 1-methyl-5-hydrazino-3-nitro-1*H*-1,2,4-triazole[269c]. The benzidine rearrangement of 2-phenylhydrazinoimidazoles has been studied[274b,c, 390a,673]. When the *para* position in the benzene ring is blocked by bromo or sulphonic acid groups no rearrangement occurs, and 2-amino-imidazole is obtained[274b,c,673]. An *ortho* substituent does not affect the rearrangement[412a]. A semidine rearrangement of 3,5-dialkyl-4-phenylhydrazinopyrazole has been noted[330k].

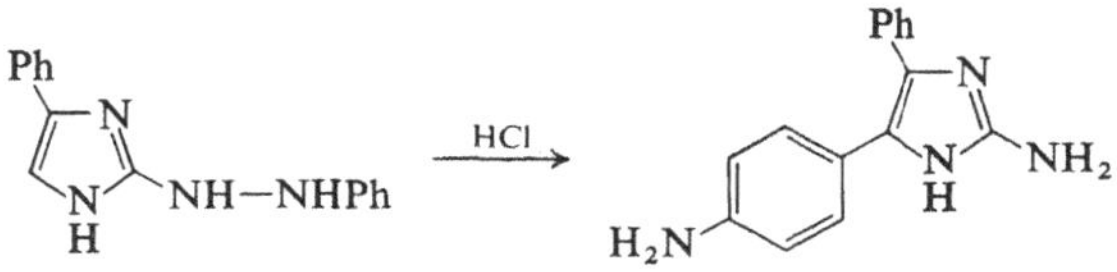

9 4.3 Nitramines
Very little study has been made of these compounds. The sole examples of nitramino compounds appear to be in the 1,2,4-triazole series (see table A.13). Zinc dust and acetic acid reduce the nitramines to the hydrazines[514a].

9.4.4 Nitrosamines
Although a reference has appeared to a nitrosaminopyrazole[771], only nitrosamino-1,2,4-triazoles[83e,f,268g,430c,913b,1193b,1224a] and -tetrazoles[83e,591b,916c,1226a] appear to be at all common. Apparently 5-alkyl-4-aryl-3-nitrosamino-4*H*-1,2,4-triazoles are more stable than those unsubstituted on ring nitrogen[913b]. Although stable at 20°C in lower concentrations, aqueous hydrochloric acid in greater than 25% concentration decomposes the nitrosamines. Depending on the acid concentration the product may be the amine, or a mixture of chlorotriazole and triazene. The nitrosamines give a positive Liebermann reaction[913b],

and may be reduced to hydrazines[83f,430c,913b]. The neutral compounds dissolve in aqueous alkali and they are stable to boiling[913b]. Coupling reactions occur only slowly in acid medium. Concentrated sulphuric or nitric acids convert nitrosamines into diazonium salts[1193b]. The explosive hydrate of 5-nitrosamino-3-phenyl-1,2,4-triazole reacts violently with HI, and with HBr at 0°C forms the 5-bromo compound [430c]. Guanazole forms a mono- and a di-nitroso amine. The former loses nitrous acid when warmed with acetic acid and rearranges to the monodiazonium salt with concentrated hydrochloric acid. Reduction of the dinitroso compound forms a monohydrazino amine[83f].

The reactions of nitrosaminotetrazoles are similar to those of the 1,2,4-triazoles. However 1-phenyl-5-nitrosaminotetrazole appears to be fairly stable to heat. Heating it under reflux with benzene induces a Gomberg-Bachman phenylation, with the compound perhaps reacting as the tautomeric diazohydroxide[916c]. The photolysis of 4-benzylnitrosamino-4H-1,2,4-triazoles has been studied[1224a].

Butler and co-workers have suggested[1202e], on the basis of previous observations[913b,1226a], that the principle factor which decides between a primary nitrosamine and a diazonium salt being formed seems to be the presence or absence of a labile hydrogen atom adjacent to the primary amino group in the starting material. The presence of such a hydrogen atom inhibits nitrosamine formation possibly because it favours the formation of a reactive diazotate form.

Table A.14 lists some compounds of this type.

9.4.5 Diazonium salts
Pyrazole-4-diazonium salts have exceptional thermal stability[330k,392g, 1193a]. Thus the salt prepared from 4-amino-3,5-dimethylpyrazole is recovered unchanged after heating for three hours at 100°C in the presence of five equivalents of hydrochloric acid[1193a]. Even after two days heating with three equivalents of hydrochloric acid the salt is not completely decomposed. A kinetic study of the decomposition reaction has established that the reaction is not unimolecular, even though the decomposition of pyrazole-4-diazonium chloride is[1193a]. It has been demonstrated that 5-amino-1-methylpyrazole-4-carboxamide also forms a stable diazonium salt[771]. Imidazole diazonium salts are not very well known. They may often only be prepared in strongly acid media[607f, 772,947g,1114m,1220], and are commonly unstable[1021d,1176c,1234-5]. Although the diazonium salt formed from the corresponding nucleotide is unusually unstable[1234-5], that formed from 4-aminoimidazole-5-carboxamide has a life of more than two years[1107,1176c], the unusual stability being attributed to resonance stabilisation of the salt. Diazonium salts of 1,2,3- and 1,2,4-triazoles are reasonably stable, but 3-methyl-1 2,4-triazole-5-diazonium chloride is unstable in the presence of hydro-

chloric acid[431b]. The action of heat[1193b] or hydrochloric acid
[431b] on 3-alkyl-1,2,4-triazole-5-diazonium chlorides results in the
formation of the 5-chloro compounds. When 1,2,4-triazole-3-diazonium
chloride is heated in boiling alcohol the unsubstituted triazole is
produced[688b]. Tetrazole-5-diazonium chloride may be isolated as a
crystalline, explosive solid[1236]. Even the process of preparation of
tetrazole diazonium salts may result in explosion if the concentration of
the solutions rises above 2%[1237], a hazard also attributed to the
1,2,4-triazole series[1192]. When tetrazole-5-diazonium chloride is
thermally decomposed in the presence of ethylene, the products are
nitrogen, methane, allene, propyne and hydrogen cyanide. It has been
proposed[1236] that the diazonium compound produces carbon atoms
which add to ethylene with the formation of the three-carbon fragments.

Nucleophilic substitution reactions of diazonium salts have been
discussed earlier (pp. 138-9, 142).

Pyrazole diazonium salts couple normally with activated aromatic
molecules[330k,392g,431a,431c,432c,957a,1019b,1020c,1193a], and also
with β-dicarbonyl compounds[419e,431c]. When the diazonium salts are
stable, coupling occurs in the imidazole series[414a,452a,1219] with
triazole diazonium salts[83f,139,431b,617i,688b,913b,1188,1193d], and
tetrazole diazonium salts[83h,1047c,1237]. Dyes formed by coupling of
1,2,4-triazole-3-diazonium salts have proved useful for polyester and pol-
acrylonitrile fabrics[601d,608a,c,617e,i]. With sulphamic acid, tetrazolyl-
5-diazonium chloride condenses to form (198) which decomposes to
5-aminotetrazole[1223]. Nitroalkane anions couple with 1H-1,2,3-triazol-
3-yldiazonium salts to form the hydrazones[1238].

Coupling power is lost or reduced when diazonium compounds are
treated with base, which converts them into diazo compounds (199).
The reaction is general for pyrazoles[432c,486,957a], but has received
little attention in the other series of azoles. Perhaps the loss of coupling
power of 3-alkyl-1,2,4-triazole-5-diazonium salts in alkaline solution[431b]

might indicate the formation of diazotriazole. Shevlin[1236] was unable to prepare 5-diazotetrazole in suitable non-aqueous solvents or in the gas phase, owing to the extreme lability of the product.

(199)

There have been a number of references to cyclisation reactions of azole diazonium salts. With excess diazomethane, pyrazole-3-diazonium chloride forms 3-(1-tetrazolyl)pyrazole[1019c]. With a 1 : 1 ratio of diazomethane to pyrazole compound a mixture of the tetrazolylpyrazole and pyrazolo[5,1-c]pyrazole is obtained[432h]. The diazonium salts prepared from 5-amino-1-methylpyrazole-4-carboxamide are readily cyclised to triazines in acid, basic or neutral aqueous solution[771]. Furthermore, the coupling products of pyrazole-3-diazonium chloride with β-ketoacids in buffered aqueous solution, are spontaneously decarboxylated and cyclised to pyrazolotriazines[419e]. The stable diazonium salt formed from 4,5-diamino-2-phenyl-2H-1,2,3-triazole cyclises with sodium acetate to form a N-phenyltriazolotriazole[688a]. Reduction of azole diazonium salts sometimes leads to the formation of hydrazines [419e,430c,1218]. However, reduction of 1,3,5-trimethylpyrazole-4-diazonium chloride with stannous chloride and acid results in the formation of some of the corresponding amine, and product in which the diazonium substituent is lost[392g].

9.4.6 *Diazo compounds*
Unlike pyrazole diazonium salts, the corresponding diazo (meso-ionic) compounds are very unstable[957a]. With hydrogen chloride in chloroform they are reconverted into the diazonium salts[888a,957a]. Coupling reactions can occur[432c,486b,1239], and the diazo group can be replaced by iodine (p. 139) and by azide (p. 137), or removed with methanolic potassium hydroxide or potassium carbonate[957a,b], or by photolysis in aqueous acetone solution[957a]. Farnum and Yates[957a,b] have noted some striking differences in the properties of isomeric diazopyrazoles (200) and (201), whose reactions are summarised in the diagram. The thermal decomposition of 3-benzoyl-5-diazo-4-phenylpyrazole (201) occurs at a much lower temperature (about 105°C) than in the case of the isomeric 4-diazo compound (35) which requires to be heated to 50°C above its melting point (148°C) before any visible decomposition occurs[957a]. When 3-benzyl-4-diazo-5-phenylpyrazole is heated at 100°C it undergoes internal cyclisation with the formation of a pyrazolo[3,4-d]-pyrazole[957a,1239]. Pure 5-diazotetrazole has not been isolated as the

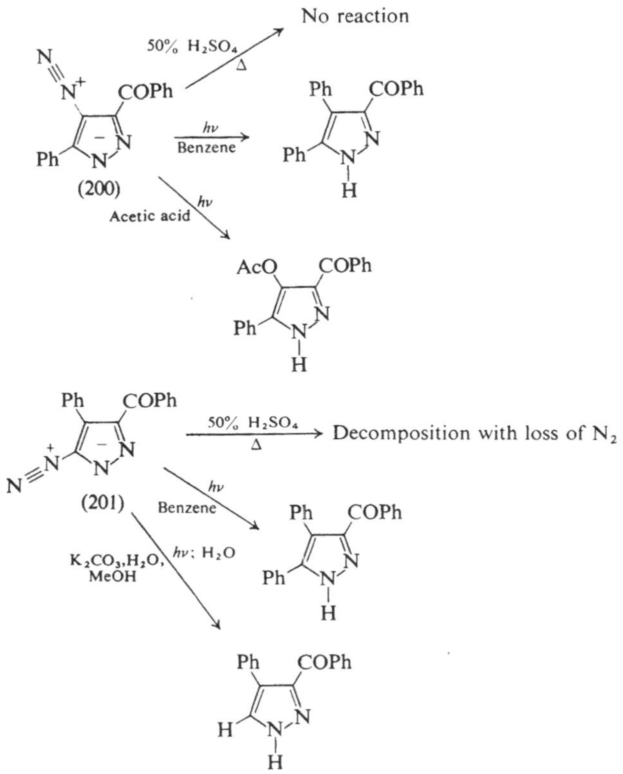

compound decomposes slowly in dilute solution, and explosively in concentrated solution [25a,b,c]. The compound can be characterised by its reduction to 5-tetrazolylhydrazine [25a,b,c], and several coupling products with substituted hydrazines have been isolated [916d,1241]. The meso-ionic diazo derivative prepared from 2-amino-4,5-dicyano-imidazole eliminates nitrogen at $80\,^\circ C$ to form a highly reactive intermediate capable of inserting the 4,5-dicyanoimidazole residue into a number of substrates [1240].

9.4.7 Azo compounds

Of the many azopyrazoles known, the 4-azo compounds are the most common. Only the arylazo derivatives of 5-azopyrazoles are known, and these are obtained by mercuric oxide oxidation of the corresponding hydrazines [735g] or by coupling of pyrazole-5-diazonium salts [330j, 1172]. Arylazoimidazoles are usually prepared by coupling imidazoles with aryldiazonium salts. The products are orange or brown crystalline compounds of high melting point. In aqueous alkaline solution the compounds are highly-coloured yellow, orange or red dyes which allow a rapid and sensitive method for locating imidazoles during chromatography (see refs. 97a, 177 for discussion). The use of azo-1,2,4-triazole dyes has been mentioned earlier (p. 215). The azo function may be reduced to hydrazino under mild conditions using zinc in alkaline medium[330k] or stannous chloride[274c]. Usually, however, when strong reducing agents are used in acid media (red P/HI; $SnCl_2$/HCl; Zn/HOAc) the product is the amine[74b,330k,414a,735h,745c,1168,1217,1242a,b]. Sodium hydrosulphite[957b] and catalytic hydrogenation[1232] also lead to the amine in good yield. Sometimes the use of stannous chloride in hydrochloric acid as reducing agent results in a benzidine-type rearrangement of the intermediate hydrazo product[274c,330k,412d]. Although 2-azoimidazoles are readily reduced to the 2-aminoimidazoles[274c,412d] ring cleavage occurs when 4-arylazoimidazoles are subjected to reducing agents[274c]. When 3-ethylmercaptoazo-4-phenylpyrazole is boiled in methanol the product is 4-phenylpyrazole[70c]. An arylazo substituent may activate the adjacent ring carbon to nucleophilic substitution[1243a]. From a study of the conformations of 2-arylazoimidazoles it has been concluded [1244] that the *s-cis–trans* form (202) is more stable than the *s-trans–trans* form (203). The demonstration of the absence of a hydrogen bond between the N–H of the ring and the N-2′ atom supports this view. When heated 4-arylazo-5-azido-2-phenyl-2*H*-1,2,3-triazoles are cyclised to triazolotriazoles[1188].

(202)   (203)

9.4.8   Azoxy compounds
Only in the pyrazole series does there appear to be any reference to this type of compound[1245].

9.5   Aryl groups
Aryl-substituted azoles are generally stable compounds with the aryl group depressing the basic characteristics of the parent heterocycle (e.g. pK_a values: imidazole, 6.95[256]; 2-phenylimidazole, 6.48[1246]; 4-phenylimidazole, 6.10[1246]; 2,4-diphenylimidazole, 5.64[1246]). The melting points of a number of arylazoles are included in table A.1. Dipole moments have been determined for arylpyrazoles[1247]. As might be expected, a *para*-nitro group on a phenyl substituent reverses the dipole (1-phenylpyrazole, $p = 6.67 \times 10^{-30}$ Cm; 1-(*p*-nitrophenyl)-pyrazole, $p = 14.68 \times 10^{-30}$ Cm). Studies of proton magnetic resonance

spectra show that the *ortho*-protons are deshielded in *N*-arylazoles[902c]. The applications of ultraviolet spectroscopy to *N*-phenylpyrazoles[97, 118], and n.m.r. spectroscopy to 1-phenylpyrazoles[114k,129,215,217, 783,1059b,1248], 5-phenyl-1,2,3-triazoles[1187] and 5-aryltetrazoles[585] indicate that a methyl substituent adjacent to the aryl group, or in an *ortho* position of the aryl group affects the coplanarity of the hetero and aryl rings. A phenyl resonance often appears as a singlet in the n.m.r. under these circumstances, although there are some exceptions [130a] (p. 22).

Aryl groups attached to a pyrazole ring are more resistant to oxidation than alkyl groups. Oxidation can, however, be facilitated by the introduction of hydroxy or amino functions in the aryl ring[2e,793a, 1026c], though under severe conditions even 1-, 3- and 5-phenyl groups can be oxidised without affecting the pyrazole nucleus[2e,1025e,1026b]. Thus, a phenyl group may be eliminated from the 1-position by acid permanganate, but it is unaffected in alkaline medium[76k,1025e]. The oxidation of 1-tolylpyrazoles may be controlled to involve only oxidation of the methyl function[321e,330g,552b,735c]. Permanganate oxidation of 3-phenylpyrazoles with heterocyclic substituents (e.g. *a*-thienyl, *β*-pyridyl) at C-5 results in the formation of 3-phenylpyrazole-5-carboxylic acid[503,1020f]. Furyl substituents on pyrazole are similarly oxidised [898a,968]. 1-Pyrazolylhydroquinones are oxidised to quinones which readily undergo nucleophilic exchange reactions[1250].

The effects of oxidising agents in alkaline medium on some aryl-imidazoles lead to products exhibiting chemiluminescence (pp. 157-8).

The hydrogenation of 1-phenylpyrazoles modifies the pyrazole ring, as does reduction with sodium and alcohol, although the aryl substituent can be modified or even lost (p. 152). Despite an early report, the imidazole ring in arylimidazoles resists hydrogenation[1255a] and sodium-alcohol reduction[1254b] (p. 153).

Electrophilic substitution of arylazoles has already been discussed (pp. 51, 63 .) The *C*- or *N*-aryl substituent attached to a pyrazole nucleus is nitrated under conditions where the pyrazole nucleus is protonated and the same is true in the imidazole and 1,2,4-triazole series[1212b]. Some bromination and chlorination[1256] can occur in the phenyl rings in the pyrazole series, and in the 1,2,3-triazoles. With 1-phenylpyrazoles some *ortho*-lithiation occurs[482].

Aryl substituents on nitrogen may sometimes be removed by nucleophilic reagents particularly if the aryl substituent carries strongly electron-withdrawing groups. This occurs with 1-(2,4-dinitrophenyl)-pyrazole reacting with hydrazine hydrate in boiling ethanol[1257], secondary amines[1258], or sodium methoxide in methanol[59c,1259]. A 1-triazino group on pyrazole is also displaced with excess amine. A dinitrophenyl group on the imidazole ring of histidine is cleaved under conditions which cleave peptides from a polymeric support (*viz.* alkaline saponification, aminolysis, or hydrazinolysis[1260]). Similar reagents are effective in the removal of nitroaryl substituents from 1,2,3-triazoles [1048b], 1,2,4-triazoles[1261a], and tetrazoles[1048b]. However, *N*-(2'-pyridyl)-imidazoles, -pyrazoles, and -1,2,4-triazoles are neither cleaved by 2M sodium hydroxide nor 2M hydrochloric acid[1262]. Even butyl-lithium lithiates the azole ring without causing nucleophilic cleavage[1262].

When 1-arylimidazoles are heated to about 600 °C they rearrange to 2-arylimidazoles (along with a small amount of the 4-isomer)[981b]. The reaction appears to be intramolecular and does not involve radicals. Although 1-(*p*-tolyl)imidazole forms high yields of 2-(*p*-tolyl)imidazole under these conditions, the corresponding *p*-nitrophenyl compound largely decomposes[981b].

The thermal decomposition of 1,5-diaryltetrazoles yields diarylcarbo-diimides and 2-arylbenzimidazoles (p. 164).

The methyl function of 1-(*p*-tolyl)pyrazole is active enough to condense with aromatic aldehydes to form stilbenes[1263]. 4-(*p*-Amino-phenyl)imidazole couples normally through the anilino amino group, and deaminates normally[390a].

The ultraviolet irradiation of 1-(*o*-nitrophenyl)-3,5-dimethylpyrazole in alcohol forms benzotriazole-1-oxides[883].

9.6   Azido groups

Azidoazoles are not well known. Table A.15 lists some of them. The compounds are crystalline, often decomposing on melting. Some are shock-sensitive[177c,1213b]. The compounds exhibit an infrared absorp-tion band in the 2300 cm^{-1} region[745a]. Reducing agents convert 5-azidopyrazoles into the 5-amino compounds (in some cases through an isolable open chain β-hydrazonitrile)[745c]. Most of the interest in azidoazoles has been focussed on their thermolysis. The azides are usually sensitive to heat and darken on standing. If heated above their melting points explosive decomposition[1213a,b] or foaming with loss of nitrogen occur[745a,b,c,1213a,b]. The structure of the thermolysis product from 5-azido-1,4-diphenyl-1*H*-1,2,3-triazole has already been discussed (p. 162). When 4-arylideneamino-3-azido-5-methyl-4*H*-1,2,4-triazoles (204) are heated, cyclisation to a triazolotriazole (205) occurs[1213a,b]. In

(204)   (205)

contrast 4-amino-3-azido-4H-1,2,4-triazoles undergo an intramolecular rearrangement during decomposition to yield 3-amino-s-tetrazines[1213b].

9.7 Cyano groups

Table A.16 lists examples of these compounds. They are hydrolysed to the amides with concentrated sulphuric acid[59d,949a] or with aqueous alkali[607b,1063b]. Although 4,5-dicyanoimidazole is resistant to boiling dilute aqueous solutions of ammonia or hydrochloric acid, partial hydrolysis to the monoamide is achieved with 1M sodium hydroxide, and stronger base at higher temperatures converts the remaining cyano group[1063b]. The dicyanoimidazoles are readily soluble in hot water and in alcohol, but only dissolve with difficulty in other solvents[1058b, c]. In the infrared the cyano compounds exhibit a stretching frequency at about 2200 cm^{-1} [1106].

The cyano group may be reduced to the aminomethyl function with sodium in ethanol[746,1058b,c], or by hydrogenation over Raney nickel [1106]. The electron-withdrawing effects of two nitrile groups in dicyanoaminopyrazoles cause the amino function to be non-basic[1270]. In fact 5-amino-3,4-dicyanopyrazole is weakly acidic (pK_a = 7.0)[1270].

There has been a recent study[1264] of the [1,5]-sigmatropic rearrangement of 2-cyanoisopyrazoles to 1-cyanopyrazoles.

9.8 Halogen substituents

Table A.17 lists some details of some haloazoles.

Although other N-halo derivatives are labile, N-iodopyrazoles are stable, probably because the iodonium cation is more stable than those of other halogens. In fact the formation of an iodonium ion from 1-iodo-4,5-dimethylpyrazole is so facile that the compound readily rearranges [77b]. Halo substituents decrease basic strength and increase acidic

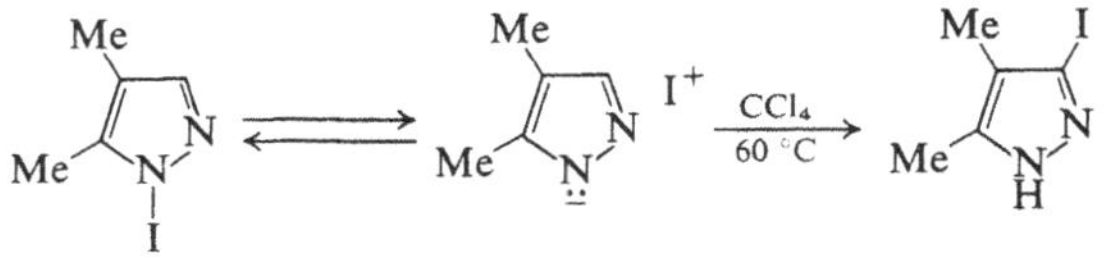

strength in azoles (e.g. imidazole, pK_a 6.95; 4-bromoimidazole, pK_a 3.60); 2,4,5-tribromoimidazole dissolves in aqueous sodium carbonate, and forms a hydrochloride which is hydrolysed by water[374a]. N-Iodoimidazoles do not dissolve in acids and alkalies, and decompose on heating with liberation of iodine[1301]. Often N-haloazoles are unstable compounds, sometimes explosively so (e.g. N-halo-1,2,3-triazoles[77c]). Infrared[77c, 495b], and nuclear magnetic resonance[212] studies of haloazoles have been made.

As in other haloaromatic compounds, the halogen atoms in *C*-halo-azoles are relatively unreactive to nucleophilic reagents (pp. 135, 143). In the pyrazole series Grignard reagents are only formed when promoted by active alkyl halides[898h,1281a,b]. In the 1-phenylpyrazole series the halogen reactivity decreases in the order 5- > 4- > 3-position[898h]. Thus 5-bromo-, 4-bromo-, and 3-bromo-1-phenylpyrazole gave the corresponding Grignard reagents in 92%, 60%, and 40% yields respectively[898h]. The ease of formation of Grignard reagents also depends on the halogen substituent. Although 4-iodo-1-phenyl-[1281b] and 4-bromo-1-phenyl-pyrazole[1281a] form Grignard reagents, particularly in the presence of methylene chloride[1281a], 3- and 4-chloro-1-phenyl-pyrazole[898h,1281b] will not react, and 5-chloro-1-phenylpyrazole only forms the reagent in the presence of 1,2-dibromoethane[898h]. Bromo groups at C-4 of pyrazole are removed in reaction with phenyl- or butyl-lithium[77g] (p. 60). The catalytic replacement of halogen by acetylenic moieties occurs with difficulty, with 5-substituents being the most reactive[1133]. In the imidazole series, acetylenic compounds react more readily with 2-iodo- than with 4-iodo-imidazoles[1144a,b]. Nucleophilic substitution reactions of haloazoles, which have been discussed in detail (pp. 135, 143) are facilitated by electron-withdrawing substituents adjacent, or in close proximity, to the halogen atom. Nucleophilic substitution of allyloxy groups for chlorine in 1-phenyl-5-chlorotetrazole occurs with some facility[598].

Halogen substituents on carbon atoms of 1,2,4-triazoles can be removed with red phosphorus and hydriodic acid[430c]. Removal or reduction of a halogen substituent is closely related to nucleophilic substitution, and may in fact involve nucleophilic attack. Although lithium aluminium hydride replaces a halogen by hydrogen in some cases [744] it does not always succeed[98]. Catalytic methods using palladium carbon[554b,1302], or Raney nickel[98,1289f,1303] are generally successful, while red phosphorus and hydriodic acid[6b,98], zinc and acetic acid[98], zinc and hydrochloric acid[321c], sodium sulphite[374a, 410b,412b,454,1291], sodium amalgam[6b,430c] and sodium and alcohol [373c,495b] have been used with varying success. Thus 5-chloro-1-methylpyrazoles are not reduced by red phosphorus with hydriodic acid, or with zinc and mineral acids[98,321c]. Furthermore, reduction with sodium and alcohols, besides removing a halogen, may partially hydro-genate the heteroring – particularly in the case of pyrazoles[321c] (p. 152).

Sulphite reacts with bromoimidazoles with either displacement of bromine by hydrogen, or by sulphonic acid, depending mainly on the structure of the substrate (p. 147). In imidazoles with a free imino nitrogen a 2-bromo substituent is readily replaced by hydrogen[374a, 412b], and a 4-bromo group less readily[374a]. Thus 2,4,5-tribromoimidazole, when treated at reflux temperatures with sulphite, yields a mixture of 4-bromoimidazole, 4,5-dibromoimidazole, imidazole, and 4-bromoimidazole-5-sulphonic acid[374a]. When there is a methyl substituent on the ring nitrogen even a 2-bromo substituent is only slowly replaced by hydrogen. However, a 5-nitro substituent activates a 2-bromo group which can now be readily replaced by sulphonic acid [412b]. It is interesting that in the iodoimidazoles it is the 2-iodo substituent which is more resistant to sulphite reduction than a 4-iodo substituent[410b]. In 1,3-dichloro-1*H*-1,2,4-triazoles the *N*-chloro group is reduced by bisulphite leaving the 3-chloro substituent unaffected[268b]. Nitrogen bound halogens are very labile when treated with reducing agents[77b,1304].

The nitration of haloazoles can result in the replacement of the halogen atom by the nitro group (p. 68)[1305]. Nitrodeiodination is common in imidazoles (p. 68).

The rearrangement of *N*-halo compounds to the *C*-halo compounds has been noted with *N*-iodopyrazoles[77b] and postulated for *N*-iodoimidazoles for the latter have been suggested as intermediates in the electrophilic *C*-iodination process[1301,1306] (*cf.* p. 56). The sodium salts of 2,4,5-triarylimidazoles react with bromine in dry ether to form *N,N'*-diimidazolyls, for, as all the ring carbon atoms are substituted, no rearrangement to *C*-haloimidazoles can occur[1301,1307b].

The nucleophilic exchange of one halogen atom for another in quaternary pyrazolium salts has already been mentioned (p. 140).

9.9   Hydroxy, alkoxy and aryloxy substituents

The tautomerism of *C*-hydroxyazoles has been discussed (p. 12). Consideration of *N*-hydroxy-*N*-oxide tautomerism is included in §9.12. Table A.18 lists physical data and references to a number of alkyl- and aryl-substituted hydroxyazoles and ethers. It should be noted that designation of compounds in this table as 'hydroxy' does not preclude '-one' or '*N*-oxide characteristics'. *N*-Alkoxy- or *N*-hydroxy-*N'*-oxides are listed in §9.12.

While 4-hydroxypyrazoles exhibit marked enol character the same is not true for 3- and 5-hydroxypyrazoles which are normally classed as pyrazolones. 4-Hydroxypyrazoles are amphoteric compounds which form salts with alkalies and with mineral acids. For example, 1-aryl-4-hydroxy-5-methylpyrazoles are stable, colourless, crystalline compounds which are sparingly soluble in cold water, but readily soluble in most organic solvents[440a]. They dissolve in dilute aqueous alkalies or ammonia or acids[440a]. Pyrazol-3-ones and pyrazol-5-ones have little enolic character, although there are cases where the hydroxy form may be the predominant tautomeric species, as in 5-hydroxy-3-methyl-1-(α-pyridyl)pyrazole

Me

(206)

(206) in which hydrogen bonding has a stabilising effect[1345].
1-Hydroxypyrazoles are often rather unstable compounds[762,1313a,b].

Both 2- and 4-hydroxyimidazoles are tautomeric systems. The 2-substituted compounds are acidic compounds which have lost the normal imidazole basic properties. They dissolve in dilute alkali with the formation of salts, are high melting crystalline compounds stable to acid hydrolysis, and although they exhibit a positive ferric chloride test most of their properties give rise to their designation as imidazol-2-ones. This is supported by their characteristic absorption in the ultraviolet region [1346]. Imidazol-4(5)-ones contain both amidine and keto–enol systems. The compounds are amphoteric, forming both silver salts and picrates [973b]. Imidazol-4-ones are less stable chemically than imidazol-2-ones. Both the 2- and 4-substituted imidazoles exist preferentially in the keto forms[208b,924a,1249] (p. 18). 1-Hydroxyimidazoles are amphoteric compounds which have some of the characteristics of N-oxides[142, 1323,1325b,c,1347]. Tautomerism in the C-hydroxy-1,2,3- and -1,2,4-triazole series has been discussed[1249]. C-hydroxy-1,2,3-triazoles are strong monobasic acids which form stable metal salts. They are readily isomerised to α-diazodicarboxylic amides[429d,1348]. Although 1,2,3-triazol-5-ones are well known[764,1093d,1326c] few 2-aryl-4-hydroxy-$2H$-1,2,3-triazoles had been reported[688a,764,1093d] until recently [1326c]. 1,2,4-triazolones or C-hydroxy-1,2,4-triazoles are also acidic compounds. Among the earliest known was 3,5-dihydroxy-1,2,4-triazole (urazole)[1179]. Extreme (often explosive) instability is a feature of the chemistry of 1-hydroxytetrazoles[1343-4,1349b]. 4-Hydroxy-$4H$-1,2,4-triazoles are amphoteric compounds[1124c].

Reducing agents such as sodium amalgam, sodium and ethanol, tin and hydrochloric acid, or hydriodic acid do not affect 4-hydroxypyrazoles, but reduction can be achieved by heating the compounds with phosphorus tribromide and yellow phosphorus in a sealed tube[440a]. Usually, however, treatment of 3- or 5-hydroxypyrazoles with phosphorus halides and oxyhalides leads to nucleophilic displacement of the oxygen function by halogen[897c,964b,1243b,1350] (p. 140). No reduction of 1,3-dimethylpyrazol-5-ones occurred on treatment with phosphorus and phosphorus tribromide, but phosphorus oxychloride replaced the oxygen function with chlorine[98]. Hydroxy groups in the 4-position are only displaced by halogen when an activating group is present in the pyrazole molecule[964b] (p. 141). The reduction of pyrazolones has been discussed (p. 152). 4-Arylidine derivatives of 3-methyl-1-phenylpyrazol-5-one are reduced at the exocyclic double bond by lithium aluminium hydride[744]. While *N*-hydroxypyrazoles are unaffected by sodium dithionite, the compounds are readily reduced by zinc in acetic acid[762, 1313a,b]. In imidazol-2-ones and -4-ones the oxygen functions are resistant to reduction. The use of sodium in ethanol[1254a], hydrogen and nickel[1255b], Raney nickel[1351], hydrogen with platinum or palladium[354,1352], or sodium amalgam in acetic acid[1290b] reduces only the ring. A hydroxyl function attached to nitrogen can be reduced by hydrogenation over Raney nickel or with phosphorus trichloride in chloroform[1347]. In the 4*H*-1,2,4-triazole series, 4-hydroxy compounds are similarly reduced with zinc and acetic acid, phosphorus trichloride or phosphorus oxychloride[1124b,c]. Sometimes chlorine substitution products are formed in the reduction[1124b,c].

Some parallels with the pyrazolones are observed in the reaction of lithium aluminium hydride with 1,2,4-triazol-5-ones[1353]. If the ring double bond is conjugated to the exocyclic double bond the reduction only affects the heteroring.

4-Hydroxypyrazoles are oxidised by Tollen's reagent, Fehling's solution [964a,b], and by cold alkaline permanganate[440a], while even air oxidised 1,4-dihydroxypyrazoles to the pyrazol-4-one-1-oxides[1313a,b]. The Cu(II) catalysed oxidation of 4-substituted pyrazol-5-ones yields 4-hydroxypyrazol-5-ones and 4,4'-bispyrazolones[801]. Peracids oxidise 1,4-dihydroxypyrazoles to 1,3,4-oxadiazin-6-one-4-oxides (207), perhaps via pyrazolone-*N*-oxides[800]. Mild air oxidation of antipyrine (and derivatives such as pyramidone (208)) affects only the *C*-methyl group [1119]. When imidazol-2-ones are oxidised ring-opening can occur with a

(207)

variety of products being formed[14a,353c,d,1314,1337] (p. 157). Oxidation of imidazol-2-one-4-carboxylic acid with chromium trioxide gives parabanic acid[602,1254c]. Oxidation of the imidazol-4-ones ruptures the ring. Permanganate in the presence of sodium ethoxide oxidises 5-hydroxy-4-methyl-1-phenyl-1H-1,2,3-triazole to pyruvic anilide [1066a].

(208)

Hydroxyazoles undergo nucleophilic substitution more or less readily depending on the substitution pattern. Thus, in the pyrazole series displacement by halogen and aryl (from Grignard reagent attack)[744] are common (pp. 137, 140). Such substitutions in imidazol-2-ones are less common[212]. C-Hydroxy-1,2,4-triazoles lose the OH function on heating with phosphorus pentasulphide, and are converted into chlorotriazoles with phosphorus pentachloride (p. 141). It is noteworthy that phosphorus pentasulphide in xylene only converts pyrazolones into pyrazolethiones[330f] (p. 148).

4-Hydroxypyrazoles are activated at C-5 to electrophilic substitution. Diazo-coupling and halogenation have been mentioned (pp. 45, 47). Nitrosation occurs at C-5[964b]. The 5,5-dichloro compounds formed in the chlorination of 4-hydroxypyrazoles can be reduced to 5-chloro-4-hydroxypyrazoles[1d,440b,c]. In pyrazol-5-ones the methylene function at C-4 is sufficiently activated for condensation to occur with ketones in acid media producing 4-alkylidenepyrazolin-5-ones (p. 37). In alkali, dipyrazolinyl compounds are formed[335e,397]. Although there is no reaction in aprotic, non-polar solvents, aldazines and ketazines react like carbonyl compounds at the activated 4-position of pyrazol-5-ones when polar solvents are employed, or in the presence of tertiary amines[1356].

Imidazol-2-ones are substituted by electrophiles at C-4 and C-5[354] (p. 34), and imidazol-4-ones couple readily with diazonium salts[398] (p. 45), as do 4-hydroxy-1,2,3-triazoles[817] (p. 45).

Acylation[964a,b,1093e] and sulphonylation[440a] of 4-hydroxypyrazoles (p. 122), pyrazol-5-ones[158,329e,707] (pp. 121, 126), imidazolones[398,602,973b] (p. 123), and 1,2,3-triazolones[605,1326c], which may involve reaction at oxygen, nitrogen, or carbon have been discussed earlier. Ready formation of acetoxy derivatives is noticed with 1-substituted 5-hydroxy-1*H*-1,2,4-triazoles[61a,b] (p. 124), but a more complicated state of affairs accompanies the reaction of acetic anhydride with 4-hydroxy-4*H*-1,2,4-triazoles[1124b,c]. Whereas 4-hydroxy-3-phenyl-4*H*-1,2,4-triazole (209) yields a *N*-acetyl triazolone (210), if the 5-position is blocked by a methyl substituent (211) is obtained (a reaction characteristic of *N*-oxides).

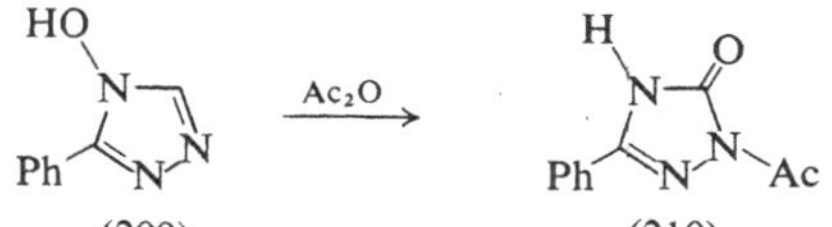

(209)   (210)

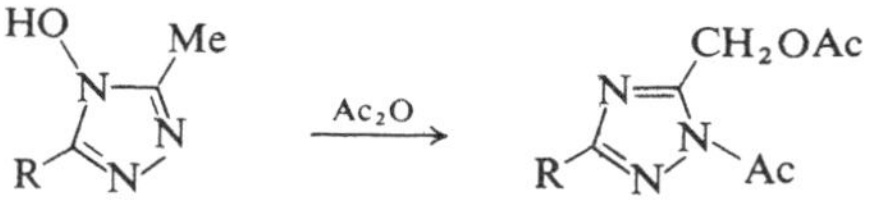

(211)

Acetoxy-pyrazoles[681] and -1,2,3-triazoles[1326c] are easily hydrolysed by either acids or bases. Some 5-acetoxy-1,2,4-triazoles are hydrolysed by boiling water[61b].

In the thermal cleavage of 5-hydroxypyrazole methyl carbonates the primary product is the 3-methyl-5-methoxy compound which rearranges to some extent to the *N*-methylpyrazolone[698].

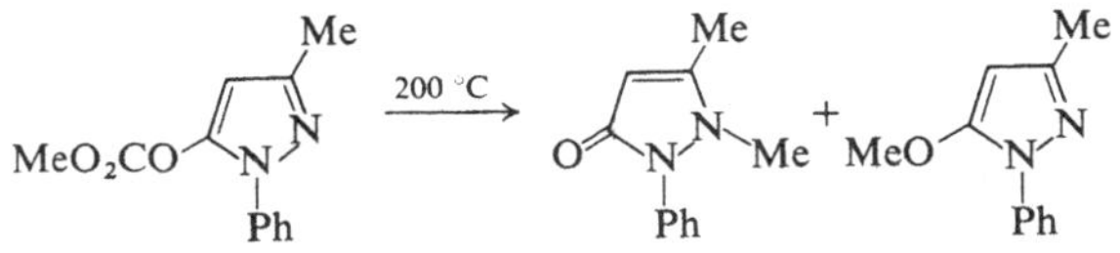

The extensive studies which have been made of the alkylation of pyrazol-3- and -5-ones have already been considered (pp. 85–9). Until recently there were few references to 4-alkoxypyrazoles, although some 4-phenoxyethers had been prepared[1357]. In the last few years 4-hydroxy-1-phenylpyrazole has been converted in basic medium into the 4-alloxy compound[598], 4-hydroxy-3,5-dimethylpyrazole into the methoxy derivative[1312], and 4-hydroxyantipyrine into the 4-*O*-alkylated product[1358]. On the basis of n.m.r. evidence, Nye and Tang [1311a] have concluded that methylation of 4-hydroxy-1-methyl-3,5-diphenylpyrazole under alkaline conditions yields the 4-methoxy compound, but under neutral conditions a *N*-methylated zwitterionic product is obtained.

In contrast to imidazol-2-thiones, which are alkylated on sulphur,

imidazol-2-ones react at nitrogen to yield the 1,3-dialkyl compounds
(p. 91). Nucleophilic displacement of halogen is the method of choice
for the preparation of 2- and 4-methoxyimidazoles[122,1289e] (p. 143).
Methylation of 1-hydroxyimidazoles in basic conditions forms the
methoxyl compounds[142].

Although the alkoxyl derivatives are more commonly prepared by
ring-synthetic methods[907c] or by nucleophilic substitution for halogen
(p. 143), alkaline diethyl sulphate reacts with *C*-hydroxy-1,2,3-triazoles
at the oxygen function (p. 90). The methylation of 1,2,4-triazol-3-ones
has been mentioned (p. 91). Ring-synthetic methods lead to 5-alkoxy-
tetrazoles, the ether functions of which may be cleaved by hot 20%
hydrochloric acid[83g].

When 3,5-diphenyl-5-hydroxypyrazol-4-one is heated, or treated with
strong acids possessing dehydrating properties, the product is 3,5-
diphenyl-4-hydroxypyrazole[1311c].

Photolysis of 4-hydroxy-4*H*-1,2,4-triazole removes the oxygen function
[1124a]. Irradiation of meso-ionic 1,4-diaryl-1*H*-1,2,4-triazol-3-ones yields
benzimidazole, arylisocyanate and arylazobenzene[887,1359]. The
photolysis of 5-phenoxy-1-phenyl-1*H*-tetrazole[892,1360] has been
discussed (p. 180).

Pyrazolyl ethers may be cleaved under vigorous conditions (e.g. by
concentrated acids at 150°C)[76k,100b,1025e], though a nitroso
function on an adjacent carbon atom may facilitate the cleavage[76k].
Ethers of pyrazolium salts are cleaved merely on heating[615c], and
also readily by alkalies[615c,1025d]. Quaternary 1,2,3-triazole ethers are
dequaternised exclusively by loss of the methyl group attached to
oxygen, giving triazol-5-ones[576k]. 5-Aryloxytetrazoles are cleaved by
catalytic hydrogenolysis[767,1140]. Alkoxyl groups may, of course, also
be removed by nucleophilic substitution procedures[212,551d,964b,
1361].

Alkoxypyrazoles form quaternary salts with alkyl halides. Salts of the
same class are also formed by the addition of alkyl halides to antipyrine
and its analogues (p. 101). Such salts regenerate antipyrine on thermal
decomposition or alkaline hydrolysis.

When 2-alkoxy-1-methylimidazoles are heated they rearrange to alkyl-
imidazol-2-ones[1362]. The *O*-allyl and *S*-allyl compounds rearrange 15–
20 times faster. Although 5-allyloxy-1-phenyl-1*H*-tetrazole rearranges on
heating for one hour at 100–150°C to 1-allyl-4-phenyltetrazol-5-one,
3-allyloxy-1-phenylpyrazole was stable under the same conditions[598].
At higher temperatures a mixture of 4-allyl-3-hydroxy-1-phenylpyrazole,
2-allyl-1-phenylpyrazol-3-one, and decomposition products were formed.

At 185 °C for one day, 2-allyl-1-phenylpyrazol-3-one rearranged to give 15–20% of 3-alloxy-1-phenylpyrazole[598].

Spontaneous thermal rearrangement of 1-methoxypyrazole-2-oxides to the isomeric 5-methoxy compounds probably proceeds through an ionic mechanism[1322a]. The compounds undergo the normal acyloxy migration of N-oxides with acetic anhydride to form 3- and 5-acetoxymethyl-pyrazoles[1322b]. With phosphorus trichloride or benzoyl chloride abnormal reactions lead to 3-chloromethylpyrazoles, and in some cases the N-oxide function is lost also[1322b].

9.10   Metallic derivatives

The azoles form metallic salts (with e.g. $NaNH_2$ or RMgBr) which are extensively hydrolysed by water. The resulting anions react very readily with electrophiles such as cyanogen bromide (p. 128), phenyl isocyanate (p. 110), and propargyl bromide[1363]. The sodium salts of 4-chloro- and 4-bromo-pyrazole give ketimines (212) and cyanogen bromide[1363].

(212)

The sodium, potassium and calcium salts of imidazole prepared in liquid ammonia are very rapidly hydrolysed[283]. Reactions with Grignard reagents are similar. Pyrazolyl-1-magnesium bromide is a grey oil, insoluble in ether, and rapidly decomposed by water to pyrazole and magnesium hydroxybromide[56]. With propionyl chloride it reacts to form 1-ethoxycarbonylpyrazole[896a]. Oddo and Mingoia reported[1364] that imidazolyl-1-magnesium bromide, heated for two hours at 55–60 °C in a sealed tube with acetyl chloride, was converted into 2-acetylimid-azole. Their reported physical data for this compound do not correspond with those of authentic 2-acetylimidazole[372], and it seems evident that they obtained 1-acetylimidazole in an impure state, and that this was hydrolysed to imidazole during their work-up procedure. Recent studies[906b] in which an imidazolyl Grignard reagent was converted by phosgene into 1,1′-carbonyldiimidazole seem to confirm this supposition. N-Imidazolyl Grignard reagents have been used in the synthesis of 1-(2′,4′-dichlorobenzoyl)-2,4,5-tribromoimidazole[909c].

There have been numerous references to the preparation and reactions of silver salts of the azoles. These salts are usually insoluble[569b], or sparingly soluble compounds, which in the case of tetrazoles detonate violently on heating. Copper salts of tetrazoles behave similarly. Silver salts of pyrazoles have been used in preparing mono- and poly-halogenated pyrazoles (pp. 46, 49). Imidazole silver salts have been used in *N*-alkylations (p. 77). Imidazolides of sulphonic acids form complexes when treated with silver nitrate[150*l*].

Complexes of azoles, particularly imidazoles, with transition metal ions have received considerable attention in recent years. The ability to form co-ordination complexes may play a major role in the biological utilisation of the imidazole moiety. Thus, complexes form readily when imidazoles react with cations such as Co^{2+}[1365-7], Cd^{2+}[273], Hg^{2+} [1368], Cu^{2+}[273,1366-7,1369], Ni^{2+}[254,1366-7], and Zn^{2+}[1365, 1370]. With Cu^{2+} and Zn^{2+} the co-ordination number is 4[254] or 6 [1370], with Ni^{2+} it is 6[254], and with Hg^{2+} the value is 2[1368]. Co-ordination polymers of imidazole and various iron carbonyls have been studied[705], as have the reactions of haemoglobins[705] with imidazoles.

Finar and co-workers[1281a,b] have made a detailed study of the reactions of the Grignard reagent prepared from 4-bromo-1-phenylpyrazole. In the carbonation of 1-phenylpyrazole-4-magnesium bromide, in addition to carboxyl product[898h,1281a] there is always some 1,1'-diphenyl-4,4'-bipyrazolyl by-product[1281a]. The yield of bipyrazolyl can be increased by the addition of anhydrous cobaltous chloride to the reaction mixture. Hydrolysis under nitrogen of the Grignard reagent also yields about 8% of the bipyrazolyl[1281a]. Attempts to prepare 4-hydroxy-1-phenylpyrazole by oxidation of the Grignard reagent failed [1281b,1372]. With equimolar amounts of aldehydes, carbinols are formed, but when two moles of benzaldehyde react with the Grignard reagent a 50% yield of 4-benzoyl-1-phenylpyrazole is obtained[1281b], formed by oxidation of the alkoxide complex by benzaldehyde. When an endeavour was made to prepare bis(1-phenylpyrazole-4-yl)methanol from the reaction of 1-phenylpyrazole-4-carboxaldehyde with the Grignard reagent the major product isolated was tris(1-phenylpyrazol-4-yl)methane (believed to be formed via a tertiary alkoxide which is reduced by a hydride ion transfer mechanism). The isolation of small quantities of bis(1-phenylpyrazol-4-yl)ketone and di[bis(1-phenylpyrazol-4-yl)methyl]-ether pointed at least to the transient formation of the required carbinol. A similar state of affairs was found in the reaction of the Grignard reagent with ethyl benzoate when the product was a substituted toluene [1281b]. Normal reactions of 1-phenylpyrazole-4-magnesium bromide occurred with acid chlorides, ketones[1281a], halogens, and some aldehydes[1281b].

N-Imidazolyl Grignard reagents which are prepared by the reaction of Grignard reagents with imidazoles unsubstituted on nitrogen[909c], or with some imidazolides of sulphur[1373], react with acid chlorides to yield ketones[906b,909c].

Rather more extensive use has been made of the organosodium or organolithium derivatives of azoles. These derivatives react smoothly with carbon dioxide to give acids[77g,480a,1303], with dimethyl sulphate [77g] or alkyl halides[484,840,981a] to form alkyl derivatives, with dimethylformamide to yield aldehydes[375,1182a], with aldehydes and ketones to form alcohols[74e,f,77g,372,451,480a,1182a], with ethyl formate to form alcohols[372], with acid halides[322,480a] and some aromatic aldehydes[74f] to form ketones, with halogens to produce halo-substituted heterocycles[1144a], and with quinoline to form the 2-quinolyl derivative[480a]. The replacement of 5-lithio by bromine in 1,2-dimethyl-5-lithioimidazole is accomplished with bromodiethylamine [451]. 1-Methyl-5-lithiotetrazole undergoes fragmentation to nitrogen and lithium methylcyanamide at temperatures in excess of –50 °C (p. 167). although below –60 °C it is fairly stable and a useful reagent for the preparation of 5-substituted tetrazoles. The compound reacts readily with halogens, cyanogen bromide, sulphur, aldehydes ketones and acid chlorides. but not appreciably with benzyl bromide or benzyl cyanide. With methyl benzoate there is formed a labile addition compound which may be hydrolysed to 5-benzoyl-1-methyltetrazole[484,840].

In many early references to pyrazoles it was noted that the compounds, whether substituted on nitrogen or not, reacted with mercuric salts to form products of varied composition[485,1025e,1026e,1033,1374-5]. 1-Arylpyrazoles react with mercuric acetate to form 4-acetoxymercury derivatives (p. 60). More recently, Grandberg[323b] has shown that mercuric salts mercurate pyrazoles in the 4-position, and also give other complexes with variable quantities of the mercuric salts. The proportion of the mercuric salt which is bound as a complex is a function of the basic strength of the pyrazole.

The mercuric group in mercurated pyrazoles behaves in a variety of reactions just like phenylmercuric chloride[59a,b,393b]. The group is removed by dilute hydrochloric acid, replaced by bromine in reaction with bromine in acetic acid, and by benzylmercapto on reaction with thiocyanogen and benzyl chloride[59a]. Nitrosyl sulphuric acid reacts with 3-phenylpyrazole-4-mercuric acetate to produce a low yield of the 4-diazo compound[486a]. A mercuric group at C-4 of pyrazole is cleaved very readily by aqueous sodium hydroxide[323b]. Thus the 1 : 1 complex of 4-chloromercuri-1,3,5-trimethylpyrazole with mercuric chloride reacts with aqueous sodium hydroxide to form bis(1,3,5-trimethyl-4-pyrazolyl) mercury in 25% yield. A similar complex of 4-chloromercuri-3,5-dimethyl-1-ethylpyrazole was converted into the 4-hydroxymercuri analogue in low yield. Aqueous alkali converted bis(3,5-dimethyl-1-phenyl-4-benzylpyrazole) mercuric chloride complex into mercuric oxide and the original pyrazole, but the 4-chloromercuri- and 4-bromomercuri-1-phenyl-pyrazole were stable to the reagent. 4-Methyl-3,5-dipropylpyrazole forms a chloromercuri derivative which is stable to alkali[323b]. Mercuric salt

complexes with 1,3-diphenylimidazolium have been studied[326].
Whereas sodium salts of imidazole, 1,2,4-triazole, etc. are largely ionic,
the mercury salts have been postulated to adopt a linear polymeric
structure. Results of recent studies of thallium and tin[703] derivatives
suggest that they too are polymeric, at least in the solid state. Stable
organotin derivatives of azoles may be prepared as long as at least two
nitrogen atoms are 1,3 to each other. The compounds are non-explosive
and often stable to boiling water[703].

9.11   Nitro and nitroso groups

Table A.19 lists examples of these compounds.

9.11.1   Nitro groups

Spectra (pp. 11, 18, 22) and ionisation constants (p. 24) of nitroazoles
have been mentioned. The nitro substituent has a strong base-weakening
effect, and also can reverse the direction of the dipole unless other substituents exert compensating effects[1247]. A nitro group exhibits a large
effect on the chemical shifts of adjacent protons in the n.m.r. spectrum
[217,489a,b,1391]. Nitroazoles are usually high-melting crystalline solids
which are sparingly soluble in most solvents. Polynitroimidazoles are strong
acids which dissolve readily in water[1072c,d,1233b,1380]. Methylation
results in loss of acidic properties and a change in solubility characteristics
[1072d]. Although nitro-pyrazoles and -imidazoles have been studied
extensively, nitro-triazoles and -tetrazoles are less well known, and may be
unstable to the point of possessing explosive properties[1114j]. Infrared
spectroscopic studies[1387,1390] indicate that salt-formation with electro-
positive metals involves the nitro group, with the formation (in the case
of nitroimidazoles) of an isoimidazole ring. Nitroimidazoles with a free
NH group dissolve in alkali hydroxides, carbonates or aqueous ammonia
to form yellow solutions containing the nitroimidazole anion[274c].

The decrease in basic strength brought about by the introduction of
a nitro group into the heterocyclic ring often prevents formation of salts
with acids. Thus 4-nitroimidazole (major tautomeric species (p. 24))

will not form a stable hydrochloride or picrate, and although it will dissolve in concentrated hydrochloric acid dilution of the solution reprecipitates the free base[274c]. 1-Methyl-5-nitro-imidazole is a strong enough base to form a stable picrate and hydrochloride whereas the 1,4-isomer, in which the nitro function is adjacent to the 'basic' ring nitrogen, will not[141,266]. The 1,4-isomer is, however, more stable since the effect of heat on 1,3-dimethyl-4(or 5)-nitroimidazolium iodide yields only methyl iodide and 1-methyl-4-nitroimidazole[497]. A similar state of affairs exists with 2-iodo-1-methyl-4- (pK_a -1.70) and -5-nitro-imidazole (pK_a -0.14)[122], but not with the corresponding 2-methoxyl isomers where the relative basic strengths are reversed, perhaps resulting from an increased importance of the differential stability of a linear conjugated system (5-nitro isomer) over a branched conjugated system (4-nitro isomer). As expected, 3-nitro-1,2,4-triazole is a monobasic acid with basic properties suppressed. The absence of significant protonation in acid solution is reflected in the absence of any downfield shift in the C-5 proton in trifluoroacetic acid solution[995a], a normal phenomenon in the n.m.r. spectra of protonated triazoles[216].

Considerable effort has been expended in the study of the effects of reducing agents on nitroazoles. Partial reduction of 4-nitropyrazoles to the corresponding nitroso compounds is possible with alkali stannites [74b,c]. Treatment with one mole of sodium stannite for 10 minutes at 100°C yields 30% of nitroso compound; two moles of stannite for 30 minutes yield mainly 4-aminopyrazole with only a trace of 4-nitroso-pyrazole; prolonged heating with three moles of stannite results in complete reduction to amino compound[74b]. Examination of the relative reaction rates for a series of substituted 4-nitropyrazoles suggests that substituents in the 3- and 5-positions have a pronounced influence on the reaction[74b]. Complete reduction of nitro to amino may be achieved using zinc and acetic acid[615a], tin and hydrochloric acid [452a,456a,615a,748,1168,1392], sodium hydrosulphite[752], aluminium amalgam with moist ether or alcohol[74c,426a,431c], catalytically using hydrogen with Raney nickel[172f,307a,544,1289d], Adams catalyst[495a 565,1101a], or Pd/C with hydrazine[59e,1023c], acetic acid[1184], or hydrogen[766,1393], and with red phosphorus and hydriodic acid[757]. 1-Nitropyrazoles have been reduced electrolytically[1295]. Selective reduction of the 3-nitro group in 3,4-dinitropyrazoles is sometimes possible. Thus 3,4-dinitro-5-(β-pyridyl)pyrazole can be reduced to the 3-amino-4-nitro product[504]. A 4-nitro group is generally more resistant to reduction than a phenylnitro substituent[59e]. The reducing action of tin and hydrochloric acid often results in complete fragmentation of the heterocyclic ring (p. 168). When 2-nitroimidazole is reduced with Adams catalyst an unstable imidazole reduction product is obtained[1379]. When 1-methyl-3,5-dinitro-1*H*-1,2,4-triazole is treated with hydrazine under differing conditions a variety of reduction products is observed [269c]. Reduction of a nitro group is sometimes accompanied by removal of a halogen atom adjacent to the nitro substituent[1289d].

The electron withdrawing effect of a nitro substituent greatly facilitates nucleophilic displacement reactions at adjacent carbon atoms (see pp. 135, 143 etc.). Thus in 4-nitropyrazoles a 5-chloro (but not a 3-chloro [1354a]) group is activated to substitution. Such activation also applies to adjacent methyl groups which can be induced to take part in base-catalysed condensations with carbonyl compounds, a property which parallels the reactivity of the methyl group in α-picolines[1394]. The formation of styrylimidazoles from the reaction of 4-methyl-5-nitro-imidazoles with benzaldehyde is an example of this type of reaction[564, 1115b]. It has, however, not proved possible to form the 2-styrylimid-azole by reaction of 1,2-dimethyl-5-nitroimidazole with benzaldehyde in basic medium[498]. The kinetics of interaction of hydroxide ion with 1-methyl-5-halo-3-nitro-1*H*-1,2,4-triazoles indicate a second-order substitution of halogen, or nitro to give the respective oxotriazolines in a 30 : 1 ratio[1389b].

Rearrangement of nitro groups from a ring nitrogen atom to ring carbon occurs in the presence of concentrated sulphuric acid[74c] or on thermolysis[489a,b,1116a]. When 1-nitropyrazole is treated with concentrated sulphuric acid 4-nitropyrazole (the normal electrophilic nitration product from pyrazole treated with mixed acid) is formed[489a]. Sometimes there is loss of the 1-nitrosubstituent, e.g. 1-nitro-3-(*p*-nitrophenyl)pyrazole reacts with sulphuric acid to yield 3-(*p*-nitrophenyl)-pyrazole[499a,b], and 1-nitro-4-iodopyrazole is converted into 4-iodopyrazole under the same conditions[74c]. On thermolysis (~ 140°C), however, the rearrangement usually leads to the 3- (or 5-)nitropyrazole [489a,b]. If C-3 is blocked the nitro group migrates to C-5; but 1-nitro-5-methylpyrazole is transformed into a mixture of 3-methyl-4- and 3-methyl-5-nitropyrazole in the ratio 93 : 7. The latter compound is presumably formed via the intermediacy of 3-methyl-1-nitropyrazole

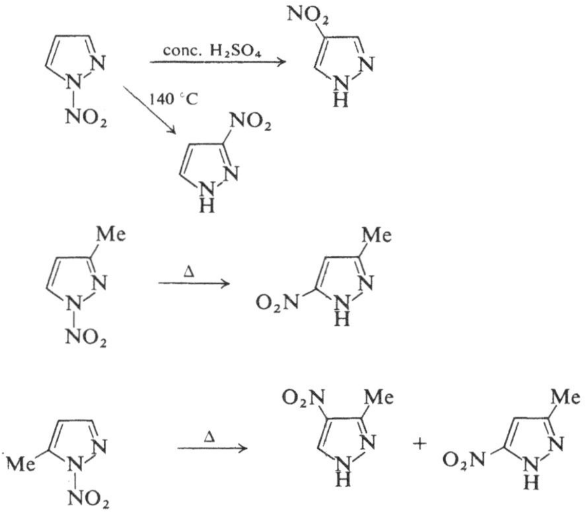

[489b]. Kinetic and other experiments suggest that the thermal rearrangements are first-order reactions which are uncatalysed, intramolecular, and do not involve radicals[98,489a,491,1116a,b]. A [1,5]-sigmatropic shift followed by an isomerisation reaction has been suggested[489b,1116a],

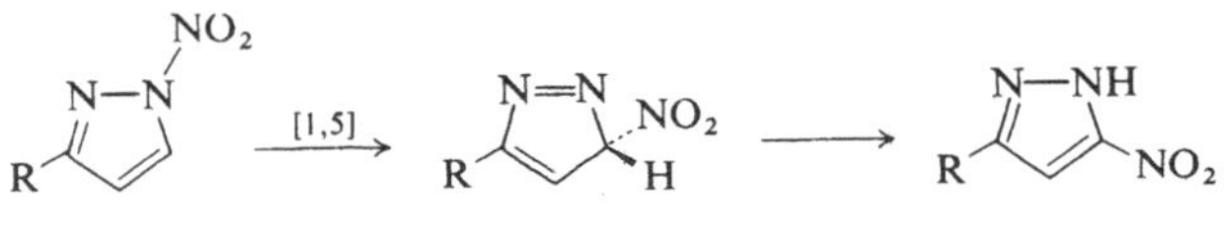

although this can be criticised on the grounds that the required orbital overlap requires an initial breaking of the aromatic system. An isomerisation to an isopyrazole (the reverse of a van Alphen rearrangement) followed by a [1,5]-sigmatropic shift of hydrogen seems to be a preferable alternative. 1-Nitro-1*H*-1,2,4-triazoles similarly rearrange to the 3-nitro isomers[1116a]. In cold concentrated sulphuric acid, 1,4(5?)-dinitroimidazole loses the *N*-nitro function; at higher temperatures 4,5-dinitroimidazole is formed[1072d].

Complexes form between 5-nitrotetrazoles and ferrous salts[292].

9.11.2 *Nitroso compounds*

Most of the studies of nitrosoazoles are related to the comparatively stable nitroso-pyrazoles and -imidazoles. Nitroso-triazoles and -tetrazoles are not well known and are unstable. 5-Nitrosotetrazole-2-carboxamide is a highly explosive red compound which liberates nitrogen and nitrous acid when boiled with water or dilute sulphuric acid, and forms the sulphate of the base[1349a]. In the monomeric state 3- and 4-nitrosopyrazoles are generally blue or green in colour (e.g. 3,5-dimethyl-4-nitroso-pyrazole - blue needles; 4-nitroso-3,5-diphenylpyrazole - green plates)[74b,530,1116b]. In contrast to nitrosobenzenes, alkyl-substituted 4-nitrosopyrazoles exist as monomers in the solid state and on solvation [74b,1116b], although there are examples of dimeric colourless or yellow crystalline forms[74b,530] which form blue solutions in polar solvents, indicating dissociation into monomers on solvation. When 4-nitroso-pyrazoles are unsubstituted on nitrogen they can be formulated as a tautomeric mixture of nitrosopyrazole and oximinoisopyrazole. The compounds are not easily decomposed in acidic medium, and their blue colour, absorption at short wavelengths and relatively high acidities point to a considerable contribution to the resonance hybrid of a dipolar form [1116b]. In alkaline solution the corresponding anions are resonance stabilised[74b].

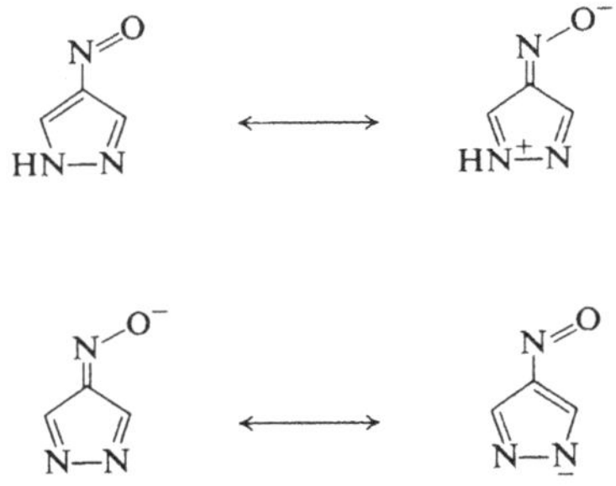

A variety of reducing agents have been employed to convert nitroso groups into amino groups. Hydrazine and alcohol[146a,898e,1166,1378] (sometimes accompanied by Raney nickel[898e,1166,1378]) is the most commonly used reagent, but tin and hydrochloric acid[1030b,1193e], ammonium sulphide[1030b,d], and zinc and acetic acid[1193e] have also proved successful. Hydrazine in acetic acid reduces nitrosopyrazoles to azopyrazoles[146a]. Dropwise treatment with phenylhydrazine results in some reduction of 5-nitroso-2,4-diphenylimidazole to the 5-amino-imidazole[414b]. Ring modification also occurs, as shown (p. 168).

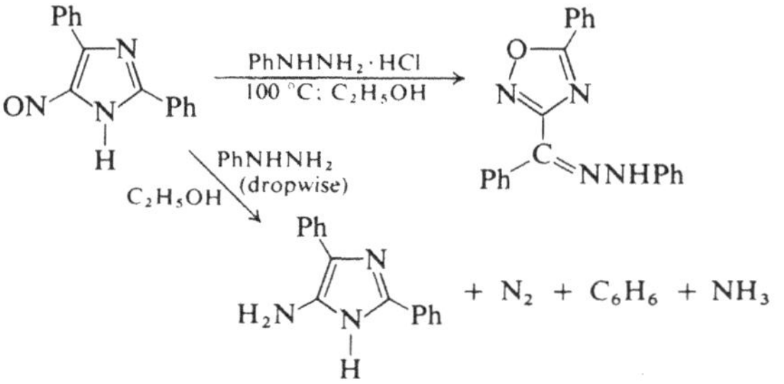

Nitrosoazoles resemble nitrosobenzenes in that they may be oxidised to nitro compounds by permanganate[537] or concentrated (or fuming) nitric acid[74b,530,1166].

The nitroso function takes part in condensation reactions with many compounds containing active methylene groups, e.g. 2,4-dinitrotoluene, 4-nitrobenzyl cyanide, isoxazolones, β-diketones, β-ketoesters etc. to form Schiffs bases[1030b,1310b,1395]. Related condensations also occur with benzoyl chloride in the presence of potassium hydroxide[1030b] and with aminopyrazoles to produce unsymmetrical azopyrazoles[74b]. The interaction in hot acetic acid of 2,4-diphenyl-5-aminoimidazole and 2,4-diphenyl-5-nitrosoimidazole forms a dark blue compound with a structure tentatively assigned as (213)[414a].

(213)   (214)

When 3-methyl-1,5-diphenyl-4-nitrosopyrazole in boiling ethanol is treated with a drop of sodium hydroxide solution an azoxybispyrazole (214) is formed[964c,1245]. 4-Nitroso-5-aminopyrazoles can be used in carboxyl group activation in peptide synthesis where they act as dinucleophiles[1174]. Mono- and di-acyl derivatives are formed in reaction with one and two moles respectively of *N*-protected amino acid in the presence of dicyclohexylcarbodiimide. The diacyl compound is a very strong acylating agent which is reported to be effective with sterically hindered amino acids and which does not appear to induce racemisation [1174]. The unusual reaction which occurs when 4-nitrosopyrazoles unsubstituted at N-1 are treated with phosphorus pentachloride has been discussed (p. 170).

When heated in aqueous or alcoholic solution, 1-nitrosopyrazoles rearrange to the 4-nitroso isomers[76k].

9.12 Oxides

Physical properties of some *N*-oxides are given in table A.20. Since *N*-hydroxyazoles are tautomeric with *N*-oxides they have been included here.

Attempts to convert azoles into *N*-oxides by treatment with peroxides in acidic conditions (or with peracids) have usually been unsuccessful (p. 126). However, the product ($C_{10}H_{11}N_3O_2$) isolated by Michaelis [330k] from the interaction of peracetic acid and 4-amino-3-methyl-1-phenylpyrazole may have been a 4-hydroxylaminopyrazole 2-oxide, since reduction with zinc in acetic acid converted it into 3-methyl-4-nitroso-1-phenylpyrazole. When the lophyl (2,4,5-triphenylimidazolyl) radical is treated with 3% hydrogen peroxide a thermochemiluminescent hydroperoxide is formed. With t-butyl peroxide the analogous t-butyl hydroperoxide results[810].

(215)   (216)

A number of studies of the tautomerism of azole N-oxides which are capable of existence as N-hydroxyazoles can be criticised on the grounds that conclusions drawn from chemical reactivity do not provide a sound basis for structure allocation. Both Allan[1323] and Akagane[1324] preferred the hydroxide form (215) for 4,5-dimethyl-2-phenylimidazole 3-oxide (216) because of the amphoteric nature of the compound, and because the derived acetate formed a well-defined hydrochloride. On comparably inconclusive evidence Cornforth[973b] had preferred the N-oxide structure. Recently, a rigorous study[142] based on spectroscopic examination of 2,4,5-triphenylimidazole 3-oxide and suitable reference compounds (1-methyl-2,4,5-triphenylimidazole 3-oxide and 1-methoxy-2,4,5-triphenylimidazole) has shown satisfactorily that the compound exists predominantly in the OH form in non-polar solvents, with increasing amounts of N-oxide as the solvent polarity increases. In aqueous solution comparable amounts of both forms exist, while in the solid state the compound forms strongly hydrogen bonded aggregates in which distinction between the two species is blurred. Two crystalline forms are apparent from a study of the infrared and mass spectra[142, 1325a], but although Zimmermann[1325a] designates these as hydroxy and N-oxide species the evidence is inconclusive. The 4-hydroxy-4H-1,2,4-triazoles appear to exist mainly as N-oxides, associated in solution and in the solid state[699,1124c].

Infrared[142,168d,1124c,1325a], ultraviolet[142,1124c,1324,1396], nuclear magnetic resonance[142,1124c,1261b,1324,1396], and mass spectral[142,873,1124b,1325a] studies have been made of N-hydroxyazoles and N-oxides. Measurements of pK values have been made[142, 1323,1397] for some examples and electron spin resonance methods have been applied to a study of N-oxide radicals[1325c].

N-Hydroxyazoles and azole N-oxides are crystalline hygroscopic compounds which are often light-sensitive. The hygroscopic nature often leads to the compounds being isolated as crystalline hydrates[1323-4]. Some of the compounds are not particularly stable to heat (some deoxygenation of 1-methyl-2,4,5-triphenylimidazole 3-oxide occurs on vacuum sublimation

[1377]) while others decompose above their melting points[1324,1340a, 1397]. Compounds which contain both hydroxyl and *N*-oxide functions are capable of forming inter- or intra-molecular hydrogen bonds. Thus 1-hydroxy-2,4,5-trimethylimidazole 3-oxide forms association complexes in chloroform solution. Even at high dilutions the complexes exist as trimers[1261b]. When the nitrogen functions are adjacent, as in 1-hydroxy-pyrazole 2-oxides, intramolecular hydrogen bonding leads to limited solubility in most solvents, and the ready formation of 2 : 1 chelate complexes between the sodium salts and Cu(II), Ni(II), Co(II) or Cd(II) [762]. 1-Hydroxypyrazole 2-oxides are relatively strong acids and are soluble in dilute aqueous alkali from which they may be reprecipitated on acidification[762]. Other *N*-hydroxy compounds and *N*-oxides also have acidic properties[142,1323,1400], but basic characteristics are not wholly suppressed, for the compounds are able to form hydrochlorides and picrates[245,1323,1340f,1399a]. Tetrazole *N*-oxides form crystalline silver salts when treated in aqueous solution with silver nitrate[1400].

Removal of *N*-oxide functions is readily achieved using zinc in acetic acid[330k,1124b,1313a,1326c,1340a,1399b], zinc and hydrochloric acid [1340a,b,c,f], hydriodic acid[1340c,f], phosphorus trichloride in chloroform solution[1124b,1322b,1347,1396], phosphorus oxychloride[1124b], catalytically[245,1347], or with sodium dithionite[1313a,1326c,1397]. Zinc in acetic acid sometimes fails as a deoxygenating reagent, particularly for bridged diimidazoles which are relatively insoluble in acetic acid. In these circumstances zinc in formic acid is usually successful[1399b]. Reduction of 1-hydroxypyrazole 2-oxides with sodium dithionite removes the *N*-oxide function only to produce 1-hydroxypyrazoles[1326c]. In analogous fashion dithionite reduces 3,4-diazacyclopentadienone dioxides (217) to 1,4-dihydroxypyrazoles.

Triethylphosphite does not appear to be a suitable reagent for the deoxygenation of 1-methoxypyrazole 2-oxides, and benzoyl chloride appears to combine deoxygenation with side chain halogenation of a 3-methyl substituent[1322b].

Dehydrogenation of 1-hydroxyimidazoles and the corresponding
3-oxides with lead dioxide gives *N*-oxides and *N,N'*-dioxides of imid-
azolyls[1325b,c]. These short-lived radical products are also formed when
alkali salts of 1-hydroxyimidazole 3-oxides are treated with halogen in
polar organic solvents[1325c]. Reaction with hydroquinone reconverts
the imidazolyl 1,3-dioxides into the 1-hydroxy 3-oxides[1325b]. On
standing in air, 1,4-dihydroxypyrazoles are transformed into 3,4-diaza-
cyclopentadienone monoxides (218)[1313b], while peracid converts them
into 1,3,4-oxadien-5-one 4-oxides (219)[800].

(218)   (219)

Because of the difficulty involved in the synthesis of azole *N*-oxides
unsubstituted on ring carbon atoms, there is little information available
concerning the directive effect of the oxide substituent on electrophilic
or nucleophilic substitution. The nitration of 1-methylpyrazole 2-oxide
in sulphuric acid at C-5 appears to involve the free base form of the
heterocycle (p. 68).

N-Hydroxy compounds, and *N*-oxides which are tautomeric with
hydroxy forms, show typical reactions of *N*-hydroxy compounds. They
form acetyl[779,1323] and benzoyl[1324] derivatives which are still
capable of forming hydrochlorides. The compounds are thus *O*-acyl or
O-aroyl in structure. Methylation of 2,4,5-triphenylimidazole 3-oxide with
diazomethane[1377] or with methyl iodide in alkaline medium[142]
yields the *N*-methoxyl product. Sometimes acylation is more complicated.
When 1-methylpyrazole 2-oxide reacts with acetic anhydride the products
are 3-acetoxy-1-methylpyrazole and 2-acetyl-1-methylpyrazolin-3-one[245].
A rearranged product (221) results when 1-hydroxy-4,5-dimethyl-3-

(220)   (221)

phenylpyrazole 2-oxide (220) is acylated[1326c] and also on acetylation
of 1-methoxypyrazole 2-oxides[1322b]. A reaction showing some similarities to this occurs when (220) is nitrosated to yield the 4-nitropyrazol-
enine 1,2-dioxide[1326c]. Oxidation apparently accompanies the
nitrosation.

Few photochemical studies have been made in this series. Irradiation of 1-benzyl-3-phenyl-1,2,4-triazole 4-oxide in methanol or methylene chloride gives a mixture of 1-benzyl-3-phenyl-Δ^2-1,2,4-triazolin-5-one and deoxygenated starting material[1224b].

Tertiary amines react with 1-hydroxy-2-chloromethylimidazole 3-oxides to form quaternary salts quaternised in the side chain[1398].

The thermal rearrangement of 1-methoxypyrazole 2-oxides results in the formation of isopyrazole *N*-oxides[1322a]. While 1-methoxy-3,4,5-trimethyl- and 1-methoxy-4,5-dimethyl-3-phenylpyrazole 2-oxides undergo spontaneous thermolysis to the isomeric 5-methoxy compounds, by what is probably an ionic mechanism, 1-methoxy-3,4-dimethyl-5-phenylpyrazole 2-oxide rearranges much more slowly to produce a mixture of 3-, 4-, and 5-methoxy isomers[1322a].

9.13 Quaternary salts

Quaternary salts of the azoles are usually hygroscopic crystalline compounds, soluble in polar solvents but rather insoluble in highly covalent solvents. The structures of the salts have merited attention for many years and it is now generally agreed (see pp. 95–8) that pyrazolium salts are 1,2-disubstituted, imidazolium are 1,3-disubstituted, 1,2,3-triazolium are usually 1,3- and sometimes[1326c] 1,2-disubstituted, while 1,2,4-triazolium salts are 1,4-disubstituted. The quaternisation of tetrazoles is rather more complex. While tetrazolium salts derived from the oxidation of formazyl derivatives are 2,3-disubstituted[791], 1,3- and 1,4-disubstitution occurs[1111b,c,1402], particularly on alkylation of *N*-substituted tetrazoles. Some azole quaternary salts are listed as derivatives in tables in this book, while more extensive tables are provided by Fusco[901], Hofmann[58], Nineham[791], and Elguero[1401].

The quaternary hydroxides of the series are strong bases. Many of the tetrazolium salts are light-sensitive. The observation that crystalline tetrazolium salts turn yellow on exposure to light has been interpreted [1403] as a photoisomerisation caused solely by a change in position of the anion in the crystal layers.

Diquaternary salts of 1,2,4-triazole have been prepared only recently [1404]. The white, crystalline compounds decompose on standing even if stored at low temperatures in dry conditions.

Quaternisation results in considerable changes in the relative activities of ring positions and substituents. The heterocyclic ring becomes more susceptible to nucleophilic attack and less affected by electrophilic reagents. The reduction in electron density at a carbon atom between two substituted nitrogen atoms is reflected in pronounced deshielding in the n.m.r. spectrum of a hydrogen atom attached to that carbon atom [222,1405]. In some imidazole quaternary salts H-2 slowly exchanges with deuterium in deuterium oxide solution (p. 41). The effects of quaternisation are felt to a lesser extent at other ring carbons. Chlorine atoms in the 5-position of pyrazolium are displaced by bromine or amino, although additional activation in the form of an aryl substituent is a normal requirement[551c,735d,1354b] (pp. 140, 134). 1,3,4-Trimethyl-imidazolium iodide can lose a proton from C-2 to form a zwitterion capable of catalysing a benzoin condensation[1406] (p. 42). Similarly, 1,3-diphenylimidazolium reacts with phenylisothiocyanate to form a betaine (p. 32), and with mercury salts to form carbene complexes[326]. The susceptibility of imidazolium to nucleophilic attack is taken advantage of in the stereospecific nucleophilic displacement of the 1,2-dimethyl-imidazolium group from optically active 1-[(diethylamino)phenylphosphino-thioyl]-2,3-dimethylimidazolium iodide in the preparation of optically active thiophosphoryl compounds[1407]. Imidazolyl anion is employed as the attacking nucleophile.

Ring substituents may also be activated in azole quaternary salts. In basic deuterium oxide the 5-methyl hydrogen atoms of 1,4,5-trimethyl-tetrazolium iodide exchange slowly. In the corresponding 1,3,5-salt the reaction is slower, but also shows some exchange of *N*-methyl hydrogens [1111a]. Methyl activation is also involved in condensation reactions leading to cyanine dyes from 1,2,4-triazole quaternary salts[540g,613c]. In 1,2,4-triazolium salts a 4-amino substituent is deprotonated in basic medium to form a stable triazole-*N*-imine (222) which can be protonated or alkylated at the exocyclic nitrogen[1408].

(222)

It is this susceptibility of azolium salts to nucleophilic attack which is responsible for the effects of basic reagents on them. This is seen in the ring opening of pyrazolium (p. 167), imidazolium (p. 168) and 1,2,4-triazolium salts (p. 170). The so-called pseudo-bases formed by treatment of imidazolium salts with alkali are probably open-chain compounds [411b,1146b,c,d,1289g,h,1409].

Sometimes the action of base induces azolium halides to eliminate hydrogen halide. Thus 5-aminopyrazolium iodides form imino compounds [330b,1354b]. Similarly, in a reaction which is reversed by treatment with perchloric acid, 3- or 5-alkyl-1,2,4-triazolium perchlorates are converted by sodium ethoxide to *C*-ylides or anhydro bases[1408].

Bases such as pyridine and piperidine have been used to induce dequaternisation of azole quaternary salts. *C*-methoxyl- and *C*-methylthio-1,2,3-triazolium salts are dealkylated in piperidine to yield the triazolone and triazoliosulphide respectively[576k,1326b]. The dequaternisation occurs exclusively from the exocyclic oxygen or sulphur function. Analogous reactions occur when 3-mercapto- and 3-methylthio-1,2,4-triazolium salts are heated above their melting points or boiled with pyridine, although the dequaternisation products are usually meso-ionic in nature[223c,613c]. In contrast to the 1,2,3-triazole series the corresponding methoxyl compounds are reported to be unreactive[223c].

When 4-amino-1-methyl-2,3-diphenylimidazolium is treated with potassium hydroxide, a Dimroth rearrangement occurs[1007].

Azole quaternary salts are usually resistant to oxidation, but the betaine of 2,3-di(*p*-hydroxyphenyl)-1,2,3,4-tetrazolium-5-carboxylic acid is reported[17c] to undergo oxidation to tetrazole. Reducing agents have a variety of effects on azole quaternary salts, forming azolines, azolidines and sometimes causing ring-opening. Hydride reductions are especially interesting[1415] (p. 153). Sometimes catalytic hydrogenation can result in dequaternisation, particularly if one of the nitrogen substituents is benzyl[149b]. *C*-Aryl substituents are reported[1403] to have a marked influence on the ease of reduction of tetrazolium salts. Powerful reducing agents cleave tetrazolium salts to formazans (p. 154). The observation that 2,3,5-triphenyltetrazolium chloride forms radicals under reducing conditions[1412b,1413] is responsible for its wide use as a staining agent for living materials to test biological redox activity (p. 154). For an e.s.r. study of the radicals the reduction was best carried out with aluminium amalgam *in vacuo* in 1,2-dimethoxyethane. The use of tetrahydrofuran as solvent resulted in too rapid decay of the radicals, while zinc-amalgam or zinc dust reduction proved too fast.

Much of the chemical interest in azole quaternary salts is a consequence of their dealkylation reactions, whether induced thermally or by the action of basic reagents. Most of the original work related to pyrazoles, but the other members of the series show similarities. Pyrolysis of pyrazole quaternary halides (usually at ~ 200°C *in vacuo*) results in loss of one of the nitrogen substituents as alkyl halide [60d,63f,66a,b,e,f, 100c,d,330a,d,392g,552b,1141b,1354c] and undoubtedly involves a nucleophilic substitution mechanism. It is found that the groups cleaved most readily from a ring nitrogen are those which most readily form cations. Thus, the ease of cleavage diminishes in the order: benzyl>allyl >methyl>ethyl>propyl>phenyl. Structural factors may alter this sequence, and preferential loss of a substituent from one of the two nitrogens may occur [66f]. Rearrangements are sometimes observed in

these thermolyses. When benzylpyrazolium salts are heated, expulsion of the benzyl halide is sometimes accompanied by simultaneous rearrangement of the residual alkyl group from one nitrogen atom to the other [66a,f,100d]. It may be that this phenomenon can be explained in terms of the transalkylation recently described by Grandberg and Kost [898b]. When 1-methyl-, 1-ethyl-, and 1-propyl-3,5-dimethylpyrazoles were heated with benzyl chloride in an open vessel at 150°C the 1-alkyl group was replaced by benzyl. Both decompositions of the quaternary salt formed

in situ are reversible but the more volatile alkyl halide is removed from the reaction mixture allowing the 1-benzyl-3,5-dimethylpyrazole to accumulate. Rearrangement of an alkyl group from nitrogen to C-4 of pyrazole has been reported[60d]. Whereas 1,2,3,5-tetramethylpyrazolium iodide (223) yields 1,3,5-trimethylpyrazole normally when heated in an open vessel, in a sealed tube at 260°C the product is 1,3,4,5-tetramethyl-pyrazole (224)[392g] probably indicating some secondary alkylation

(223)   (224)

under conditions where the alkyl halide is not removed from the site of reaction. Transalkylation also occurs when 1-benzylpyrazole is heated with methyl iodide in a sealed tube at 100°C[1401]. The reaction products are 1-benzylpyrazole methiodide and 1-methylpyrazole methiodide. At 140°C the 1,3-dibenzylpyrazolium iodide is also formed[1401]. Occasionally a *N*-alkyl function can be transferred to an exocyclic heteroatom, usually sulphur or selenium[552b,1354c]. This process probably proceeds through an intermediate salt quaternised exocyclically.

(some)

Similar considerations apply to the thermolysis of quaternary salts of imidazoles[66h,373a,c,374b,981a,1177,1289b]. Electronegative substituents in the 4- or 5-positions of 1,3-dimethylimidazolium weaken the adjacent *N*-alkyl bond. Thus 4-chloro-, 4-bromo- or 4-nitro-1,3-dimethylimidazolium iodides decompose with the preferential formation of the 1-methyl-4-substituted imidazoles[373c,374b,1289b]. Even though the influence of a 4-phenyl group is less pronounced the 1,4-isomer still predominates when the quaternary salt is decomposed[373c]. The order of firmness of attachment of alkyl groups to nitrogen is similar to the case of pyrazolium salts except that ethyl, butyl and propyl substituents show some seemingly inconsistent behaviour[66b] which merits further study. The thermal decompositions of 1,2,3-triazolium[576d,k,1326b,c], 1,2,4-triazolium[223c,613c,624,1414], and tetrazolium salts[1111b,c] have been largely neglected. In mixed alkyl tetrazolium iodides, methyl iodide is eliminated more readily than ethyl iodide[1111c]. The thermal stability of 1,3-dimethyl-5-phenyltetrazolium iodide is less than that of the 1,4-isomer. When the former isomer is heated the 1-methyl group is preferentially eliminated[1111c]. Likewise, at 130°C 1,3,5-trimethyltetrazolium iodide eliminates the 1-methyl group predominantly to yield 2,5- and 1,5-dimethyltetrazoles (85 : 15). At higher temperatures this selectivity begins to be lost[1111b]. The 1,4,5-trimethyl salt decomposes at 270°C to yield 1,5- and 2,5-dimethyltetrazoles (74 : 21), a phenomenon which can be accounted for if some intermolecular methylation occurs[1111b]. A report[624] that thermal dequaternisation of 1,2,4-triazole ethiodide produces 1,2,4-triazole and ethyl iodide merits confirmation. Dealkylation of 1,4-diethyltetrazolium occurs specifically at N-4[1414]. In 1,3-dimethyl-1,2,3-triazolium iodides a benzyl group is eliminated in preference to methyl[576d], but in 1,2-dialkyl salts most dequaternisation occurs from N-1[1326c]. However, in meso-ionic 1,2,3-triazoles dequaternisation from nitrogen is accompanied by alkylation of an exocyclic sulphide. A benzyl group at N-1 appears to be most labile[1326d]. Dequaternisation of methoxyl-substituted 1,2,3-triazolium salts occurs exclusively from the oxygen function to form the triazol-5-one[576k]. Similarly 3-mercapto- or 3-methylthio-1,4,5-trisubstituted-1,2,4-triazolium halides form meso-ionic triazoles when heated alone, or with pyridine[223c,613c]. 5-Mercapto-3-methyl-1,4-diphenyl-1,2,4-triazolium chloride is an exception in that it forms the triazoline-5-thione[223c]. Quaternary fluoroborates of azoles in which the nitrogen substituents are alkyl and acyl are converted into *N*-alkylazoles on treatment with water or alcohol[621] (see p. 99).

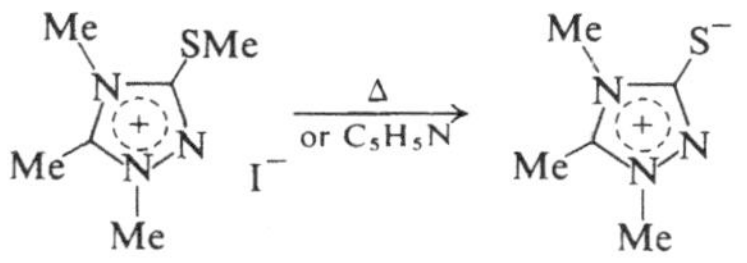

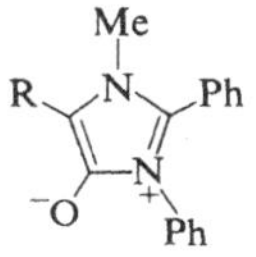

R = H, COMe, COEt

Some meso-ionic imidazoles quaternised on a ring nitrogen have been studied. The 5-acyl-substituted derivatives are quite stable due to delocalisation of the negative charge on the exocyclic oxygen, but the unsubstituted meso-ionic system (R=H) is very susceptible to electrophilic attack [1007]. Analogous meso-ionic 1,2,3-triazoles are dequaternised and *O*-benzoylated when heated with benzoyl chloride[576d]. The meso-ionic 1,4-diphenyl-1,2,4-triazolium compounds mentioned above did not take part in 1,3-dipolar reactions[223c], nor was it possible to brominate the nucleus under conditions effective for sydnones.

9.14 Radicals

The reactions of azoles with radicals have been discussed (p. 149). In this section, the properties and reactions of azole radicals will be considered.

The reduction of tetrazolium salts to form radicals[1413] has already been mentioned (p. 154). From a study[1412b] of the e.s.r. spectrum the stable tetrazolinyl radicals, formed by oxidation of formazans, reduction of tetrazolium salts, or by disproportionation of a mixture of both[790,1413,1416], have been assigned a resonance-stabilised acyclic structure[1412b]. The relative stabilities of these radicals increase with electron-withdrawing power of the substituents[1412a].

In recent years e.s.r. spectroscopy has been brought to bear on a number of studies of azole radicals[1325b,c,1417-18] and radical anions [1419]. It has, for example, allowed the recognition of pyrazole and imidazole anion radicals as the tautomeric α-pyrrolenine forms[1419].

rather than

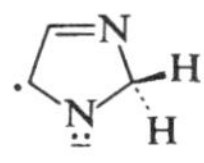

The chemiluminescence generated by the oxidation of arylimidazoles [1307a,b,1420-4] has been discussed (p. 157).

Triarylimidazolyl radicals oxidise electron-rich substrates by rapid electron abstraction from tertiary amines, iodide and metal ions[1423b]. The results of a kinetic study[1425] show that the rate-determining step in the reaction of the lophyl radical (L·) with an aromatic tertiary amine is electron exchange at the amino nitrogen (*viz.* L· + $>$N- → L:⁻ +$>$N-). The rate constant varies with substituents which increase the electron density at the amine nitrogen[1425]. A further example of this type of reaction is the photo-oxidation of leuco-triphenylmethane dyes by these radicals[1426-7]. Phenols, mercaptans, primary and secondary amines, and active methylene and methine compounds are oxidised by hydrogen abstraction[1423b,1425]. The rate constant for the oxidation of hydroquinone with triarylimidazolyl has a large kinetic isotope effect (k_H/k_D = 5.6) in benzene saturated with water or deuterium oxide. The abstraction of a hydrogen radical is therefore rate limiting[1423a]. Both types of oxidation reaction normally lead to the neutral triarylimidazole as product with the hydrogen being abstracted from the solvent if necessary[1423b]. The radicals react with nitric oxide to give *N*-nitrotriarylimidazoles[1307a], but do not react with aromatic hydrocarbons, aliphatic alcohols or vinyl monomers at rates which compete detectably with the dimerisation or oxidation reactions[1423b].

The oxidation of 2,4-diaryl-5-thienylimidazoles produces deeply coloured substances (green in solution) which resemble lophyl radicals in their reactions, except that they exhibit lower dehydrogenating capacity [1429a]. When oxidised with lead dioxide diarylnaphthylimidazoles also produce radicals with paramagnetic susceptibility[1428b]. Further examples of lophyl-type radicals with e.g. furyl substituents have been studied recently[1428-30].

When 1-triphenylmethylimidazoles are heated they form deep blue melts which probably contain trityl and imidazolyl radicals. The products isolated from these melts show that rearrangement to 2- and 4-triphenylmethylimidazoles has occurred[569a,b,c]. Radical intermediates have also been implicated in the photoaddition of acetone to 1- and 2-methyl-, and 1,2-dimethyl-imidazole[493], and they almost certainly are involved in some of the thermolyses of 1-substituted imidazoles[981b].

In contrast to nitroxide radicals, the radicals which are formed on dehydrogenation of 1-hydroxyimidazoles or 1-hydroxyimidazile 3-oxides are unstable in solution[1325b,c]. In the γ-irradiation of 1-substituted-5-aminotetrazoles radicals of the type R–CH$_2$· (where R = 5-substituted tetrazole) are formed[1202d].

9.15 Sulphur-containing compounds

This section will comprise the reactions of thiols (or thiones), thioethers, sulphonic acids, sulphones and other sulphur containing derivatives. Tables A.21 and A.22 list some representative examples.

9.15.1 Substituents on ring carbon

The dequaternisation of sulphur-substituted azole quaternary salts has been discussed (§9.13). Reference has been made to the ultraviolet (p. 14) and infrared (p. 18) spectra of mercaptoazoles.

Many of the mercapto derivatives of azoles may exist in the tautomeric thione forms. A study[1452b] of 3-a-pyridyl-1,2,4-triazoline-5-thione and its 1-, 2- and 4-methyl derivatives has shown that the thione modification predominates in ethanol solution. Pyrazole-3- and -5-thiones are acidic compounds which are readily soluble in alkali (4-mercapto-pyrazoles are less well known). Imidazole-2-thiones, similarly, are high melting solids which dissolve in aqueous alkali with the formation of stable salts. Unlike other imidazoles they are too weakly basic to form picrates or hydrochlorides. Compounds in which the oxygen atoms of triazolone, urazole etc. are replaced by sulphur are also very strongly acidic.

The alkylation of azole thiones[1446,1455-6] has been discussed (p. 94).

When thiones are treated with oxidising agents a variety of products arise depending on the reaction conditions. Mild oxidation with air[506a, 1446], iodine[330f,1354d,1446], hydrogen peroxide[374c,1446,1457] N-bromosuccinimide[1447], bromine in alkali[1198], and sometimes electrolytically[1118c], converts them into disulphides. More vigorous conditions have occasionally allowed isolation of a sulphinic acid, as in the carefully controlled oxidation with hydrogen peroxide of 4-methyl-imidazole-2-thione. Analogously, some thiol ethers of the series have been oxidised to sulphoxides. Peroxyphthalic acid transforms 3-benzylthio-1 2,4-triazole into a mixture of sulphoxide and sulphone[1118c], while controlled treatment with hydrogen peroxide converts the corresponding methylthio compound into the sulphoxide[1191a]. Sulphonic acids are the commonest products when azole thiones are oxidised (with hydrogen peroxide, chlorine, bromine, chromic acid or nitric acid)[24a,83d,126c, 321j,330a,d,506a,552b,1118b,1270,1279,1344,1446,1461]. The oxidation process is complicated with 1,2,4-triazoline-3-thione which is not converted into the sulphonic acid with alkaline hydrogen peroxide; neutral or acid peroxide or bromine yield traces only, and the last named reagent produces the acid sulphate of 1,2,4-triazole[1118b]. When pyrazolethiones carry substituents on both nitrogen atoms as in thiopyrine the oxidation product is a betaine[330a]. Disulphides, rather than

sulphonic acids, are the usual oxidation products from 1,2,3-triazolethiones[1458]. When oxidation is more vigorous, desulphurisation to the parent azole results. Nitric acid[12a,1440], hydrogen peroxide[1191a], ferric chloride[1459], and oxygen in the presence of platinum catalyst [1120b] have proved effective. Bromine causes some oxidative cleavage of the carbon–sulphur bond, but also leads to products in which the mercapto group is replaced by bromine[1118c]. Although the sulphur-containing groups may also be removed by treatment with concentrated hydrochloric acid at high temperatures[552b], and with nickel boride [1460], the commonest procedure employs Raney nickel which smoothly cleaves thiol (and thioether) groups[453a,674a,969c,971a,1120b,1227b, 1446]. It is probable that the oxidative desulphurisation reactions proceed *via* the unstable sulphinic acids, which lose sulphur dioxide with great facility. The high resistance of sulphonic acids to acid hydrolysis makes their intermediacy in the reaction unlikely. Chromic acid oxidation of 2-mercapto-4,5-diphenylimidazole forms *N,N'*-dibenzoylurea[425b]. When 1,2,4-triazoline-3-thiones are treated with chlorine or bromine unstable sulphonyl halides result[1101a,1118c].

Alkylmercaptoazoles, which are basic compounds, are oxidised to sulphones with normal oxidising agents[24c,83d,321j,506a,552b,607b, 735h,1048b,1279], and to sulphoxides by careful oxidation with hydrogen peroxide[1191a].

Removal of the sulphur function is achieved with Raney nickel[453a, 1187,1227b]; or by high temperature treatment with acids[552b]. Cleavage to thiols or thiones is accomplished with hydriodic acid[24c, 1326d], acetyl iodide in acetic acid[1442], sodium in liquid ammonia [1114b], or aluminium bromide in benzene[1447]. Alkylthio groups activate adjacent ring positions to electrophilic substitution[321j,552b]. An alkylmercapto group at C-5 of tetrazole resists displacement by ammonia or amines, while bromine in acetic acid induces ring fission [590b]. When 2-alkylthio-1-methyl- or 2-allylthio-1-methyl-imidazoles are heated they undergo Claisen rearrangement[1362].

Azole disulphides are high melting, crystalline compounds which may be cleaved to thiones with hydrogen sulphide[1446], or by the action of sulphur dioxide on an aqueous acidic solution of the disulphide[374c, 1447]. More drastic oxidation can lead to sulphones or sulphonic acids [126c,1445a,1446], while Raney nickel treatment produces two mole-cules of azole[1445a]. With chlorine in carbon tetrachloride bis(1,3,5-tri-substituted pyrazol-4-yl) disulphides are converted into pyrazole sulphenyl chlorides[1256].

Sulphinic acids (and derivatives) are not very common in the azole series. The few pyrazole-sulphinic acids known have been prepared by reduction of sulphonyl chlorides with amalgamated sodium[446], while a small number of unstable imidazole-sulphinic acids are available from the careful oxidation of thiones (p. 249). Reduction of 3-methyl-1,5-diphenylpyrazole-4-sulphinic acid with sodium and alcohol gives a mixture of the corresponding thiol and a pyrazoline[506a]. Treatment with dilute acid results in loss of sulphur dioxide, while bromine replaces the sulphinic acid group[506a]. Oxidation of sulphoxides to sulphones occurs readily[598,1191a].

Sulphonic acids and sulphones are much more common. The former have high melting points and probably exist as zwitterions. Although amphoteric they mainly display weakly acidic properties. The free acids are rather variable in both stability and reactivity, for, while pyrazole-3-, -4- and -5-sulphonic acids react with phosphorus pentachloride to form sulphonyl chlorides[330f,431c,446,1354d], imidazole-4-sulphonic acids [543-4], and 1,2,4-triazole-3-sulphonic acids[1118b] resist esterification and reaction with thionyl chloride and phosphorus halides. The com-pounds are often highly stable to hydrolysis, e.g. 1,2,4-triazole-3-sulphonic acids are stable to boiling aqueous hydrobromic acid or alkali at 80-90°C [1118b]; pyrazole-sulphonic acids are only hydrolysed at high tempera-tures[330f,431c,446,1354d]; imidazole-4-sulphonic acids[431c,446,1025f, 1462] are hydrolysed by concentrated hydrochloric acid at 170°C, while the 2-isomers are even more resistant[543]. The free acids are converted into hydroxy derivatives when subjected to aqueous alkali treatment[24c, 83d]. Sulpho groups may be displaced by bromine[446] or cyano[607b].

Halogen derivatives include sulphenyl and sulphonyl halides. The former are rather labile compounds which have received little attention. Pyrazolyl-4-sulphenyl chlorides[1256] are unstable in the atmosphere with respect to the disulphides. In dry nitrogen they form sulphides with acetylacetone, adducts with cyclohexene, sulphenamides with amines, sulphenate esters with alcohols, and derivatives with acetone and with dimethylaniline. The reaction product with acetone is probably a sulphide since it forms a sulphone on oxidation. Excess chlorine results in decomposition of the sulphenyl chloride [1256]. 4-Bromosulphenyl-1,2,3-triazolium salts react in a similar fashion [1326d].

Sulphonyl halides are much more stable and may be converted readily into sulphonamides [544,1101a,1118c,1463], and (with some exceptions) into sulphonic acids [1118b,c]. Reduction to the thiols is possible [446, 1463]. The corresponding amides are less reactive than the halides. A sulphonamido group can be removed from C-4 of imidazole with diethylamine [1392]. Catalytic reduction with Adams' catalyst converted a 5-sulphonamidoimidazole into the corresponding 5-sulphamylimidazole [1101a].

9.15.2 Substituents on ring nitrogen
Azoles with oxysulphur substituents on nitrogen react in much the same way as *N*-acylazoles, and so they may be classified as azolides. At room temperature imidazolides of sulphonic acids are relatively stable to hydrolysis, but are hydrolysed rapidly when heated in water [150e]. In alkaline conditions hydrolysis is rapid even at low temperature [150e,1434]. Tosyl groups are rapidly hydrolysed from pyrazole [172b,c,1159a] and

triazole[1338] ring nitrogen atoms under alkaline conditions. This is in contrast to other sulphonamides which are very stable to alkaline hydrolysis.

In chloroform solution some 5-alkoxyl-1-arylsulphonyl-1*H*-1,2,3-triazoles appear to exist in equilibrium with an acyclic isomer[1464].

Normal azolide reactions of the compounds include aminolysis to sulphonamides[150e], alcoholysis to sulphonic esters[150e*l*,1436], and ether formation with sodium alkoxides[150e,933b]. The compounds are generally less reactive than the corresponding carboxylic derivatives, but high yields of products can be obtained. In reaction with the sodium salt of benzyl alcohol, 1-*p*-toluenesulphonylimidazole forms 63% of the dibenzyl ether, but 24% of 1-benzyl-imidazole is produced in a competing reaction[150e]. Attempts to formylate 1-benzenesulphonylpyrazole remove the substituent[59d]. Transacylation reactions are important, and 1,1'-thionyldiimidazole has similar applications to those of 1,1'-carbonyldi-imidazole[150k,933c,1373,1435,1465]. Reaction of 1-*p*-toluenesulphonyl-imidazole with diphenylmethanol in the presence of the sodium derivative of imidazole yields three isomeric diphenylmethylimidazoles[150*l*]. Tri-fluoromethanesulphonic imidazolide is a convenient reagent for intro-ducing the 'triflate' group[1432].

9.16 Miscellaneous substituents

Silicon-containing substituents have recently been studied in some detail. A *N*-trialkylsilyl group is readily removed by hydrolysis[190d,1467] or alcoholysis[190d], and reacts smoothly in acylations[190b,f,1465] (p. 117). In similar reactions thionyl chloride, phosphorus trichloride and phosphorus oxychloride produce 1,1'-thionyldiimidazole, tris(1-imidazolyl)-phosphine, and the corresponding phosphine oxide respectively. The *N*-trimethylsilyl derivatives of pyrazole, imidazole and 1,2,4-triazole are thermally stable, but the silylated tetrazoles are decomposed by heat [190c,906d,e,1466]. In the n.m.r. spectrum 1-trimethylsilylpyrazoles show non-equivalent 3- and 5-substituents which become equivalent at high temperatures. An intramolecular rearrangement of the silyl substi-tuent is adduced to explain the phenomenon[1309]. The dipole moments of some 2-trimethylsilyl-2*H*-1,2,3-triazoles have been measured[190d].

The 2-trimethylsilylimidazoles are of value synthetically in that the substituent is replaced exothermically when treated with carbonyl com-pounds[939,980,1467]. Selective desilylation of 3- and 4-trimethylsilyl-pyrazoles is possible. The silyl groups at C-3, but not at C-4, can be removed by nucleophilic reagents, while concentrated sulphuric acid displaces a 4-silyl substituent[906c].

Many *N*-trimethylstannyl-imidazoles and -1,2,4-triazoles are more stable to hydrolysis than the foregoing silyl compounds, although there are some exceptions e.g. diethyldi(1,2,4-triazolyl)tin is quickly converted by water into the dialkyltin oxide. The observation that only azoles with 1,3-heteroatoms can form stable organotin derivatives suggests that some form of association is operating[703].

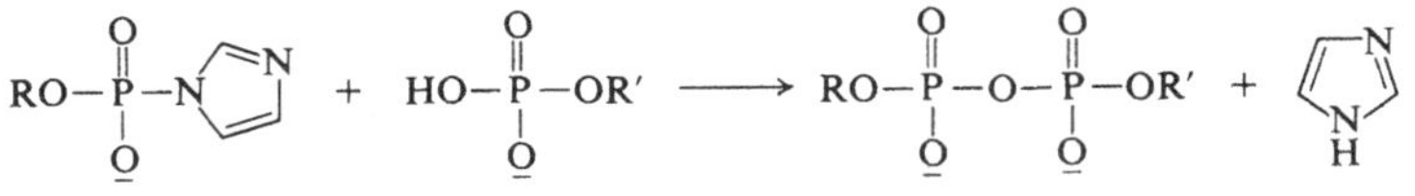

Imidazolides of phosphoric acid have found a number of synthetic applications, particularly as phosphorylating agents[918a,933a,934a,b,935, 1468]. Thus ammonolysis[935] and aminolysis[918a] form phosphor-amides in high yield, alcoholysis leads to phosphate esters[935] (in particular, alcoholysis of imidazolides of monoesters of phosphoric acid yields diesters of phosphoric acid), and reaction with carboxylic acids forms primarily acylphosphates[935]. Adenosine diphosphate has been prepared by converting the monophosphate into its imidazolide and then treating the product with orthophosphate[1468]. Rapid hydrolysis in water is a characteristic of the imidazolides of phosphoric acid, with di- and tri-imidazolides being especially sensitive to moisture[934a].

Appendix 1
The orientations of 1,3- and 1,5-dimethylpyrazoles

One of the crucial orientations in the pyrazole series is that which distinguishes 1,3- from 1,5-dimethylpyrazole. Both isomers form the same methiodide which is cleaved thermally to produce equal quantities of the isomers[66f].

von Auwers and Hollmann[66d] established the structures of the isomeric dimethylpyrazoles using a method based on the decarboxylation of isomeric pyrazole carboxylic acids. The deduction of the structures of these acids and their esters depends on a comparison of the esterification and hydrolysis rates which are affected by steric hindrance at the carboxyl functions. Only 4-bromo-1,5-dimethylpyrazole-3-carboxylic acid can form an ester, since in the other isomer both methyl and bromo groups hinder reaction at the carboxyl substituent. Similar considerations apply to hydrolysis of the esters[66f]. In a similar fashion Rojahn[62a] came to the same conclusion regarding the orientation.

Criticism[97] of this orientation assignment has since been negated [98-9]. Burness[97] had considered that the reaction between hydrazine and β-ketobutyraldehyde acetal proceeds through an intermediate hydrazone which, on cyclisation, can only form 1,3-dimethylpyrazole. However, the irreproducibility[98-9] of this synthesis coupled with an unequivocal preparation of 1,3-dimethylpyrazole from 1,3-dimethylpyrazolin-5-one[98] seems finally to have settled the problem. N.m.r. spectroscopic methods are now able to be applied to such orientation problems. Chemical shift values for 1,3- and 1,5-dimethylpyrazoles, tabulated below, indicate that in the 5-methyl isomer H-3 is at lower field and CH_3-5 is at higher field than the corresponding H-5 and CH_3-3 protons in 1,3-dimethylpyrazole.

Chemical shifts (δ) for isomeric pyrazoles[98]

Pyrazole	CH_3-3	CH_3-5	H-3	H-5	H-4
1,3-dimethyl-	2.22			7.13	5.93
1,5-dimethyl-		2.19	7.27		5.90

Appendix 2
Properties of azoles

Table A.1. The azoles and their alkyl- and aryl-derivatives

Pyrazoles	M.p./°C	B.p./°C (mmHg)	Derivatives (m.p./°C)
Pyrazole[2a,b,c]	70	187	Picrate (160) Nitrate (148) 1-Acetyl (b.p. 140°C at 52 mmHg[56]) 1-Benzoyl (b.p. 220-5 °C at 60 mmHg[56])
1-Me[63b]		127	Picrate (148)
3-Me[1c,65a]		204	Picrate (142)
4-Me[17d]		205 (730)	Picrate (142)
1-Ph[1d,59a,60d]	11	246; 58-62 (0.01)	Methiodide (179)
3-Ph[1e,2e]	78	313	Picrate (170-1)
4-Ph[61c]	230		Picrate (155)
1,3-Me$_2$[a]		136	Picrate (138)
1,4-Me$_2$[62c]			Picrate (165)
1,5-Me$_2$[a]		153	Picrate (172)
3,4-Me$_2$[63b]	58	111 (10-11)	Picrate (153)
3,5-Me$_2$[64,65b]	107	218 (758)	Picrate (167)
1-Me-3-Ph[66b,g]	56	138-9 (12)	Picrate (132-3)
1-Me-5-Ph[66b,g]		118 (12)	Picrate (143-4)
3-Me-1-Ph[67a,68]	37	255 (753)	
4-Me-1-Ph[60d]	264-6		
5-Me-1-Ph[69b,d,329c]		263.5 (762)	Picrate (97-8)
3-Me-4-Ph[70a]	141		
3-Me-5-Ph[66e]	128	191-3 (14)	Picrate (159)
1-PhCH$_2$-3-Me[66a,f]		141 (14)	Methiodide (154) Picrate (114)
1,3-Ph$_2$[66b]	84-5		
1,5-Ph$_2$[66b]	55-6		
1,3,4-Me$_3$[100g]		160	Picrate (163.5-164.5)
1,3,5-Me$_3$[1c,66a]	37	170 (755)	Picrate (147)
1,4,5-Me$_3$[100g]		176-7	Picrate (175-6)
3,4,5-Me$_3$[1c]	137-8	232 (753)	Picrate (237-9)
1,3,4-Ph$_3$[71][b]	185		
1,3,5-Ph$_3$[72]	140		
1,4,5-Ph$_3$[73]	212		
3,4,5-Ph$_3$[70b]	265		

	M.p./°C	B.p./°C (mmHg)	Derivatives (m.p./°C)
1,3,4,5-Me$_4$[1c,63b,f]		190–3	Picrate (194)

Imidazoles[c]

	M.p./°C	B.p./°C (mmHg)	Derivatives (m.p./°C)
Imidazole	90	256	Picrate (212)
			Nitrate (118)
			1-Acetyl (102)
			1-Benzoyl (19)
1-Me	–6	198	Picrate (158)
2-Me	141	267	Picrate (213)
4-Me	56	264	Picrate (162)
1-Ph	13	277	Picrate (155)
2-Ph	148	340	Picrate (238)
4-Ph	133		Picrate (216)
1-PhCH$_2$	71–2		
2-PhCH$_2$	126		
4-PhCH$_2$	84–5		
1,2-Me$_2$		205	Picrate (181)
1,4-Me$_2$		200	Picrate (167)
1,5-Me$_2$		220	Picrate (168)
2,4-Me$_2$	92		Picrate (142)
4,5-Me$_2$	120		Hydrochloride (305)
1-Me-4-Ph	111		Picrate (245)
1-Me-5-Ph	97		Picrate (138)
4-Me-2-Ph	181		
4-Me-5-Ph	185		
2,4-Ph$_2$[d]	193		
4,5-Ph$_2$	249		Picrate (235)
2,4,5-Me$_3$	183	271	Picrate (163)
2,4,5-Ph$_3$	275		Picrate (163)
1,2,4,5-Me$_4$	58		Picrate (189)
1,2,4,5-Ph$_4$	221		

1,2,3-Triazoles

	M.p./°C	B.p./°C (mmHg)	Derivatives (m.p./°C)
1,2,3-Triazole[18h,19a,b, 74d,75a]	23	203 (739)	N-Benzoyl (111)
			N-Acetyl (62)
1-Me[76d]	15–16	228 (752)	
2-Me[77c]	21.5–22	89–90 (714)	
		50.5–51 (165)	
1-Ph[19a,b,78]	56	172 (185)	
2-Ph[79]		223–4 (716)	
4-Ph[80a,81]	144		
1,5-Me$_2$[76d]	–4	255 (751)	
4,5-Me$_2$[17e]	70[e]		
4-Me-2-Ph[79]		150 (60)	
5-Me-1-Ph[19a,b]	64		
4-Me-1-Ph[78]	81		
1,5-Ph$_2$[19c]	114		
2,4-Ph$_2$[82]	57		
4,5-Ph$_2$[83b]	139		

	M.p./°C	B.p./°C (mmHg)	Derivatives (m.p./°C)
1,2,4-Triazoles			
1,2,4-Triazole[76,22d,84b]	121	260	Nitrate (138) Hydrochloride (168–9)
1-Me[22f]	20	178	
3-Me[20c,22g]	95	265	
4-Me[24b]	90		
1-Ph[20c,22d,g]	47	266	Picrate (159)
3-Ph[22g,81,85,86a]	121		
4-Ph[22a,75b]	122		Picrate (172)
3,5-Me$_2$[87]	143		
3-Me-1-Ph[22g]	87	274	Picrate (171)
3-Me-4-Ph[22g]	112^f		Picrate (134)
5-Me-1-Ph[22g]		275	Picrate (146)
4-Me-3-Ph[85]	112–3		
1,3-Ph$_2$[88]	96–7		Picrate (148)
1,5-Ph$_2$[89]	91		Picrate (139)
3,4-Ph$_2$[12c]	142		Picrate (174)
3,5-Ph$_2$[90d]	192		*N*-acetyl derivative (105)
1,3,5-Ph$_3$[91]	104		Ethiodide (145)
3,4,5-Ph$_3$[83c,92]	292		
Tetrazoles			
Tetrazole[18f,93]	157.5–158		
1-Me[80j,94]	38–9	110 (5)	
2-Me[80d,94]	9–10	146	
5-Me[93]	148–148.5		
1-Et[80d]		162 (30)	
2-Et[80d]		70–1 (35); 152–5	
1-Ph[19g,24c]	66		
2-Ph[18b]		An oil which explodes on heating	
5-Ph[93]	217–18		
1,5-Me$_2$[95]	71.8–72.8		
2,5-Me$_2$[95]		57.0–57.2 (13)	
5-Me-1-Ph[19g]	97.5		
5-Me-2-Ph[19f]	40	140 (15)	
1-Me-5-Ph[96]	102–3		

(a) Despite some recent doubts[97] it is now certain that these assignments of structure[62a,66d,f] are correct[98,99]. See Appendix 1.
(b) This compound is most probably 2,4,6-triphenylpyrimidine[57].
(c) The data for imidazoles are from the comprehensive tabulation by Hofmann[58].
(d) Also, m.p. 168 °C; dimorphic.
(e) M.p. of trihydrate, 97 °C.
(f) M.p. of monohydrate, 68 °C.

Table A.2. A selection of ultraviolet absorption data for azoles

Pyrazoles	Organic solvent λ_{max}/nm ($\log_{10}\epsilon$)	H_2O λ_{max}/nm ($\log_{10}\epsilon$)	Acid media λ_{max}/nm ($\log_{10}\epsilon$)	Basic media λ_{max}/nm ($\log_{10}\epsilon$)
Pyrazole[74d,113,116,118]	210 (3.45)[a] 210 (3.53)[c] 212 (3.45)[d]	211 (3.61)[113]	217 (3.67)[116][b]	
1-Me[118]	216 (3.63)[c]		219 (3.72)[e]	
3-Me	214 (3.58)[116][c]		218 (3.79)[118][e]	
4-Me	220 (3.47)[116][c] 219 (3.49)[74d][d]		226 (3.65)[118][b]	
1,3-Me$_2$[118]	221 (3.67)[c]		223 (3.84)[e]	
1,5-Me$_2$[118]	217 (3.64)[c]		221 (3.77)[e]	
3,4-Me$_2$	222 (3.63)[118]		229 (3.78)[116][b]	
3,5-Me$_2$	214 (3.65)[118][c] 214 (3.61)[74d][d]		220 (3.88)[118][e]	
1,3,4,5-Me$_4$[118]	228 (3.67)[c]		232 (3.87)[e]	
1-Ph	~206 (4.21), 255 (4.15)[118][c] 253 (4.18)[125b][f]		206 (4.03), 246 (3.98)[118][e]	
3-Ph[118]	<210 (>4.15), 249 (4.23)[a] ~206 (4.20), 249 (4.20)[c]		~208 (4.08), 252 (4.23)[e]	
4-Ph[119]	249 (4.11)[f]		236 (4.11)[e]	
3-Me-1-Ph[125b]	256 (4.15)[f]			
5-Me-1-Ph[125b]	240 (4.00)[f]			
1-Me-3-Ph[118]	253 (4.26)[a]		257 (4.25)[e]	
1-Me-4-Ph[119]g	251 (4.18)[f]		236 (4.04)[e]	
1,3-Me$_2$-4-Ph[119]g	245 (4.09)[f]			
1,5-Me$_2$-4-Ph[119]g	243 (4.05)[f]			

Table A.2. Continued.

Pyrazoles	Organic solvent λ_{max}/nm (log$_{10}\epsilon$)	H$_2$O λ_{max}/nm (log$_{10}\epsilon$)	Acid media λ_{max}(nm/log$_{10}\epsilon$)	Basic media λ_{max}/nm (log$_{10}\epsilon$)
1,3,5-Me$_3$-4-Ph[119][g]	240 (4.00)[f]			
3,5-Me$_2$-4-Ph[119][g]	240 (3.99)[f]			
1-Me-4-Ph[118]	249 (4.10)[a]			
1-C$_6$H$_4$.NO$_2$(p)[114d]	313 (4.24)[c]		246 (4.12)[e]	
3-Et-1-C$_6$H$_4$.NO$_2$(p)[97]	224 (3.89), 323 (4.23)[f]			
5-Et-1-C$_6$H$_4$.NO$_2$(p)[97]	216 (3.98), 299 (4.03)[f]			
3-Et-4-Me-1-C$_6$H$_4$.NO$_2$(p) [97]	232 (3.89), 335 (4.28)[f]			
5-Et-4-Me-1-C$_6$H$_4$.NO$_2$(p) [97]	217 (4.05), 306 (4.03)[f]			
1-C$_6$F$_5$	239 (3.96)[121][c]			
3,5-Me$_2$-1-C$_6$F$_5$	225 (3.87), 240 (3.84) [121][c]			
3-CO$_2$H[118]	214 (3.96)[c]		214 (3.90), 235 (3.75)[e]	
3-CO$_2$Me[118]	217 (3.94)[c]		214 (3.91), 235 (3.73)[e]	
1-MeCO[74d][h]	239 (4.06)[d]			
5-NH$_2$-3-Me-1-i-Pr[126h]	228 (3.86)[f]			
3-NH$_2$-1-Ph[125b,126h]	284 (4.23) (4.32)[f]			
4-NH$_2$-1-Ph[125b]	282 (4.16)[f]			
5-NH$_2$-1-Ph[125b]	240 (4.17)[f]			
3,5-(NH$_2$)$_2$-1-Ph hydro-chloride[127]	249 (4.26)[f]			
3,5-(NH$_2$)$_2$-1-PhCH$_2$CH$_2$CH$_2$CH$_2$ hydrochloride[127]	237 (4.12)[f]			

Table A.2. Continued

Pyrazoles	Organic solvent λ_{max}/nm ($\log_{10}\epsilon$)	H_2O λ_{max}/nm ($\log_{10}\epsilon$)	Acid media λ_{max}/nm ($\log_{10}\epsilon$)	Basic media λ_{max}/nm ($\log_{10}\epsilon$)
3,5-Me$_2$-4-PhN$_2$[114c]	330 (4.26), 410 (2.98)^s			
1,3,5-Me$_3$-4-PhN$_2$[114c]	332 (4.36), 430 (3.18)^s			
3,5-Me$_2$-4-PhN$_2$-1-Ph [114c]	334 (4.41), 428 (3.15)^s			
4-Cl[74d]	220 (3.44)^d			
4-Br[74d]	221 (3.39)^d			
4-I[74d]	226 (3.39)^d			
3,4,5-Br$_3$[74d]	222.5 (3.69)^d			
4-NO$_2$	269 (3.92)[74d]^d 275 (3.91)[128]^c	274 (3.92)[120]	238 (3.89)[128]^i	320 (4.07)[128]^j
1-Me-4-NO$_2$[120]		280–2 (3.96)		
3-Me-4-NO$_2$[120]		280–4 (3.91)	281–4 (3.91)^k	322 (4.08)^l
3,5-Me$_2$-4-NO$_2$[120]		284 (3.94)	284 (3.93)^k	321 (4.10)^l
1,3,5-Me$_3$-4-NO$_2$[120]		290 (3.97)		
4-NO$_2$-1,3-Ph$_2$[129]	256 (4.24), 300 (3.95)^c			
4-NO$_2$-1,5-Ph$_2$[129]	245 (4.24), 280 (3.85)^c			
3-Me-4-NO$_2$-1-Ph[129]	300 (3.95)^c			
Pyrazol-3-ones and some O-alkyl derivatives^o				
1,2,5-Me$_3$[130a]	257.5 (3.97)^a	246.5 (3.98)^m	227.5 (3.94)^n	
3-EtO-1,5-Me$_2$[130a]	225.5 (3.74)^a	221.5 (3.80)^m	229.5 (4.00)^n	
1,5-Me$_2$[130a]	227.5 (3.74)^a	245.5 (3.83)^m	226.5 (3.95)^n	
1-Ph[131]	272 (4.28)^c			239 (3.74)^j

Table A.2. Continued

Pyrazol-3-ones and some O-alkyl derivatives[o]	Organic solvent λmax/nm (log₁₀ε)	H₂O λmax/nm (log₁₀ε)	Acid media λmax/nm (log₁₀ε)	Basic media λmax/nm (log₁₀ε)
3-MeO-1-Ph[131]	272 (4.24)[c]			
4-Me-1-Ph[131]	278 (4.37)[c]			
4-NO₂-1-Ph[131]	326, 270, 238 (3.98; 3.88; 4.15)[c]			
Pyrazol-5-ones and some O-alkyl derivatives[t]				
1,2,3-Me₃[130a]	257.5 (3.97)[a]	246.5 (3.98)[q]	227.5 (3.94)[n]	
1,3,4,4-Me₄[130a]	250 (3.63)[a] p	248 (3.56)[q]	260.5 (3.59)[r]	
5-EtO-1,3-Me₂[130a]		—pq	224 (3.89)[n]	
1,3-Me₂[130a]	251 (3.61)[a]	241 (3.90)[q]	222.5 (3.89)[n]	233 (3.81)[j]
3-Me-4-PhN₂[114c]	410 (4.16)[s]			
4-PhN₂-1-Ph[132b]	250 (4.37), 398 (4.34)[c]			
3-Me-4-PhN₂-1-Ph	251 (4.37); 393 (4.38)[132b][c] 393 (4.35)[114c][s]			
5-MeO-3-Me-4-PhN₂-1-Ph	399 (4.32); 400 (3.72)[114c][s]			
Imidazoles				
Imidazole	207.5 (3.70)[133][cp]	206 (3.54), 265 (0.18)[113][u]	207.5 (3.69)[133][e]	213.5 (3.66)[133][v]
1-Me[133]	212 (3.63)[c]	211 (2.63)[139][aa]	210 (3.63)[e], <215 (>3.55)[139][aa]	

Table A.2. Continued

Imidazoles	Organic solvent λ_{max}/nm ($\log_{10}\epsilon$)	H$_2$O λ_{max}/nm ($\log_{10}\epsilon$)	Acid media λ_{max}/nm ($\log_{10}\epsilon$)	Basic media λ_{max}/nm ($\log_{10}\epsilon$)
2-Me	209.5 (3.61)[c]			
4-Me	215 (3.67)[133][c] ~225 (3.79), 243 (3.84) ~266 (3.27)[a]	214.5 (3.69)[44b]	216 (3.68)[133][e] 218 (3.94), ~230 (3.85), ~261 (3.08)[x]	290 (1.44)[134][w]
1-Ph[135]	236 (3.93), ~264 (3.27)[c]			
2-Ph[133]	271 (4.20)[c]		265 (4.20)[e]	300 (4.1)[y]
4-Ph	260 (4.17)[136][f]	257 (4.2)[134]		
4-CHO	256 (4.07)[124][c] 253 (3.4)[z]		237 (3.85)[137b][x]	280 (4.17)[137b][v]
1-MeCO[138]	270.5 (2.99), 277 (2.91)[c]			
4-CO$_2$H[124]				
4-CONHPh[139]		260 (4.19)[aa]	263 (4.00)[aa]	280 (4.06)[aa]
2-Br[139]		<220[aa]	220 (3.86)[aa]	<220[aa]
4-Br[139]		211 (3.61)[aa]	217 (3.71)[aa]	
2-Br-1-Me[139]		<220[aa]	221 (3.81)[aa]	
5-Br-1-Me[139]		216 (3.68)[aa]	219 (3.68)[aa]	
2-NO$_2$[140]		325 (3.95)[aa]	298 (3.91)[aa]	372 (4.13)[aa]
4-NO$_2$[139,140,141]		298 (3.80)[aa]	267 (3.85)[aa]	227 (3.52); 352 (4.01)[aa]
2,4-$(NO_2)_2$[140]		304 (4.05)[aa]		354 (4.09)[aa]
1-Me-2-NO$_2$[140]		325 (3.93)[aa]	300 (3.89)[aa]	
1-Me-4-NO$_2$[139,140,141]		225 (3.55), 301 (3.83)[aa]	268 (3.85)[aa]	
1-Me-5-NO$_2$[139,140,141]		227 (3.53), 305 (3.91)[aa]	367 (3.80)[aa]	
1-Me-2,4-$(NO_2)_2$[140]		305 (4.06)[aa]		
1-Me-4-NO$_2$-5-NC$_5$H$_{10}$[139]		232.5 (3.90), 285 (3.63) 406 (3.81)[aa]		
1-Me-5-NO$_2$-2-NC$_5$H$_{10}$[139]		223 (3.93), 398 (4.02)[aa]	229 (4.11), 350 (3.84)[aa]	356 (4.01)[aa]
4-Cl-5-NO$_2$[140]		304 (3.82)[aa]	267 (3.86)[aa]	

Table A.2. Continued

Imidazoles	Organic solvent λ_{max}/nm (log$_{10}\epsilon$)	H$_2$O λ_{max}/nm (log$_{10}\epsilon$)	Acid media λ_{max}/nm (log$_{10}\epsilon$)	Basic media λ_{max}/nm (log$_{10}\epsilon$)
2-I-5-NO$_2$[122]	*235 (3.66), 314 (3.89)*[f]	*240 (3.66), 325 (3.83)*[aa]	*307 (3.80)*[aa]	*273 (3.53), 363 (3.97)*[aa]
4-Cl-1-Me-5-NO$_2$[140]		312 (3.93)[aa]	272 (3.82)[aa]	
5-Cl-1-Me-4-NO$_2$[140]		308 (3.85)[aa]	272 (3.88)[aa]	
2-Br-1-Me-4-NO$_2$[139]		309 (3.86)[aa]	220 *(3.76),* 280 *(3.88)*[aa]	
2-Br-1-Me-5-NO$_2$[139]		316 (3.96)[aa]	223 *(3.73),* 282 *(3.83)*[aa]	
5-Br-1-Me-4-NO$_2$[139]		221 *(3.68),* 314 *(3.85)*[aa]	221 *(3.69),* 288 *(3.87)*[aa]	
2-I-1-Me-4-NO$_2$[122]	240 *(3.69),* 315 *(3.88)*[f]	243 *(3.65),* 327 *(3.89)*[aa]	308 *(3.79)*[aa]	
2-I-1-Me-5-NO$_2$[122]	258 (3.61), 323 (3.84)[f]	262 (3.59), 331 (3.83)[aa]	307 (3.75)[aa]	
2-MeO-5-NO$_2$[122]	233 (3.58), 330 (4.07)[f]	237 (3.56), 330 (4.12)[aa]	225 *(3.54),* 307 *(3.89)*[aa]	237 (3.59), 285 (3.52), 366 (4.15)[aa]
2-MeO-1-Me-4-NO$_2$[122]	235 *(3.56),* 307 *(3.83)*[f]	240 *(3.52),* 320 *(3.87)*[aa]	230 *(3.52),* 294 *(3.84)*[aa]	
2-MeO-1-Me-5-NO$_2$[122]	247 (3.61), 327 (4.06)[f]	248 (3.63), 339 (4.09)[aa]	225 *(3.56),* 305 *(3.96)*[aa]	
2-SH[143b]		258 (4.16)		
Imidazole oxides				
2,4,5-Ph$_3$ 3-oxide[142]		294 (4.35)[aa]	282 (4.38)[aa]	305 (4.35)[aa]
1-Me-2,4,5-Ph$_3$ 3-oxide[142]		245 *(4.27),* 275 *(4.18)*[aa]	263 (4.31)[aa]	
1-MeO-2,4,5-Ph$_3$[142]		292 (4.36)[aa]	285 (4.43)[aa]	
1,2,3-Triazoles				
1,2,3-Triazole	210 (3.63)[144][c] 213 (3.54)[74d][d]		211 (3.66)[145][b]	
1-Me	213 (3.64)[145][c]			
4-Me	216 (3.61)[74d][d]			

Table A.2. Continued

1,2,3-Triazoles	Organic solvent λ_{max}/nm (log$_{10}\epsilon$)	H$_2$O λ_{max}/nm (log$_{10}\epsilon$)	Acid media λ_{max}/nm (log$_{10}\epsilon$)	Basic media λ_{max}/nm (log$_{10}\epsilon$)
4-n-C$_7$H$_{13}$	215 (3.68)[144][c]			
4,5-Me$_2$	221 (3.68)[74d][d]			
1-Ph	243 (4.01)[145][c]		215 (3.90), 247 (3.90)[145][b]	
2-Ph	262 (4.21)[145][c]		258 (4.15)[145][bb]	
4-Ph	209 (4.02), 245 (3.98)[146b][c] 245 (4.18)[144][c]			
4-CHO[139]		237 (3.87)[aa]	214 (3.67), 235 (2.75)[aa]	260 (4.09)[aa]
4-CO$_2$H[139]		<220[aa]		<215 (>3.85)[aa] (mono-anion) 226 (3.94)[aa] (di-anion)
4-CHO-1-Me[139]	240 (4.13)[74d][d]	240 (3.91)[aa]	215 (3.67)[aa]	
4-CO$_2$H-1-Me[139]	247.5 (4.14)[74d][d]	216 (3.95)[aa]		<210 (3.91)[aa]
N-MeCO 1-MeCO-4-Me 1-MeCO-4,5-Me$_2$	253 (4.22)[74d][d]			
4-NH$_2$			245 (3.50)[147][cc]	228 (3.73)[147][dd]
4-NHPh	251 (4.20)[148d][c]			
4-NH$_2$-5-CO$_2$H[149c]		226 (3.85), 261 (3.86)[aa]	221 (3.91), 280 (3.41)[aa]	217 (3.75), 256 (3.75)[aa]
4-NH$_2$-5-CONH$_2$[149c]		225 (3.87), 260 (3.85)[aa]	224 (3.94)[aa]	223 (3.67), 265 (3.90)[aa]
4-NH$_2$-5-CONH$_2$-2-Me[149c]		217 (3.81), 273 (3.80)[aa]		
4,5-Br$_2$[74d]	231 (3.65)[d]			
4-Br-1-Me[139]		226 (3.49)[aa]	229 (3.59)[aa]	
4-Br-2-Me[139]		229 (3.71)[aa]		
5-Br-1-Me[139]		222.5 (3.71)[aa]	230 (3.67)[aa]	
5-Cl-1-Ph[148e]	228 (4.03)[c]			

Table A.2. Continued

1,2,3-Triazoles	Organic solvent λ_{max}/nm ($\log_{10}\epsilon$)	H_2O λ_{max}/nm ($\log_{10}\epsilon$)	Acid media λ_{max}/nm ($\log_{10}\epsilon$)	Basic media λ_{max}/nm ($\log_{10}\epsilon$)
5-HO-1-Ph[148e]	235 (4.03)[c]			
4-CO$_2$H-5-Cl-1-Ph[148e]	231 (4.03)[c]			
4-CO$_2$Me-5-Cl-1-Ph[148e]	218 (4.24)[c]			
1,2,4-Triazoles				
1,2,4-Triazole	205 (2.30)[150a][z]	187 (3.52)[113]		
1-Ph[117]	239 (4.04)[c]			
3-Ph[117]	241.5 (4.15)[c]			257 (4.13)[ee]
4-Ph[117]	224.5 (4.04)[c]			
1-Me-3-Ph[117]	243 (4.18)[c]			
1-Me-5-Ph[117]	235 (4.07)[c]			
3-Me-1-Ph[117]	244 (4.19)[c]			
3-Me-4-Ph[117]	272, 267 (2.60)[c]			
4-Me-3-Ph[117]	270 (2.74)[c]			
5-Me-1-Ph[117]	224.5 (3.88)[c]			
3-Me-5-Ph[117]	244 (4.19)[c]			261 (4.18)[ee]
1- or 4-MeCO[150a]	221.5 (3.89)[z]			
1- or 4-MeCO-3,5-Me$_2$[117]	222 (3.84)[a]			
3-NH$_2$		<220[139]	<220[139]	<220[139]
5-NH$_2$-1-Me[139]		<215	<215	
5-NH$_2$-4-Me[139]		<210 (>3.57)	<210 (>3.58)	
5-Br-1-Me[139]		<220		
5-Br-4-Me[139]		<220		
3-HO-1-Ph[117]	282 (3.96)[c]			
2-Me-1-Ph-3-one[117]	280 (3.87)[c]			284 (4.06)[ee]

Table A.2. Continued

Tetrazoles[z]	Organic solvent λ_{max}/nm ($\log_{10}\epsilon$)	H_2O λ_{max}/nm ($\log_{10}\epsilon$)	Acid media λ_{max}/nm ($\log_{10}\epsilon$)	Basic media λ_{max}/nm ($\log_{10}\epsilon$)
Tetrazole				
1,5-$(CH_2)_5$[113]	(End absorption only, above 220 nm)	188 (3.51), 245 (1.0)		
1-Ph[152]	236 (3.97)[c]			
2-Ph[153d]	252 (4.1)[c]			
5-Ph[148h,151b,154]	240 (4.19)[c]			
1-Me-5-Ph[154]	232 (>4.0)[c]			
2-Me-5-Ph[154]	240 (*ca.* 4.04)[c]			
5-NH_2	218 (3.49)[148h,155][c]	(End absorption only, above 220 nm)[151a]		
5-NH_2-1-Me[155]	222 (3.47)[c]			
5-NH_2-1-Ph[155]	229 (3.81)[c]			
5-MeNH[155]	225 (3.49)[c]			
1-Me-5-MeNH	227 (3.52)[155][c]	232 (3.48)[157][aa]	225 (3.27)[157][aa]	
1-Me-5-Me_2N[155]	232 (3.54)[c]			
5-NH_2-1-$C_6H_4NO_2$(*p*)[155]	217 (3.93)[c]			
5-NH_2-1-$C_6H_4NO_2$(*m*)[155]	225 (4.07)[c]			
1,4-Me_2-5-NH[155]	260 (3.24)[c]			
1,4-Me_2-5-MeN	267 (3.02)[155][c]	272 (3.20)[157][aa]	226 *(3.33)* [157][aa]	
1-Me-5-NC_5H_{10}[139]		232 *(3.45)*[aa]	211 (3.70)[aa]	
5-HO[156]			213 (3.51)[ff]	226 (3.52)[gg]
5-HO-1-Me[156]			218.5 (3.50)[ff]	228 (3.57)[gg]
5-HO-1-Ph[156]			244.5 (3.98)[ff]	229 (3.80), 253 (3.97)[gg]
5-SH[148f]	245 (4.13)[c]			
5-SMe[148f]	230 *(3.52)*[c]			

Table A.2. Continued

Results in italics indicate shoulders or inflections.

(a) In cyclohexane, hexane, or octane.
(b) In 6M HCl.
(c) In ethanol.
(d) In dioxan.
(e) In 1M HCl.
(f) In methanol.
(g) Results for solutions in n-heptane and dioxan have also been recorded[119].
(h) For several *N*-acetylpyrazoles see ref. 74d.
(i) In 71% sulphuric acid.
(j) In 0.1M NaOH.
(k) In 0.05M HCl.
(l) In 0.05M NaOH.
(m) For these 3 compounds at pH 5, 7, and 5.5, respectively.

(n) In 10M H_2SO_4.
(o) Numerous examples besides those given have been reported[131].
(p) End absorption only.
(q) In aqueous buffers.
(r) In 14.5M H_2SO_4.
(s) In chloroform.
(t) Numerous examples besides those listed have been reported[114c, 132a,b].
(u) For the significance of the long-wavelength absorption see the text.
(v) In 0.01M NaOH.
(w) In 0.025M NaOH.
(x) In 0.1M HCl.

(y) In 1M NaOH.
(z) In tetrahydrofuran.
(aa) At pH appropriate for the neutral molecule, cation, or anion, as the case may be. For pK_a values see table A.4.
(bb) In 3M HCl.
(cc) Hydrochloride in neutral solution.
(dd) Hydrochloride in bicarbonate solution.
(ee) KOH in alcohol.
(ff) In 0.01M HCl-MeOH.
(gg) In 0.01M NaOH-MeOH.

Table A.3. A selection of proton magnetic resonance data for azoles[a]

Pyrazoles	Solvent[b]	Position 1	2	3	4	5	Ref.
Pyrazole[c]	A			2.39(d), $J_{3,4} = 1.9$, 2.13	3.69(t)		114k, 214–16
1-Me[d]	B	-3.60			3.44		216
	C	6.19		2.26(d)	3.60(t)		218
	C			2.70(d), $J_{3,4} = 2.0$	3.90(t)	2.78(d)	114k
3-Me	A	6.12		2.51(d)	3.78(t), $J_{4,5} = 2.3$	2.65(d)	114f, k
	D	5.74		1.89(d)	3.17(t)	1.95(d)	114k
	A			7.68	3.94(d)	2.52(d)	114k
4-Me[e]	C			2.74	8.05, $J_{4,5} = 1.7$		214
1-Et	A	8.53(t), 5.82(q), $J = 7.1$		2.50(d)	3.77(t)	2.62(d)	114k
1,3-Me$_2$	A	6.20		7.77, $J_{3,4} = 2.3$	4.05(d), $J_{4,5} = 2.3$	2.78(d)	114k
	E	6.72		7.76	4.11(d), $J_{4,5} = 2.0$	3.28(d)	114k
1,4-Me$_2$	D	5.80		7.44	3.42(d), $J_{4,5} = 2.0$	2.11(d)	114k
	A	6.20		2.72	7.96	2.88	114k
	E	6.70		2.82	8.10	3.44	114k
	D	5.80		2.11	7.74	2.16	114k
1,5-Me$_2$	A	6.27		2.64(d), $J_{3,4} = 2.0$	4.02(d)	7.78	114k
	E	6.82			4.15(d)	8.32	114k
	D	5.88		2.02(d)	3.37(d)	7.45	114k

Table A.3. Continued

Pyrazoles	Solvent[b]	Position					Ref.
		1	2	3	4	5	
1,3,4-Me_3	A	6.24		7.83	8.04	2.98	114k
	E	6.74		7.82	8.16	3.48	114k
	D	5.92		7.59	7.86	2.32	114k
1,3,5-Me_3	A	6.34		7.83	4.22	7.83	114k
	E	6.83		7.76	4.27	8.32	114k
	D	6.00		7.54	3.64	7.54	114k
1,4,5-Me_3	A	6.26		2.81	8.04	7.84	114k
	E	6.77			8.20	8.40	114k
	D	5.93		2.22	7.84	7.59	114k
1-Ph	A			2.28(d) $J_{3,4}=1.9$	3.54(q) $J_{4,5}=2.5$	2.13(d)	114k, 215, 217
3-Ph	A				3.40(d) $J_{4,5}=2.5$		114k
	D				2.92(d) $J=2.2$ $J_{4,5}=2.2$	1.80(d)	114k
4-Ph	D	6.08		1.60			114k
1-Me-4-Ph[f]	A	6.40		2.25		2.42	119
1,3-Me_2-4-Ph[f]	A			7.63	2.68	2.67	119
3-Me-1-Ph	A	2.30–2.90		7.63	3.80	2.22	215, 217
5-Me-1-Ph	A	2.60		2.46	3.85	7.71	215, 217
1-C_6H_4·NO_2(p)[g]	A	2.10(d,t), 1.64(d,t)		2.17(d) $J_{3,4}=1.8$	3.43(q) $J_{4,5}=2.7$	1.94(d)	114k, 217
3-Me-1-C_6H_4·NO_2(p)	A	2.20(d,t), 1.71(d,t)		7.60	3.66(d) $J_{4,5}=2.5$	2.09(d)	114k, 217
4-Me-1-C_6H_4·NO_2(p)	A	2.18(d,t), 1.71(d,t)		2.39	7.82	2.20(m)	114k

Table A.3. Continued

Pyrazoles	Solvent[b]	Position 1	2	3	4	5	Ref.
5-Me-1-C_6H_4.NO_2(p)	A	2.30(d,t), 1.64(d,t)		2.36(d)	3.70(d)	7.52	114k, 217
1-C_6H_3(NO_2)$_2$(2,4)g	A	2.16(q), 1.43(q), 1.28(q)		2.20(q), $J_{3,4}=1.8$, 2.36(q)	3.42(q), $J_{4,5}=2.7$	2.14(q)	114k
1-COMe	C			$J_{3,4}=0.6$, $J_{3,5}=0.6$	3.60(q), $J_{4,5}=2.9$	1.78(q)	218
1-COMe-3-tBu	C				3.71(d), $J_{4,5}=2.9$	1.88(d)	218
3-CO_2Et	A			8.60(t), 5.49(q)	3.14(d), $J_{4,5}=2.3$	2.20(d)	114k
1-$CONH_2$h	A			2.34(q), $J_{3,4}=1.5$, $J_{3,5}=0.7$, 7.71	3.60(q), $J_{4,5}=2.7$	1.74(d)	114k
1-$CONH_2$-3-Me	A			2.58(d), $J_{3,4}=1.5$	3.81(d)	1.89(d)	114k
1-$CONH_2$-5-Me	A			2.12(q), $J_{3,4}=1.7$, $J_{3,5}=0.6$	3.91(d), $J_{4,5}=2.6$	7.41	114k
1-O_2S.C_6H_3Me(p)h	F	7.64(Me)			3.42(q), $J_{4,5}=2.8$	1.55(q)	114k
3-NH_2	G			3.70(d), [NH(3 protons)$\tau = 4.88$(broad)]$J_{4,5} = 2.2$	4.48(d)		172f
3-Cl	A			2.47(d), $J_{4,5}=1.9$	3.76(d)		114k
4-Cl	A			2.43			114k
4-Br	A			2.42			114k

Table A.3. Continued

Pyrazoles	Solvent[b]	Position 1	2	3	4	5	Ref.
4-I	A			2.35			114k
4-Br-1-Me	A	6.12		2.54		2.60	114k
4-Cl-1-Ph	A			2.34		2.08	114k
4-Br-1-Ph	A			2.34		2.09	114k
1-COMe-4-Cl	C			2.44 $J_{3,5} = 0.7$		1.81	218
3-Me-4-NO$_2$[i]	A			7.33		1.77	114k
4-NO$_2$-1-Ph[i]	A	2.14–2.60		1.72		1.34	217
5-Me-3-NO$_2$-1-Ph	A	2.52			3.20	7.64	217
5-Me-4-NO$_2$-1-Ph	A	2.50		1.76		7.36	59k
Pyrazolium ions							
1-Me-2-Ph		5.72	2.25	1.50 $J_{3,4} = J_{4,5} = 2.9$	2.91 $J_{3,5} = 1.1$	0.77	121
2-H-1-Ph	H		−2.87				121
Pyrazolones							
1-Ph-5-one[l-] [CH-form]	A			2.56 $J_{3,4} = 1.3$	6.55		197a
1-C$_6$H$_4$.NO$_2$(p)-5-one[i] [CH-form]	A			2.50	6.53		197a
4-CO$_2$Et-1-Ph-5-one[i]	A		0.75 (broad; under aromatic protons)	2.25 $J_{3,4} = 1.4$			197a
4-CO$_2$Et-1-C$_6$H$_4$.NO$_2$(p)-5-one [NH-form]	A			2.21			197a

Table A.3. Continued

Pyrazole N-oxides	Solvent[b]	Position 1	2	3	4	5	Ref.
1-Me-2-oxide	A	6.29		2.84(dd)	3.92(dd)	2.92(dd)	245
1-Me-5-NO_2-2-oxide	A	5.87		2.77(q)	2.77(q)		246
Imidazoles							
Imidazole[j]	A		2.14		2.75		150p, 216, 219
	D		1.18		2.34		216
	I		2.20		2.79		216
	H		1.4	$J = 1.4$ and 2.4	2.5		219
1-Me[j]	A	6.30	2.53		2.92 $J_{4,5} = 1.0$	3.12	216
2-Me	D	5.88	1.27		2.40	2.48	216
	A		7.56		3.03		220
	H		7.2		2.6		219
4-Me	A		2.44 $J_{2,4} = J_{2,5} = 1.1$		7.73	3.25	221
1-Et	A		2.50		3.06		219
1-COMe	A		1.82		2.86	2.49	219
4-CO_2H	D		0.97 $J = 0.8$ and 1.5			1.48	216
2-Cl	F				2.93		212
4-Cl	F		2.35			2.77(d) $J = 1.25$	212
2-Br	D		2.39		2.36		216
4-Br	F					2.77	212

Table A.3. Continued

Imidazoles	Solvent[b]	Position					Ref.
		1	2	3	4	5	
4-Cl-2-Me	D		1.10 $J_{2,5} = 1.5$ 2.54			2.33	216
	J		$J_{2,5} = 1.3$ 7.76			2.98	219
2-Br-1-Me	F					3.0	212
	A	6.36			2.96	2.96	216
	D	5.97			2.35	2.35	216
5-Br-1-Me	A	6.37	2.41		2.93		216
	D	5.96	1.17		2.39		216
4-Br-2-Me	F		7.74			2.92	212
	J					3.13	219
5-Cl-2,4-Me$_2$	F		7.83		7.95		212
4,5-Cl$_2$	F		2.25				212
4,5-Br$_2$	F		2.20				212
4,5-Cl$_2$-2-Me	F		7.77				212
4,5-Br$_2$-2-Me	F		7.74				212
4-NO$_2$[k]	D		0.89			1.49	216
	F		2.12			1.66	150p
	H		1.30			1.70	150p
1-Me-4-NO$_2$	A	6.10	2.46			2.13	150p, 216
	D	5.78	1.06			1.62	150p, 216
1-Me-5-NO$_2$	A	5.95	2.36		1.91		150p, 216
	D	5.55	0.97		1.46		216
2-Me-4-NO$_2$	F		7.64			1.80	150p
5-Me-4-NO$_2$	F		2.36				150p
1,2-Me$_2$-5-NO$_2$	C	6.10	7.58		2.24		150p

Table A.3. Continued

Imidazoles	Solvent[b]	Position					Ref.
		1	2	3	4	5	
2-Br-1-Me-4-NO$_2$	A	6.20				2.09	216
	D	6.04				1.74	216
5-Br-1-Me-4-NO$_2$	A	6.21	2.32				216
	D	5.89	1.08				216
Imidazolium salts							
1,3-Me$_2$	K		1.22 (broad s) $J_{NMe,H2} = 0.45$		4.7(d)		222
Imidazole N-oxides							
1,4,5-Me$_3$-3-oxide	A	6.45	2.03		7.86(s)	7.86(s)	246
1-PhCH$_2$-5-Me-3-oxide	A	2.70(m), 4.92	1.90		3.01	7.89	246
1-PhCH$_2$-4,5-Me$_2$-3-oxide	A	2.75(m), 5.01	2.19		7.82	7.93	246
2-CHO-1,5-Me$_2$-3-oxide	A	6.15	−0.03		2.96	7.73	246
2-CHO-1,4,5-Me$_3$-3-oxide	A	6.14	−0.10		2.77	7.80	246
1,2,3-Triazoles							
1,2,3-Triazole[l]	A	(τ_{NH} = −2.05)			2.10 (2.25)		216, 219
	D				1.31		216
	I				2.04		216
1-Me[m]	A	5.87			2.20	2.82	216
	D	5.50			1.41	1.53	216
2-Me[m]	D	5.52			1.56(d)	1.51(d)	114f
	A	5.82			2.43		114f
	F	5.83			2.23		114f
1-C$_6$H$_4$.NO$_2$(p)	F	1.76(d,t), 1.56(d,t) J = 10.0 and 2.5			1.93(d) $J_{4,5}$ = 1.2	0.97(d)	114f

Table A.3. Continued

1,2,3-Triazoles	Solvent[b]	Position 1	2	3	4	5	Ref.
2-C6H4·NO2(p)	F	1.80(d,t), 1.62(d,t) J = 10.2 and 2.3			1.79		114f
4-CHO	A					1.72	216
	D					0.90	216
1-COMe	C	7.13			2.32	1.72	190f
2-COMe	C	7.24			2.12, $J_{4,5} = 1.3$	1.79	190f
4-CHO-1-Me	A	5.72				1.10	216
	D	5.48				0.97	216
4-CO2H	D					1.65	216
4-CO2H-1-Me	A	5.76				1.09	216
	D	5.44					216
4-Br-1-Me[m]	A	5.78				2.25	216
	D	5.48				1.51	216
5-Br-1-Me	A	6.00			2.02		216
	D	5.52			1.42		216
4-Br-2-Me	A	5.84				2.35	216
	D	5.58				2.10	216
1,2,4-Triazoles							
1,2,4-Triazole[n]	A			1.70			216, 219, 225a
	D			0.60			216
	I			1.81			216
1-Me	A	6.00		2.01		1.83	216, 225a
	D	5.66		1.20		0.42	216
	F	6.12		2.05		1.53	225a

Table A.3. Continued

1,2,4-Triazoles	Solvent[b]	Position 1	2	3	4	5	Ref.
3-Me	A			7.46		1.91	225a
	F			7.65		1.99	225a
	G			7.58		2.08	224
4-Me	A			1.99	6.37		225a
	F			1.52	6.34		225a
3-Et	G			8.71, 7.20		2.05	224
1,3-Me$_2$	A	6.16		7.62		2.06	225a
	F	6.21		7.77		1.73	225a
1,5-Me$_2$	A	6.18		2.26		7.54	225a
	F	6.25		2.26		7.54	225a
3,4-Me$_2$	A			7.56	6.36	1.93	225a
	F			7.69	6.45	1.67	225a
3-Ph	G			1.8–2.7		1.78	224
1-C$_6$H$_4$·NO$_2$(p)^o	F			1.64		0.45	225b
4-C$_6$H$_4$·NO$_2$(p)^o	F			0.67			225b
1,3-Ph$_2$	L					1.65	223a
1,5-Ph$_2$	L			2.08			223a
1-COMe	A			1.94		1.02	219
	L			1.97		1.05	223a
3-CO$_2$Me	G			6.13		1.52	224
3-NH$_2$	A					2.05	216
	D					1.58	216
	G					2.53	224
4-NH$_2$	L			1.73			223a
3-NHMe	G			7.12		2.48	224
3-NMe$_2$	G			7.03		2.48	224
5-NH$_2$-1-Me	A	6.33		2.55			216

Table A.3. Continued

1,2,4-Triazoles	Solvent[b]	Position 1	2	3	4	5	Ref.
5-NH$_2$-4-Me	D	6.01		1.71			216
3-Cl	D			1.62	6.12		216
3-Br	G					1.57	224
3-I	G					1.59	224
	G					1.69	224
5-Br-1-Me	A	6.05		2.05			216
	D	5.72		1.20			216
5-Br-4-Me	A			1.66	6.28		216
	D			0.35	5.88		216
3-OMe	G			6.15		2.03	224
3-SMe				7.44		1.75	224
3-SEt				8.67, 6.90		1.75	224
Tetrazoles							
Tetrazole	D					0.05	216
	I					1.27	216
	L					0.75	216
1-Me	A	5.73				1.02	216
	D	5.46				0.12	216
2-Me	A	5.54				1.40	216
	D		5.40			0.88	216
5-Br-1-Me	A	5.87					216
	D	5.74					216
5-Br-2-Me	A		5.60				216

Table A.3. Continued

(a) Values of τ are recorded for nuclear protons, or for the protons present in substituent groups. Coupling constants are in hertz. Multiplicities are recorded in obvious ways by letters in parenthesis.

(b) A = CDCl$_3$, B = D$_2$O, C = CCl$_4$, D = CF$_3$CO$_2$H, E = C$_6$H$_6$, F = (CD$_3$)$_2$SO, G = (CD$_3$)$_2$CO, H = H$_2$SO$_4$, I = 2M-NaOD, J = tetrahydrofuran, K = H$_2$O, L = MeCN.

(c) Ref. 214 gives $\tau_{NH} = -3.13$ (δ values were reported). Refs. 114k and 216 give results for a variety of solvents.

(d) Ref. 114k gives results for several solvents. See also refs. 214 and 215.

(e) Ref. 214 gives $\tau_{NH} = -3.04$ (δ values were reported).

(f) Ref. 119 reports δ values, and gives phenyl signals only when they are 'essentially singlets'; see text.

(g) Ref. 114k gives results for a number of 1-(p-nitrophenyl)- and 1-(2,4-dinitrophenyl)-pyrazoles.

(h) Ref. 114k gives results for several 1-carboxamidopyrazoles and 1-p-toluenesulphonylpyrazoles.

(i) The original reports δ values, and also results for solutions in (CD$_3$)$_2$SO.

(j) Ref. 216 gives results for a variety of solvents.

(k) Ref. 150p also records spectra for some nitroimidazoles in H$_2$SO$_4$.

(l) Figures in parenthesis are from ref. 114f, which also gives results for other solvents.

(m) Ref. 114f gives results for several solvents.

(n) For other solvents see refs. 216, 223a, 224, 225a.

(o) Chemical shifts for the benzene ring are also reported.

Table A.4. Acidity constants of azoles

Pyrazoles	pK_a as base	pK_a as acid
Pyrazole	2.47[84a] [ad], 2.53[128] [ae] 2.52[114h] [afg], 2.50[247] [be] 2.52[247] [bg], 2.61[248a] [afgo]	14.21[115] [aeh]
1-Me	2.04[84a] [ad]	
1-t-Bu	2.00[247] [be], 1.95[247] [bg]	
3-Me	3.55[84a] [ad]	
4-Me	3.09[114h] [afg]	
1,5-Me_2	3.11[84a] [ad]	
3,5-Me_2	4.37[84a] [ad], 4.12[114h,l] [afg]	
3,4,5-Me_3	4.63[114h,l] [afg]	
3-MeCO		11.85[249] [ce]
3-CO_2H		3.74[250] [cg]
1-Me-5-CO_2H		3.27[250] [cg]
3-Me-5-CO_2H		3.79[250] [cg]
1,3-Me_2-5-CO_2H		3.31[250] [cg]
1,5-Me_2-3-CO_2H		4.24[250] [cg]
1-CH_2CO_2H		3.31[250] [cg]
3,5-Me_2-1-CH_2CO_2H		3.90[250] [cg]
1-Ph-3-CO_2H		3.60[125b] [bg]
1-Ph-4-CO_2H		4.40[125b] [bg]
1-Ph-5-CO_2H		2.70[125b] [bg]
3-NH_2-1-Ph	2.96[125b] [bg]	
4-NH_2-1-Ph	4.80[125b] [bg], 3.22[251a] [ag]	
5-NH_2-1-Ph	3.14[125b] [bg]	
3-Cl	-0.48[1141] [ae]	
4-Cl	0.60[1141] [ae]	
4-Br	0.64[1141] [ae]	
4-I	0.82[1141] [ae]	
3-Br-4-Me	0.24[1141] [ae]	
3-Cl-5-Me	0.30[1141] [ae]	
4-Cl-3-Me	1.42[1141] [ae]	
4-Br-3-Me	1.46[1141] [ae]	
4-I-3-Me	1.59[1141] [ae]	
4-Cl-3,5-Me_2	2.22[1141] [ae]	
4-I-3,5-Me_2 [gg]	2.36[1141] [aeg]	
4-OH-1-Ph		9.05[125b] [bg]
4-NO_2	-2.0[128] [ae]	9.64[128] [ae], 9.67[120] [ce]
3-Me-4-NO_2		10.06[120] [ce]
3,5-Me_2-4-NO_2		10.65[120] [ce]
Pyrazol-3-ones and related alkoxypyrazoles		
1,5-Me_2	2.60[130a,b]	8.91[130a,b]
3-EtO-1,5-Me_2	2.05[130a,b]	
1,2,5-Me_3	2.22[130a,b]	
1-Ph		7.57[125b] [bg]
5-Me-1-Ph	1.79[130a,b]	8.23[130a,b]
3-MeO-5-Me-1-Ph	1.17[130a,b]	
2,5-Me_2-1-Ph	1.66[130a,b]	
Pyrazol-5-ones and related alkoxypyrazoles		
Pyrazol-5-one	2.5[252] [i], 2.32[114g] [ej]	7.94[130a,b]

Table A.4. Continued

	pK_a as base	pK_a as acid
Pyrazol-5-ones and related alkoxypyrazoles		
3-Me-5-MeO	2.62[114g][ej]	
3-Me	2.70[114g][ej]	
4-Me	2.33[114g][ej]	
1,3-Me$_2$	2.35[130a,b], 2.31[114g][ej]	
1,4-Me$_2$	1.97[114g][ej]	
3,4-Me$_2$	2.67[114g][ej]	
5-EtO-1,3-Me$_2$	3.51[130a,b]	
1,2,3-Me$_3$	2.22[130a,b], 2.14[114g][ej]	
1,2,3,4-Me$_4$	2.05[114g][ej]	
1,3,4,4-Me$_4$	–3.79[130a,b]	
1-Ph		6.56[125b][bg]
3-Me-1-Ph	1.42[130a,b], 1.30[114g][ej]	7.17[130a,b]
5-EtO-3-Me-1-Ph	2.34[130a,b]	
2,3-Me$_2$-1-Ph	1.40[130a,b], 1.38[253][ae]	
3,4-Me$_2$-1-Ph	1.39[130a,b]	7.38[130a,b]
5-EtO-3,4-Me$_2$-1-Ph	2.55[130a,b]	
2,3,4-Me$_3$-1-Ph	1.24[130a,b], 1.28[253][ae]	
3,4,4-Me$_3$-1-Ph	–4.02[130a,b]	
4-NH$_2$-2,3-Me$_2$-1-Ph	4.21, –1.26[253][aek]	
4-NMe$_2$-2,3-Me$_2$-1-Ph	4.87, –1.85[253][aek]	
4-Br-2,3-Me$_2$-1-Ph	–0.42[253][ael]	
4-I-2,3-Me$_2$-1-Ph	–0.34[253][ael]	
Imidazoles		
Imidazole	7.31[m], 7.12[a], 6.93[254][ngo]	14.52[255a][ae], 14.44[258][afo],
	7.05[255b][aef], 7.07[84a][d]	14.17[115][aeh]
	6.95[256][afp], 7.00[247][bg]	
	7.103[bq]	
1-Me	7.20[273][a], 7.32[84a][ad]	
	7.33[139][b]	
2-Me	7.86[256][afp]	
4-Me	7.52[256][afp], 7.37[259][r]	
	8.06[260][afg]	
2-Et	8.00[256][afp]	
2,4-Me$_2$	8.36[256][afp], 8.91[260][fgs]	
4-CH$_2$OH	6.38[256][afp]	
2-Ph	6.39[256][afp], 6.48[255b][aef]	13.32[255b][aef]
4-Ph	6.00[256][afp], 6.10[255b][aef]	13.42[255b][aef]
1-Me-4-Ph	5.78[141][aef]	
2,4-Ph$_2$	5.64[255b][aef]	12.53[255b][aef]
4,5-Ph$_2$	5.90[255b][aef]	12.80[255b][aef]
1-MeCO	3.6[261][at]	
4-CO$_2$H[u]		
4-CO$_2$Et[u]		
4,5-(CO$_2$H)$_2$	–1.55[139][b]	3.43, 8.27, 13.6[139][b]
2-NH$_2$[263]	8.46[agv], 8.39[agw]	
2-NH$_2$-1-Me[263]	8.65[agv], 8.44[agw]	
2-NH$_2$-4,5-Me$_2$[263]	9.21[agv], 8.88[agw]	
2-NH$_2$-4,5-Ph$_2$[263]	7.04[agw]	
4-Cl-1-Me	3.10[140][g]	

Table A.4. Continued

Imidazoles	pK_a as base	pK_a as acid
5-Cl-1-Me	4.75[140] [gx]	
2-Br	3.85[139] [b]	11.03[139] [b]
4-Br	3.88[139] [b]	12.32[139] [b]
2-Br-1-Me	3.88[139] [b]	
5-Br-1-Me	5.26[139] [b]	
2-NO$_2$	−0.81[140] [j], −0.20[264]	7.15[140] [y], 6.40[264]
1-Me-2-NO$_2$	−0.48[140] [j]	
4-NO$_2$ [z]	−0.05[141] [aef], −0.16[140] [j]	9.30[141] [aef], 9.20[140] [gaa]
1-Me-4-NO$_2$	−0.53[141] [aef], −0.58[140] [j]	
1-Me-5-NO$_2$	2.13[141] [aef], 2.12[140] [j] 1.95[264]	
2-Me-4-NO$_2$	0.50[265]	
2,4-(NO$_2$)$_2$	−7.33[140] [j]	2.85[140] [y]
1-Me-2,4-(NO$_2$)$_2$	−7.47[140] [j]	
4-NH$_2$-1,2-Me$_2$-5-NO$_2$	2.50[266]	
5-NH$_2$-1,2-Me$_2$-4-NO$_2$	0.33[266]	
4-Cl-5-NO$_2$	−3.62[140] [ej]	5.85[140] [gy]
2-I-4-NO$_2$	−0.85[122] [ej]	6.82[122] [e]
4-Cl-1-Me-5-NO$_2$	−1.42[140] [ej]	
5-Cl-1-Me-4-NO$_2$	−3.49[140] [ej]	
5-Br-2-Me-4-NO$_2$	−0.55[266]	
2-Br-1-Me-4-NO$_2$	−3.07[139] [b]	
2-Br-1-Me-5-NO$_2$	−0.75[139] [b]	
5-Br-1-Me-4-NO$_2$	−1.77[139] [b]	
2-I-1-Me-5-NO$_2$	−0.14[122] [ej]	
2-I-1-Me-4-NO$_2$	−1.70[122] [ej]	
2-MeO-5-NO$_2$	−0.90[122] [ej]	6.06[122] [e]
2-MeO-1-Me-4-NO$_2$	−0.44[122] [ej]	
2-MeO-1-Me-5-NO$_2$	−1.03[122] [ej]	
4-MeO-1,2-Me$_2$-5-NO$_2$	2.65[266]	

1,2,3-Triazoles		
1,2,3-Triazole	1.17[149d] [bfg]	9.42[149d] [bg] 9.26[248b] [afg]
1-Me	1.25[149d] [bfg]	
2-Me	<1[149d] [bg]	
4-Ph		6.32[80i] [b]
4-CHO	−0.10[139] [b]	6.63[139] [b]
4-CHO-1-Me	−0.59[139] [b]	
4-CO$_2$H		3.22, 8.73[248b] [afg]
1-Me-4-CO$_2$H		3.16[139] [b]
1-Ph-4-CO$_2$H		2.88[248b] [afg]
5-Me-1-Ph-4-CO$_2$H		3.73[248b] [afg]
4,5-(CO$_2$H)$_2$		1.86, 5.90, 9.30[248b] [afg]
1-Ph-4,5-(CO$_2$H)		2.13, 4.93[248b] [afg]
4-CN-5-Me		6.13[80i] [b]
4,5-(CN)$_2$		1.47[267] [a]
4-Br-1-Me	−1.67[139] [b]	
5-Br-1-Me	−0.47[139] [b]	
4,5-Br$_2$		5.37[248b] [afg]
4-NH$_2$-5-CONH$_2$	−0.23[149c] [be]	7.79[149c] [be]

Table A.4. Continued

1,2,3-Triazoles	pK_a as base	pK_a as acid
4-NH$_2$-5-CONH$_2$-2-Me	0.10[149c] [be]	
4-NH$_2$-5-CO$_2$H		4.27, 9.43[149c] [bg]
4-NH$_2$-5-CO$_2$H-2-Me	-0.28[149c] [be]	3.76[149c] [be]

1,2,4-Triazoles

	pK_a as base	pK_a as acid
1,2,4-Triazole	2.19[84a,268a] [ad], 2.27[268a] [b] 2.98[bb], 2.45[248a] [afgo]	10.26[268a] [b], 10.04[248a] [afgo]
1-Me	3.20[bb]	
3-Me	3.28[268a] [b], 3.68[bb]	10.73[268a] [b]
4-Me	3.40[bb]	
3-Et	3.20[268a] [b]	10.69[268a] [b]
3,5-Me$_2$	3.68[84a,268a] [ad]	
3,5-Et$_2$	3.64[84a,268a] [ad]	
3-Ph	2.05[268a] [b]	9.29[268a] [b]
3-NH$_2$	4.04[268a] [b], 4.17[270] [a] 4.50[bb]	11.08[268a] [b], 11.25[270] [a]
4-NH$_2$	3.23[bb]	
3-NH$_2$-5-Me	4.68[bb]	
5-NH$_2$-1-Me	4.29[139] [b], 4.20[bb]	
5-NH$_2$-4-Me	5.38[139] [b]	
3-NH$_2$-5-Ph	3.93[bb]	
4-NH$_2$-3,5-Me$_2$	3.66[bb]	
3,5-(NH$_2$)$_2$	4.43[268a] [b]	12.12[268a] [b]
3-Me-5-NHNO$_2$		4.80, 11.30[148a]
3-Cl		8.13[268a] [b], 8.06[270] [a]
3-Br		8.00[268a] [b]
3-I		8.11[268a] [b]
3,5-Cl$_2$		5.22[268a] [b]
3,5-Br$_2$		5.23[268a] [b]
3,5-(CF$_3$)$_2$		3.00[271] [cc]
5-OH		9.11[270] [a]
5-OH-3-Me		9.61[270] [a]
3-MeO		9.96[268a] [b]
3-MeS		9.20[268a] [b]
3,4-Me$_2$-5-S:		8.19[163b] [ag]
4-Me-3-Ph-5-S:		7.66[163b] [ag]
3-NH$_2$-5-OH		8.84[270] [a]
3-Cl-5-OH		6.02[270] [a]
5-OH-3-NO$_2$		3.63[270] [a]
3,5-(MeS)$_2$		8.20[268a] [b]

Urazoles[161] [j]

	pK_a as base	pK_a as acid
1-Ph	-4.1	4.85
2-Me-1-Ph	-4.7	6.97
3-MeO-1-Ph	-3.4	6.93
4-Me-1-Ph	-4.2	4.73
2,4-Me$_2$	-4.8	
3-MeO-4-Me-1-Ph	-3.1	
3,5-(EtO)$_2$-1-Ph	-0.55	
3-EtO-1-Ph	-2.9	6.96

Table A.4. Continued

Tetrazoles	pK_a as base	pK_a as acid
Tetrazole		4.86[80g,h][b], 4.79[93][ag]
		4.89[148a][cg]
		4.90[248a][afgo]
1-Me[ff]	4.3[80f][d]	
2-Me[ff]	3.58[80f][d]	
5-Me		5.56[93][ag],
		5.63[248a][afgo]
5-Ph		4.53[93][acc],
		4.38[248a][afgo]
5-CO$_2$Et		4.35[80g,h][b]
5-CONH$_2$		2.37[80g,h][b]
5-NH$_2$	1.82[272][ag]	6.03[149a][bgdd],
		6.00[148a][cg]
5-NH$_2$-1-Me	1.82[272][ag]	
5-NH$_2$-1-Ph	1.12[272][ag]	
5-:NH-1,4-Me$_2$	>4[148f][agdd], 9.57[157][e]	
1-Me-5-MeNH	0.55[157][e]	
5-Cl		2.07[148a][cg]
5-Br		2.13[148a][cg]
5-I		2.85[148a][cg]
5-CF$_3$		1.7[248a][afgo]
5-OH		5.40, 10.26[248a][afgo]
5-OH-1-Ph		5.53[156][gddee]
5-PhO		3.49[248a][afgo]
1-Me-5-SH		3.65[148f][agdd]
1-MeS		4.07[148f][agdd],
		4.00[248a][afgo]
1-Ph-5-SH		3.86[148f][agdd]

(a) 25 °C.
(b) 20 °C.
(c) Temperature not given ('room temperature').
(d) 'Classical' values from measurements of conductivity, solubility and ester hydrolysis.
(e) Spectrophotometric.
(f) 'Thermodynamic' value.
(g) Potentiometric.
(h) H_- scale.
(i) Extrapolated from titrations with perchloric acid in acetic acid.
(j) H_0 scale.
(k) The second pK_a was evaluated using H_A and H_0. Correlation with H_A was the better.
(l) Evaluated using H_A and H_0. Correlation with H_0 was the better.
(m) 15 °C.
(n) 35 °C.
(o) Thermodynamic quantities for the ionisation are given in the reference quoted.

(r) 30 °C.
(s) 4 °C.
(t) From rate of hydrolysis.
(u) Measurements on solutions in 23.3% w/w ethanol at 30 °C have been reported[262].
(v) In 0.1M KCl.
(w) In 1 : 1 (v/v) EtOH–0.1 M KCl.
(x) In MeOH–H$_2$O (1 : 3).
(y) In MeOH–H$_2$O (1 : 1).
(z) Protonation follows H_0 in perchloric acid but deviates from it in hydrochloric acid[141].
(aa) In HCONMe$_2$–H$_2$O (1 : 2).
(bb) The original paper (ref. 269a) was not available. The value of pK_a tabulated is derived from the reported pK_b values assuming measurements to have been made at 25 °C (pK_w = 14.00).
(cc) 50% aqueous dioxan.
(dd) 50% EtOH.
(ee) 27 °C.
(ff) See text.

Table A.4. Continued

(p) E.m.f. measurements.

(q) Thermodynamic value using the cell Pd, H_2(1 atmos.)|$C_3H_4N_2$, HCl, KCl|AgCl, Ag. Values of pK_a are given for intervals of $5\,^\circ$C over the range 0–$50\,^\circ$C[257].

(gg) In ref. 247, table 5, this compound is wrongly described as 3,5-dimethyl-pyrazole.

Table A.5. The nitration of arylazoles[a]

Aryl group	Other substituents in azole ring	Conditions[b]	Products	Ref.
Pyrazoles				
1-Ph[a]		A	4-NO$_2$-1-Ph	118, 251, 499b
		B or C	83% 1-(p-O$_2$N.C$_6$H$_4$.)	118, 502
		C (12 °C)	86% 1-(p-O$_2$N.C$_6$H$_4$.)	59c
		C (22 °C)	76% 4-NO$_2$-1-(p-O$_2$N.C$_6$H$_4$.)	59c
		D	4-NO$_2$-1-Ph	129
1-Ph	3-Me[a]	A	3-Me-4-NO$_2$-1-Ph + some 3-Me-4-NO$_2$-1-(p-O$_2$N.C$_6$H$_4$.)	499b
		B	3-Me-1-(p-O$_2$N.C$_6$H$_4$.)[c]	1c
1-Ph	5-Me[a]	B	mixture[c]	1c
1-Ph	4-Et	A	4-Et-3-NO$_2$-1-Ph	491
1-Ph	3,5-Me$_2$[a]	A	3,5-Me$_2$-4-NO$_2$-1-Ph	499b
		C	3,5-Me$_2$-1-(p-O$_2$N.C$_6$H$_4$.)	499b
		C	3,5-Me$_2$-4-NO$_2$-1-(p-O$_2$N.C$_6$H$_4$.)	503
1-Ph	3-CO$_2$H	C	3-CO$_2$H-1-(p-O$_2$N.C$_6$H$_4$.)	499b
1-Ph	3-CO$_2$H-5-Me	B or C	3-CO$_2$H-5-Me-1-(O$_2$N.C$_6$H$_4$.)[c]	1c
1-Ph	5-Cl-3-Me	B	5-Cl-3-Me-1-(p-O$_2$N.C$_6$H$_4$.) + some of the dinitrophenyl compound	321c
		C	5-Cl-3-Me-1-(p-O$_2$N.C$_6$H$_4$.)	321c
1-Ph	5-Br-3-Me		Behaviour analogous to that of chloro compound	321c
1-Ph	4-Br-3-Me	C (12 °C)	4-Br-3-Me-1-(p-O$_2$N.C$_6$H$_4$.)	59h
1-Ph	4,5-Br$_2$-3-Me	C (100 °C)	4,5-Br$_2$-3-Me-1-(p-O$_2$N.C$_6$H$_4$.)	59h
1-Ph	4-NO$_2$	C	85% 4-NO$_2$-1-[2,4-(O$_2$N)$_2$.C$_6$H$_3$.]	59c, 118
1-Ph	3-Me-4-NO$_2$	C	3-Me-4-NO$_2$-1-(p-O$_2$N.C$_6$H$_4$.)	499b
1-Ph	3,5-Me$_2$-4-NO$_2$	C	3,5-Me$_2$-4-NO$_2$-1-(p-O$_2$N.C$_6$H$_4$.)	499b
1-o-O$_2$N.C$_6$H$_4$.		C (22 °C)	66% 4-NO$_2$-1-(o-O$_2$N.C$_6$H$_4$.)	59c

Table A.5. Continued

Aryl group	Other substituents in azole ring	Conditions[b]	Products	Ref.
Pyrazoles				
$1\text{-}o\text{-}O_2N.C_6H_4\cdot$	$4\text{-}NO_2$	C (100°C)	$73\%\ 4\text{-}NO_2\text{-}1\text{-}[2,4\text{-}(O_2N)_2.C_6H_3\cdot]$	59c
$1\text{-}m\text{-}O_2N.C_6H_4\cdot$	$4\text{-}NO_2$	C (100°C)	$70.5\%\ 4\text{-}NO_2\text{-}1\text{-}[2,4\text{-}(O_2N)_2.C_6H_3\cdot]$	59c
		C (22°C)	$73\%\ 4\text{-}NO_2\text{-}1\text{-}(m\text{-}O_2N.C_6H_4\cdot)$	59c
$1\text{-}m\text{-}O_2N.C_6H_4\cdot$	$4\text{-}NO_2$	C (100°C)	$57\%\ 4\text{-}NO_2\text{-}1\text{-}[3,4\text{-}(O_2N)_2.C_6H_3\cdot]$	59c
$1\text{-}p\text{-}O_2N.C_6H_4\cdot$	$4\text{-}NO_2$	C (100°C)	$72\%\ 4\text{-}NO_2\text{-}1\text{-}[3,4\text{-}(O_2N)_2.C_6H_3\cdot]$	59c
		C (22°C)	$74\%\ 4\text{-}NO_2\text{-}1\text{-}(p\text{-}O_2N.C_6H_4\cdot)$	59c, 118
$1\text{-}p\text{-}O_2N.C_6H_4\cdot$	$4\text{-}NO_2$	C (100°C)	$84\%\ 4\text{-}NO_2\text{-}1\text{-}[2,4\text{-}(O_2N)_2.C_6H_3\cdot]$	59c
$1\text{-}[2,4\text{-}(O_2N)_2.C_6H_3\cdot]$		C (100°C)	$84\%\ 4\text{-}NO_2\text{-}1\text{-}[2,4\text{-}(O_2N)_2.C_6H_3\cdot]$	59c
		C (22°C)	$64\%\ 4\text{-}NO_2\text{-}1\text{-}[2,4\text{-}(O_2N)_2.C_6H_3\cdot]$	59c
1-Picryl		C	$4\text{-}NO_2\text{-}1\text{-}picryl$	490b
$1\text{-}(p\text{-}Ph.C_6H_4\cdot)$		A	$4\text{-}NO_2\text{-}1\text{-}(p\text{-}Ph.C_6H_4\cdot)$	251
3-Ph		A	$N\text{-}Ac\text{-}(p\text{-}O_2N.C_6H_4\cdot) + N\text{-}NO_2\text{-}Ph$[d]	499a
		B or C	$3\text{-}(p\text{-}O_2N.C_6H_4\cdot)$	426e, 499a
3-Ph	5-Me	C (100°C)	$5\text{-}Me\text{-}4\text{-}NO_2\text{-}3\text{-}(p\text{-}O_2N.C_6H_4\cdot)$	74b, 503
3-Ph	5-Cl	C	$5\text{-}Cl\text{-}3\text{-}(O_2N.C_6H_4\cdot)$	330d
3-Ph	$4\text{-}Br\text{-}5\text{-}Cl$	B	$4\text{-}Br\text{-}5\text{-}Cl\text{-}3\text{-}(O_2N.C_6H_4\cdot)$	330d
$3\text{-}(\beta\text{-}C_5H_4N)$		C	$4\text{-}NO_2\text{-}3\text{-}(\beta\text{-}C_5H_4N)$	504
$3\text{-}(\beta\text{-}C_5H_4N)$	$5\text{-}NO_2$	C (100°C)	$4,5\text{-}(NO_2)_2\text{-}3\text{-}(\beta\text{-}C_5H_4N)$	504
4-Ph		B	Mixture including a mono and a dinitro compound	505
4-Picryl	1-Me	B	$1\text{-}Me\text{-}3\text{-}NO_2\text{-}4\text{-}picryl$	490c
		C	$1\text{-}Me\text{-}3,5\text{-}(NO_2)_2\text{-}4\text{-}picryl$	490c
$1,3\text{-}Ph_2$		A (0°C)	$4\text{-}NO_2\text{-}1,3\text{-}Ph_2 + 1\text{-}(p\text{-}O_2N.C_6H_4\cdot)\text{-}3\text{-}Ph +$ $4\text{-}NO_2\text{-}1\text{-}(p\text{-}O_2N.C_6H_4\cdot)\text{-}3\text{-}Ph$	129
		A (20°C)	$4\text{-}NO_2\text{-}1\text{-}(p\text{-}O_2N.C_6H_4\cdot)\text{-}3\text{-}Ph$	129
		C (0°C)	$25\%\ 1,3\text{-}(p\text{-}O_2N.C_6H_4\cdot)_2$	129

Table A.5. Continued

Aryl group	Other substituents in azole ring	Conditions[b]	Products	Ref.
Pyrazoles				
1,3-Ph$_2$	4-NO$_2$	C (0°C)	43% 4-NO$_2$-1-(p-O$_2$N.C$_6$H$_4$.)-3-Ph	129
1-(p-O$_2$N.C$_6$H$_4$.)-3-Ph		C (0°C)	87% 1 3-(p-O$_2$N.C$_6$H$_4$.)$_2$	129
1,5-Ph$_2$		A	4-NO$_2$-1,5-Ph$_2$	129, 251
		C	1,5-(p-O$_2$N.C$_6$H$_4$.)$_2$	129, 251
5-Ph	1-(p-O$_2$N.C$_6$H$_4$.)	A	4-NO$_2$-1-(p-O$_2$N.C$_6$H$_4$.)-5-Ph	251a
		C	1,5-(p-O$_2$N.C$_6$H$_4$.)$_2$	251a
1,5-Ph$_2$	3-Me	A	3-Me-4-NO$_2$-1,5-Ph$_2$	506b
		C	3-Me-4-NO$_2$-1,5-(p-O$_2$N.C$_6$H$_4$.)$_2$	503, 506b
			3-Me-1-Ph-5-(p-O$_2$N.C$_6$H$_4$.)	506b
			3-Me-1,5-(p-O$_2$N.C$_6$H$_4$.)$_2$	506b
1-[2,4-(O$_2$N)$_2$.C$_6$H$_3$.]-5-Ph	3-Me	C	3-Me-5-(p-O$_2$N.C$_6$H$_4$.)-1-[2,4-(O$_2$N)$_2$.C$_6$H$_3$.]	503
3,5-Ph$_2$		C (110°C)	4-NO$_2$-3,5-(p-O$_2$N.C$_6$H$_4$.)$_2$	74b
1,3,5-Ph$_3$		C	1,3,5-(p-O$_2$N.C$_6$H$_4$.)$_3$ + a dinitro compound	503
Imidazoles				
1-Ph		C, E	1-(p-O$_2$N.C$_6$H$_4$.)	374b, 453d
		D	5-NO$_2$-1-Ph	374b
2-Ph		E (100°C)	50% 2-(p-O$_2$N.C$_6$H$_4$.) + 1.5% 2-(o-O$_2$N.C$_6$H$_4$.) + 0.2% 2-(m-O$_2$N.C$_6$H$_4$.)	412c, 453d
		E (room temp.)	57% 2-(p-O$_2$N.C$_6$H$_4$.) + <5% 2-(m-O$_2$N.C$_6$H$_4$.)	
2-Ph	1-Me	E (100°C)	<43.4% 1-Me-2-(p-O$_2$N.C$_6$H$_4$.)	453d
2-Ph	4-Me	B	4-Me-5-NO$_2$-2-(p-O$_2$N.C$_6$H$_4$.)	320d

Table A.5. Continued

Aryl group	Other substituents in azole ring	Conditions[b]	Products	Ref.
Imidazoles				
2-Ph	4-CO$_2$H	E	52% 4-CO$_2$H-2-(p-O$_2$N.C$_6$H$_4$.) + 19% 4-CO$_2$H-2-(m-O$_2$N.C$_6$H$_4$.)	412c
2-Ph	4,5-(CO$_2$H)$_2$	C	52% 4,5-(CO$_2$H)$_2$-2-(m-O$_2$N.C$_6$H$_4$.) + 19% 4,5-(CO$_2$H)$_2$-2-(p-O$_2$N.C$_6$H$_4$.)	274b, 412c
2-Ph	4-Br	F	Only product isolated was 4-Br-2-(p-O$_2$N.C$_6$H$_4$.) (31%)	453c
2-Ph	4,5-Br$_2$	F	63% 4,5-Br$_2$-2-(p-O$_2$N.C$_6$H$_4$.) + 1.8% of an isomer	453c
4-Ph		E	69% 4-(p-O$_2$N.C$_6$H$_4$.) + 25% 4-(o-O$_2$N.C$_6$H$_4$.)	507
		B or C	4-(p-O$_2$N.C$_6$H$_4$.)	320d
4-Ph	1-Me	E	1-Me-4-(p-O$_2$N.C$_6$H$_4$.)	497
4-(3,4-Cl$_2$.C$_6$H$_3$.)	H or 1-Me	E	Mainly nitrated at C-5	508b
4-(p-O$_2$N.C$_6$H$_4$.)		E	5-NO$_2$-4-(p-O$_2$N.C$_6$H$_4$.)	507
4-(p-O$_2$N.C$_6$H$_4$.)	1-Me	E	1-Me-5-NO$_2$-4-(p-O$_2$N.C$_6$H$_4$.)	497
5-Ph	1-Me	E	1-Me-5-(p-O$_2$N.C$_6$H$_4$.)	497
5-(p-O$_2$N.C$_6$H$_4$.)	1-Me	E	1-Me-4-NO$_2$-5-(p-O$_2$N.C$_6$H$_4$.)	497
2,4,5-Ph$_3$		B	2,4,5-(p-O$_2$N.C$_6$H$_4$.)$_3$	509
1,2,3-Triazoles				
1-Ph-1*H*	4-Me	C	1-(p-O$_2$N.C$_6$H$_4$.)	455
1-Ph-1*H*	5-Me	C	1-(p-O$_2$N.C$_6$H$_4$.)	455
1-Ph-1*H*		C	1-(p-O$_2$N.C$_6$H$_4$.)	455
1-Ph-1*H*	4-CO$_2$H	C	4-CO$_2$H-1-(O$_2$N.C$_6$H$_4$.)[c]	19b
1-Ph-1*H*	5-CO$_2$H	C	5-CO$_2$H-1-(O$_2$N.C$_6$H$_4$.)[c]	19b
			5-CO$_2$H-1-(p-O$_2$N.C$_6$H$_4$.)	455
1-Ph-1*H*	4-CO$_2$H-5-Me	C	4-CO$_2$H-5-Me-(1-p-O$_2$N.C$_6$H$_4$.)	455
2-Ph-2*H*		A	2-(p-O$_2$N.C$_6$H$_4$.)	456a

Table A.5. Continued

Aryl group	Other substituents in azole ring	Conditions[b]	Products	Ref.
1,2,3-Triazoles				
2-Ph-2H		B	2-(p-O$_2$N.C$_6$H$_4$.)	79
2-Ph-2H		C	2-(p-O$_2$N.C$_6$H$_4$.) + some 4-NO$_2$-2-(p-O$_2$N.C$_6$H$_4$.)	456a, 510a
2-Ph-2H	4-Me	C	4-Me-2-(p-O$_2$N.C$_6$H$_4$.)	510b
2-Ph-2H	4-CHO	B	4-CHO-2-(p-O$_2$N.C$_6$H$_4$.)	511
2-Ph-2H	4-CO$_2$H	B	4-CO$_2$H-2-(O$_2$N.C$_6$H$_4$.)[c]	275
2-(p-O$_2$N.C$_6$H$_4$.)-2H		C (25°C)	4-NO$_2$-2-(p-O$_2$N.C$_6$H$_4$.)	456a
2-(p-O$_2$N.C$_6$H$_4$.)-2H		C (80°C)	4-NO$_2$-2-[2,4-(O$_2$N)$_2$.C$_6$H$_3$.]	456a
2-(p-O$_2$N.C$_6$H$_4$.)-2H	4-NO$_2$	C (25 or 80°C)	4-NO$_2$-2-[2,4-(O$_2$N)$_2$.C$_6$H$_3$.]	456a
1,2,4-Triazoles				
1-Ph-1H	3-CO$_2$H	C	3-CO$_2$H-1-(O$_2$N.C$_6$H$_4$.)[c]	18e
1,2,3,4-Tetrazoles				
1-Ph-1H		B	1-(p-O$_2$N.C$_6$H$_4$.)	24c
2-Ph-2H	5-CO$_2$H	B	5-CO$_2$H-2-(O$_2$N.C$_6$H$_4$.)[c]	18f
5-Ph		C	5-(m-O$_2$N.C$_6$H$_4$.)[e]	512b

(a) Some thienylazoles have also been nitrated[503]. See also table 11, p. 65.

(b) A, nitric acid and acetic anhydride; B, fuming nitric acid; C, nitric and sulphuric acids; D, nitronium borofluoride in sulpholan or chloroform; E, nitrate of base added to sulphuric acid; F, solution of base in sulphuric acid treated with potassium nitrate.

(c) The historical importance of these cases is mentioned on pp. 1-4.

(d) See text.

(e) The *m*-orientation, not fully substantiated in the original work, is supported by later syntheses[513a].

Table A.6. The *N*-arylation of azoles by halogenobenzenes

	Halogenobenzene[a]	Conditions[b]	Products[c]	Refs.
Pyrazoles				
Pyrazole	A	1		114b
	B	2		114b
	B	3		432d
	C	2		114b
	C	3, 4		432d
	D, E, F	17		632
	G, H, I	17		632
	J, K, L	17		632
	M, N, O	17		632
3-Me	A	5	3-Me	114b
	B	2	3-Me	114b
	C	2	3-Me	114b
4-Me	A	5		114b
	B	2		114b
	C	6		114b
3-Et	B	2	3-Et	114b
3,4-Me$_2$	B	2	85% 3,4- + 15% 4,5-Me$_2$	114b
	C	2	3,4-Me$_2$	114b
3,5-Me$_2$	B	2		114b
	C	6		114b
3,4,5-Me$_3$	B	2		114b
	C	6		114b
3-Ph	B	2	3-Ph	114b
4-Ph	B	2		114b
3-Me-4-Ph	B	2	90% 3-Me-4-Ph + 10% 5-Me-4-Ph	114b
4-Me-3-Ph	B	2	90% 4-Me-3-Ph + 10% 4-Me-5-Ph	114b
3-Me-5-Ph	B	2	5-Me-3-Ph	114b
4,5-Me$_2$-3-Ph	B	2	35% 4,5-Me$_2$-3-Ph + 65% 3,4-Me$_2$-5-Ph	114b
3,5-Ph$_2$[d]	B	7		114b
4-Br-3-Me	B	2	4-Br-3-Me	114b
	C	6	4-Br-3-Me	114b
4-Br-3-Ph	B	2	4-Br-3-Ph	114b
4-Br-3-Me-5-Ph	B	2	90% 4-Br-5-Me-3-Ph + 10% 4-Br-3-Me-5-Ph	114b
4-NO$_2$	P	8		59c
Imidazoles				
Imidazole	B	9		633
	Q	9		633
	L, R, S, T, U, V, W	10		411c
	D, E, F	17		632
	G, H, I	17		632
	J, K, L	17		632
	M, N, O	17		632
2-Me	A, B	2		634
4-Me[e]	B	9		633
	A, B	2	1-Ar-4-Me	634
2,4-Me$_2$	A, B	2	1-Ar-2,4-Me$_2$	634

Table A.6. Continued

	Halogenobenzene[a]	Conditions[b]	Products[c]	Refs.
Imidazoles				
4,5-Me$_2$	A, B	2		634
2,4,5-Me$_3$	A, B	2		634
4,5-(CH$_2$)$_4$	P	11		411b
4,5-Ph$_2$	P	12		411b
2-Me-4,5-Ph$_2$	P	13		411b
1,2,3-Triazoles				
1,2,3-Triazole	Y	14	20% 1-*o*-nitrophenyl-1,2,3-triazole + 62% 2-*o*-nitrophenyl-1,2,3-triazole	177a
	A	5	Both possible isomers	114f
	B	2	1-(2,4-dinitrophenyl)-1,2,3-triazole	114f
	C	6	1-(2,4,6-trinitrophenyl)-1,2,3-triazole	114f
	X	4	1-Substituted compound	635
1,2,4-Triazoles				
1,2,4-Triazole	A	1	80% 1-*p*-nitrophenyl-1,2,4-triazole + 20% 4-*p*-nitrophenyl-1,2,4-triazole	225b
	B	15	1-(2,4-dinitrophenyl)-1,2,4-triazole	225b
	C	16	1-(2,4,6-trinitrophenyl)-1,2,4-triazole	490a
	D, E, F	17	1-Aryl compound	632
	G, H, I	17	1-Aryl compound	632
	J, K, L	17	1-Aryl compound	632
	M, N, O	17	1- and 4-aryl compound	632
3-Me	A	1	3-Methyl-1-*p*-nitrophenyl-1,2,4-triazole and on one occasion a trace of the 4-aryl isomer	225b
	B	2	Product of unknown orientation (10%)	225b
3,5-Me$_2$	A	1	The 1-aryl isomer	225b
	B	2	The 1-aryl isomer	225b
Tetrazoles				
Tetrazole	D, Y		Reaction failed	177c

(a) A, 4-nitrofluorobenzene; B, 2,4-dinitrofluorobenzene; C, 2,4,6-trinitrochlorobenzene; D, 2-nitrochlorobenzene; E, 3-nitrochlorobenzene; F, 4-nitrochlorobenzene; G, 2-chlorobenzonitrile; H, 3-bromobenzonitrile; I, 4-bromobenzonitrile; J, 2-bromoacetophenone; K, 3-bromoacetophenone; L, 4-bromoacetophenone; M, 2-bromopyridine; N, 3-bromopyridine; O, 4-chloropyridine; P, 2,4-dinitrochlorobenzene; Q, 4,4'-difluoro-3,3'-dinitrodiphenyl sulphone; R, 2-bromoanisole; S, 3-bromoanisole; T, 4-bromoanisole; U, 1,4-dibromobenzene; V, 4-bromodimethylaniline; W, 4-bromobenzaldehyde; X, several α-chloro-heterocycles (see p. 105); Y, 2-nitrofluorobenzene.

Table A.6. Continued

(b) 1, Anhydrous KF at 170 °C; 2, boiling ethanol; 3, ether; 4, neat; 5, anhydrous
    KF at 200 °C; 6, boiling benzene; 7, benzene in sealed tube at 180 °C;
    8, ethanolic potassium hydroxide; 9, aqueous sodium bicarbonate; 10, cuprous
    bromide and potassium carbonate in boiling nitrobenzene; 11, sodium acetate in
    ethanol; 12, fuse at 165–170 °C; 13, fuse at 185–190 °C; 14, sodium carbonate
    in dimethylformamide; 15, ethanol at room temperature; 16, γ-butyrolactone
    at 25 °C; 17, CuO and pyridine.

(c) Where only one product can be formed it is not mentioned. In other cases
    substituents in the product are numbered with reference to the *N*-aryl group.

(d) 3,5-Diphenyl- 4-bromo-3,5-diphenyl-, and 4,5-dibromo-3-methyl-pyrazole did not
    react with 2,4-dinitrofluorobenzene in boiling ethanol.

(e) Use of an excess of 2,4-dinitrofluorobenzene led to ring-opening consequent upon
    quaternisation.

Table A.7. The addition of azoles to unsaturated compounds

	Unsaturated compound	Conditions	Ref.
Pyrazoles			
Pyrazole	Acrolein	Steam bath	432d
	Methyl acrylate	Steam bath	432d
	Acrylonitrile	Steam bath	432d
	Ethyl α-bromoacrylate	Steam bath	432d
	1-Chloro-1,2,2-trifluoro-ethylene[a]	Steam bath	432d
	Ethyl crotonate	Steam bath	432d
	Diethyl fumarate	Steam bath	432d
	Dimethyl fumarate	Steam bath	432d
	Diethyl maleate	Steam bath	432d
	Methyl methacrylate	Steam bath	432d
	Methyl vinyl ketone	Steam bath	432d
	2-Nitropropene	Steam bath	432d
3,5-Me$_2$[c]	Acrylonitrile	Dioxan, 90–100 °C	126a
	Acrylonitrile	Various conditions[b]	126e
	Maleic anhydride	Dioxan, 90–100 °C	126a
	Methyl methacrylate	3h/200 °C	126a
3-Me-5-Ph	Acrylonitrile[d]	5h/160 °C	126e
	Maleic anhydride	Dioxan, 90–100 °C	126a
3,5-Ph$_2$	Acrylonitrile	Sealed tube/4h/150 °C	126e
	Maleic anhydride	Dioxan, 90–100 °C	126a
4-Cl-3,5-Me$_2$	Acrylonitrile	Sealed tube/150 °C	126e
4-Cl-3,5-Ph$_2$	Acrylonitrile	Sealed tube/150 °C	126e
4-Br-3,5-Me$_2$	Acrylonitrile	Sealed tube/150 °C	126e
4-Br-3,5-Ph$_2$	Acrylonitrile	Sealed tube/150 °C	126e
4-I[e]	Ethyl acrylate	Dioxan/Triton B	473
	Ethyl α-bromoacrylate	Dioxan/Triton B	473
	CH$_2$:C(NHAC)CO$_2$Et	Dioxan/Triton B	473
4-I-3,5-Me$_2$	Ethyl acrylate	Dioxan/Triton B	473
3,4-I$_2$[f]	Ethyl acrylate	Dioxan/Triton B	473
3,4-I$_2$-5-Me[f]	Ethyl acrylate	Dioxan/Triton B	473
4-NO$_2$	Acrylic acid	Pyridine catalyst	623c
	Benzalacetone	Triton B	623c
	Chalcone	Pyridine catalyst	623c
	Dibenzalacetone	Triton B	623c
	p-Methoxybenzaceto-phenone	Triton B	623c
3,5-Me$_2$-4-NO$_2$	Acrylonitrile	Sealed tube/150 °C	126e
	Acrylic acid	Pyridine catalyst	623c
	Benzalacetone	Triton B	623c
	Chalcone	Pyridine catalyst	623c
	Dibenzalacetone	Triton B	623c
	p-Methoxybenzaceto-phenone	Triton B	623c
Imidazoles			
Imidazole	Acrylonitrile	90–100 °C	641
	1-Chloro-1,2,2-trifluoro-ethylene	Tetrahydrofuran/autoclave/100 °C	642e
	2-Vinylpyridine	Acetic acid catalyst, 110 °C	643
	4-Vinylpyridine	Acetic acid catalyst, 110 °C	643
	2-Methyl-6-vinylpyridine	Acetic acid catalyst, 110 °C	643
	2,4-Dimethyl-6-vinylpyridine	Acetic acid catalyst, 110 °C	643

Table A.7. Continued

	Unsaturated compound	Conditions	Ref.
Imidazoles			
2-Me	2-Vinylpyridine	Acetic acid catalyst, 110°C	643
	4-Vinylpyridine	Acetic acid catalyst, 110°C	643
	2-Methyl-6-vinylpyridine	Acetic acid catalyst, 110°C	643
	2,4-Dimethyl-6-vinylpyridine	Acetic acid catalyst, 110°C	643
2-Et	2-Vinylpyridine	Acetic acid catalyst, 100°C	643
	4-Vinylpyridine	Acetic acid catalyst, 100°C	643
	2-Methyl-6-vinylpyridine	Acetic acid catalyst, 100°C	643
	2,4-Dimethyl-6-vinylpyridine	Acetic acid catalyst, 100°C	643
2-Ar	2-Vinylpyridine		508a
	4-Vinylpyridine		508a
1,2,3-Triazoles			
1,2,3-Triazole[g]	Acrylic acid	Pyridine catalyst	623a
	Acrylonitrile	Dioxan/Triton B	235
	Benzalacetone	Triton B	623a
	Chalcone	Triton B	623a
	Dibenzalacetone	Triton B	623a
	Nitroethylene	Chloroform	235
1,2,4-Triazoles			
1,2,4-Triazole[g]	Acrylic acid	Pyridine catalyst	623c
	Benzalacetone	Triton B	623c
	Chalcone	Pyridine catalyst	623c
	Dibenzalacetone	Triton B	623c
	p-Methoxybenzalaceto-phenone	Triton B	623c
Tetrazoles			
1,2,3,4-Tetrazole[g]	Acrylic acid	Pyridine catalyst	623c
	Benzalacetone	Triton B	623c
	Chalcone	Pyridine catalyst	623c
	Dibenzalacetone	Triton B	623c
	p-Methoxybenzalaceto-phenone	Triton B	623c

(a) The product was 1-chloro-1,2,2-trifluoro-2-(1-pyrazolyl)ethane.

(b) The reactants were heated together under various conditions, and also in the presence of sodium alkoxides or acetic acid. The best yield (87%) resulted when no catalyst was used, the reactants being heated in a sealed tube at 130°C for 4h.

(c) No reaction occurred with benzalacetone[270b].

(d) Both possible products were formed in the ratio 30 : 1. The major product was thought to be β-1-(3-methyl-5-phenylpyrazolyl)propionitrile.

(e) 4-Iodopyrazoles are said not to react with acrylonitrile, being resinified under severe conditions[270c].

(f) Both possible products were formed.

(g) Reaction is usually assumed to occur at N-1, but in the case of the addition of acrylonitrile to 1,2,3-triazole n.m.r. spectroscopy suggested that N-2 was the site of addition.

Table A.8. *N*-acyl- and *N*-carboxylazoles[a]

	M.p. (b.p./mmHg)/° C	Ref.
Pyrazole		
1-Ac	(151–2); picrate, 160	74d
1-COEt	(148–50/48)	896a
1-Ac-3,5-Me$_2$	(70/12)	897c
1-COCH$_2$CN-3,5-Me$_2$	118–21	
1-COCH$_2$SH-3,5-Me$_2$	118.0–9.5	
1-COCH$_2$Cl-3,5-Me$_2$	68–70	
1-COCH$_2$OPh-3,5-Me$_2$	85–7	
1-CONH$_2$-3,5-Me$_2$	112–13	
1-COCH$_2$Ph-3,5-Me$_2$	56.5–8.0	
1-COPh-3,5-Me$_2$	(158/12) 123.5–4.5	897c, 898g
1-CO(C$_6$H$_4$-*p*-NO$_2$)-3,5-Me$_2$	122.5–3.5	897c
1-CO(C$_6$H$_4$-*p*-NH$_2$)-3,5-Me$_2$	95.5–6.5	897c
1-Ac-5-NHAc	190.5–1.5	670b
1-Ac-4-Br	93–4	74d
1-Ac-4-Cl	73	
1-Ac-4-I	114	
1-Ac-3,4,5-Br$_3$	106	
1-Ac-4-NO$_2$	58	
1-Ac-4-Et	(82/12)	
1-Ac-3,5-Et$_2$	(94/12)	
1-Ac-4-Me	(68/12)	74d
1-Ac-3 4,5-(C$_6$H$_4$-*p*-OMe)$_3$	136–7	899
1-COPh-3,4,5-(C$_6$H$_4$-*p*-Me)$_3$	166–7	
1-COPh-3,4,5-(C$_6$H$_4$-*p*-OMe)$_3$	210–12	
1-COPh-3,4,5-(2-naphthyl)$_3$	255	
1-COPh-3,4,5-(C$_6$H$_4$-*p*-Cl)$_3$	191–3	
1-COPh-3.4,5-(C$_6$H$_4$-*p*-CHMe$_2$)$_3$	208–10	
1-CO(*p*-Br-C$_6$H$_4$)-3,4,5-(C$_6$H$_4$-*p*-OMe)$_3$	177–8	899
1-Ac-4-Ph	81.5–2.5	900a
1-COCH$_2$NH$_2$-3,5-Me$_2$	101–2	896b
1-COCH$_2$Cl	(170–2/52)	896a
1-Ac-3,4-Me$_2$-5-OAc	54	158
1-Ac-5-COAc-3-Me	38	158
1-COPh-5-OCOPh-3-Me	128	158
1-CONH$_2$	140–1	902a
1-CONHPh	108.5–9.5	581a
1-CONHPh-3-Me-5-OEt	89–90	581a
1-CONHPh-3,5-Ph$_2$	108	581a
1-CONHPh-3,5-Me$_2$	66.5–7.0	581a
1-CONHPh-5-Me-3-Ph	158–60	581a
1-CONH$_2$-4-Br-3,5-Me$_2$	152–3	902a
1-CONH$_2$-4-Br	187–9	114k
1-CONH$_2$-3,4-Me$_2$	173–5	114k
5-Pyrazolones		
1-Ac-3-Me	170–2	158
1-Ac-3-Ph	126	158
Imidazole		
1-COH	54–5	150m
1-Ac	104	150m
1-COEt	38	150m
1-CO-n-Pr	19	150m

Table A.8. Continued

	M.p. (b.p./mmHg)/°C	Ref.
Imidazole		
1-CO-n-Bu	(100/18)	150m
1-CO-n-pentyl	35	150m
1-CO-n-heptyl	47	150m
1-CO-n-nonyl	63	150m
1-CO-n-undecyl	70.5–71.0	150m
1-CO-i-Pr	(92/18)	150m
1-CO-t-Bu	56	150m
1-COCH$_2$Ph	53	150m
1-COCHPh$_2$	126–8	150m
1-COCPh$_3$	148	150m
1-COCH:CHPh	133–4	150m
1-CO-(CH:CH)$_2$Me	86	150m
1-COCF$_3$	(137–8)	150m
1-COCCl$_3$	39–40	150m
1-CO(CH$_2$)$_4$CO$_2$Me	63–4	150m
1-COCH(OAc)Ph	100–2	150m
1-COCH:PPh$_3$	184–6	150m
1-COC(Me):PPh$_3$	172–5	150m
1-COPh	204	150m
1-CO(C$_6$H$_4$-2-Cl)	68–9	150m
1-CO(C$_6$H$_4$-2-Br)	(107–10/10^{-3})	150m
1-CO(C$_6$H$_4$-2-Ac)	89	150m
1-CO(C$_6$H$_4$-2-CO$_2$Me)	110.0–11.5	150m
1-CO(C$_6$H$_4$-3-OMe)	(100/10^{-2})	150m
1-CO(C$_6$H$_4$-3-Cl)	44–5	150m
1-CO(C$_6$H$_4$-3-Br)	59–61	150m
1-CO(C$_6$H$_4$-3-NO$_2$)	89	150m
1-CO(C$_6$H$_4$-3-CO$_2$CH$_3$)	97–9	150m
1-CO(C$_6$H$_4$-4-Me)	75	150m
1-CO(C$_6$H$_4$-4-Et)	22–4	150m
1-CO(C$_6$H$_4$-4-i-Pr)	53.5–4.5	150m
1-CO(C$_6$H$_4$-4-t-Bu)	110	150m
1-CO(C$_6$H$_4$-4-Ph)	120.0–1.5	150m
1-CO(C$_6$H$_4$-4-Ac)	68–9	150m
1-CO(C$_6$H$_4$-4-Cl)	86.5–7.5	150m
1-CO(C$_6$H$_4$-4-Br)	94	150m
1-CO(C$_6$H$_4$-4-NHAc)	179–80	150m
1-CO(C$_6$H$_4$-4-NMe$_2$)	109	150m
1-CO(C$_6$H$_4$-4-NO$_2$)	123	150m
1-CO(C$_6$H$_4$-4-CO$_2$Et)	77–8	150m
1-CO(C$_6$H$_4$-4-C$_5$H$_4$N)	95	150m
1-CO(1-imidazolyl)	116–18	150m
1-COCH(Me)CMe$_3$	48–50	905
1-COCH(Me)Ph	(147–8/12)	905
1-COCH$_2$CH(Me)Ph	(170–2/20)	905
1-COCH(Me)(C$_6$H$_4$-4-NO$_2$)	97–9	905
1-COCH(OPh)Me	86–7	905
1-COCEt$_3$	(102.5/1.5)	653
1-CONHPh	114.5–15.5	581a
1-CONHPh-4,5-Ph$_2$	210.5	581a
1-CONPh$_2$	123–5	657
1-CONPh$_2$-4-Ph	195–7	657
1-CO$_2$Et	(99–100/12)	150g

Table A 8. Continued

	M.p. (b.p./mmHg)/°C	Ref.
Imidazole		
1-CONHBu	75.0–6.5	150i
1-CONH-cyclohexyl	95.5–7.0	150i
1-CONEt$_2$	48.0–9.5	150i
1-CONMePh	72.5–3.5	150i
1 3-diAc-2-imidazolone	105–6	602
1,2,3-Triazole		
1-Ac	62	74d, 190f
1-Ac-4,5-Br$_2$	122	74d
1-Ac-4-Me	67–9 (91–3/12)	74d, 190f
1-Ac-4,5-Me$_2$	36	74d, 190f
1-Ac-4-n-Bu	35	190f
1-Ac-4-n-Pr	28	190f
1-Ac-4-Ph	80–5	190f
1-CONH-n-Bu	71–2	150i
1-CONHPh	110.5–11.0	150i
1-CO$_2$Et-4-Ph	80.5–2.0	907a
1-CO$_2$Et-5-(C_6H_4-p-NO$_2$)	156–8	907a
2-Ac	39–41 (110/50)	190f
2-Ac-4,5-Me$_2$	39	190d
2-Ac-4-Ph	110	190f
2-Ac-4-Me	(94–5/13)	190f
2-Ac-4-n-Bu	(126–8/13)	190f
2-Ac-4-n-Pr	(113–15/13)	190f
2-COPh-4-Me	66.5–8.0	907b
2-COPh-4-Ph	191.0–3.5	907b
2-CO(C_6H_4-m-NO$_2$)-4,5-Me$_2$	140.5–2.0	907b
2-CO(C_6H_4-p-NO$_2$)-4-Me	147–8	907b
2-CO(C_6H_4-m-NO$_2$)-4-Me	110.5–13.0	907b
2-CO(C_6H_4-p-Cl)-4-Me	110.5–12.0	907b
2-CO(C_6H_4-p-OMe)-4-Me	51.5–3.0	907b
2-CO(C_6H_4-p-NO$_2$)-4-Ph	154–5	907b
2-CO(C_6H_4-m-NO$_2$)-4-Ph	150–2	907b
2-CO(C_6H_4-p-Cl)-4-Ph	118.5–19.5	907b
2-CO(C_6H_3-m,p-Cl$_2$)-4-Ph	136–7	907b
2-CO(C_6H_4-p-OMe)-4-Ph	103–4	907b
2-CO(C_6H_4-m-NO$_2$)-4-(C_6H_4-p-NO$_2$)	190–2	907b
2-CO(C_6H_4-p-Cl)-4-(C_6H_4-p-NO$_2$)	166.5–8.0	907b
2-CO$_2$Et-4-Ph	59.0–9.5	907b
2-CO$_2$Et-4-(C_6H_4-p-NO$_2$)	149–50	907b
1,2,4-Triazole[b]		
1-Ac	40–2	150a, 906a
1-Ac-3,5-Me$_2$	90–1	577a
1-Ac-3,5-Ph$_2$	107–8	577a
1-COCMe$_3$	(50/18)	150b
1-CO$_2$Et	(105–6/11)	906a
1-CONHPh	112.0–12.5; 122–3	581a, 150g
1(4)-Ac-3-NH$_2$	151–4	675b
1(4)-Ac-3,5-(NH$_2$)$_2$	~240	675b
1(4)-Ac-3,5-(NHAc)$_2$	225	675b
1(4)-Ac-3-NHAc	191–3	675b
1(4)-CONMe$_2$-5(3)-NH$_2$	144.0–4.5	607g

Table A.8. Continued

	M.p. (b.p./mmHg)/°C	Ref.
1,2,4-Triazole[b]		
1(4)-CONMe$_2$-3-NHAc	140.5–2.0	607g
1(4)-CONMe$_2$-5(3)-NHCOPh	145–6	607g
1(4)-COPh-3-NH$_2$-5-Me	171–2	514e
1(4)-Ac-3-NH$_2$-5-(C$_6$H$_4$-p-Br)	213	514e
1(4)-COPh-3 NH$_2$-5-(C$_6$H$_4$-p-Br)	214–15	514e
1(4)-COPh-3-NH$_2$-5-(C$_6$H$_4$-p-NO$_2$)	234	514e
1(4)-CONH$_2$-3-NH$_2$	122–3	911a
1(4)-Ac-3-picrylamino	215	490a
1(4)-COPh-3-NH$_2$	193	150d
1(4)-COPh-3-NHPh-5-Me	169–72	676d
1(4)-COPh-3-NHPh-5-Et	115–16	676d
1(4)-COPh-3-NHPh-5-n-Pr	111–13	676d
1(4)-COPh-3-NPhCOPh-5-Me	121–4	676d
1(4)-COPh-3-NPhCOPh-5-n-Pr	141–6	676d
1-CONHMe	126.0–41.5	911a
1-CSNHMe	116–19	911a
1-CSNHCH$_2$Ph	122.0–3.5	911a
1-CONH$_2$	130–8	911a
1-CONH(CH$_2$)Cl	102.0–3.5	911a
1(4)-CONH(CH$_2$)Cl-5(3)-NH$_2$	129.0–30.5	911a
1(4)-CONHMe-5(3)-NH$_2$	188–90	911a, 914a

R		
CO	138–9	906a
C·O—⟨ ⟩—CO	232	150c
CO(CH$_2$)$_4$CO	169–70	150c
COCH=CHCO*(trans)*	115–18	150c
Tetrazole		
2-Ac	58–9	150a, 153a
2-Ac-5-Ph	95–7	153a

(a) See also table 21 (p. 113); a more comprehensive table has been published[901].
(b) The presumed orientation of substituents based on recent studies (p. 118) has been included in the table. As there is dispute as to whether 3-amino-1,2,4-triazoles are acylated at N-4[514c,e] or N-1[911,912a] compounds of this type are listed as 1(4)-acyl-1,2,4-triazoles.

Table A.9. Some C-acylazoles

	m.p. (b.p./mmHg)/°C	Derivative[b] (m.p./°C)
Pyrazole[a]		
3-CHO[895d]	149–50	
5-Me-3-CHO[895d]	190	
4,5-Ph$_2$-3-CHO[895b]	160–1	
1,4-Me$_2$-3CHO[943a]	126–7	S (216)
1,5-Me$_2$-3-CHO[943a]	56 (115–20/13)	S (201); O (177–8)
5-Me-1-Ph-3-CHO[915a,943c]	(185/22)	P[943c] (173); S[943c] (183); O[943c] (166)
4-Br-5-Me-1-Ph-3-CHO[943c]	71	O (194); S (217)
4-CHO[944a]	82–3	
1-Me-4-CHO[59f]	(106–8/20)	DNP (264–5)
1,3-Me$_2$-4-CHO[946]	47 (129–31/18)	
1,5-Me$_2$-4-CHO[946]	58 (129/18)	
1-Ph-4-CHO[59a]	85	O (167.0–7.5)
1,3,5-Me$_3$-4-CHO[946]	79	S[943a] (213–14)
3-Me-1-Ph-4-CHO[59g,947b]	53; 60–1	
5-Me-1-Ph-4-CHO[943b]	76.0–6.5	P (145)
1,3-Ph$_2$-4-CHO[947b]	142	
3,5-Me$_2$-1-Ph-4-CHO[59g,947b]	128	S[943b] (191–2); P[943b] (140–1)
3-Me-1,5-Ph$_2$-4-CHO[59g]	136–7	
5-Me-1,3-Ph$_2$-4-CHO[59g,943b]	131–2; 101.0–1.5	S[943b] (207)
1,3,5-Ph$_3$-4-CHO[59g]	160.0–1.5	
1-Ph-5-CHO[943c]	Oil	S (168); O (177)
1,4-Me$_2$-5-CHO[943a]	126–7	
1,3-Me$_2$-5-CHO[943a]	(80–3/12)	Pi (133); S (206); O (148)
3-Me-1-Ph-5-CHO[943c]	Oil	S (166, 195); O (146, 179)
1-Et-3-Me-5-CHO[895d]	116–17	
3,4-(CHO)$_2$[944b,948]	203–5	
3-Ac[949c,950]	100–1	S[960b] (204–5); NP[960b] (249)
4-Me-3-Ac[951]	102–3	P (135–6)
5-Me-3-Ac[946]	97	S (229)
5-Me-1-Ph-3-Ac[952]	90	O (175); S (228); O (151) [953]
5-Me-1-(p-MeOC$_6$H$_4$)-3-Ac[953]	84	
1,5-Ph$_2$-3-Ac[73]	88	
1,4-Me$_2$-3-Ac[954]	19–20 (63–5/0.1)	
1,5-Me$_2$-3-Ac[946]	56	
4-Me-1-Ph-3-Ac[955]	72–4	O (149)
3-COEt[950]	207–10	
3-COPh[950,956]	98–9	S (185–7)
4-Ph-3-COPh[957a]	193–4	
5-Ph-3-COPh[957a]	170.0–1.5	
1-Ph-3-COPh[958]	52	
4,5-Ph$_2$-3-COPh[957a]	177.0–7.5	
5-Me-1-Ph-3-COPh[958]	91	DNP (210)
1,4,5-Ph$_3$-3-COPh[959a]	155	DNP (210–11)
4-Ac[946,950]	111	
3-Me-4-Ac[960b]	63.5	O (156–8)
1-Ph-4-Ac[59f,945a]	127.5–8.5	
5-Me-1-Ph-4-Ac[961a]	107–8	H (94–5); P (175)
3,5-Me$_2$-4-Ac[646]	128	Pi[898g] (146–7)
1-Et-3,5-Me$_2$-4-Ac[126d]	30 (125–7/8)	Pi (83)

Table A.9. Continued

	m.p. (b.p./mmHg)/°C	Derivative[b] (m.p./°C)
Pyrazole[a]		
3,5-Me₂-1-Ph-4-Ac[126d]	59–60	
1.3,5-Me₃-4-Ac[126d]	69–70	Pi (90)
3-Me-1,5-Ph₂-4-Ac[950]	93–5	
5-Me-1,3-Ph₂-4-Ac[962c]	88	P (182)
4-COEt[950]	207–10	
1,3,5-Me₃-4-COEt[126d]	43–4 (137–9/10)	Pi (83–4)
3,5-Me₂-1-Ph-4-COPr[323a]	62.5–3.0	
1,3,5-Me₃-4-COAm[323a]	36.5	
5-Cl-3-Me-1-Ph-4-Ac[323a]	76–7	
3-Cl-1-Ph-4-Ac[323a]	125	
1,3,5-Ph₃-4-Ac[323a]	124	
5-COEt-3-Ac[963]	129.5–31.0	
5-COPh-3-Ac[963]	125.5–6.5	
5-Me-1-Ph-3,4-Ac₂[901]	136	
3,5-Ac₂[963]	148.0–9.5	
4-Me-3,5-Ac₂[964c]	114	di-O (217)
4-Ph-3,5-Ac₂[964c]	134	
5-COPh-4-Me-3-Ac[964c]	97	
3,5-Me₂-1,4-Ac₂[646]	50	
3,5-Me₂-4-COPh[323a]	123.0–3.5	
1-Ph-4-COPh[960c]	122–3	DNP[958] (264)
1,5-Ph₂-4-COPh[100e]	144–5	
1,3,5-Me₃-4-COPh[126b]	(201/13)	Pi (135–6)
3-Me-1-Ph-4-COPh[126b]	136	
1-Et-3,5-Me₂-4-COPh[126b]	(198–200/13)	Pi (106)
1-CH₂Ph-3,5-Me₂-4-COPh[126b]	95	Pi (84–5)
1-Ph-3,5-Me₂-4-COPh[126b]	99–100	
3-Me-1,5-Ph₂-4-COPh[959a]	115–16	DNP (207)
1,3,5-Ph₃-4-COPh[962c]	174	
3,5-Me₂-1,4-(COPh)₂[965d]	124.0–5.5	
3,4-(COPh)₂[96b]	169	
1-Ph-5-COPh[945a,967a]	119–20	DNP[945a] (194–5)
3-Me-1-Ph-5-COPh[968]	77–9	
3-COEt-5-COPh[963]	137–9	
3,5-(COPh)₂[963]	150.5–2.0	
Imidazole		
2-CHO[803]	204	DEA[970b] (116)
1-Me-2-CHO[108,375]	41–2; 34–7	O[375] (176)
1-PhCH₂-2-CHO[375]	(120/0.1)	O (170)
		P[803] (169)
4-CHO[971b]	174	DNP (291–2)
		O (183–4)
5-Me-4-CHO[972]	167	Pi (180–1)
5-Me-2-Ph-4-CHO[325]	108	DNP (310)
1-Me-5-CHO[974b]	54	Pi (172–3)
1,4-Me₂-5-CHO[972]	70	Pi (212–13)
1-i-Pr-5-CHO[974b]		P (116–18)
1-Ph-5-CHO[974b]	120–1	P (207–9)
1-Cyclohexyl-5-CHO[974b]	70–1	P (186–8)
2-NH₂-4-Me-5-CHO[975a]		DMA–Pi (229–30)
2-SMe-4-Me-5-CHO[975a]		DMA–Pi (163–5)
4-Me-2-Ph-5-CHO[975a]		DMA–Pi (239–40)

Table A.9. Continued

	m.p. (b.p./mmHg)/°C	Derivative[b] (m.p./°C)
Imidazole		
		DMA (144–5)
4,5-(CHO)$_2$[969e]	237	
1-Me-4,5-(CHO)$_2$[969e]	88	
1-Et-4,5-(CHO)$_2$[969e]	93–4	
1-PhCH$_2$-2-i-Pr-4,5-(CHO)$_2$[976a]	74–5	
1-PhCH$_2$-4,5-(CHO)$_2$[969e]	124–6	
2-Ac[372]	137	Pi (224–6)
2-Ac-4-Me[977b,869]	109; 111–13	Pi[977b] (157)
		DNP[977b] (275)
5-Ac-4-Me[978b]	151	S (212)
5-Ac-1-Me[979]	54–6 (84/2.5)	P (163–4)
1-Me-2-COPh[980,981a]	(110/0.5)	Pi[981a] (175–8)
4-Me-2-COPh[869]	192–4	
2-COPh[982b]	161–2	
4,5-Br$_2$-2-COPh[982b]	218–20	
4-Ph-2-COPh[969a]	204–5	Pi (153–4)
		DNP (272–3)
1-Ph-2-COPh[981a]	(210/2)	Pi (149)
1-PhCH$_2$-2-COPh[981a]	66	Pi (146–7)
4-Ph-2-CO-cyclohexyl[969a]	146–7	Pi (156–8)
		DNP (296–7)
4-(p-CH$_3$O-C$_6$H$_4$)-2-CO(C$_6$H$_4$-p-OCH$_3$)[983]	212–15	
4-(p-MeC$_6$H$_4$)-2-CO(C$_6$H$_4$-p-Me)[983]	203.5–5.5	DNP (277–81)
1,2,3-Triazole		
4-CHO[984a]	141–2	O[985a] (207–8)
1-Ph-4-CHO[984b]	98.5–9.5	
5-Ph-4-CHO[984b]	186.5–7.5	
2-Ph-4-CHO[986]	68–9	DNP[987] (198–200)
		S[987] (225–6)
1-Ph-5-Me-4-CHO[943d]	52	S (226–7); O (162)
1-PhCH$_2$-4-CHO[985a]	89–90	DNP (270–1)
1,5-Ph$_2$-4-CHO[943d]	104–5	S (224); O (176)
1,5-Ph$_2$-4-CHO[984b]	112–3	O (194–5)
1,4-Ph$_2$-5-CHO[984b]	171–2	O (160.5–1.5)
1-Ph-5-CHO[984b]	76.5–7.0	
2-(m-F-C$_6$H$_4$)-4-CHO[988a]	65–6	
2-(p-F-C$_6$H$_4$)-4-CHO[988a]	89–90	
2-(m-Cl-C$_6$H$_4$)-4-CHO[988a]	90	
2-(p-Cl-C$_6$H$_4$)-4-CHO[988a]	121	
2-(m-Br-C$_6$H$_4$)-4-CHO[988a]	96	
2-(p-Br-C$_6$H$_4$)-4-CHO[988a]	114	
2-(4′,3′-Cl,NO$_2$-C$_6$H$_3$)-4-CHO[988a]	113–14	
2-(4′ 3′-Cl,Me-C$_6$H$_3$)-4-CHO[988a]	105	
2-(4′,2′-Br,Me-C$_6$H$_3$)-4-CHO[988a]	99	
2-(4′,3′-Br,Me-C$_6$H$_3$)-4-CHO[988a]	106	
1-Ph-4,5-(CHO)$_2$[948]	107	bis-DEA (59)
1-PhCH$_2$-4,5-(CHO)$_2$[948]	89	
1-Hexyl-4,5-(CHO)$_2$[948]	(104–6/0.1)	
1-Ph-4,5-(CHO)$_2$[989]	145–7	bis-DNP (304–7)
2-Ph-4-Me-5-Ac[990a]		DNP (254)

Table A.9. Continued

	m.p. (b.p./mmHg)/°C	Derivative[b] (m.p./°C)
1,2,3-Triazole		
5-Me-4-Ac[991]	173–4	O (202)
1-Ph-4-Ac[959b]	113	DNP (251–2)
5-Me-1-Ph-4-Ac[959b]	99–100	DNP (211)
1-Me-4-COCO$_2$H[895d]		O (155)
3-Me-5-COCO$_2$H[895d]	178	O (196.5)
4-COPh[949b]	123–4	
1-Ph-4-COPh[959b]	125	DNP (254–5)
5-Me-1-Ph-4-COPh[959b]	Oil	DNP (246)
4-Me-2-Ph-5-COPh[992]	74	NP (231.0–1.5)
4,5-(COPh)$_2$[994]	167–9	
1-Ph-4-PhCOCH$_2$CO[959b]	169–70	
2-Ph-4-Me-5-MeCOCH$_2$[994a]	85	P (101–2)
4-(p-CH$_3$O-C$_6$H$_4$)CO[949b]	121–3	
4-(p-Cl-C$_6$H$_4$)CO[949b]	146	
4-(p-Br-C$_6$H$_4$)CO[949b]	135	
1,2,4-Triazole		
3-CHO[995b,c]	175–7	DMA (66–7); DEA (63–5); DNP (274–5)
5-Ph-3-CHO[995b,c,d]	202–4[c]	DEA (106–8); O (218–19); S (172–3); TS (242–3); DMA (102–3); DNP (286–8)
5-Me-3-CHO[995b,c,d]	188–90[c]	DMA (96–101); DNP (254–6); DEA (126–7); TS (214–15)
5-PhCH$_2$-3-CHO[995b,c,d]	169–71[c]	DMA (102–3); DNP (243–5); TS (169–71)
5-SH-3-CHO[995b,c]		DMA (178–9); DEA (161–2); DNP (281–2)
5-OH-1-Ph-3-CHO[996]		P (275)
5-(p-MeC$_6$H$_4$)-3-CHO[995b,c,d]	274–6[c]	DMA (127–8); DNP (279–81)
5-(p-ClC$_6$H$_4$)-3-CHO[995b,c,d]	242–4[c]	DMA (121–2); DNP (284–5)
5-(p-NO$_2$C$_6$H$_4$)-3-CHO[995b,c,d]	241–3[c]	DMA (168–9); DNP (290–3)
5-(2'-pyridyl)-3-CHO[995b,c,d]	202–4[c]	DNP (324–5)
5-(3'-pyridyl)-3-CHO[995b,c,d]	239–41[c]	DMA (151–2); DNP (314–15)
5-(4'-pyridyl)-3-CHO[995b,c,d]	241–3[c]	DMA (121–2); DNP (327–8)
1,5-Ph$_2$-3-CHO[175a]	144–5	O (162–70); S (222–3); DNP (247–9)
1-(p-ClC$_6$H$_4$)-5-Me-3-CHO[995d]	143–6	
1-(o-MeC$_6$H$_4$)-5-Ph-3-CHO[175a]	80–2	DNP (230–2, 256–8)
1-(m-MeC$_6$H$_4$)-5-Ph-3-CHO[175a]	98–100	
1-(p-MeC$_6$H$_4$)-5-Ph-3-CHO[175a]	92–3	S (226–7); DNP (235–6)
1-(p-MeOC$_6$H$_4$)-5-Ph-3-CHO[175a]	93–4	DNP (227–8)
1-(p-BrC$_6$H$_4$)-5-Ph-3-CHO[175a]	128–30	O (173–5, 235–6); DNP (223–4)
1-Ph-3-Ac[997]	121–3	O (185–6); TS (203–4)
(1,4,5-Ph$_3$-3-Ac)$^+$[962g]		Perchlorate (176)
1-Ph-5-Me-3-Ac[997]	88–9	O (206–8)
1-Ph-5-Et-3-Ac[997]	73–4	O (171–3)
1-(p-ClC$_6$H$_4$)-5-Me-3-Ac[997]	150–2	O (213–15)

Table A.9. Continued

	m.p. (b.p./mmHg)/°C	Derivative[b] (m.p./°C)
1,2,4-Triazole		
1-(*p*-EtOC$_6$H$_4$)-5-Me-3-Ac[997]	108–10	
1-(*p*-ClC$_6$H$_4$)-5-Et-3-Ac[997]	64–6	O (195–7)
1-(*p*-EtOC$_6$H$_4$)-5-Et-3-Ac[997]	52–4	
1-(*p*-ClC$_6$H$_4$)-3-Ac[997]	145–7	O (184–6); TS (213–14)
1-(*p*-EtOC$_6$H$_4$)-3-Ac[997]	115–17	O (173–5); TS (214–15)
3-COPh[995e]	206–8	DNP (280–2)
1-Ph-3-COPh[995e]	105–7	O (138–41, 184–7)
5-Ph-3-COPh[995e]	199–200	DNP (294–5); TS (232–3)
5-(3'-pyridyl)-3-COPh[995e]	224–5	
5-(4'-pyridyl)-3-COPh[995e]	251–3	DNP (323–4)
5-(4'-pyridyl)-3-CO(C$_6$H$_4$-*p*-Cl) [995e]	299–300	
5-Ph-3-CO(C$_6$H$_4$-*m*-NO$_2$)[995e]	221–2	
5-Ph-3-CO(C$_6$H$_4$-*p*-NO$_2$)[995e]	232–3	
4-Ph-3-COPh[327]	115–18	
1-Ph-5-Me-3-COPh[998b]	55.5	
4 5-Ph$_2$-3-COPh[327]	179–81	O (208–11)
4,5-Ph$_2$-3-CO(C$_6$H$_4$-*p*-Cl)[327]	194–6	
4,5-Ph$_2$-3-CO(C$_6$H$_4$-*p*-NO$_2$)[327]	229–31	
4-Ph-3,5-(COPh)$_2$[327]	140–2	bis-O (254–5)
4-Ph-3,5-(CO–C$_6$H$_4$-*p*-Cl)$_2$[327]	263–5	
4-Ph-3,5-(CO-C$_6$H$_4$-*p*-NO$_2$)$_2$[327]	347–50	
Tetrazole		
1-Me-5-CHO[1001]	76–9	DNP (230–7)
1-Et-5-CHO[1001]	(61/0.5)	DNP (182–90)
1-Bu-5-CHO[1001]	(75/0.3)	DNP (143–4)
1-C$_6$H$_{11}$-5-CHO[1001]	47–9	DNP (239–40)
1-CH$_2$Ph-5-CHO[1001]	(87–9/0.01)	DNP (216–17)
1-Ph-5-CHO[1001]	97	DNP (244)
1-(*p*-Me-C$_6$H$_4$)-5-CHO[1001]	90–7	DNP (229–30)
1-(*p*-NO$_2$-C$_6$H$_4$)-5-CHO[1001]	134–7	DNP (252)
3-Me-5-CHO[999]		S (225–30)
5-Ac[1000]	89–90	
3-Me-5-Ac[999]	41–2 (84–6/2)	H (173–5); P (118–20); S (230–5)
5-COPh[1000]	140–1	O (219–20)
3-Me-5-COPh[999]	45.5–6.0	S (203–4)
1-PhCH$_2$CO-5-COPh[1002]	114–15	
2-PhCH$_2$CO-5-COPh[1002]	126.0–6.5	
5-CO(C$_6$H$_4$-*p*-Cl)[1000]	174–5	
5-CO(C$_6$H$_4$-*p*-NO$_2$)[1000]	161–3	
5-CO(C$_6$H$_4$-*m*-NO$_2$)[1000]	131–2	
5-CO(C$_6$H$_4$-*p*-Br)[1000]	176–7	

(a) Much of this information on pyrazoles has been taken from the comprehensive tables by R. Fusco[901].

(b) Picrate = Pi; 2,4-dinitrophenylhydrazone = DNP; oxime = O; *p*-nitrophenyl-hydrazone = NP; phenylhydrazone = P; hydrazone = H; semicarbazone = S; thiosemicarbazone = TS; diethylacetal = DEA; dimethylacetal = DMA.

(c) m.p. of hemiaminal form.

Table A.10. Some azole–carboxylic acids and their derivatives

	m.p. (b.p./mmHg)/°C	Derivative (m.p./°C)
Pyrazole[a]		
3-CO$_2$H[77g]	210–12	Me ester[432b] (141); Et ester [432b] (160); amide[1020e] (147–9); [1021a] (159–60)
4-Me-3-CO$_2$H[17d]	218–20	Me ester (170–1); Et ester (156–8)
4-Et-3-CO$_2$H[1023b]	214–15	Et ester (102–3)
4-Ph-3-CO$_2$H[1022]	253–4	Me ester (188–90); Et ester [1024] (164–5)
5-Me-3-CO$_2$H[1025e]	236–8	Et ester (82–3); Me ester[100f] (75–8); amide[426b] (178–9)
5-Ph-3-CO$_2$H[1026d]	235–7	Me ester (181–2); Et ester (140)
4,5-Me$_2$-3-CO$_2$H[100f]	276–7	Me ester (140–2)
4-Me-5-Ph-3-CO$_2$H[900d]	248–50	Me ester[554b] (102–3)
4,5-Et$_2$-3-CO$_2$H[70a]	194	Et ester (b.p. 135–40/5 mm)
4-Ph-5-Me-3-CO$_2$H[70a]	265	Et ester (122); Me ester[895c] (154)
4-Ph-5-Et-3-CO$_2$H[70a]	241	Et ester (127–8)
4,5-Ph$_2$-3-CO$_2$H[782]	261	Me ester (218)
1-Me-3-CO$_2$H[100i]		Me ester (b.p. 120/9 mm)
1-PhCH$_2$-3-CO$_2$H[363d]		Et ester (75–6)
1-Ph-3-CO$_2$H[1028]	143–6	Me ester (77)
1,4-Me$_2$-3-CO$_2$H[100f]	171	Me ester (37–8); amide[943a] (158–60); acid chloride[943a] (73–4)
1-Me-4-Ph-3-CO$_2$H[66j]	132	Me ester (122–3)
1-Ph-4-Me-3-CO$_2$H[955]	170	
1-Ph-4-Et-3-CO$_2$H[962f]	129	Et ester (69)
1,4-Ph$_2$-3-CO$_2$H[960a]	227–8	
1,5-Me$_2$-3-CO$_2$H[62c]	176–8	Me ester[118] (71.5–2.5); acid chloride[943a] (60)
1-Me-5-Ph-3-CO$_2$H[66g]	143–4	Et ester (62–3)
1-Et-5-Me-3-CO$_2$H[66d]	136–7	Et ester (b.p. 202/12 mm)
1-Ph-5-Me-3-CO$_2$H[1029a]	134–6	Amide[1029a] (146); Me ester [901] (55–6); Et ester[59e] (35–6); acid chloride[943c] (85)
1,5-Ph$_2$-3-CO$_2$H[1030a]	185	Et ester (90)
1,4,5-Me$_3$-3-CO$_2$H[100g]	204.5–5.5	Me ester (49.0–52.5)
1,4,5-Ph$_3$-3-CO$_2$H[959a]	245	Et ester (158); acid chloride (155)
4-CO$_2$H[1031]	278–9	Et ester[974e] (76–8)
1-Me-4-CO$_2$H[77g]	208–9	
1-PhCH$_2$-4-CO$_2$H[363b]	151–2	Et ester (62–3)
1-Ph-4-CO$_2$H[1032a]	221–2	Me ester (128–9); Et ester (99–100)
3-Me-4-CO$_2$H[900e]	228	Et ester[961a] (54)
3-Ph-4-CO$_2$H[961b]	260	Et ester (85–6); Me ester[900b] (111.5–2.5)
3,5-Me$_2$-4-CO$_2$H[1033]	290	Et ester (96)
3-Me-5-Pr-4-CO$_2$H[1034]	228	Me ester (b.p. 179/10 mm)
3-Me-5-Ph-4-CO$_2$H[1035]	260–5	Et ester (b.p. 145–50/0.5 mm)
3,5-Ph$_2$-4-CO$_2$H[895c]	283–5	Me ester (180–1)
1-Ph-3-Me-4-CO$_2$H[59g, 1032b]	196–7	Me ester (70–1)

Table A.10. Continued

	m.p. (b.p./mmHg)/°C	Derivative (m.p./°C)
Pyrazole[a]		
1,3-Ph$_2$-4-CO$_2$H[900f]	202–3	
1,5-Me$_2$-4-CO$_2$H[77g]	184–5	
1-Ph-5-Me-4-CO$_2$H[59g, 1029b]	166–8	Me ester[915a] (71); Et ester[1029b] (55–6); acid chloride[943b] (147); amide[1036] (137–40)
1,5-Ph$_2$-4-CO$_2$H[960a]	180	Et ester (112–14)
1,3,5-Me$_3$-4-CO$_2$H[62c]	176–7	
1,5-Me$_2$-3-Ph-4-CO$_2$H[62c]	197–8	Me ester (63–4)
1,3-Me$_2$-5-Ph-4-CO$_2$H[62c]	176–7	
1-Ph-3-OMe-4-CO$_2$H[1037]		Et ester (144–6)
1-Ph-3-OEt-4-CO$_2$H[1037]		Et ester (101)
1-Ph-3,5-Me$_2$-4-CO$_2$H[59g, 943b]	200–1	Et ester[1038] (68–70); acid chloride[943b] (64)
1,5-Ph$_2$-3-Me-4-CO$_2$H [1039a]	205	Amide (193.0–4.5); acid chloride[1040] (78–80)
1,5-Ph$_2$-3-Et-4-CO$_2$H[1020a]	192	Et ester (119)
1,3-Ph$_2$-5-Me-4-CO$_2$H[59g]	193–4	Me ester[900f] (100–1); acid chloride[943b] (82); amide [962a] (232)
1,3,5-Ph$_3$-4-CO$_2$H[72,59g, 1041]	238–9; 205–6	Et ester (145.0–6.5); Me ester [895c] (140–1); amide[962a] (199–200)
1-Me-5-CO$_2$H[443]	222	Me ester[100f] (b.p. 73/9 mm)
1-PhCH$_2$-5-CO$_2$H[77g]	171–2	
1-Ph-5-CO$_2$H[1029a]	179–81	Me ester (67); acid chloride [943c] (53)
1,3-Me$_2$-5-CO$_2$H[118]	207	Me ester (b.p. 91/11 mm); acid chloride[943a] (b.p. 75–80/ 12 mm); amide[943a] (165)
1-Me-3-Ph-5-CO$_2$H[66g]	183–4	
1-Et-3-Me-5-CO$_2$H[66d]	141–2	Et ester (b.p. 101.5/12 mm)
1-Et-3-Ph-5-CO$_2$H[66g]	162–3	Et ester (b.p. 193–4/12 mm)
1-PhCH$_2$-3-Me-5-CO$_2$H[66f]	156	Et ester (b.p. 170–1/9 mm)
1-Ph-3-Me-5-CO$_2$H[154]	189–90	Me ester (65–6); amide (181); Et ester[59e] (40–1); acid chloride[943c] (39–41)
1,3-Ph$_2$-5-CO$_2$H[968]	217–18	
1,4-Me$_2$-5-CO$_2$H[100f]	171	Me ester (25–6); amide[943a] (40); acid chloride[943a] (40)
1-Me-4-Ph-5-CO$_2$H[66j]	210–11	Me ester (69)
1,3,4-Me$_3$-5-CO$_2$H[100g]	179	Me ester (b.p. 113/16 mm)
1,4-Ph$_2$-3-Me-5-CO$_2$H[1042]	201–2	Et ester (138–40)
1,3,4-Ph$_3$-5-CO$_2$H[1042]	222	
3,5-(CO$_2$H)$_2$[1043]	290	bis-Acid chloride (86–7); bis-Me ester[1026e] (153)
4-Me-3,5-(CO$_2$H)$_2$[1044]	315	bis-Et ester (106–7); bis-Me ester[1045] (128–9)
4-Ph-3,5-(CO$_2$H)$_2$[1026a]	243–6	bis-Et ester (96)
1-Me-3,5-(CO$_2$H)$_2$[77g]	250	bis-Me ester[900a] (72.0–3.5)
1-Et-3,5-(CO$_2$H)$_2$[1043]	243	bis-Acid chloride (b.p. 125/ 12 mm)
1-Ph-3,5-(CO$_2$H)$_2$[1029c]	266	bis-Me ester (127–9); bis-amide (190)

Table A.10. Continued

	m.p. (b.p./mmHg)/°C	Derivative (m.p./°C)
Pyrazole[a]		
4,5(3)-(CO$_2$H)$_2$[432b,944a]	244; 260–1	bis-Me ester (141); bis-Et ester [974f] (69–70); bis-acid chloride[944a] (310–15); bis-amide[1046] (327)
3-Me-4,5-(CO$_2$H)$_2$[900d]	229–30	bis-Me ester[432b] (106)
3-Ph-4,5-(CO$_2$H)$_2$[66i]	235	
1-Ph-3,4-(CO$_2$H)$_2$[965c]	234	bis-Me ester (97–8)
1-Ph-5-Me-3,4-(CO$_2$H)$_2$ [1047a]	247	bis-Et ester[901] (51.5)
1,5-Ph$_2$-3,4-(CO$_2$H)$_2$[1d]	217–18	bis-Me ester[900f] (97.0–7.5); bis-Et ester[1048a] (89)
1-Me-4,5-(CO$_2$H)$_2$[1043]	(.2H$_2$O) 233–5 (.H$_2$O) 179–80	bis-Me ester (b.p. 204–7/20 mm)
1-Ph-4,5-(CO$_2$H)$_2$[1029b]	215–16	bis-Amide (253–5); bis-Me ester (75–6); bis-Et ester[974f] (b.p. 168–73/0.7 mm)
1-Ph-3-Me-4,5-(CO$_2$H)$_2$ [1039b]	202–3	
1,3-Ph$_2$-4,5-(CO$_2$H)$_2$[1469]	197	bis-Me ester[900f] (150.5–1.5)
3,4,5-(CO$_2$H)$_3$[1031,1049]	213; 233	tri-Me ester[2f] (118); tri-Et ester[974d] (91)
1-Ph-3,4,5-(CO$_2$H)$_3$[1d,974f]	184; 213–14	tri-Et ester[974f] (74.0–5.5)
Imidazole[b]		
2-CO$_2$H[363a]	163–4	
1-Me-2-CO$_2$H[1050a]	97	Et ester[980] (44–6); amide [375] (170)
1-Ph-2-CO$_2$H[480a]		Et ester (94.0–4.5)
1-PhCH$_2$-2-CO$_2$H[480a]	103–4	Amide[375] (160)
4-Me-2-CO$_2$H[284]	175	
4,5-Ph$_2$-2-CO$_2$H[208c]	143–4	Et ester (197–200); anilide (221–2)
4-CO$_2$H[974a]	283	Et ester (157–8); Me ester[1051] (154–6); amide[374f] (215)
1-Me-4-CO$_2$H[1052]		Me ester (97–8)
2-Me-4-CO$_2$H[274c]	262	Et ester[274d] (156)
5-Me-4-CO$_2$H[1050a,274c]	222–3	Et ester[973e] (204–5)
2-Ph-4-CO$_2$H[274d]	239	Et ester (189)
2-PhCH$_2$-4-CO$_2$H[1053a]		Me ester (214–15)
2-Et-4-CO$_2$H[274d]	252	Et ester (129)
2,5-Me$_2$-4-CO$_2$H[973e]	247	Et ester (165–7)
2-Ph-5-Me-4-CO$_2$H[973a]	177	Et ester[1054] (198–9)
5-Ph-4-CO$_2$H[453b]		Et ester (225)
2-PhCH$_2$-5-Me-4-CO$_2$H[973e]		Et ester (167–8)
1-Me-2,5-Ph$_2$-4-CO$_2$H[915c]		Et ester (124–5)
1-Me-5-CO$_2$H[480a]	275–7	Me ester (39–40)[480a] (56–7) [974a,1052]
1-Ph-5-CO$_2$H[1055a]	203	Et ester[974a] (80–1)
1-PhCH$_2$-5-CO$_2$H[1055a]		Et ester (64); Me ester (63–4)
1-PhCHMe-5-CO$_2$H[1056]	188–9	Me ester (173–4)
1-Ph(CH$_2$)$_2$-5-CO$_2$H[1056]		Me ester (63–4)
1,4-Me$_2$-5-CO$_2$H	205–6[972] (b.p. 75/ 0.2 mm)[1050a]	

Table A.10. Continued

	m.p. (b.p./mmHg)/°C	Derivative (m.p./°C)
Imidazole[b]		
4,5-$(CO_2H)_2$[274c]	288	bis-Amide[570] (>360); bis-Me ester[570] (200-3); bis-Et ester[974c,1059a] (151-2)
1-Me-4,5-$(CO_2H)_2$[974c, 1057]	262-3	bis-Amide[570] (263-6); bis-Me ester[974c] (46.5-7.0)
2-Me-4,5-$(CO_2H)_2$[274c]	320-2	bis-Amide[1058b] (325); bis-Et ester[1058a] (88)
1-Ph-4,5-$(CO_2H)_2$[974c]	199-200	bis-Me ester (86-7); bis-Et ester (84-5)
2-Ph-4,5-$(CO_2H)_2$[274c]	271	bis-Et ester[1058a] (190)
2-Et-4,5-$(CO_2H)_2$[274c]	259	bis-Amide[1058c] (258); bis-Et ester[1058c] (94)
1-Et-4,5-$(CO_2H)_2$[1021b]		bis-Amide (215)
1-PhCH$_2$-4,5-$(CO_2H)_2$ [1059a,1060]		bis-Et ester[1060] (131.0-1.5) [1059a] (b.p. 200.2/0.1 mm)
2-PhCH$_2$-4,5-$(CO_2H)_2$[982b]	248	
1,2,3-Triazole[c]		
4-CO_2H[77c 82,276b,915e, 1061b,1062,1063a]	211-15; 220-4 [276b,1063a]	Me ester[915e] (135-8); amide [1061a,1063a] (256-7); Et ester[1063a] (117-18)
1-Me-4-CO_2H[77c,276b, 895d]	224	Et ester[1071] (93.0-4.5); amide[1071] (150-220)
1-Me-5-CO_2H[76c]	188	
2-Me-4-CO_2H[1064a]	142	
5-Me-4-CO_2H[77c,962h, 1065]	213-14[77c,962h]; 220[1065]	Et ester (161-2); anilide[962h] (191)
1-Et-5-CO_2H[76c]	178-82	
1-Ph-4-CO_2H[915e,1055c, 1066a,1067]	148-50	Me ester[915e,1066a,1067] (120-2); Et ester[959b] (88); amide[1055c] (229-31)
1-Ph-5-CO_2H[915e,1066a, 1067]	176	Me ester (101); Et ester[1066a] (54-5); amide[1066a] (146)
4-Ph-5-CO_2H[949a,1019a, 1064e,1068]	205-6	Et ester[949a] (92-3)
2-Ph-4-CO_2H[1069,1070]	191	Me ester[1070] (85-6); Ph ester [1070] (111-12); amide[1070] (142-3)
1-PhCH$_2$-4-(or 5)-CO_2H [1072a]	173-4	
1-PhCH$_2$-5-CO_2H[1061a]	196-7	
1,5-Me$_2$-4-CO_2H[1061a]	203	
1-Et-5-Me-4-CO_2H[1061a]	184	
2-Et-4-Me-5-CO_2H[1064a]	131	
1-Ph-5-Me-4-CO_2H[943d, 1073]	140-1; 148[1066a]	Me ester[1066a] (73-4); Et ester[1066a] (60); amide[943d] (217); acid chloride[943d, 1074] (133)
2-Ph-4-Me-5-CO_2H[994a,b]	95[994a]; 202[994b]	
1-PhCH$_2$-5-Me-4-CO_2H [1061a]	168-9	
1,5-Ph$_2$-4-CO_2H[943d, 1066a,1075]	164-5[1066a]; 176-8 [1075]	Me ester[1066a,1075] (135); Et ester[1066a] (134-5); amide[943d] (174); acid chloride[943d] (101)

Table A.10. Continued

	m.p. (b.p./mmHg)/°C	Derivative (m.p./°C)
1 2,3-Triazole[c]		
1,4-Ph$_2$-5-CO$_2$H[984b]	177.5	
2,5-Ph$_2$-4-CO$_2$H[1076]	208–9	
1-PhCH$_2$-5-Ph-4-CO$_2$H[1077]	187–8	
1-Ph-5-PhCH$_2$-4-CO$_2$H[959a]	150–1	
4,5-(CO$_2$H)$_2$[1063a 1078, 1079]	198–200	bis-Me ester[190c,1063a] (131); mono-amide[1063a] (246)
1-Me-4,5-(CO$_2$H)$_2$[985a, 1063a,1080]	174–5[1080]; 179–80[985a]	bis-Me ester[1063a] (64–5)
1-Et-4,5-(CO$_2$H)$_2$[1010, 1061a]	(.H$_2$O) (108–10)	
1-Bu-4,5-(CO$_2$H)$_2$[985a]	136.0–6.5	
1-iso-Bu-4,5-(CO$_2$H)$_2$[1010]	136–7	
2-Me-4,5-(CO$_2$H)$_2$[1064e]		bis-Me ester (55–60)
1-PhCH$_2$-4.5-(CO$_2$H)$_2$ [985b,1072a]	183	bis-Me ester (48–9); bis-amide (199)
2-Ph-4,5-(CO$_2$H)$_2$[989, 994b]	250[994b]; 259–61[989]	bis-Me ester[1081] (89); bis-Et ester[1081] (41–2); bis-amide [1081] (279); bis-acid chloride [1081] (122)
1-Ph-4,5-(CO$_2$H)$_2$[1067]	147–8	bis-Me ester[959b,1067] (127); bis-amide[943d] (228); bis-acid chloride[959b] (40)
1,2,4-Triazole		
3-CO$_2$H[363c,1082,1083]	135–7	Me ester[910] (198); Et ester [363c,1082] (178); amide [910] (312)
1-Ph-3-CO$_2$H	197[997]; 181–2 [1083]	Et ester[915d] (66–7); [1084] (72)
3-Ph-5-CO$_2$H[1085]	114	
3-Me-5-CO$_2$H[1085]	131–2	
1,3-Ph$_2$-5-CO$_2$H[828b]		Et ester (118.5–9.5)
1,5-Ph$_2$-3-CO$_2$H[175a, 1086]	177–8	Me ester[175a] (158–9); amide [175a] (198–9); Et ester[915d] (169.5–70.5)
1-Ph-5-Me-3-CO$_2$H[1087a,b, 1088]	176	Amide[1087a] (170)
1-Ph-5-PhCH$_2$-3-CO$_2$H [1087a b]	144	
4,5-Ph$_2$-3-CO$_2$H[1089]	135–8	Et ester (151–2); Me ester[962i] (205–6)
3 5-(CO$_2$H)$_2$[1090]		
5-Me-1-(p-NO$_2$-C$_6$H$_4$)-3-CO$_2$H[1088,1091a]		Et ester (158)
5-Ph-1-(p-NO$_2$-C$_6$H$_4$)-3-CO$_2$H[175a 1091a]		Me ester (176–8)
5-Me-1-(o-Cl-C$_6$H$_4$)-3-CO$_2$H[1091a]		Et ester (102–4)
5-Ph-1-(o-Me-C$_6$H$_4$)-3-CO$_2$H[175a]	171–2	Me ester (92–4)
5-Ph-1-(m-Me-C$_6$H$_4$)-3-CO$_2$H[175a]	182–3	Me ester (127–8)
5-Ph-1-(p-Me-C$_6$H$_4$)-3-CO$_2$H[175a]	177–8	Me ester (132–3); amide (155–6)

Table A 10. Continued

	m.p. (b.p./mmHg)/°C	Derivative (m.p./°C)
1,2,4-Triazole		
5-Ph-1-(*p*-OMe-C$_6$H$_4$)-3-CO$_2$H[175a]	176–7	Me ester (106–7); amide (183–4)
5-Ph-1-(*p*-Br-C$_6$H$_4$)-3-CO$_2$H[175a]	179–80	Me ester (138–9); amide (176–8)
1-Ph-5-(*p*-Br-C$_6$H$_4$)-3-CO$_2$H[175a]		Me ester (157–9)
Tetrazole		
5-CO$_2$H		Et ester[1016a,1064b] (87–8); amide[80d] (234)
1-Et-5-CO$_2$H[80c]	124–5	Amide (125–6)
2-Me-5-CO$_2$H[80d,1092]	204–5	Me ester[999] (b.p. 104.5–5.5/2 mm)
2-Ph-5-CO$_2$H[276a]	138	Amide (167–8); hydrazide [1093f] (200)
1,5-(CO$_2$H)$_2$[888a]		bis-Me ester (77–80)

(a) The majority of these data are taken from ref. 14.
(b) Some of these data come from tables published by Hofmann[58].
(c) 2-substituted-2*H*-1,2,3-triazoles are tabulated as 2-substituted-1,2,3-triazoles.

Table A.11. Alkenyl and alkynyl azoles

Pyrazoles

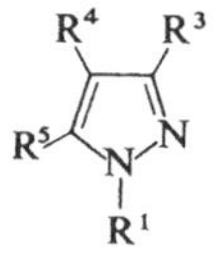

R^1	R^3	R^4	R^5	M.p. (b.p./mmHg)/°C	Ref.
-CH=CH$_2$	H	H	H	139–40 (62/52)	898k, 1128
H	-C≡CH	H	H	55; 45–6 (86/4)	1019g, 1049
H	-C≡CH	H	Me	82	1129
H	-C≡CH	H	Ph	103.0–4.5	1138
Me^a	-C≡CH	H	CO$_2$Me	114–17	1019g
COMe^a	-C≡CH	H	CO$_2$Me	106–7	1019g
H	-C≡CH	H	H	78	1019g
H	-C≡CMe	H	H	71–2 (112–14/6)	1130
H	-C≡CEt	H	H	38–9 (120–2/5)	1130
-CH=CH$_2$	Me		Me	(67–8/11)	898k
Ph	H	-CH=CH$_2$	H	39–40 (123–8/2)	898k, 945c
-CMe=CH$_2$	H	H	H	154–5	1128
-CH=CHMe	H	H	H	167–9^b	1128
-CEt=CHMe	H	H	H	(108/14)^b	1128
-CH=CMe$_2$	H	H	H	(75–6/19)	1128
H	-CH=CHPh	H	Ph	157–9	1131a
H	-CH=CPh$_2$	H	Ph	160–2	1132
Me	H	H	-CH=CHMe	100–1	902b
Me	H	H	-CH=CMe$_2$	35–7	902b
Me	H	-CH=C(CN)$_2$	H	137	946
Me	H	-CH=C(CN)$_2$	Me	107	946
Ph	Me	-CH=CH-CO$_2$H	H	166.5–7.0	59g

Table A.11. Continued

R¹	R³	R⁴	R⁵	M.p. (b.p./mmHg)/°C	Ref.
Ph	H	–CH=CH–CO₂H	Me	199–200	59g
Ph	Me	–CH=CH–CO₂H	Me	163–4	59g
Ph	Ph	–CH=CH–CO₂H	Me	211.5–12.5	59g
Ph	Ph	–CH=CH–CO₂H	Ph	246.0–7.5	59g
Ph	H	–CH=CH–COPh	Me	110.5–11.0	59g
Ph	Me	–CH=CH–COPh	Me	102.0–2.5	59g
Ph	Me	–CH=CH–COPh	Ph	169–70	59g
Ph	Ph	–CH=CH–COPh	Me	139.5–40.5	59g
4-NO₂C₆H₄	Me	H	–CH=CHMe	138–40	902b
2,4-di-NO₂C₆H₃	Me	H	–CH=CHMe	138–40	902b
4-NO₂C₆H₄	Me	H	–CH=CMe₂	90–3	902b
2,4-di-NO₂C₆H₃	Me	H	–CH=CMe₂	74–6	902b
2,4-di-NO₂C₆H₃	Ph	H	–CH=CHMe	160–1	902b
4-NO₂C₆H₄	Ph	H	–CH=CMe₂	138–40	902b
2,4-di-NO₂C₆H₃	Ph	H	–CH=CMe₂	165–7	902b
4-NO₂C₆H₄	–CH=CHMe	H	Me	85–6	902b
2,4-di-NO₂C₆H₃	–CH=CMe₂	H	Me	143–4	902b
2,4-di-NO₂C₆H₃	–CH=CHMe	H	Me	164–6	902b
2,4-di-NO₂C₆H₃	–CH=CHMe	H	Ph	146	902b
2,4-di-NO₂C₆H₃	–CH=CMe₂	H	Ph	113–14	902b
H	Me	–CH₂–CH=CH₂	Me	(128–32/7)	898j
Me	Me	–CH₂–CH=CH₂	Me	(98–100/10)	898j
–CH₂–CH=CH₂	Me	–CH₂–CH=CH₂	Me	(115–19/11)	898j
–CH₂–CH=CH₂	H	H	Me	(188–92/751)	898b
–CH₂–CH=CH₂	H	H	Me	(181–2)	898b
–CH₂–CH=CH₂	Me	H	H	(171)	100d
Ph	Me	–CH₂–CH=CH₂	Me	(205/40)	900c
Ph	H	–CH=CHMe	H	39–40	59f

Table A.11. Continued

R¹	R³	R⁴	R⁵	M.p. (b.p./mmHg)/°C	Ref.
Me	H	H	-C≡C-C(OH)Me₂	54.5-5.5 (103-4/0.5)	1133
Me	-C≡CH	-C≡CH	H	54-5	1133
Me	H	-C≡CH	-C≡CH	65-6	1133
Me	-C≡CH	-C≡CH	-C≡CH	Polymerises	1133

Imidazoles

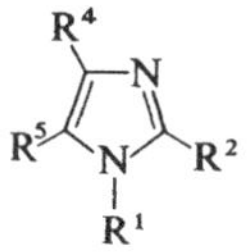

(Imidazole ring with substituents R⁴ and R⁵ at the 4- and 5-positions, R¹ on the 1-nitrogen, and R² at the 2-position.)

R¹	R²	R⁴	R⁵	M.p. (b.p./mmHg)/°C	Ref.
-CH=CH₂	H	H	H	(100-13/20)	1134a
H	-CH=CH₂	H	H	(80/10)	1136
H	H	-CH=CH₂	H	127-9	1114f
-CH=CH₂	Me	H	H	83.0-4.5	1137
-CH=CH₂	Me	Ph	Ph	(101-4/2)	1134a, 1139
-CH=CH₂	H	Me	H	83-4	1140
-CH=CH₂	Me	Me	H	(67-9/4)	1141a
-CH=CH₂	Ph	H	H	(67-8/2)	1141a
Me	-CH=CH₂	H	H		1139
H	Ph	-C≡CH	H	(75/2)	1114d
-CH₂-CH=CH₂	H	H	H	(40-2/0.1)	1142
H	-CH₂-CH=CH₂	H	H	142	1143
H	H	-CH₂-CH=CH₂	H	Liquid	981c
					981c

Table A.11. Continued

R¹	R²	R⁴	R⁵	M.p. (b.p./mmHg)/°C	Ref.
H	-CH=CHPh	H	H	178-9	973d
H	-CH=CHPh	Me	H	232-3	973d
-CH₂CH=CHMe	H	H	H	(52-4/0.15)	1143
-CH₂C≡CMe	H	H	H	(59.0-9.5/0.1)	1143
H	H	NO₂	-CH=CHPh	303	564
Me	H	NO₂	-CH=CHPh	150-1	564
Me	H	-CH=CHPh	NO₂	214-15	564
Me	NO₂	H	-CH=CH₂	106-8	1145
Me	NO₂	H	-CH=CHPh	178-80	1145
Et	NO₂	H	-CH=CH₂	45-7	1145
Me	-C≡CH	H	Cl	76-7	1144a
Me	-C≡CH	-C≡CH	Cl	90.5-1.5	1144a
Me	-C≡CH	-C≡CH	Cl	149.5-50.5	1144a
Ph	-C≡CH	H	H	60	1146a
Me	-C≡CH	H	H	Syrup	1144b
Me	H	-C≡CH	-C≡CH	129.0-9.5	1144b

1*H*- or 2*H*-1,2,3-Triazole

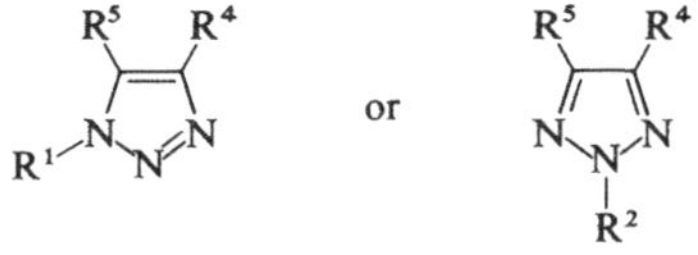

R¹	R²	R⁴	R⁵
-CH=CH₂		H	H
Ph		H	-CH=CH₂
-CPh=CH₂		CO₂H	Me

M.p.	Ref.
(103-4/15)	1055b
64-5; 67-9	1075, 1147
159-60	1148a

Table A.11. Continued

R¹	R²	R⁴	R⁵	M.p. (b.p./mmHg)/°C	Ref.
-CH=CHPh		CO₂H	Me	180ᶜ	1148a
-CPh=CHMe		CO₂Me	CO₂Me	88–90ᵈ	1148b
-CMe=CHMe		CO₂Me	CO₂Me	58–60ᵈ	1148b
				36.5–7.5ᶜ	1148b
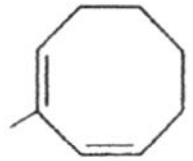		CO₂Me	CO₂Me	76–8	1148b
-CH=CH(t-Bu)		CO₂H	Me	148.5–9.0ᶜ	1148a
-CH=CH(t-Bu)		CO₂H	Ph	155ᶜ	1148a

1,2,4-Triazoles

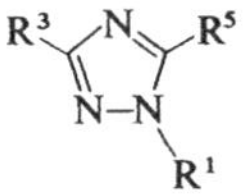

R¹	R³	R⁵	M.p. (b.p./mmHg)/°C	Ref.
-CH=CH₂	H	H	(112–13/50)	1149
Ph	Me	-CH=CHPh	74	117

Tetrazoles

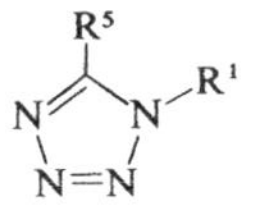

R^5 structure or R^5 structure

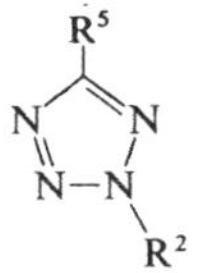

R^1	R^2	R^5	M.p. (b.p./mmHg)/°C	Ref.
-CH=CH$_2$		H	(94/1)	1021g, 1150, 1151
Me	-CH=CH$_2$	-CH=CH$_2$	15-20	1021g, 1150
	Me	H	(66-8/60)	1021g, 1150, 1151
		-CH=CH$_2$	(c. 80)	1021g, 1150
-CH$_2$-CH=CH$_2$		H	(101/1)	1021g, 1150, 1151
	-CH$_2$-CH=CH$_2$	H	(80-1/20)	1021g, 1150, 1151
-CH=CHPh		Ph	75-7[d]	1152
-CH=CHPh		Me	99-101[d]	1152
-CH=CH(furyl-2)		Ph	97-9[d]	1152
-CH=CH(thienyl-2)		Ph	58-60[d]	1152
-CH=CH(thienyl-2)		Me	99-100[d]	1152
-CH=CHC$_6$H$_4$-p-NO$_2$		Me	152-4[d]	1152
		Me	174-6[c]	1152

(a) Orientation uncertain.
(b) Mixture of *cis* and *trans* forms.
(c) *Trans* form.
(d) *Cis* form.

Table A.12. Aminoazoles

	M.p. (b.p./mmHg)/°C	Derivative (m.p./°C)
Pyrazoles		
3-NH$_2$[432c 1158,1159a]	38–40 (140–2/7)	Picrate[1127b] (218–20)
3-NHAc[670c]	223–4	
4-NH$_2$[1160]	80–2	Picrate[1161] (195–6)
4-NHAc[473]	199–201	
1-Me-3-NH$_2$[902d]	(80–2/0.1)	
4-Me-3-NH$_2$[1141b]	(136–40/6)	
3-Me-4-NH$_2$[74b,307a,898l]	97[74b]; 90[898l]	Dihydrochloride[307a] (195–200)
1-Me-5-NH$_2$[902d,700,1158, 1162]	68–9	Monotosyl[1159a] (188–90); Ditosyl[1159a] (151–2)
3-Me-5-NH$_2$	47–8 (146–7/4)[1158] 95–6 (153–4/4) [1055d]	
1-Et-5-NH$_2$[1055d]	(75–6/25)	
1-Ac-5-NHAc[670c]	190.5–1.5	
1-Ph-3-NH$_2$[700,898c,1159b]	88–9	
1-Ph-4-NH$_2$[59e,747,1163]	104–5	Hydrochloride (245–50)
1-Ph-5-NH$_2$[700,1162,1164a]	42–3 (142–4/0.15)	
3-Ph-5-NH$_2$[323f,1158]	121–8	Picrate[1127b] (198–9)
4-Ph-3-NH$_2$[1027]	172–3	Picrate (202–3)
1-Ph-3-NMe$_2$[902d]		
1-PhCH$_2$-3-NH$_2$[1159b]		
1-PhCH$_2$-5-NH$_2$[453a]	79–81	
3-(3'-pyridyl)-4-NH$_2$[694,757]	176	Dipicrate (205)
3,5-Me$_2$-4-NH$_2$[74b,1166]	203–4	Benzylidene[431c] (139–40)
1,3-Me$_2$-5-NH$_2$[1158]	77–8	
1-Et-3-Me-5-NH$_2$[1055d]	100–1	
1-Ph-4-Me-3-NH$_2$[898c]	111–12	
1-Ph-5-Me-3-NH$_2$[898c]	82–3	
1-Ph-3-Me-5-NH$_2$[323f,745c]	110–11	
1-Ph-3-Me-4-NH$_2$[330k]	88 (312–13)	Picrate (138); Benzylidene (113)
1-Ph-5-Me-4-NH$_2$[59e]	32–3 (114–16/0.08)	
1-i-Pr-3-Me-5-NH$_2$[323f]	111–12 (131–2/21)	
3-PhCH$_2$-5-Ph-4-NH$_2$[957a,b]	139.5–40.0	
1-PhCH$_2$-4-Me-5-NH$_2$[1055d]	129–30	
1-PhCH$_2$-3-Ph-5-NH$_2$[1055d]	125–6	
1-PhCH$_2$-3-Me-5-NH$_2$[323f,i]	69–71 (188–93/8)	Picrate[323i] (145–6)
1-PhCH$_2$CH$_2$-3-Me-5-NH$_2$ [323f,i]	108.5–9.5 (184–9/6)	Picrate[323i] (180.0–0.5)
1,3-Ph$_2$-5-NH$_2$[323f]	123	
1,5-Ph$_2$-3-NH$_2$[898c]	135–7	
1-Ph-3-Me-5-PhCH$_2$NH[1005a]		
1-Ph-3-(p-NH$_2$-C$_6$H$_4$)-5-NH$_2$ [323f]	210–11	
1-Ph-3,5-Me$_2$-4-NH$_2$[1168]	38–40	Hydrate (59); p-nitrobenzylidene (155–7)
1,5-Ph$_2$-3-Me-4-NH$_2$[1168]	102–4	p-nitrobenzylidene (158–9)
1,3,5-Me$_3$-4-NH$_2$[392g,1166]	100–1	
1-(o-MeO-C$_6$H$_4$)-3-Me-5-NH$_2$ [323f]	87	
1-(p-NO$_2$-C$_6$H$_4$)-3-Me-5-NH$_2$ [323f]	161–2	

Table A.12. Continued

	M.p. (b.p./mmHg)/°C	Derivative (m.p./°C)
Pyrazoles		
1-heptyl-3,5-Me$_2$-4-NH$_2$[656]	(143–4/2)	
1-heptyl-3,5-Me$_2$-4-NHAc[656]	73–5	
1-Me-3,4-Ph$_2$-5-NH$_2$[1169]	168–70	
1-(2′-C$_{10}$H$_7$)-3-Me-5-NH$_2$[323f]	149–50	
1-Ph-3-Me-4-COPh-5-NH$_2$[321l]		Hydrazone (276–7)
1-(p-CH$_3$-C$_6$H$_4$)-3,4-Ph$_2$-5-NH$_2$ [745c]	183–5	
1,3-Ph$_2$-4-(p-Cl-C$_6$H$_4$)-5-NH$_2$ [745c]	149–50	
1-i-Pr-3-Et-4-Me-5-NH$_2$[323f]	62–3 (137–8/19)	
1-Ph-3-Me-4-(p-CH$_3$O-C$_6$H$_4$)-5-NH$_2$[745c]	137–9	
1-Ph-3-Et-4-Me-5-NH$_2$[323f]	60–1	
1-(o-MeO-C$_6$H$_4$)-3-Et-4-Me-5-NH$_2$[323f]	57–8 (174–6/8)	
1-(2′-C$_{10}$H$_7$)-3-Et-4-Me-5-NH$_2$ [323f]	85–6	
1,3,4-Ph$_3$-5-NH$_2$[745c 1170]	168–9	
Diaminopyrazoles		
3,4-(NH$_2$)$_2$[172f,1171a]		Sulphate (227–9; 236)
1-Me-3 4-(NH$_2$)$_2$[1171b][a]		Sulphate (231–2)
1-Me-3-AcNH-4-NH$_2$[1171b][a]		Picrate (202–3)
1-PhCH$_2$-3,5-(NH$_2$)$_2$[127]		Hydrochloride (160–1)
1-Ph-3,4-(NH$_2$)$_2$[1171b][a]		Hydrochloride (231–3)
1-Ph-3-Me-4,5-(NH$_2$)$_2$[537]		Dihydrochloride[1020d] (250)
1.5-(NH$_2$)$_2$-3,4-(CN)$_2$[1114l]	224–5	
Haloaminopyrazoles		
1-NH$_2$-4-Cl[1114l]	67–8	
3-NH$_2$-4-Cl[172f]	121–2	
4-NH$_2$-3-Me-5-Cl[426a]	182–3	
4-NH$_2$-3-Me-1-Ph-5-Cl[552e]	49	Picrate (95); benzylidene (72); hydrochloride (222)
4-NMe$_2$-3-Me-1-Ph-5-Cl[552e]		Hydriodide (147)
3-NH$_2$-4-Br[172f]	138–9	
5-NH$_2$-3-Me-4-Br[426c]	226	
1-NH$_2$-3,4,5-Br$_3$[1114l]	127–8	
Aminopyrazole carboxylic acids and derivatives		
5-NH$_2$-4-CO$_2$Et[1158]	104.5–5.5	
5-NH$_2$-1-Me-4-CO$_2$H[771]		
5-NH$_2$-1-Me-4-CONH$_2$[771]	99–101	
5-NH$_2$-3-Me-4-CO$_2$H[1173]	124–5	
5-NH$_2$-3-Me-4-CO$_2$Et[1158, 1173]	109–10	
5-NH$_2$-3-Me-4-CO$_2$Me[1173]	145.0–6.5	
5-NH$_2$-3-Et-4-CO$_2$Me[1173]	102–3	
5-NH$_2$-3-n-Pr-4-CO$_2$Me[1173]	80–1	
3-NH$_2$-1-PhCH$_2$-4-CO$_2$Et[1159b]		
5-NH$_2$-1,3-Me$_2$-4-CO$_2$Et[1158]	108–9	

Table A.12. Continued

	M.p. (b.p./mmHg)/°C	Derivative (m.p./°C)
Nitro- and nitroso-substituted aminopyrazoles		
1-NH$_2$-4-NO$_2$[1114l]	100-1	
3-NH$_2$-4-NO$_2$[1159a,1171a]	242	
3-NH$_2$-1-Me-4-NO$_2$[1171b]	265-6	
3-NH$_2$-1-Ac-4-NO$_2$[1171b]	167-70	
3-NHAc-4-NO$_2$[172f]	242-4	
5-NH$_2$-1-Me-4-NO$_2$[1162]	200	
5-NH$_2$-1-Ac-4-NO$_2$[1171b]	184-6	
5-NAc$_2$-1-Me-4-NO$_2$[1171b]	130-1	
5-NHAc-1-Me-4-NO$_2$[1171b]	181-2	
5-NH$_2$-1-Ph-4-NO$_2$[1162]	154-6	
5-NH$_2$-1-Ph-3-Me-4-NO[1174]		
5-NH$_2$-1-Ph-3-Me-4-NO$_2$[537]	167-8	
Imidazoles		
1-NH$_2$-4,5-Me$_2$-2-Pr[1134d]		
1-NH$_2$-2-Me-4-Ph[1181]	140-1	
1-NH$_2$-2-PhCH$_2$-4-Ph[1181]	137-8	
1-NH$_2$-2,4-Ph$_2$[1181]	163	
2-NH$_2$		Picrate[274] (236); nitrate [412d] (135-6); hydrochloride (152)
1-Me-2-NH$_2$[1182b]		
4-Me-2-NH$_2$[673]		Picrate (186-7)
4,5-Me$_2$-2-NH$_2$[673]		Picrate (245); hydrochloride (289)
1-Ph-2-NH$_2$[1182b]		
4,5-Ph$_2$-2-NH$_2$[390a]	243	
4-(p-NH$_2$-C$_6$H$_4$)-2-NH$_2$[274c]	148	Dipicrate (256)
4-(p-NH$_2$-C$_6$H$_4$)-5-Me-2-NH$_2$ [274c]		Picrate (255)
4-(p-NH$_2$-C$_6$H$_4$)-5-Ph-2-NH$_2$ [390a]	265	Dihydrochloride (310)
4-NH$_2$[274a]		Dihydrochloride (184)
4-NHAc[174]	225	
1-Me-4-NHAc[174]	249-52	
5-Me-4-NH$_2$[274a,452a]		Picrate (195); hydrochloride (189)
1 5-Me$_2$-4-NH$_2$[20c]		Picrate (220); hydrochloride (225)
5-(p-NH$_2$-C$_6$H$_4$)-4-NH$_2$[274c, 507]		Picrate (>300); hydrochloride (>300)
2-PhCH$_2$-5-Ph-4-NH$_2$[674b]	199	Picrate (215); hydrochloride (200)
2-Me-5-Ph-4-NH$_2$[414a,674b]		Picrate (219); hydrochloride (238)
1-PhCH$_2$-4-NHAc[174]	180-1	
2-Ph-4-NHAc[1184]	196-203	
2-Ph-4-NHCO$_2$Et[1184]	170	
2-Ph-5-Me-4-NH$_2$[1184]	230	
2-Ph-5-Me-4-NHAc[1184]	220	
2-(p-NH$_2$C$_6$H$_4$)-4-NHAc[1184]	280	Picrate (265); hydrochloride (282)
2,5-Ph$_2$-4-NH$_2$[414a,b]	162[414b]	Picrate[674b] (220); hydrochloride[414a] (250)

Table A.12. Continued

	M.p. (b.p./mmHg)/°C	Derivative (m.p./°C)
Imidazoles		
1-Me-5-NH$_2$[1176a]		Picrate (189)
1-Et-5-NH$_2$[1176a]		Picrate (185)
1,2-Me$_2$-5-NH$_2$[1176a]		Picrate (193)
1,4-Me$_2$-5-NH$_2$[20c]		Picrate (209)
4-Ph-1,2-(NH$_2$)$_2$[1183]	242–3	
Halogen substituted aminoimidazoles		
1-Me-5-Cl-4-NH$_2$[373c]		
1-Me-4-Cl-5-NH$_2$[373c]		
1-Me-2-Br-4-NHAc[174]	192–3	
1-Me-2-Br-5-NHAc[174]	223–4	
1-Et-2-Br-5-NHAc[174]	175.0–6.5	
1-Bu-2-Br-5-NHAc[174]	138–40	
1-Ph-2-Br-5-NHAc[174]	200–2	
1-PhCH$_2$-2-Br-4-NHAc[174]	184–5	
1,4-Ph$_2$-2-Br-5-NH$_2$[174]	195–210	
Aminoimidazole carboxylic acids and derivatives		
5-NH$_2$-4-CONH$_2$	254–5[1063b]	
	170[1176a]	Picrate (240)
1-Me-5-NH$_2$-4-CONH$_2$[1176a]	260	Picrate (249)
2-Me-5-NH$_2$-4-CONH$_2$[1176a]		Picrate (240); hydrochloride (246–7)
1-Et-5-NH$_2$-4-CONH$_2$[1176a]	230–2	Picrate (249)
1-cyclohexyl-5-NH$_2$-4-CONH$_2$ [1176a]	209	
1,2-Me$_2$-5-NH$_2$-4-CONH$_2$ [1176a]	286	
1-Et-2-Me-4-NH$_2$-5-CONH$_2$ [495b]	167–8	
1-Pr-2-Et-4-NH$_2$-5-CONH$_2$ [495b]	172.0–2.5	
1-Bu-2-Pr-4-NH$_2$-5-CONH$_2$ [495b]	150–1	
1-i-Bu-2-i-Pr-4-NH$_2$-5-CONH$_2$ [495b]	156.0–6.5	
1-cyclohexyl-2-Me-5-NH$_2$-4-CONH$_2$[1176a]		Hydrate (240)
2-Me-1 5-(NH$_2$)$_2$-4-CONH$_2$ [1185]	272–3	
1 2,3-Triazoles		
5-Me-1-NH$_2$[964d]	70	Hydrochloride (138)
2-NH$_2$[17e]		Picrate (130); hydrochloride (114)
2-NHCOPh[17e]	151	
4,5-Me$_2$-2-NH$_2$[17e]	95	
4-NH$_2$[147,1141d,1186]	74–5	Picrate[1141d] (178); hydrochloride[147,1141d] (142)
4-NHCH$_2$Ph[723b]	110	
1-Me-4-NH$_2$[276b]	88–90	Hydrochloride (181–4)
1-Me-5-NH$_2$[276b]		Hydrochloride (181–2)
1-Ph-4-NH$_2$[1178]	110	
1-Ph-4-NHAc[1178]	143	

Table A.12. Continued

	M.p. (b.p./mmHg)/°C	Derivative (m.p./°C)
1,2,3-Triazoles		
2-Ph-4-NH$_2$[688a]	70	
2-Ph-4-NHAc[688a]	166	
1-Ph-5-NH$_2$[148c,816b]	110	
4-Ph-5-NH$_2$[988c]	125	
1-PhCH$_2$-5-NH$_2$[149e]	129	
1-CH$_2$SMe-4-Ph-5-NH$_2$[1187]	134–5	
2-Ph-4-N$_2$Ph-5-NH$_2$[1188]	206–8	
1,4-Ph$_2$-5-NH$_2$[723b,1189]	173; 179	
1-Me-5-CONH$_2$-4-NH$_2$[149b]	174	
5-CONH$_2$-4-NH$_2$[149c]		
5-CONH$_2$-4-NHCH$_2$Ph[149e]	232–4	
5-CO$_2$H-4-NHCH$_2$Ph[149e]	154	
2-Me-5-CONH$_2$-4-NH$_2$[149c]	193	
1-Ph-4-CONH$_2$-5-NH$_2$[723b]	167	
1-Ph-4-CO$_2$H-5-NH$_2$[723b]	140–2	
1-Ph-4-CO$_2$Me-5-NH$_2$[723b]	170	
1-PhCH$_2$-4-CONH$_2$-5-NH$_2$ [149b,e]	233–5	
1-PhCH$_2$-4-CO$_2$H-5-NH$_2$[149e]	173	
2-Ph-4-Me-5-NH$_2$[998c]	83	
2-Ph-4,5-(NH$_2$)$_2$[688a]	143	Picrate (153); hydrochloride (210)
1,2,4-Triazoles		
3-NH$_2$[688b]	159	Picrate[688b] (227–8); nitrate[688b] (174)
3-NHAc[607g,675b]	289–91	
3-NHPh[1191a]	172–4	Picrate (174–6)
4-NH$_2$[1072b]	76–7	
4-NHCH$_2$Ph[630a]	109–11	
5-Me-3-NH$_2$[431b,586]	146–8	
1-Ph-3-NH$_2$[430c]	150	
4-Ph-3-NH$_2$[1191a]		Picrate (221–3)
5-Ph-3-NH$_2$[1192]	188–9	
5-Ph-3-NHAc[675b]	209–12	
5-Pr-3-NH$_2$[1193b]	143	Nitrate (153); picrate (152)
5-i-Pr-3-NH$_2$[1193b]	112	Nitrate (176); picrate (193–4)
3,5-Me$_2$-4-NH$_2$[1072b,1195]	195–6	
3-Et-5-Me-4-NH$_2$[1195]	122–4	
3,5-Et$_2$-4-NH$_2$[1072b]	165–6	
3,5-Ph$_2$-4-NHAc[1093a]	180	
3,5-Ph$_2$-4-NH$_2$[1195]	268–9	
3-PhCH$_2$-5-Ph-4-NH$_2$[1195]	190–2	
3,5-(PhCH$_2$)$_2$-4-NH$_2$[1195]	163–4	
Polyamino-1,2,4-triazoles		
3,5-(NH$_2$)[1196a]	206	Picrate[1197] (249)
3-NHAc-5-NH$_2$[675b]	292–3	
3,5-(NHAc)$_2$[675b]	327–30	
3-NHPh-5-NH$_2$[1198]	160–2	Picrate (230–2)
3 4-(NH$_2$)$_2$[913a]	213	Benzylidene (246–8)
5-Me-3,4-(NH$_2$)$_2$[913a]	213	Benzylidene (119)
5-Et-3,4-(NH$_2$)$_2$[913a]	195	

Table A.12. Continued

	M.p. (b.p./mmHg)/°C	Derivative (m.p./°C)
Polyamino-1,2,4-triazoles		
5-Pr-3,4-$(NH_2)_2$[913a]	171	Benzylidene (153)
5-i-Pr-3,4-$(NH_2)_2$[913a]	204	Benzylidene (149)
5-i-Bu-3,4-$(NH_2)_2$[913a]	204	
5-PhCH$_2$-3,4-$(NH_2)_2$[913a]	192	
4,5-$(NH_2)_2$[676b]	233	
4,5-$(NHAc)_2$[676b]	254	
3-Me-4,5-$(NH_2)_2$[676b]		
3,4,5-$(NH_2)_3$[1196c]	257	Picrate[1197] (282–4)
3-Ph-4,5-$(NH_2)_2$[1201]	224	
Aminotriazoles with other functional groups		
1-Me-3-NO_2-5-NH_2[269c]	253–5	
5-OH-3-NH_2[1191b,1198]	286–90	Picrate (210–12)
5-SH-3-NH_2[1198]	300–2	
5-SCH$_2$Ph-3-NH_2[1198]	109–11	
3-SMe-5-NHPh[1191a]	187–8	
3,5-$(OH)_2$-4-NH_2[1199]	270	
3-CO_2H-5-NH_2[688b]	182	Picrate (176); HCl (171-2)
Tetrazoles		
5-NH_2[515b]	206	Nitrate (178–9)
5-NHAc[697b]	271–3	
5-NHCOPh[25c,697b]	277–82	
5-NMe$_2$[1194a]	235–6	
5-NEt$_2$[1194a]	124–5	
5-N(i-Pr)$_2$[1194a]	184	
5-NBu$_2$[1194a]	132.5–3.5	
5-NHCH$_2$Ph[590d]	180–1	
5-N(CH$_2$Ph)$_2$[1194a]	158–9	
1-Me-5-NH_2[83e,155,1194b]	222[83e] ; 232[155]	Benzylidene[83e] (157)
1-Me-5-NHAc[83e]	164	
1-Me-5-NHMe[155]	172	Hydrochloride[515e] (209)
1-Me-5-NMe$_2$[155,579a]	43–4[299] (114–16/3) [155]	Hydrochloride[155] (152–4)
2-Me-5-NH_2[123,591b]	102	
2-Me-5-NHAc[1203a]	152–3	
1-Et-5-NH_2[1194b]	147.5–8.5	
1-Et-5-NHEt[515e]	92–3	Hydrochloride (161–3)
1-Pr-5-NH_2[515e]	150–2	
1-Pr-5-NHPr[515e]	70–2	Hydrochloride (141–2)
1-i-Pr-5-NH_2[515e]		Hydrochloride (192–4)
1-i-Pr-5-NH-i-Pr[515e]	160–1	
1-Bu-5-NH_2[515e]	148–9	Hydrochloride (156–7)
1-Bu-5-NHBu[515e]	72–4	
1-i-Bu-5-NH_2[1194b]	212.0–12.5	
1-$(CH_2)_2$CN-5-NH_2[640]	115–16	
1-Ph-5-NH_2[155,1194b]	160	
1-Ph-5-NHMe[579a]	133.5–6.5	
1-Ph-5-NMe$_2$[579a]	110–11	
1-(p-NO_2-C_6H_4)-5-NH_2[1194b]	221–3	
1-(p-NH_2-C_6H_4)-5-NH_2[1194b]	200–1	
1-PhCH$_2$-5-NH_2[590d,1194b]	191–2	
1-PhCH$_2$-5-NHCH$_2$Ph[515e]	167–8	Hydrochloride (160–1)

Table A.12. Continued

	M.p. (b.p./mmHg)/°C	Derivative (m.p./°C)
Tetrazoles		
1-Ph(CH$_2$)$_2$-5-NH$_2$[515e]	175-7	
2-Ph-5-NH$_2$[1093f]	142	Benzylidene (123)
2-Ph-5-NHAc[1093f]	177	
2-PhCH$_2$-5-NH$_2$[590d]	86	
1-(p-MeC$_6$H$_4$)CH$_2$-5-NH$_2$[590d]	194-5	
1-(p-ClC$_6$H$_4$)CH$_2$-5-NH$_2$[590d]	194-5	
1-(p-BrC$_6$H$_4$)CH$_2$-5-NH$_2$[590d]	209-10	
2-(p-MeC$_6$H$_4$)CH$_2$-5-NH$_2$[590d]	124	
2-(p-ClC$_6$H$_4$)CH$_2$-5-NH$_2$[590d]	107	
2-(p-BrC$_6$H$_4$)CH$_2$-5-NH$_2$[590d]	130-1	
5-NH(CH$_2$C$_6$H$_4$-p-Me)[1138]	206	
5-NH(CH$_2$C$_6$H$_4$-p-Cl)[590d]	210	
5-NH(CH$_2$C$_6$H$_4$-p-Br)[590d]	217-18	

(a) These compounds were originally assigned[1171b] as 4,5-diamino, but recent n.m.r. studies[1162] would suggest that the orientation is 3,4-diamino.

Table A.13. Nitramino-1,2,4-triazoles

Compound	M.p./°C	Ref.
3-NHNO$_2$	216–17	514a, 903, 1212b
3-Me-5-NHNO$_2$	212–13	514a
3-Ph-5-NHNO$_2$	211–12	514a
3-Et-5-NHNO$_2$	193–4	514a
3-(o-HO-C$_6$H$_4$)-5-NHNO$_2$	215–19	514a
3-(4'-pyridyl)-5-NHNO$_2$	284–5	514a

Table A.14. Nitrosaminoazoles

	M.p./°C	Ref.
Pyrazole		
1-Me-4-CO$_2$H-5-NHNO	189–91	771
1,2,4-Triazoles		
3-Ph-5-NHNO	Explodes	430c
3-NMeNO	154	268g
4-NH$_2$-3-Me-5-NMeNO	174	268g
3-N(CH$_2$Ph)NO	176	533
3-N(CH$_2$CH$_2$Ph)NO	135	533
4-Ph-3-Pr-5-NHNO	141–4	913b
4-(p-MeC$_6$H$_4$)-3-Pr-5-NHNO	145–8	913b
4-(p-BrC$_6$H$_4$)-3-Pr-5-NHNO	142–4	913b
4-(p-ClC$_6$H$_4$)-3-Pr-5-NHNO	145–6	913b
4-(p-ClC$_6$H$_4$)-3-Et-5-NHNO	167–8	913b
3,5-(NHNO)$_2$	187	83f
5-N$_3$-3-NHNO	134	83f
4-N(CH$_2$Ph)NO	108	1224a
4-N(CH$_2$-C$_6$H$_4$-p-Cl)NO	128	1224a
Tetrazoles		
1-PhCH$_2$-5-NHNO	56	1226a
1-Ph-5-NHNO	110	1226a
2-PhCH$_2$-5-NHNO	Oil	1226a

Table A.15. Azidoazoles

	M.p./°C	Ref.
Pyrazoles		
$3-N_3$	55-7	1019d
3-Me-1-Ph-5-N_3	Oil	745c
1,4-Ph_2-5-N_3	68-9 (dec.)	745c
1,4-Ph_2-3-Me-5-N_3	73-4 (dec.)	745c
3-Me-1-Ph-4-(p-$NO_2C_6H_4$)-5-N_3	108 (dec.)	745c
1-(p-$CH_3C_6H_4$)-3,4-Ph_2-5-N_3	110 (dec.)	745c
1,3,4-Ph_3-5-N_3	93-6	745c
1,2,3-Triazoles		
1,4-Ph_2-5-N_3	70	745c
1-(p-$CH_3C_6H_4$)-4-Ph-5-N_3	75 (dec.)	745a
5-NH_2-2-Ph-4-N_3		693h
2-Ph-5-$N_2C_6H_5$-4-N_3	Softens 159-61	1188
1,2,4-Triazoles		
5-NHNO-3-N_3	134 (dec.)	83f
1-Me-3-NO_2-5-N_3	126-7	269c
4-Et-3-Me-1-Ph-5-N_3(fluoroborate)	99-101	826
3-Me-4-N=CH-Ph-5-N_3	113-16	1213a,b
3-Me-4-(N=CH-C_6H_4-p-Me)-5-N_3	130-1	1213b
3-Me-4-(N=CH-C_6H_4-p-OMe)-5-N_3	122-3	1213b
3-Me-4-(N=CH-C_6H_4-p-Cl)-5-N_3	134-6	1213b
3-Me-4-(N=CH-C_6H_4-p-NO_2)-5-N_3	152	1213b
3-Me-4-(N=CH-C_6H_4-o-NO_2)-5-N_3	140-1	1213b
3-Me-4-(N=CH-C_6H_4-o-N_3)-5-N_3	153	1213b
4-(N=CHPh)-5-N_3	149-52	1213b
4-(N=CH-C_6H_4-p-Me)-5-N_3	158-61	1213b
4-(N=CH-C_6H_4-p-OMe)-5-N_3	156-7	1213b
4-(N=CH-C_6H_4-p-Cl)-5-N_3	159	1213b
4-(N=CH-C_6H_4-p-NO_2)-5-N_3	168-71	1213b
3-Et-4-(N=CHPh)-5-N_3	68.5-70.0	1213b
3-Et-4-(N=CH-C_6H_4-p-Me)-5-N_3	95-6	1213b
3-Et-4-(N=CH-C_6H_4-p-OMe)-5-N_3	107-8	1213b
3-Et-4-(N=CH-C_6H_4-p-NO_2)-5-N_3	160-1	1213b
3-Ph-4-(N=CHPh)-5-N_3	142	1213b
3-Ph-4-(N=CH-C_6H_4-p-NO_2)-5-N_3	168	1213b
Tetrazoles		
1-Ph-5-N_3	96-7; 99	83e, 177c
1-(p-Cl-C_6H_4)-5-N_3	101-4	608d
1-(p-F-C_6H_4)-5-N_3	86.0-6.5	177c
1-(m-F-C_6H_4)-5-N_3	86-8	177c
1-(p-NO_2-C_6H_4)-5-N_3	128-9	177c
1-(m-NO_2-C_6H_4)-5-N_3	131-2	608d
1-(2',4'-CL_2-C_6H_3)-5-N_3	90-2	608d
3,3'-(p-C_6H_4)-5,5'-diN_3-bistetrazole	205 (dec.)	177c

Table A.16. Cyanoazoles

	M.p./°C	Ref.
Pyrazoles		
3,4-Me$_2$-1-CN	99	1264
3-CN	149.5–50.0	1242d
4-Me-3-CN	130	897f
1-Ph-3-CN		915b
1-Ph-4-CN	95	1141c, 1265b,c
3-Ph-4-CN	134	1266
1-PhCH$_2$-4-CN	63–4	363b
1-(o-NO$_2$C$_6$H$_4$)-4-CN	175–7	1265b
1-(m-NO$_2$C$_6$H$_4$)-4-CN	162–3	1265a
1-(p-NO$_2$C$_6$H$_4$)-4-CN	190–1	1141c, 1265a
1-(o-NH$_2$C$_6$H$_4$)-4-CN		1265b
1-(p-NH$_2$C$_6$H$_4$)-4-CN	174–5	1141c, 1265a
1-(o-MeC$_6$H$_4$)-4-CN	105–7	1265c
1-(p-MeC$_6$H$_4$)-4-CN	128-30	1265c
3-(p-BrC$_6$H$_4$)-4-CN	195	1266
1-Ph-3-Me-4-CN	93–4	1039e
1-Ph-3,5-Me$_2$-4-CN	88–90	1020b
1-Ph-3-Me-5-CH$_2$Cl-4-CN	130·5	1039g
1-Ph-3-Me-5-CH$_2$OH-4-CN	114–16	1039g
1-Ph-3-Me-5-CH$_2$OEt-4-CN	68.5	1039g
1-Ph-3-Me-5-CH$_2$OPh-4-CN	73–4	1039g
1-Ph-3-Me-5-CH$_2$NH$_2$-4-CN	58–9	1039g
1,3-Ph$_2$-4-CN	133–5	915f
1,5-Ph$_2$-4-CN	182	1266
3,5-Ph$_2$-4-CN	230	1267a
1-Ph-5-(p-BrC$_6$H$_4$)-4-CN	212	1266
1-Ph-3,5-Me$_2$-4-CN	89–90	1020b
1,3-Ph$_2$-5-Me-4-CN	134	962a
1,3-Ph$_2$-5-CH$_2$OPh-4-CN	158	1039f
1,3-Ph$_2$-5-CMe$_3$-4-CN	163–4	962a
1,3-Ph$_2$-5-(CH=CH-Ph)-4-CN	205	1039f
1,5-Ph$_2$-3-Me-4-CN	189	1020b, 1267b, 1268
1-Ph-3-(p-MeC$_6$H$_4$)-5-CH$_2$OPh-4-CN	167	1039f
1,3,5-Ph$_3$-4-CN	189; 197	1030c, 1267a, 1269
1,4-Ph$_2$-5-CN	182	843a
Aminocyanopyrazoles		
1-Me-5-NH$_2$-3,4-(CN)$_2$	131.5-3.0	947e, 1270
1-Me-3-NH$_2$-4-CN	135–6	947d
1-i-Pr-3-NH$_2$-4-CN	97	947d
1-cyclohexyl-5-NH$_2$-4-CN	108.5–10.0	1271
1-Ph-5-NH$_2$-4-CN	135–7	1021c
3-CF$_3$-5-NH$_2$-4-CN	172	1021e
4-(p-tosyl)-5-NH$_2$-3-CN	230–1	1021f
1,4-(p-tosyl)$_2$-5-NH$_2$-3-CN	207.5–9.0	1021f, 1270
1-(p-tosyl)-3-Cl-5-NH$_2$-4-CN	190–2	1021f, 1270
1,3-Ph$_2$-5-NH$_2$-4-CN	172–3	1270
3,5-(NH$_2$)$_2$-4-CN	170–1	1272
1-CONH$_2$-5-NH$_2$-4-SO$_2$Me-3-CN	203–4	1021f, 1270
1-CONH$_2$-4-(p-tosyl)-5-NH$_2$-3-CN	191.0–2.5	1021f, 1270
1-CONH$_2$-3-Ph-5-NH$_2$-4-CN	200–1	1270
1-CONH$_2$-3-Me-5-NH$_2$-4-CN	206	1270
1-CONH$_2$-5-NH$_2$-4-CN	253	1270
1-CONH$_2$-3-SO$_2$Me-5-NH$_2$-4-CN	190–3	1021f, 1270

Table A.16. Continued

	M.p./°C	Ref.
Aminocyanopyrazoles		
1-CONH$_2$-3-SMe-5-NH$_2$-4-CN	214.5–15.0	1021f, 1270
1-CONH$_2$-3-OEt-5-NH$_2$-4-CN	214–15	1270
1-CONH$_2$-4-SO$_2$Ph-5-NH$_2$-3-CN	194–5	1270
3-SO$_2$Me-5-NH$_2$-4-CN	200–3	1021f
5-NHAc-3,4-(CN)$_2$	269–71	947e
1-Ph-5-NH$_2$-3,4-(CN)$_2$	195	1270
1-(p-NO$_2$C$_6$H$_4$)-5-NH$_2$-3,4-(CN)$_2$	252–3	1270
1-Ac-5-NH$_2$-3,4-(CN)$_2$	203–7	1270
1-PhCO-5-NH$_2$-3,4-(CN)$_2$	>200	1270
1-(p-tosyl)-5-NH$_2$-3,4-(CN)$_2$	214.5–16.0	1270
1-BrCH$_2$CO-5-NH$_2$-3,4-(CN)$_2$	210.0–11.5	1270
1-CONH$_2$-5-NH$_2$-3,4-(CN)$_2$	>240	1270
1-CSNH$_2$-5-NH$_2$-3,4-(CN)$_2$	Dec.	1270
1-CO$_2$Et-5-NH$_2$-3,4-(CN)$_2$	207–8	1270
Acyl- and carboxy-pyrazoles		
1-Ph-3-Me-4-CO$_2$H-5-CN	250–1	1039d
1-Ph-3-Me-5-CO$_2$H-4-CN	212	1039e
1-Ph-3-Me-4-CO$_2$Et-5-CN	88–9	1039d
1-Ph-3-Me-5-CONHPh-4-CN	167–8	1039e
1,3-Ph$_2$-5-CO$_2$H-4-CN	217	1039c
1-Ph-3-(p-MeC$_6$H$_4$)-5-CO$_2$H-4-CN	208–9	1039c
3-Ac-4-Me-5-CN	122.5	1021f, 1265b
3-COEt-4-Me-5-CN	106	1021f, 1265b
3-CO(2'-furyl)-4-Me-5-CN	184.5	1021f
3-(COC$_6$H$_4$-p-OMe)-4-Me-5-CN	147	1265b
Nitropyrazoles		
1,3-Me$_2$-4-NO$_2$-5-CN	98–100	693g
1-(p-NO$_2$C$_6$H$_4$)-3-Me-4-NO$_2$-5-CN	169	62d
Imidazoles		
1-CN	59.5–60.5	1109
2-CN		1273
1-Me-2-CN	Picrate 160	375
1-PhCH$_2$-2-CN	Picrate 100	375
2-PhCH$_2$-4-CN	192.5–4.0	973c
2-Ph-4-CN		1142
5-t-octyl-4-CN		1274
1-PhCH$_2$-2-Me-5-CN	123–4	746
1-PhCH$_2$-2-CH$_2$Cl-5-CN		1126
4,5-(CN)$_2$	174–5	1275
2-Me-4,5-(CN)$_2$	225	1058b
2-Et-4,5-(CN)$_2$	185	1058c
1-Me-5-Cl-2-CN		1009
5-CONH$_2$-4-CN	272–3	1063b
Aminoimidazoles		
5-NH$_2$-4-CN	129.0–9.5	1063b
1-Me-4-NH$_2$-5-CN	178	748, 914b
1-Me-5-NH$_2$-4-CN	195–6	1176b
1,2-Me$_2$-5-NH$_2$-4-CN	242	1176b
1-Et-2-Me-4-NH$_2$-5-CN	162	748

Table A.16. Continued

	M.p./°C	Ref.
Nitroimidazoles		
1-Me-5-NO$_2$-2-CN		607b
1-Me-4-NO$_2$-5-CN	141–3	565, 751
1-Et-2-Me-4-NO$_2$-5-CN	78–9	332a, 748
1-Pr-2-Et-4-NO$_2$-5-CN	62.5–3.5	495b
1-Bu-2-Pr-4-NO$_2$-5-CN	36.5–7.5	495b
1-i-Bu-2-i-Pr-4-NO$_2$-5-CN	131.5	495b
1,2,3-Triazoles		
2,4-Ph$_2$-5-CN	133–4	1135
4-Ph-2-(*p*-BrC$_6$H$_4$)-5-CN	140	1135
5-Ph-4-CN	166–7	949a
5-Me-1-Ph-4-CN	98	897f
2-Ph-4-CN	88–9	510b
2-Me-4-CONH$_2$-5-CN		1106
Amino-1,2,3-triazoles		
2-Me-4-NH$_2$-5-CN	115	1106
1-Me-4-NH$_2$-5-CN	187	1106
1-Me-5-NH$_2$-4-CN	229–30	1106
1-PhCH$_2$-5-NH$_2$-4-CN	182	147, 1106, 1131b
5-NH$_2$-4-CN	226–8	147
1,2,4-Triazoles		
5-CN	187	910
5-Me-3-CN	135–6	1021h, 1276
5-Ph-3-CN	123–4	1021h, 1276
1-(*p*-NO$_2$C$_6$H$_4$)-5-Me-3-CN	138.5–9.5	1091b
1-(*p*-ClC$_6$H$_4$)-5-Me-3-CN	106–8	1091b
1-(*p*-MeOC$_6$H$_4$)-5-Me-3-CN	80–2	1091b
1-(*o*-Me-*p*-NO$_2$C$_6$H$_3$)-5-Me-3-CN	126–8	1091b
1-(*o*-OMe-*p*-NO$_2$C$_6$H$_3$)-5-Me-3-CN	152–4	1091b
Tetrazoles		
5-CN	103–4	1276

Table A.17. Halogen substituted azoles

	M.p. (b.p./mmHg)/°C	Ref.
Pyrazoles[a]		
1-Halo		
3,4-Me$_2$-1-I	175–80	77b
3,4,5-Me$_3$-1-I	174–6	77b
3-Halo		
3-F	(77–8/12)	1019b
3-Cl	40 (76–7/0.5)	1019b
1,5-Me$_2$-3-Cl	46–7 (210–12)	62b, 66j, 100a
1-Me-5-Ph-3-Cl	76	100b, 330d
1-Et-5-Me-3-Cl	(216–18)	62b, 100a
1-Et-5-Ph-3-Cl	(152–3/13)	100b
1-PhCH$_2$-5-Me-3-Cl	(295–300)	62b, 100a
1-PhCH$_2$-5-Ph-3-Cl	67–8	100b
1-Ph-5-Me-3-Cl	(170/15)	330a
1,5-Ph$_2$-3-Cl	64	330e
1,4,5-Me$_3$-3-Cl	30–1	100d
1-Ph-4,5-Me$_2$-3-Cl	34	330c
1-Ph-3-Cl	31.5 (120–1/3)	898h
1-Ph-4-Et-5-Me-3-Cl	92	330c
5-Me-3-Cl	115; 119	100a, 321j
5-Ph-3-Cl	142	100b, 330d
4,5-Me$_2$-3-Cl	121.5–2.5	62b, 100d
4-Me-5-Ph-3-Cl	115–16	100b
3-Br	70 (70–2/0.1)	1019b
4-Me-3-Br	127–8	77f
4-Ph-3-Br	146–7	70c
1-Ph-3-Br	(115–16/1)	898h
3-I	72–3	432c, 1019b
4-Me-3-I	159	77b
4,5-Me$_2$-3-I	140–1	77b
4-Halo		
4-Cl	77 (220)	77d, 1277
3-Me-4-Cl	65–6	77d, 100d
3-Ph-4-Cl	102	100i
1-Me-4-Cl	(167/756)	433
1-Ph-4-Cl	75 (106/1)	898h, 1277, 1278
3,5-Me$_2$-4-Cl	117.5–18.5	100d, 431c, 898l
3-Me-5-Ph-4-Cl	108–10	66e
1,5-Me$_2$-4-Cl	178–9	100d
1-Me-5-Ph-4-Cl	(153–6/12)	100i
1-PhCH$_2$-5-Me-4-Cl	(288)	100d
1,3-Me$_2$-4-Cl	(177–8)	77d
1-Me-3-Ph-4-Cl	(163–5/9)	100i
1,3,5-Me$_3$-4-Cl	(113–15/35)	100d
1-Ph-3,5-Me$_2$-4-Cl	(162–5/10)	63b
1-PhCH$_2$-3,5-Me$_2$-4-Cl	(138–9/1–2)	898i
1,5-Ph$_2$-3-Me-4-Cl	50	1279
4-Br	93–4; 96–7	1277, 1280, 1281b
3-Me-4-Br	76–7	63b, 100d, 1025g
3-Ph-4-Br	116–17	1e, 66b,c
1-Me-4-Br	(76–8/18)	77f,g
1-PhCH$_2$-4-Br	44–5	363b

Table A.17. Continued

	M.p. (b.p./mmHg)/°C	Ref.
4-Halo		
1-Ph-4-Br	81	447, 898h, 1277, 1282
3,5-Me$_2$-4-Br	123	77d, 898g,i
3-Me-5-Ph-4-Br	93–6	1242c
3,5-Ph$_2$-4-Br	198–9	990b, 1266
1,5-Me$_2$-4-Br	38.0–9.5	100d
1-Me-5-Ph-4-Br	53–4; 65–6	66b,g
1-Ph-5-Me-4-Br	(~130/25)	943c
1,3-Me$_2$-4-Br	(76–7/10)	66d
1-Me-3-Ph-4-Br	(175–6/12)	66b
1-Ph-3-Me-4-Br	(311–13)	59h
1,3,5-Me$_3$-4-Br	32–4	77g
1-Ph-3,5-Me$_2$-4-Br	(173)	60e
1,5-Ph$_2$-3-Me-4-Br	75	1b, 1283
1,3,5-Ph$_3$-4-Br	142	392d
4-I	108–9	392d, 473, 615a
3-Me-4-I	185–7	431c, 473
1-Me-4-I	64–5	77b
1-Ph-4-I	76.5	1284
3,5-Me$_2$-4-I	137	77d
1,3-Me$_2$-4-I	113–15	473
1,5-Me$_2$-4-I	Oil	473
3-Ph-4-I	136	1285
1,3,5-Me$_3$-4-I	65	392g, 473, 1133
5-Halo		
1-Ph-5-Cl	(250)	330i, 898h
1,3-Me$_2$-5-Cl	(157–8)	62b, 66j, 98, 100a, 322
1-Me-3-Ph-5-Cl	62	100b, 330d
1-Et-3-Me-5-Cl	(167)	100a, 322
1-Et-3-Ph-5-Cl	(160–1/18)	100b
1-Pr-3-Me-5-Cl	(83–4/22)	322
1-Me-3-Et-5-Cl	(82–3/28)	322
1-Me-3-Pr-5-Cl	(78–9/10)	322
1-Pr-3-Et-5-Cl	(104–5/30)	322
1-Me-3-i-Pr-5-Cl	(72–4/10)	322
1-PhCH$_2$-3-Me-5-Cl	(146/13)	100a
1-PhCH$_2$-3-Ph-5-Cl	53–4	100b
1-Ph-3-Me-5-Cl	(146–8/20)	126b, 321b,c, 551d, 735c
1,3-Ph$_2$-5-Cl	56	62a
1,3,4-Me$_3$-5-Cl	(68/13)	100d
1,3-Me$_2$-4-Ph-5-Cl	(100–4/0.5)	98
1-Ph-3,4-Me$_2$-5-Cl	26 (147/12)	735c, 1243b
1-Ph-3-Me-4-Et-5-Cl	40 (175/50)	735c
1-Ph-3-Me-4-PhCH$_2$-5-Cl	50	735c
1-Ph-5-Br	55.5–6.0 (98–100/2)	898h
1-Ph-3-Me-5-Br	(153/15)	321c
1-Ph-3,4-Me$_2$-5-Br	81	552b
1-Me-5-I	76.0–6.5	1133
1-Ph-3-Me-5-I	Oil	321b, 735c
1-Ph-3,4-Me$_2$-5-I	78	735c

Table A.17. Continued

	M.p. (b.p./mmHg)/°C	Ref.
1,4-dihalo		
1-I-4-Cl	127–8	77d
1-I-4-Br	127–8	77d
3,5-Me$_2$-1-I-4-Cl		77d
3,5-Me$_2$-1-I-4-Br	222–3	77d
3,5-Me$_2$-1,4-I$_2$	228–39	77d
3,4-dihalo		
5-Me-3,4-Cl$_2$	115–17; 128	77d, 321j, 426a
5-Ph-3,4-Cl$_2$	95–6	330d
1-Ph-5-Me-3,4-Cl$_2$	Oil	330a
1,5-Me$_2$-3-Cl-4-Br	56	62b
5-Ph-3-Cl-4-Br	90	330d
1-Et-5-Me-3-Cl-4-Br	39	62b
1-PhCH$_2$-5-Me-3-Cl-4-Br	62–3	100a
1-Ph-5-Me-3-Cl-4-Br	(194/15)	330a
5-Me-3-Cl-4-Br	140	321j
5-Me-3-Cl-4-I	152	321j
3,4-Br$_2$	135–6	77f, 898i
5-Me-3,4-Br$_2$	145	77f
1-Ph-3,4-Br$_2$	83.5–4.0	965a, 1284
1,5-Me$_2$-3,4-Br$_2$	74	443
1-PhCH$_2$-3,4-Br$_2$		898h
3,4-I$_2$	162–3	77b, 473
1-Me-3,4-I$_2$	56 (120–5/2)	473
5-Me-3,4-I$_2$	177–9	473
3,5-dihalo		
1-Ph-3,5-Cl$_2$	25–6 (170–2/16)	551a
1-Ph-4-Me-3,5-Cl$_2$	(155/16)	551a
1-Ph-4-Et-3,5-Cl$_2$	81	321i
3,5-Br$_2$	151	77g
1-Ph-3,5-Br$_2$	50	330i
4-Ph-3,5-Br$_2$	157.0–7.5	945b
4,5-dihalo		
1-Ph-4,5-Cl$_2$	48	330i
1-Me-3-Ph-4,5-Cl$_2$	25.5 (317)	330d
1-Ph-3-Me-4,5-Cl$_2$	56	321b, 551d, 1243c
1,3-Ph$_2$-4,5-Cl$_2$	87–8	1286
1-Ph-5-Cl-4-Br	65	330i
1,3-Me$_2$-5-Cl-4-Br	35–6	100a, 322
1-Me-3-Ph-5-Cl-4-Br	65–7	330d
1-PhCH$_2$-3-Me-5-Cl-4-Br	(183/13)	100a
1-Et-3-Me-5-Cl-4-Br	(93–4/10)	322
1-Pr-3-Me-5-Cl-4-Br	(93–5/6)	322
1-Me-3-Pr-5-Cl-4-Br	(105–7/17)	322
1-Me-3-i-Pr-5-Cl-4-Br	(98–100/16)	322
1-Me-3-Bu-5-Cl-4-Br	(118–20/16)	322
1-Ph-3-Me-5-Cl-4-Br	56 (93–5/0.2)	321b, 322
1,3-Me$_2$-4,5-Br$_2$	74–5	447
1-Ph-3-Me-4,5-Br$_2$	92	59h, 321c

Table A.17. Continued

	M.p. (b.p./mmHg)/°C	Ref.
4,5-dihalo		
1-Me-3-Ph-5-Cl-4-I	105	330d
1-Me-4,5-I$_2$	104–5 (105–12/2)	1133
3,4,5-trihalo		
1-Ph-3,4,5-Cl$_3$	82	330i
1-Ph-3,5-Cl$_2$-4-Br	85	330i
3,4,5-Br$_3$	180–1; 184; 193–4	77f, 756, 898i
1-Ph-3,4,5-Br$_3$	106–7; 122	330i, 965a
1-PhCH$_2$-3,4,5-Br$_3$	81–2	898h
1,3,4-I$_3$	122	77b
3,4,5-I$_3$	200; 221	77f, 473
1-Me-3,4,5-I$_3$	152–3	1133
Imidazoles		
1-Halo		
1-I-2,4,5-Me$_3$	134	410d
2-halo		
2-F	Dec.; HCl salt 215–16	1287
2-Cl	165–6	212
4,5-Ph$_2$-2-Cl	217	1254d
2-Br	207	753
4-Me-2-Br	124–5	412b
1,4-Me$_2$-2-Br	51–2	449
4,5-Ph$_2$-2-Br	205–6	454
4-Ph-2-Br	153	453b
2-I	135–6	410b
4-Me-2-I	170–1	410b
1-Me-2-I	89–90	1144b
4 (and 5)-Halo		
4-F	101–4	1287
4-Cl		212
1-Me-4-Cl	(154–5/44) picrate 166–7	373a, 1289b
1-Et-4-Cl	Picrate 146–7	373a
2-Me-4-Cl		212
1-Me-5-Cl	(119–20/51)	179, 373a, 976b, 1289b
1-Et-5-Cl	(286–8)	373a
1-MeCH:CH-5-Cl	Picrate 110–3	373a
1,2-Me$_2$-4-Cl	93–4	373c
1-Et-2-Me-4-Cl	(157–8/27) Picrate 154–5	373c, 1289b
1-Pr-2-Me-4-Cl	Picrate 152–3	373c
1-Pr-2-Et-4-Cl	(122/6)	1289b
1-i-Bu-2-Me-4-Cl	Picrate 141–2	373c
1,2-Me$_2$-5-Cl	Picrate 175–6	982b
1-Et-2-Me-5-Cl	(228–32/620) picrate 155	373c, 495b, 1289i, 1290a
1-Ph-2-Me-5-Cl	Picrate 176	1290a
1-Me-2-Ph-5-Cl	106–7	976b
2,4-Me$_2$-5-Cl		212
1-Et-2-Ph-5-Cl	65–7	1291
1-Pr-2-Et-5-Cl	(228–32/620) picrate 155.0–6.5	495b

Table A.17. Continued

	M.p. (b.p./mmHg)/°C	Ref.
4 (and 5)-Halo		
1-Bu-2-Pr-5-Cl	(252–6/620) picrate 100–1	495b
1-Bu-2-Me-5-Cl	Picrate 151–2	495a
1-PhCH$_2$-2-Ph-5-Cl	79–80	1289i
1,2-Ph$_2$-5-Cl	133	1291
4-Br	130–1	374a, 1292
1-Me-4-Br	Picrate 179	374b
2-Me-4-Br	162–3	11
5-Me-4-Br		410b, 412b
2-Ph-4-Br	206–7	453c
5-Ph-4-Br	242–5	320d, 453b
5-Me-2-Ph-4-Br		924b
1-Me-5-Br	45–6	374b, 1289j, 1292
1,2-Me$_2$-5-Br	91–2	451
1,4-Me$_2$-5-Br	Oxalate 96	412b, 1293
1-Me-4-Ph-5-Br	80–90	453b
2-Me-4-I	144–5	410b
Dihalo		
2,4-Cl$_2$	184–5	212, 753
1-Me-4,5-Cl$_2$	54–5	1289a, j
2-Me-4,5-Cl$_2$		212
1-Bu-2-Pr-4,5-Cl$_2$	(152–4/0.5)	1289a
1-Me-2-I-5-Cl	99–100	1144a
1-Me-4-I-5-Cl	75.5–6.0	1144a
2,4-Br$_2$	193	374a, f
4,5-Br$_2$	225	374a, 1292
1-Me-4,5-Br$_2$	79–80	374b, 1289a, 1292
2-Me-4,5-Br$_2$	239–40	11, 1289a
4-Me-2,5-Br$_2$	214–15	412b
2-Et-4,5-Br$_2$	157–8	1289a
4-Ph-2,5-Br$_2$	198–9	453b
1,4-Me$_2$-2,5-Br$_2$	41–3	412b, 453b
1,5-Me$_2$-2,4-Br$_2$	125	412b, 453b
4-Br-2-I	174	410b
5-Me-4-Br-2-I	147–8	410b
2,4-I$_2$	182	474a
4-Me-2,5-I$_2$	191–2	410b
1 2-Me$_2$-4,5-I$_2$	142	122
2-Ph-4,5-I$_2$	194–5	1294
2-Me-4,5-I$_2$	199	410d
1-Me-4,5-I$_2$	141.5–2.5	1144b
Tri- and tetrahalo		
1-Me-2,4,5-Cl$_3$	93–4	1289j
1-Me-5-Cl-2,4-I$_2$	162.5–3.5	1133
1,4,5-Br$_3$		264
2,4,5-Br$_3$	221	374a, 1292
1-Me-2,4,5-Br$_3$	93.0–4.5	374b, 1289a
1-Bu-2,4,5-Br$_3$		909a
2,4-Br$_2$-5-I	181	410b
4,5-Br$_2$-2-I	215.5	410b
1,4,5-I$_3$		264
2,4,5-I$_3$	182–3	452b, 1296

Table A.17. Continued

	M.p. (b.p./mmHg)/°C	Ref.
Tri- and tetrahalo		
2-Me-1,4,5-I_3	160 (dec.)	410d
1-Me-2,4,5-I_3	149.5–50.5	1144b
1-Et-2,4,5-I_3	141–2	410d
1,2,4,5-I_4	160 (dec.)	410d, 1296
1,2,3-Triazoles[b]		
1-Me-4-Cl	69.5–70.0	77c
2-Me-4-Cl (?)	(62–5/39)	1061b, 1297
4-Me-5-Cl	77–8	77c
1-Ph-4-Cl	47–8	148c
1-Ph-5-Cl	50	816b
4,5-Me_2-1(?)-Cl	39–40 (dec.)	77c
1,4-Ph_2-5-Cl	137–8	1298
2-Et-5-Me-4-Cl (?)	(86–8/40)	1061b, 1297
1-(p-MeC$_6$H$_4$)-4-Ph-5-Cl	124–5	1298
4-Me-1(?),5-Cl_2	43–4 (dec.)	77c
1-Me-4-Br	98.5–9.0	77c, 276b
1-Me-5-Br	41–2	276b
2-Me-4-Br (?)	(62–5/22)	1061b, 1297
4-Me-5-Br	128–9	77c
1-Ph-4-Br	121.5–2.5	542
4,5-Me_2-1(?)-Br	75 (dec.)	77c
5-Me-2-Et-4-Br (?)	(84–5/30)	1061b, 1297
4,5-Br_2	194 (dec.)	77a, c
2-Me-4,5-Br_2	66.5–7.5	77c
4-Me-1(?)-5-Br_2	89–90 (dec.)	77c
1(?),4,5-Br_3		77c
1(?)-I	140	77c
4-I	110–11	77c
1-Me-4-I	123	77c
4-Me-5-I	152.5	77c
4,5-Me_2-1(?)-I	199.5–200.0	77c
2-Me-4,5(?)-I_2	127–8	77c
4-Me-1(?),5-I_2	135 (explodes 163–4)	77c
1,2,4-Triazoles		
Monohalo		
3-Ph-1(?)-Cl	108–9	268b
3,5-Ph_2-1-Cl	62–3	268b
3-Cl	167–8; 54–6(?)	268b, 688b, 908b
5-Me-3-Cl	151	268b
3-Et-5-Cl	101–2	1299
3-Pr-5-Cl	90	1193b
3-i-Pr-5-Cl	133	1193b
3-C_5H_{11}-5-Cl	(134–5) ~45	908b
1-Ph-3-Cl	76–8	1300
4-Ph-3-Cl	118	1255a, d
5-Ph-3-Cl	182	908b
1-Ph-3-Me-5-Cl	84	826
4-Ph-5-Pr-3-Cl	108	913b
4-(p-ClC$_6$H$_4$)-5-Et-3-Cl	143	913b
4-(p-ClC$_6$H$_4$)-5-Pr-3-Cl	77	913b
4-(p-BrC$_6$H$_4$)-5-Pr-3-Cl	86	913b

Table A.17. Continued

	M.p. (b.p./mmHg)/°C	Ref.
Monohalo		
1,3-Ph$_2$-5-Cl	68–70	1134c
3-Br	189	268a, 430c
3-Ph-5-Br	186–8; 90–1	908b, 1190
1-Et-3-Ph-5-Br	54–5	908a
1-Et-5-Ph-3-Br	(~130/0.001)	908a
4-Et-5-Ph-3-Br	86–7	908a
3-I	208	430c
Di- and trihalo		
1,3-Cl$_2$	87	268b, 908b
3,5-Cl$_2$	148	83f
5-Me-1,3-Cl$_2$	69	268b, 908b
1-Me-3,5-Cl$_2$	(96–7/14)	1134c
1-Ph-3,5-Cl$_2$	97–8	1134c
1-(p-ClC$_6$H$_4$)-3,5-Cl$_2$	103.0–4.5	1134c
1-(o-ClC$_6$H$_4$)-3,5-Cl$_2$	72.5–4.0	1134c
1-(m-NO$_2$C$_6$H$_4$)-3,5-Cl$_2$	108.5–10.0	1134c
1,3,5-Cl$_3$	82–3	268b, 908b
1,3-Br$_2$	211–12	268c
3,5-Br$_2$		268a
Tetrazoles		
5-Cl	73	1093b
1-(m-NO$_2$C$_6$H$_4$)-5-Cl	132	608d
1-(p-ClC$_6$H$_4$)-5-Cl	70–2	608d
1-(o,p-Cl$_2$C$_6$H$_3$)-5-Cl	78–80	608d
1-Ph-5-Cl	124	83e, 598
5-Br	156	1093b
1-Me-5-Br	71.0–1.5	999
2-Me-5-Br	(48.0–8.5/5)	999
1-Ph-5-Br	157	83e
5-I	~190 (dec.)	1093b
1-Ph-5-I	140 (dec.)	83d

(a) A more extensive series of tables with references up to 1962 has appeared[901].
(b) Some assignments of orientation are uncertain, and are shown with a question mark in parenthesis.

Table A.18. Hydroxyazoles (azolones)

	M.p. (b.p./mmHg)/°C	Ref.
3-Hydroxypyrazoles		
Unsubstituted	166	1159c
1-Ph	153–5	131, 1308
1,5-Me$_2$	180.5–183.0	130a
5-Me-1-Ph	171.0–173.5	130a
	165.0–166.5	1308
1,5-Ph$_2$	252–3	1308
4-Hydroxypyrazoles		
Unsubstituted	118.0–18.5	964a
1-Me	Oil	964a
1-Ph	119–20	964b
3,5-Me$_2$	173–5	1310a, 1311a, 1312
5-Et-3-Me	179–81	947a
3-Me-5-Ph	188; 194–5	1310a, 1313a
3-Et-5-Ph	178–9	1313a
5-Me-1-Ph	136.5–7.5	440a
3,5-Ph$_2$	235–7	1311c, 1313a
5-Et-3-Me-1-Ph	108–10	947a
1-Me-3,5-Ph$_2$	175–7	1311a
5-Hydroxypyrazoles		
1-Me	112.5–13.5	1159c
1-cyclohexyl	119–21	1159c
1,3,4-Ph$_3$	158 (dec.)	1315
3-Ph	236	158
1-Hydroxypyrazoles		
3,4,5-Me$_3$	183–4	762
4-Me-3,5-Ph$_2$	204–6	762
4,5-Me$_2$-3-Ph	167–8	762
1,4-Dihydroxypyrazoles		
3-Me-5-Ph	171–2	1313a
3-Et-5-Ph	165–6	1313a
3,5-Ph$_2$	171–2	1313a
3-Alkoxypyrazoles		
5-Me-3-OMe	49–50	330a, 1316
5-Me-3-OEt	66–7	76k, 158, 1316
1-Ph-3-OMe	(96–8/0.35)	131
5-Ph-3-OMe	106.0–6.5	100b
1,5-Me$_2$-3-OEt	(110/15)	130a
4,5-Me$_2$-3-OMe	85	1316
4,5-Me$_2$-3-OEt	93; 98	76k, 1316
4-Et-5-Me-3-OMe	106–7	1316
4-Et-5-Me-3-OEt	86	1316
4-Pr-5-Me-3-OEt	80.5	1316
5-Me-1-Ph-3-OMe	(150–60/16; 92.0–2.5/0.05)	130a, 330a
4-Et-1-Ph-5-Me-3-OMe	(175/16)	1317
4,5-Ph$_2$-3-OEt	157–8	1320
4-Alkoxypyrazoles		
4-OEt	68–9	1321
4-O-i-Pr	60–1	1321

Table A.18. Continued

	M.p. (b.p./mmHg)/°C	Ref.
4-Alkoxypyrazoles		
1-Ph-4-OEt		1318
3,5-Ph$_2$-4-OMe	160.5-2.5	1311b
1-Me-3,5-Ph$_2$-4-OMe	88.5-9.5	1311b
5-Alkoxypyrazoles		
1-Ph-5-OEt	34-5	329a, 1319
1,3-Me$_2$-5-OEt	(110/15)	130a, 1114c
3-Me-1-Ph-5-OMe	(277-82/725)	17b, 392f
3-Me-1-Ph-5-OEt	38.5-40.0 (300-2/752)	130a, 392f, 466, 898f
3,4-Me$_2$-1-Ph-5-OMe	(244-5/225)	392f
3,4-Me$_2$-1-Ph-5-OEt	60 (73/2)	130a, 329e
1,3-Ph$_2$-5-OEt	67-9	915f
1-Me-3,4-Ph$_2$-5-OEt	157-8	1320
4-CH$_2$Ph-3-Me-1-Ph-5-OCH$_2$Ph	139-40	900c
1-Alkoxypyrazoles		
4,5-Me$_2$-3-Ph-1-OMe[a]	(115/0.3)	762
1-Hydroxyimidazoles		
4,5-Me$_2$-2-Ph		1323
4,5-Me$_2$-2-Et	53.0-3.5	1324
2,4,5-Me$_3$	62-3	1324
2-i-Pr-4,5-Me$_2$	141.0-1.5	1324
4,5-Me$_2$-2-CH=CH$_2$	90-4	1324
4,5-Ph$_2$	210-11	1324
2-Me-4,5-Ph$_2$	247-8	1324
2-CH=CH$_2$-4,5-Ph$_2$	180-4	1324
2,4,5-Ph$_3$	229-31	142, 1325a
1-Alkoxyimidazoles		
2,4,5-Ph$_3$-1-OMe	131-2	142
4-Hydroxy-1,2,3-triazoles		
1-Ph	160; 170-3	1178; 605
2-Ph	123-4	688a, 1326c
1,5-Ph$_2$	188-93	605
2,5-Ph$_2$(?)	229	1093b
4-Alkoxy-1,2,3-triazoles		
1-Me-4-OMe	Oil; 27-8	576a, j
2-Me-4-OMe	(152)	576a
1-Ph-4-OEt	77-8	605
1-PhCH$_2$-4-OMe	50-1	687
1-Me-5-Ph-4-OMe	79	576c
5-Hydroxy-1,2,3-triazoles		
1-Me	166-7	276a, 576a
1-Ph	124-5	526
1-PhCH$_2$	157-8	276a
4-Ph	176-8	1326a
1-Me-4-Ph	203	1326a
4-Me-1-Ph	148; 133-4	1066b, 1326a
4-Me-2-Ph	330	892

Table A.18. Continued

	M.p. (b.p./mmHg)/°C	Ref.
5-Hydroxy-1,2,3-triazoles		
1,4-Ph$_2$	157–8; 150–1; 166	429a, 1298, 1326a
4-Ph-1-(Me-*p*-C$_6$H$_4$)	173–4	1298
5-Alkoxy-1,2,3-triazoles		
1-Me-5-OMe	(122–37/19)	576a
1-Ph-5-OMe	49	576c
1-Ph-5-OEt	65–6	526, 604
1-PhCH$_2$-5-OMe	61–3	526
1-Me-4-Ph-5-OMe	58	526
2-Me-4-Ph-5-OMe (?)	30	526
4-Me-1-Ph-5-OMe	61	526
4-Me-1-Ph-5-OEt	40–1 (147–9/0.6)	604
1,4-Ph$_2$-5-OMe	86–7	526
1,4-Ph$_2$-5-OEt	88–90	604
3-Alkoxy-1,2,3-triazoles		
1-Ph-3-OEt (?)	64–5	816b
3-Hydroxy-1,2,4-triazoles		
Unsubstituted	226–7; 234	430a, 1191b
1-Ph	274	1328
4-Ph	184	1225e
5-Ph	321–2	86b, 1225c, 1329, 1330, 1332
5-PhCH$_2$	226–7	897e, 1330, 1333
5-(*p*-MeC$_6$H$_4$)	372	1330
5-Me-4-Bu	80	1225b
5-Me-4-Ph	155	1225e
5-Me-4-PhCH$_2$	150	1225e
5-Et-1-Ph	191–2	1328
1,5-Ph$_2$	288–9	1334
2,5-Ph$_2$	234–5	516a
4,5-Ph$_2$	259	516a, 1225b, c, 1335
4-Ph-5-PhCH$_2$	159	1225b
4-Hydroxy-1,2,4-triazoles		
Unsubstituted		1124b
3-Ph	184; 152	1124c, 1336
3-PhCH$_2$	110	1124c
3-Me-5-Ph	223	1124c
3-Me-5-PhCH$_2$	141	1124c
3,5-Ph$_2$	189	1124c
5-Ph	184	1336
5-Hydroxy-1,2,4-triazoles		
1-Ph	183	61a, 1338
3-Ph-1-Me	218–19	85, 577b
1-Ph-3-Me	163–4; 167	61a, 577b, 1338
3-Et-1-Ph	Acetoxyderiv. 62–3	61a
1-Ph-3-Pr	146	61b
3-PhCH$_2$-1-Ph	187	61a, 1330
1-PhCH$_2$-3-Ph	228	1339
1-Ph-3-(*p*-MeC$_6$H$_4$)	267	1330

Table A.18. Continued

	M.p. (b.p./mmHg)/°C	Ref.
5-Hydroxy-1,2,4-triazoles		
1,3-Ph$_2$	231–2	1340e
3,4-Ph$_2$	256	1335
3,5-Dihydroxy-1,2,4-triazoles		
Unsubstituted (urazole)	244	1179
1-Me	216	1341
1,2-Me$_2$	167	1341
1-Ph	262–3	1342
4-Ph	203	1225a, d
Alkoxy-1,2,4-triazoles		
5-Ph-3-OMe	147	676a
5-Ph-3-OEt	121; 116–17	86b, 676a
5-PhCH$_2$-3-OMe	143	676a
5-PhCH$_2$-3-OEt	135	676a
5-PhCH$_2$-3-OPr	108	676a
4,5-Ph$_2$-3-OMe	180	1335
Hydroxy- and alkoxy-tetrazoles		
N-(1)-OH (?)	145 (dec.)	1343
5-Ph-1-OH	124 (dec.)	1344
5-OH	254	24c, 1093b, 1193d
1-Me-5-OH	122	83d
1-Ph-5-OH	188	24a, 770b
5-OMe	159	83g
5-OEt	98	83g
5-OAr (various)		767

(a) Boyle[1322b] suggests that this assignment should be 3,4-Me$_2$-5-Ph-1-OMe.

Table A.19. Nitro- and nitroso-azoles

	M.p. (b.p./mmHg)/°C	Ref.
1-Nitropyrazoles		
Unsubstituted	93	74c, 489a
3-Me	54–5	489a, 1295
4-Me	42.5	489b
5-Me	Oil	489a, 1295
3-t-Bu	66	489b
3,5-Me$_2$	66–7	74c
4-Et	Oil	489a
3,5-Et$_2$	Oil	74c
3,4,5-Me$_3$		74c
3-Ph	122	489b, 499a, b
3-C$_6$H$_4$-p-NO$_2$	184–6	489b
4-Br	50	74c, 1295
4-Cl	26–7	74c
4-I		74c, 1295
3-Nitropyrazoles		
Unsubstituted	174–5	489a, b, 530
4-Me	187	489b
4-Et	135–43	489a
1,4-Me$_2$	97	491
1-Me-4-Et	103	491
1-Me-4-CO$_2$H	179–80	491
4-SiMe$_3$	262–3	530
5-Me-4-CN		1134e
1-Ph	97	700, 962j
4-Ph	209–10	70c, 1027
5-Ph	198	489b
5-Me-1-Ph	96	217, 962j
5-Et-1-Ph	78	962j
1-picryl	225–6	1248
1-Me-4-picryl	197	490c
1,4-Ph$_2$	79	491
5-t-Bu	190.0–0.5	489b
5-C$_6$H$_4$-p-NO$_2$	260	489b
4-Nitropyrazoles		
Unsubstituted	162 (323)	74b, c, 1025d, e
1-Me	91–2	488
3-Me	134 (325/748)	898l, 1025e
1,3-Me$_2$	72	100d
1,5-Me$_2$	112	100d
3,5-Me$_2$	124–7 (325/743)	898l, 1020c, 1166
1,3,5-Me$_3$	56–7; 70	392g, 1025f, 1166
1-CH$_2$OH	99.5	898d
1-CH$_2$OH-3,5-Me$_2$	133	898d
3-Et-5-Me	144–5	1020d
3,5-Et$_2$	83	74b, c
1-CH$_2$CH$_2$OH	92–4	488
1-CH$_2$CH$_2$NO$_2$	51–2	488
3-t-Bu	118.5–19.0	489b
3-t-Bu-5-Me	148	1116b
1-C$_7$H$_{15}$-3,5-Me$_2$	(156/2)	898e
1-Ph	126–7	59c, 118, 1162

Table A.19. Continued

	M.p. (b.p./mmHg)/°C	Ref.
4-Nitropyrazoles		
3-Me-1-Ph	109.0–10.5	59e, 129
5-Me-1-Ph	113–14	59e
3-Me-5-Ph	144–5	74b
3,5-Me$_2$-1-Ph	103–4	499a, b, 964c, 1020c
3-Ph	185.0–5.5	489b
1,3-Ph$_2$	123	129
1,5-Ph$_2$	150	129, 1162
3,5-Ph$_2$	196–7	74b
3,5-di-t-Bu	205–8	1116b
1-(o-NO$_2$-C$_6$H$_4$)	148	59c, 118
1-(m-NO$_2$-C$_6$H$_4$)	155–6	59c
1-(p-NO$_2$-C$_6$H$_4$)	149–50	59c, 118, 129
1-(o-Me-C$_6$H$_4$)	69–70	1023c
1-(m-Me-C$_6$H$_4$)	76–7	1023c
1-(p-Me-C$_6$H$_4$)	91–2	1023c
1-picryl	219	490b
1-(p-NH$_2$-C$_6$H$_4$)	190–1	59c
1-(p-NO$_2$-C$_6$H$_4$)-5-Me	155–7	59e
1-(p-NH$_2$-C$_6$H$_4$)-5-Me	146–8	59e
1-(p-NO$_2$-C$_6$H$_4$)-3-Me	203–5	59e, 129, 499a, b
1-(p-NH$_2$-C$_6$H$_4$)-3-Me	167–8	59e
1-(p-NO$_2$-C$_6$H$_4$)-3,5-Me$_2$	154–6	499a, b, 503
5-(p-NO$_2$-C$_6$H$_4$)-3-Me	196	74b, 503
3,5-(p-NO$_2$C$_6$H$_4$)$_2$	279–80; 285–7	74b, 1020g
1,5-(p-NO$_2$C$_6$H$_4$)$_2$-3-Me	176–8; 181–4	1d, 503
1-(p-NO$_2$C$_6$H$_4$CH$_2$)	116–17	1114h
3,5-Me$_2$-1-(p-NO$_2$C$_6$H$_4$CH$_2$)	124–6	1114h
3-Picrylamino	267 (dec.)	490b
1-Picryl-3-picrylamino	340 (dec.)	490b
3-NH$_2$	242	1171a
3-NHAc	242–4	1171a
1-NH$_2$	100–1	1114l
5-NH$_2$-3-Me-1-Ph	167–8	537
3-Me-5-CHO	208 (dec.)	1376
1-Ph-5-CO$_2$H	190 (dec.)	909b
1-(p-MeC$_6$H$_4$)-5-CO$_2$H	162–3 (dec.)	909b
1-(m-NO$_2$C$_6$H$_4$)-5-CO$_2$H	151–2	909b
1-(p-HSO$_3$C$_6$H$_4$)-5-CO$_2$H	135–40	909b
1-(3′,4′-Cl$_2$C$_6$H$_3$)-5-CO$_2$H	148–50	909b
3,5-(CO$_2$H)$_2$	205 (dec.)	74b
3-Cl-5-Me	112	426a, 1247
5-Cl-1,3-Me$_2$	68; 77–8	426a, 693g
5-CN-1,3-Me$_2$	98–100	693g
5-OH-3-Me	267	527b
3-C$_6$H$_4$-p-NO$_2$	210–12	489b
3-C$_6$H$_4$-p-NH$_2$	187.0–7.5	489b
5-Nitropyrazoles		
3-Me	156–7	489a
1-Me 2-oxide	109	528, 700
1-Me-4-CO$_2$H	161–3	771
1-Ph	98–9	700

Table A.19. Continued

	M.p. (b.p./mmHg)/°C	Ref.
5-Nitropyrazoles		
1-picryl-4-picrylamino	243 (dec.)	490b
Dinitropyrazoles		
1,3-$(NO_2)_2$	67	489b
1,4-$(NO_2)_2$	54	489b
3,5-$(NO_2)_2$	173–4	489b
3,4-$(NO_2)_2$	87.5–8.5	489b
3-Me-1,4-$(NO_2)_2$	48	74c
3-Ph-1,4-$(NO_2)_2$	163 (dec.)	489b
3,5-Me_2-1,4-$(NO_2)_2$	91.0–1.5	74c
5-Ph-3,4-$(NO_2)_2$	149–50	489b
5-(β-pyridyl)-3,4-$(NO_2)_2$		504
5-Me-3,4-$(NO_2)_2$	120–1	489b
1-Me-3,5-$(NO_2)_2$	60–1	490d
4-Et-3,5-$(NO_2)_2$	170–1	489b
4-NH_2-1-Me-3,5-$(NO_2)_2$	162	490d
4-NHPh-1-Me-3,5-$(NO_2)_2$	158–9	490d
4-Br-1-Me-3,5-$(NO_2)_2$	110	490d
1-Me-4-picryl-3,5-$(NO_2)_2$	164; 215	490c, d
1-Me-3,5-$(NO_2)_2$ 2-oxide	186–8	528
1-Me-4-(2′,4′-di$NO_2C_6H_3$)-3,5-$(NO_2)_2$	183	490c
3-Nitrosopyrazoles		
Unsubstituted	159–62	530
4-$SiMe_3$	200–2	530
4-Nitrosopyrazoles		
Unsubstituted	232–5	530
3-C_9H_{19}	67	942
3,5-Me_2	128	74b, 964c, 1166, 1378
3,5-Et_2	76	74b
1,3,5-Me_3	80–1	1114a, 1166, 1245
1-i-Pr-3,5-Me_2	106	1378
3-t-Bu-5-Me	164	1116b
3,5-$(t-Bu)_2$	206–8	1116b
1-Cyclopentyl-3,5-Me_2	79–80	1114f
1-Cyclohexyl-3,5-Me_2	144–5	1114f
1-Cycloheptyl-3,5-Me_2	89–90	1114f
3-Me-5-Ph	153	74b, 964c
3,5-Me_2-1-Ph	94; 95–7	964c, 1166, 1378
3,5-Me_2-1-(m-MeC_6H_4)	Oil	1114a
3,5-Me_2-1-(p-MeC_6H_4)	109.5	1114a
3,5-Me_2-1-(o-Cl-C_6H_4)	117	1114a
3,5-Me_2-1-(m-Cl-C_6H_4)	84	1114a
3,5-Me_2-1-(p-Cl-C_6H_4)	118	1114a
3,5-Me_2-1-(p-F-C_6H_4)	116–17	1114a
3,5-Me_2-1-(p-Br-C_6H_4)	122	1310b
3,5-Me_2-1-(p-I-C_6H_4)	112	1114a
3,5-Me_2-1-(p-NO_2-C_6H_4)	131–2 (dec.)	1114a
3,5-Me_2-1-(p-Ph-C_6H_4)	165 (dec.)	1114a
3,5-Me_2-1-$PhCH_2$	54	1378
3,5-Me_2-1-Ph$(CH_2)_2$	64	1378

Table A.19. Continued

	M.p. (b.p./mmHg)/°C	Ref.
4-Nitrosopyrazoles		
5-NH$_2$-3-Me-1-Ph	199–200	537
3-Me-5-PhCH$_2$-1-Ph	65–6	1114e
5-Me-3-PhCH$_2$-1-Ph	100	1114e
3-Me-5-Ph-1-(*p*-Br-C$_6$H$_4$)	130	1310b
3-Me-5-Ph-1-(*p*-NO$_2$-C$_6$H$_4$)	135	1310b
3,5-Ph$_2$	196	74b
3-Me-1,5-Ph$_2$	137.5	992, 1114a, 1245
1,3,5-Ph$_3$	183	392d
2-Nitroimidazoles		
Unsubstituted	284	772, 1379
1-Me	100–2	772, 947g, 1387
4-Me	206; 211–2	947g, 1055e, 1233c
1-Et	43.5–4.5	1114n
4-Et	152–4	1055e
1-Pr	(101–3/0.1)	1114n
4-Pr	140–1	1055e
1-i-Pr	88.0–9.5	1114n
4-i-Pr	138–9	1055e
1-Bu	(115–16/0.1)	1114n
1-i-Bu	53	1114n
1-CH$_2$-CH=CH$_2$	(105/0.1)	1114n
1,5-Me$_2$	107–8	1055e, 1233a, c
1,4-Me$_2$	119–21	1233a, c
4,5-Me$_2$	215–17	947g, 1233c
4-Et-1-Me	49–50	1233b
1,5-Et$_2$	61–2	1233b
5-Et-1-Me	84–5	1233b
5-Bu-1-Me	17–18	1233b
1,4,5-Me$_3$	178–9	1055e, 1233c
1-Me-5-CH=CH$_2$	106–8	1145
1-Et-5-CH=CH$_2$	45–7	1145
1-Me-5-CH=CHPh	178–80	1145
1-Me-5-CHO	114–15	1145
1-Et-5-CHO	38–40	1145
1-Me-5-CH$_2$OH	142–4	1145
1-Me-5-CHOHCH$_2$OH	120–1	1145
1-Me-5-CO$_2$Et	65–6	1145
1-Et-5-CH$_2$CH$_2$Cl	36–7	1145
1-(*p*-Cl-C$_6$H$_4$CH$_2$)	108.0–9.5	1114m
1-(*o*-NO$_2$-C$_6$H$_4$CH$_2$)	175.5–7.0	1114m
1-(*m*-NO$_2$-C$_6$H$_4$CH$_2$)	115.0–16.5	1114m
1-(*p*-NO$_2$-C$_6$H$_4$CH$_2$)	130.5–2.0	1114m
4-Me-5-C(NO$_2$)$_3$	112	1387
1,5-Me$_2$-4-C(NO$_2$)$_3$	110.0–11.5	1387
4-Nitroimidazoles		
Alkyl- and aryl-substituted 4-nitroimidazoles		
Unsubstituted	312–13	368b, 1380b, 1385
1-Me	135	373c, 374b, 453b 497
2-Me	254	274c, 1381, 1385
5-Me	248; 253	274c, 564, 1385

Table A.19. Continued

	M.p. (b.p./mmHg)/°C	Ref.
Alkyl- and aryl-substituted 4-nitroimidazoles		
1,2-Me$_2$	182–3	498, 1381
1,5-Me$_2$	160–1	20c, 498, 564
2,5-Me$_2$	312	564
1,2,5-Me$_3$	170–1	496
1-Et	40	1381
2-Et	161	947c, 1381, 1385
2-Et-5-Me	183–5	1385
1,5-Et$_2$	99–100	1382
1-Et-2,5-Me$_2$	84.5–5.5	1382
1-Pr	21	1381
2-Pr	161–3	1385
2-i-Pr	180	1381, 1385
1-Bu	42	1381
2-Bu	130–1	1385
2-i-Bu	170–1	1385
1-t-Bu	250–1	1385
2-Ph	235–42	1221
5-Ph	160–70	377b
2-(o-F-C$_6$H$_4$)	196–9	1221
2-(p-F-C$_6$H$_4$)	224–5	1221
2-(p-Cl-C$_6$H$_4$)	264–5	1221
2-(p-NH$_2$-C$_6$H$_4$)	270	1184
2-(p-NHAc-C$_6$H$_4$)	315–20	1184
2-(o-NO$_2$-C$_6$H$_4$)	242–3	1221
2-(p-NO$_2$-C$_6$H$_4$)	290–1	1221
2-(o-NO$_2$-C$_6$H$_4$)		1221
2-(p-Ph-C$_6$H$_4$)	246.5–7.5	1221
5-(p-Cl-C$_6$H$_4$)	285	508b
5-(p-NH$_2$-C$_6$H$_4$)	.HCl 245–6	508b
5-(p-NO$_2$-C$_6$H$_4$)	293	320d, 507
5-(p-HO-C$_6$H$_4$)	>300	507
5-(p-HSO$_3$-C$_6$H$_4$)	>300	508b
1-Me-5-(p-Br-C$_6$H$_4$)	170	508b
1-Me-5-(p-Cl-C$_6$H$_4$)	238–40	508b
5-Me-2-(p-NO$_2$-C$_6$H$_4$)	248–9	320d
1-Me-5-(p-NO$_2$-C$_6$H$_4$)	187	497
1-PhCH$_2$	76	1381
2-Me-1-PhCH$_2$	106	1381
2,5-Me$_2$-1-PhCH$_2$		1123
5-CH=CHPh	303	564, 565, 1115b
1-Me-5-CH=CHPh	143–4; 150–1	564, 565
5-CH=CH-C$_6$H$_4$-OMe-p	296	1115b
1,2-Me$_2$-5-(p-Cl-C$_6$H$_4$)	200	508b
2,5-Ph$_2$	232	146c
1-CH$_2$OMe	66.5–7.0	947f, 1381
1-CH$_2$OAc	83.5–4.5	947f
1-CH$_2$CO$_2$Et	61–2	947f
1-CH$_2$OMe-2-Me	69.5–71.0	947f, 1383
1-CH$_2$CH$_2$OH-2-Me		1164b
1-CH$_2$CH$_2$Cl-2-Me	290	1078, 1164b
1-CH$_2$CH$_2$Br-2-Me		1164b
1-CH$_2$CH$_2$I-2-Me		1164b
1-(CH$_2$)$_3$CO$_2$H-2-Me	**143–4**	1110

Table A.19. Continued

	M.p. (b.p./mmHg)/°C	Ref.
Amino-4-nitroimidazoles		
1-Me-5-NH$_2$	303	1162, 1289d
1-Me-5-NHMe	156–7	969b
1-Me-5-NHEt	161–2	969b
1-Me-5-NHPr	114–18	969b
1-Me-5-NHBu	101–6	969b
1-Me-5-NHCH$_2$Ph	132–3	969b
1-Ph-5-NH$_2$	154–6	1162
1-Et-2-Me-5-NH$_2$	214–15	1289d
2-Et-1-Pr-5-NH$_2$	146–7	1289d
1-Bu-2-Pr-5-NH$_2$	127–8	1289d
1-i-Bu-2-i-Pr-5-NH$_2$	187–8	1289d
2-Me-1-CO$_2$Me-5-NH$_2$	252–5	1110
2-Me-1-(CH$_2$)$_2$CO$_2$H-5-NH$_2$	262–4	1110
Alkoxyl-4-nitroimidazoles		
2-OMe	219	122, 318j
1-Me-2-OMe	144–5	122, 318j
1-Et-2-OMe	87	122
2-OBu		122
2-OPh	209–10	122
1-Me-2-OPh	117	318j
Halogen substituted 4-nitroimidazoles		
5-Cl	214–16	1233b
1-Me-5-Cl	146–8	373c, 374b, 1233b 1289c
1-Et-2-Me-5-Cl	91; 88	373c, 495b, 1289c
1-Pr-2-Et-5-Cl	58–9	495b, 1289c
1-Bu-2-Me-5-Cl	180–2	495a
1-Bu-2-Pr-5-Cl	34.5–6.0	495b, 1289c
1-i-Bu-2-i-Pr-5-Cl	109.5–10.5	495b, 1289c
5-Br	279	374a
1-Me-2-Br	155	139
1-Me-5-Br	178–9	139, 374b
2-Me-5-Br	267–8	11, 1289e, k
5-Me-2-Br	220–1	412b
1,2-Me$_2$-5-Br	161.0–1.5	1289e, k
1,5-Me$_2$-2-Br	179–80	412b
2,5-Br$_2$	>270	374a
2-I	281	122, 1072c, 1387
1-Me-2-I	240	122, 318j
1-Et-2-I	152	318j
2-Me-5-I	271–3	122
1,2-Me$_2$-5-I	204–5	122
2,5-I$_2$	255.0–6.5	1072c, 1387
1,2,5-I$_3$	350	1072c
Sulphur-containing 4-nitroimidazoles		
5-SMe	187–9	544
5-SEt	196–8	544
5-SPr	159–61	544
5-SBu	215–16	544
5-SCH$_2$Ph	212–13	544

Table A.19. Continued

	M.p. (b.p./mmHg)/°C	Ref.
Sulphur-containing 4-nitroimidazoles		
1-Me-5-SMe	119–20	544
1-Me-5-SEt	68–70	544
1-Me-5-SPr	43–4	544
1-Me-5-SBu	45–6	544
1-Me-5-SPh	77–8	544
1-Me-5-SCH$_2$Ph	113–15	544
5-SO$_2$Me	276–7	544
5-SO$_2$Et	246.5–7.0	544
5-SO$_2$Pr	215–17	544
5-SO$_2$CH$_2$Ph	252–6	544
1-Me-5-SO$_2$Me	160–1	544
1-Me-5-SO$_2$Et	158–60	544
1-Me-5-SO$_2$Pr	165–6	544
1-Me-5-SO$_2$Bu	82–4	544
1-Me-5-SO$_2$CH$_2$Ph	161.0–2.5	544
5-SO$_2$NH$_2$	261–2	1101a
1-Me-5-SO$_2$NH$_2$	176–7	1101a
Carboxamido 4-nitroimidazoles		
5-CONH$_2$	290	565
1-Me-5-CONH$_2$	258–60	565
1-Et-2-Me-5-CONH$_2$	278.5–9.5 (dec.)	495b, 1289c
2-Et-1-Pr-5-CONH$_2$	221.0–1.5	495b, 1289c
1-Et-2-Pr-5-CONH$_2$	210	495a
1-Bu-2-Me-5-CONH$_2$	181	495a
1-Bu-2-Pr-5-CONH$_2$	184–6	495a, b, 1289c
1-i-Bu-2-i-Pr-5-CONH$_2$	227.0–7.5	495b, 1289c
Cyano-substituted 4-nitroimidazoles		
1-Me-5-CN	139–40	565
1-Et-2-Me-5-CN	78–9	495b, 1289c
2-Et-1-Pr-5-CN	62.5–3.5	495b, 1289c
1-Bu-2-Pr-5-CN	36.5–7.5	495b, 1289c
1-i-Bu-2-i-Pr-5-CN	131.5	495b, 1289c
Miscellaneous substituents		
5-CO$_2$Me		1384
5-Me-2-N$_3$	126–8	1233c
5-Nitroimidazoles		
Alkyl- and aryl-substituted		
1-Me	55	453b, 497, 564, 1380b, 1385
1,2-Me$_2$	138–40	498, 1381, 1385
1,4-Me$_2$	57–8	20c, 498, 564, 1385
1-Me-2-Et	78	947c, 1381, 1385
1-Me-2-Pr	Hydrochloride 195	1385
1-Me-2-i-Pr	60	947h
1-Me-2-Bu	Hydrochloride 190–3	1385
1-Me-2-i-Bu	Hydrochloride 190–2	1385
1-Me-2-t-Bu	Hydrochloride 192	1385
1,4-Me$_2$-2-Et	Oil	1385
1,2,4-Me$_3$	50–1	496

Table A.19. Continued

	M.p. (b.p./mmHg)/°C	Ref.
Alkyl- and aryl-substituted		
1-Me-2-Ph	160–1	1221
1-Me-2-(p-Cl-C_6H_4)	137–8	1221
1-Me-2-(o-NO_2-C_6H_4)	143–5	1221
1-Me-2-(o-F-C_6H_4)	163–5	1221
1-Me-2-(p-AcNH-C_6H_4)	246–7	1221
1-Me-2-(p-PhC$_6H_4$)	208–10	1221
1-Me-2-(o-AcNH-C_6H_4)	233–5	1221
1-Me-2-(p-F-C_6H_4)	187–8	1221
1-Me-2-(p-NO_2-C_6H_4)	135–6	1221
1-Me-4-(p-NO_2-C_6H_4)	208–9	497
1-Me-4-(p-NH_2-C_6H_4)	141–3	508b
1-Me-4-(p-AcNHC_6H_4)	214–15	508b
1-Me-4-(p-Br-C_6H_4)	99	508b
1-Me-4-(p-Cl-C_6H_4)	107–8	508b
1,2-Me_2-4-(p-Cl-C_6H_4)	145–7	508b
1-Me-4-CH=CH-Ph	214–15	564, 565
1-CH=CH_2-2-Me	48	1381
1-CH=CH_2-4-CH=CH-Ph (*trans*)		1156
1-Et	150	1381
1-Et-2-Me	145	947c, 1381
1-Et-4-Me	160	1385
1,2-Et_2	102	909c, 1381
1,4-Et_2	126–7	1382
1-Et-2-i-Pr	Picrate 118	1385
1,2-Et_2-4-Me		1381
1-Et-2,4-Me_2	35 (121–2/2)	1382
1-Pr	135	1381
1-Pr-2-Me	61	947c, 1381
1-Pr-4-Me	Tosylate 148–50	1385
1-Pr-2-Et	90	947c, 1381
1-Pr-2-i-Pr	Picrate 138–9	1385
1-Bu	52	909c, 1381
1-Bu-2-Me	120	947c, 1381
1-Bu-4-Me	Tosylate 166–7	1385
1-Bu-2-Et	90	947c, 1381
1-Bu-2-i-Pr	Picrate 172–3	1385
1-Bu-2-Et-4-Me	.HCl 135–7	1385
1-i-Bu-2-Me	82–3	947c
1-(C_5H_{11})-2-Me	68–9	947c
1-(C_6H_{13})-2-Me	63	947c
1-Ph	160–70	909c
1-PhCH$_2$-2-Me	112	1381
1-CH_2-CH=CH_2-2-Me	90	1381
1-CH_2CH_2OBu-2-Me	Oil; .HCl 158–60	1386
1-Me-2-CH_2OH		1055f
1-CH_2CH_2OEt	.HCl 172–5	909c
Amino-5-nitroimidazoles		
1-Me-4-NH_2	222.5–3.0	1289d
1-Et-2-Me-4-NH_2	198–9	1289d
2-Et-1-Pr-4-NH_2	160–1	1289d
1-Bu-2-Pr-4-NH_2	130–1	1289d
1,2-Me_2-4-NH_2	248–55	733

Table A.19. Continued

	M.p. (b.p./mmHg)/°C	Ref.
Alkoxyl 5-nitroimidazoles		
2-OMe-1-Me	138	122, 318j
4-OMe-1-Me		1122
4-OEt-1-Me		1122
2-OBu-1-Me	59	318j
4-OMe-1,2-Me$_2$	127–8	733
4-OEt-1,2-Me$_2$	97–8	733
Halogen substituted 5-nitroimidazoles		
1-Me-4-Cl	78.5–9.5	373c, 565, 1233b, 1289c
1-Et-2-Me-4-Cl	67	373c, 1289c
2-Et-1-Pr-4-Cl	58–9	1289c
1-Bu-2-Pr-4-Cl	(164/3)	1289c
1-Me-2-Br	117.0–18.5	139
1-Me-4-Br	105	374b
2-Et-4-Br	180–1	1289e, k
1,2-Me$_2$-4-Br	101–2	1289k
1,4-Me$_2$-2-Br	67–8	412b, 449
1-Me-2-I	152	122, 318j
1-Et-2-I	98–100	318j
1,2-Me$_2$-4-I	150–1	122
1-Me-2,4-I$_2$	205–7	1072c, 1387
1,2,4-I$_3$	350	1072c
Sulphur containing 5-nitroimidazoles		
1-Me-2-SH		607b
1-Me-2-SO$_2$Me	91.0–2.5	607b
1-Me-2-SO$_2$Ph	164–5	607b
1—Me— 2 —SO$_2$—⬠	88–9	607b
1-CH$_2$CH$_2$OAc-2-SO$_2$Me	87–9	607b
1-Me-4-SH	NH$_4^+$ salt 140–1	1101a
1-Me-4-SO$_3$H	260–5 (dec.)	1101a
1-Me-4-SO$_2$NH$_2$	158–9; 149–50	1101a
Miscellaneous substituents		
1-Me-2-CN		607b
1-CH$_2$CH$_2$OAc-2-CN	55–7	607b
1-Me-2-CHO		1012
1-Me-2-CO$_2$H		1055f
1-Me-2-CO$_2$Me		1055f
1-Me-2-CONH$_2$	215–16	607b
1-Me-4-CONH$_2$	290	565
1-Me-4-C(NO$_2$)$_3$	123–30	1387
Dinitroimidazoles and trinitroimidazoles		
1,4(5?)-(NO$_2$)$_2$	91.5–2.5	1072d, 1387
2-Me-1,4(5?)-(NO$_2$)$_2$	121.5–2.0	1072d, 1387
4,5-(NO$_2$)$_2$	187–8	1072c, d, 1387
1-Me-4,5-(NO$_2$)$_2$	73–5	140, 1072d, 1387
2-Me-4,5-(NO$_2$)$_2$	207–8	1072d, 1387
1,2-Me$_2$-4,5-(NO$_2$)$_2$	49–50	1387
2,4-(NO$_2$)$_2$	273–4; 266–8	140, 1387
1-Me-2,4-(NO$_2$)$_2$		140

Table A.19. Continued

	M.p. (b.p./mmHg)/°C	Ref.
Dinitroimidazoles and trinitroimidazoles		
1-Me-2,5-$(NO_2)_2$-4-I	111–13	1387
2,4,5-$(NO_2)_3$	136–8	1072c, d
1-Me-2,4,5-$(NO_2)_3$	81.5–2.5	1387
1-Me-2,5-$(NO_2)_2$-4-C$(NO_2)_3$	136–7	1387
2.5(4)-$(NO_2)_2$-4(5)-C$(NO_2)_3$	K salt 142	1387
Nitrosoimidazoles		
5-Ph-4-NO	195; 199–200	414c, 532c
2,5-Ph$_2$-4-NO	203	414a, b, c, d, 845a
2-Me-5-Ph-4-NO		1030b
1-Ac-2,4-Ph$_2$-5-NO		1030d
2-Me-5-(o-Me-C$_6$H$_4$)-4-NO		1388
1,2,3-Triazoles		
2-(o-Me-C$_6$H$_4$)-4-NO_2	68–9	1134b
2-(p-Me-C$_6$H$_4$)-4-NO_2	116	1134b
2-(o-Cl-C$_6$H$_4$)-4-NO_2	82	1134b
2-(m-Cl-C$_6$H$_4$)-4-NO_2	73–5	1134b
2-(p-Cl-C$_6$H$_4$)-4-NO_2	106–8	1134b
2-(o-MeO-C$_6$H$_4$)-4-NO_2	75	1134b
2-(m-MeO-C$_6$H$_4$)-4-NO_2	87–9	1134b
2-(p-MeO-C$_6$H$_4$)-4-NO_2	112	1134b
2-(p-NH_2-C$_6$H$_4$)-4-NO_2	217–18	1134b
2-(m-NO_2-C$_6$H$_4$)-4-NO_2	119	1134b
2-(p-NO_2-C$_6$H$_4$)-4-NO_2	143; 140	995a, 1134b
2-(2′,4′-$(NO_2)_2$C$_6$H$_3$)-4-NO_2	104	995a
2-(1,2,4-triazol-3-yl)-4-NO_2	226–7	995a
1,2,4-Triazoles		
3-NO_2	213–15	175a, 766, 1212b, 1389a
1-Me-3-NO_2	63–4	269c, 1389a
5-Me-3-NO_2	195–8	995a, 1212b, 1389a
1-Me-5-NO_2		1389a
4-Me-3-NO_2		1389a
5-Et-3-NO_2	129–31	1212b
5-Ph-3-NO_2	215	1116a
5-(p-NO_2-C$_6$H$_4$)-3-NO_2	260	1116a
3-Br-5-NO_2	157–9	766
1-Me-5-NH_2-3-NO_2	253–5	269c
1-Me-5-NHNH$_2$-3-NO_2	162–3	269c
1-Me-5-NMeNH$_2$-3-NO_2	160–1	269c
1-Me-5-N_3-3-NO_2	126–7	269c
1-Me-5-Cl-3-NO_2		1389a
3,5-$(NO_2)_2$		1114j, 1390
1(or 4)-Me-3,5-$(NO_2)_2$	98.0–8.5	1114j, 1389a
1(or 4)-Et-3,5-$(NO_2)_2$	78	1114j
1(or 4)-CH$_2$CH=CH$_2$-3,5-$(NO_2)_2$	54–6	1114j
Tetrazoles		
5-NO_2		292
Nitrosotetrazoles		
2-CONH$_2$-5-NO	180–2 (dec.)	1349a

Table A.20. Azole N-oxides and N-hydroxyazoles

	M.p. (b.p./mmHg)/°C	Ref.
1-Hydroxypyrazoles		
3,4,5-Me$_3$	183–4; p-nitrobenzoate	762
	137–8	
4-Me-3,5-Ph$_2$	204–6	762
4,5-Me$_2$-3-Ph	167–8; p-nitrobenzoate	762
	152–4	
	p-toluenesulphonyl 105	
	OMe deriv. (115/0.3)	
Pyrazole 2-oxides		
1-Me	65.9 (105–7/0.3)	245
1-Me-5-NO$_2$	109	528, 700
1-Me-3,5-(NO$_2$)$_2$	186–8	528
1-OH-3,4,5-Me$_3$	169–70	762
1-OH-4,5-Me$_2$-3-Ph	200	762
1-OH-5-Me-3,4-Ph$_2$	213–14	762
1-OH-3,4,5-Ph$_3$	230–2	762
1-OMe-3,4,5-Me$_3$	78–9	1322a
1-OMe-3,4-Me$_2$-5-Ph	92–3	1322a
1,4-Dihydroxypyrazoles		
3,5-Me$_2$	149–50	762
3-Me-5-Ph	171–2	762
3-Et-5-Ph	165–6	762
3,5-Ph$_2$	171–2	762
3-CO$_2$Et-5-Ph	180–1	1313a
3,5-(CO$_2$Me)$_2$	167–8	1313a
3,5-(CO$_2$Et)$_2$	165	1313a
1-Hydroxyimidazoles		
2,4,5-Me$_3$	62–3	1324
2-Et-4,5-Me$_2$	53.0–3.5	1324
2-i-Pr-4,5-Me$_2$	141.0–2.5	1324
2-MeOCH$_2$-4,5-Me$_2$	234–6	1324
2-Me(CH$_2$)$_2$CH(Me)-4,5-Me$_2$	125–7	1324
2-CH$_2$=CH-4,5-Me$_2$	90–4	1324
2-CCl$_3$-4,5-Me$_2$	175–80 (dec.)	1324
4,5-Ph$_2$	210–11	1324
2-Me-4,5-Ph$_2$	247–8	1324
2-CH$_2$=CH-4,5-Ph$_2$	180–4	1324
2-Me(CH$_2$)$_2$CH(Et)CH$_2$-4,5-Ph$_2$	234–5	1324
2,4,5-Ph$_3$	229–31; 240–1	142, 1325a
	OMe deriv. 131–2	142
2,5-Me$_2$-4-Ph	149–50	1396
2,4-Me$_2$-5-Ph	162–3	1396
Imidazole 3-oxides		
2,4,5-Ph$_3$(?)	240	1325a
1-Me-2,4,5-Ph$_3$	168–72	142
1-Bu-4,5-Me$_2$-2-Ph	Hydrochloride 174–6	1399a
1-CH$_2$Ph-4,5-Me$_2$-2-Ph	Hydrochloride 165–7	1399a
1-CH$_2$Ph-4,5-Me$_2$-2-(m-NO$_2$-C$_6$H$_4$)	Hydrochloride 120–5	1399a
1-CH$_2$Ph-5-Me-2,4-Ph$_2$	96–8	1399a
1-CH$_2$Ph-2,4,5-Ph$_3$	Hydrochloride 188–90	1399a
1-Bu-5-Me-4-Ph-2-(m-NO$_2$C$_6$H$_4$)	Hydrochloride 186–8	1399a

Table A.20. Continued

	M.p. (b.p./mmHg)/°C	Ref.
Imidazole 3-oxides		
1-Bu-5-Me-4-(p-ClC$_6$H$_4$)-2-(m-NO$_2$C$_6$H$_4$)	Hydrochloride 225.0–7.5	1399a
1-Bu-2,4,5-Ph$_3$	Hydrochloride 179–82	1399a
1-i-Bu-2,4,5-Ph$_3$	Hydrochloride 223–4	1399a
1-CH$_2$Ph-2-(m-NO$_2$C$_6$H$_4$)-4-(p-ClC$_6$H$_4$)	188–9	1399a
1-CH$_2$Ph-2-Ph-4-(p-ClC$_6$H$_4$)-5-Me	Hydrochloride 132–4	1399a
1,4-Ph$_2$-5-Me-2-(o-HO-C$_6$H$_4$)	184–5	1399a
1,2,4-Ph$_3$-5-Me	249–51	1399a
1-(p-ClC$_6$H$_4$)-5-Me-2,4-Ph$_2$	229–31	1399a
1,2,4,5-Ph$_4$	247–8	1399a
1-(p-MeC$_6$H$_4$)-2,4,5-Ph$_3$	235–7	1399a
1-(p-ClC$_6$H$_4$)-2,4,5-Ph$_3$	241–2	1399a

Compounds

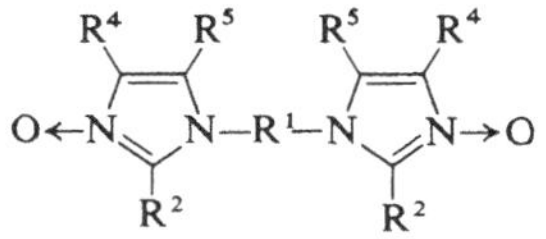

$$O{\leftarrow}N{\overset{R^4\quad R^5}{\underset{R^2}{\bigcirc}}}N{-}R^L{-}N{\overset{R^5\quad R^4}{\underset{R^2}{\bigcirc}}}N{\rightarrow}O$$

R^1	R^2	R^4	R^5		
(1)-C$_6$H$_4$-(4)	Ph	Me	Ph	251–5	1399a
(1)-C$_6$H$_4$-(4)	Ph	Ph	Ph	211–14	1399a
(4)-C$_6$H$_4$- C$_6$H$_4$-(4)	Ph	Me	Ph	297–9	1399a
(4)-C$_6$H$_4$- C$_6$H$_4$-(4′)	Ph	Ph	Ph	276–8	1399a

Compounds

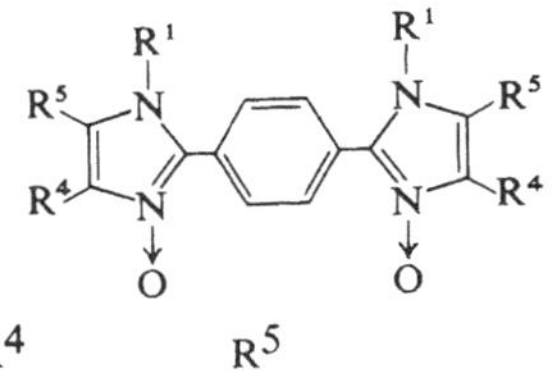

R^1	R^4	R^5		
CH$_2$Ph	Me	Ph	269.0–70.5	1399a
Bu	Ph	Ph	284.5–6.0	1399a
CH$_2$Ph	Ph	Ph	265–6	1399a

1-Hydroxyimidazole 3-oxides		
2,4,5-Me$_3$	198 (dec.); 203	1347, 1397
2-Et-4,5-Me$_2$	195 (dec.)	1397
4-Me-5-Ph	188–90	1347
2-Me-4,5-Ph$_2$	155–7	1347
4,5-Ph$_2$	239	1347
2,4,5-Ph$_3$	225; 231–2 (dec.)	1261b, 1325c 1347

1,2,3-Triazole 1-oxides		
4-Me-2-Ph	67.0–7.5	168d, 1340a
4-Me-2-(p-NO$_2$C$_6$H$_4$)	136	168d, 1340a
4,5-Me$_2$-2-Ph	92–3	168d, 1340f
4,5-Me$_2$-2-(p-BrC$_6$H$_4$)		168d

Table A.20. Continued

	M.p. (b.p./mmHg)/°C	Ref.
4,5-Me$_2$-2-(p-MeC$_6$H$_4$)	92–3	1340d
4,5-Me$_2$-2-(p-NO$_2$C$_6$H$_4$)		168d
4-Et-5-Me-2-Ph	43–4	1340b
4-Et-5-Me-2-(p-NO$_2$C$_6$H$_4$)	131	1340b
5-Et-4-Me-2-Ph	liquid	168d, 1340b
5-Et-4-Me-2-(p-NO$_2$C$_6$H$_4$)	156–7	168d, 1340b
5-Me-2.4-Ph$_2$	83	1340c
2,4,5-Ph$_3$	169	1340c

1,2,4-Triazole 4-oxides

These compounds are listed as 4-hydroxy-4H-1,2,4-triazoles in §9.9. References 873, 1124b, c, 1224b.

Tetrazole 4-oxides

Unsubstituted	153	1400
5-Me	151	1400
5-Et	140	1400

Table A.21. Azoles with sulphur substituents on nitrogen

Pyrazoles	M.p. (b.p./mmHg)/°C	Ref.
1-SO$_2$Ph	103.5–4.0	952
1-SO$_2$Ph-5-NH$_2$	158–61	172b, c
1-SO$_2$Ph-3,5-Me$_2$	67–71; 78–9	318i, 897b
1-SO$_2$Ph-3,5-Ph$_2$-4-(CH$_2$)$_2$Ph	145–6	1431
1-SO$_2$Me-3,5-Me$_2$	56–8	318i
1-SO$_2$NMe$_2$-3,5-(CN)$_2$-5-NH$_2$	166–8	1270
1-SO$_2$C$_6$H$_4$-p-Me	146–8	691
1-SO$_2$C$_6$H$_4$-p-Me-3-Me	111–12	691
1-SO$_2$C$_6$H$_4$-p-Me-5-NH$_2$	180–1	172b,c
1-SO$_2$C$_6$H$_4$-p-Me-3,5-Me$_2$	93–4; 96.5–7.5; 112–13	691, 897a, c
1-SO$_2$C$_6$H$_4$-p-Me-5-Et-3-Me	77–8	691
1-SO$_2$C$_6$H$_4$-p-Me-3,5-Et$_2$	49	691
1-SO$_2$C$_6$H$_4$-p-Me-5-Pr	87.5–8.0	691
1-SO$_2$C$_6$H$_4$-p-Me-5-i-Pr-3-Me	108–9	691
1-SO$_2$C$_6$H$_4$-p-Me-3,5-Ph$_2$	118–19	690
1-SO$_2$C$_6$H$_4$-p-Me-5-NHSO$_2$C$_6$H$_4$-p-Me	216–18	172b, c
1-SO$_2$C$_6$H$_4$-p-Me-5-N(SO$_2$C$_6$H$_4$-p-Me)$_2$	201–2	172b, c
1-SO$_2$C$_6$H$_4$-p-Me-3,4-(CN)$_2$-5-NH$_2$	214.5–6.0	1270
1-SO$_2$C$_6$H$_4$-p-Me-3-Cl-4-CN-5-NH$_2$	166–8	1270
1,4-(SO$_2$C$_6$H$_4$-p-Me)$_2$-5-NH$_2$-3-CN	207.5–9.0	1270
1-SO$_2$(CH$_2$)$_2$C$_6$H$_4$-p-Me	61–4	318i
1-SO$_2$C$_6$H$_4$-p-NH$_2$-3,5-Me$_2$	182	1200
1-SO$_2$C$_6$H$_4$-p-Br-3,5-Me$_2$	130–2	318i
1-SO$_2$C$_6$H$_4$-p-Cl-3,5-Me$_2$	121–2	318i
1-SO$_2$C$_6$H$_3$-m,p-Cl$_2$-3,5-Me$_2$	105–6	318i
1-SO$_2$C$_6$H$_4$-p-OMe-3,5-Me$_2$	122.5–4.5	318i
1-SO$_2$C$_6$H$_4$-(β-naphthyl)-3,5-Me$_2$	106.0–6.5	318i
1-SO$_2$C$_6$H$_4$-o-NO$_2$-3,5-Me$_2$	115–17	318i
1-SCCl$_3$-3-Me	(62–3/0.1)	1114g
1-SCCl$_3$-3,5-Me$_2$	59–60	1114g
1-SCCl$_3$-3-CH$_2$CHMe$_2$	(97/0.1)	1114g
1-SCCl$_3$-3-Ph	98–9	1114g
<u>Imidazoles</u>		
1-SO$_3$H	(K$^+$ salt) 221–2	1100b
1-SO$_3$H-4-Ph	> 300	452b, 1100b
1-SO$_2$CF$_3$	22 (46/11)	1432
1-SO$_2$Ph	83–4	150e
1-SO$_2$C$_6$H$_4$-p-Me	78.0–8.5	150c, e
1-SO$_2$C$_6$H$_4$-p-Me-2-Me-4-NHAc	179–80	1433
1-SO$_2$C$_6$H$_4$-p-Me-2-Et-4-NHAc	138–40	1433
1-SO$_2$C$_6$H$_4$-p-Me-4-NHAc	186	1434
1-SO$_2$C$_6$H$_4$-p-NH$_2$	182–3	150e,l
1-SO$_2$C$_6$H$_4$-p-NHAc	167	150e
1-SO$_2$C$_6$H$_4$-p-N$_2$Ph	168.0–8.5	150e
1-SO$_2$Ph-p-(NH)$_2$Ph	140–1	150e
1-SO$_2$-(1'-imidazoyl)	140–1	150l
1-SO-(1'-imidazolyl)	78–9	150l, 933c, 1373, 1435
1-S-(1'-imidazolyl)	110–11; 80–5	190b, 649a
<u>1,2,3-Triazoles</u>		
1-SO$_2$C$_6$H$_4$-p-Me-5-Me	119.5–21.0	1436
1-SO$_2$C$_6$H$_4$-p-Me-5-Ph	114–15	1436

Table A.21. Continued

1,2,3-Triazoles	M.p. (b.p./mmHg)/°C	Ref.
1-SO$_2$C$_6$H$_4$-p-Me-5-C$_6$H$_4$-p-Br	139.5	1436
1-SO$_2$C$_6$H$_4$-p-Me-5-C$_6$H$_4$-p-NO$_2$	172–3	1436
1,2,4-Triazoles		
1-SO$_2$Ph	109–10	693j
1-SO$_2$C$_6$H$_4$-p-Me	143–5	693j
1-SO$_2$Ph-3,5-Me$_2$	97.0–7.5	693j
1-SO$_2$C$_6$H$_4$-p-Cl	127–8	693j
1-SO$_2$C$_6$H$_4$-p-F	81–3	693j
1-SO$_2$C$_6$H$_4$-p-F-3,5-Me$_2$	109.0–9.5	693j
1-SO$_2$C$_6$H$_4$-p-OMe	94–5	693j
4-SO$_2$C$_6$H$_4$-p-OMe	128–30	693j
1-SO$_2$C$_6$H$_4$-p-OMe-3,5-Me$_2$	107–8	693j
1-SO$_2$C$_6$H$_3$-o,p-(OMe)$_2$	80–1	693j
1-SO$_2$C$_6$H$_3$-o,p-(OMe)$_2$-3,5-Me$_2$	51–3	693j
1,2,4-Triazol-5-ones		
1-Ph-4-SO$_2$C$_6$H$_4$-p-Me	154–5	1338
1-Ph-4-SO$_2$C$_6$H$_4$-p-Me-3-Me	162–3	1338

Table A.22. Azoles with sulphur substituents on carbon

	M.p. (b.p./mmHg)/°C	Ref.
Pyrazoles		
Thiols[a] and thioethers		
4-SH-5-Ph-3-SMe	132	1227b
5-SH-3-Me-1-Ph	109–10	330f
5-SH-3-Me-4-N_2Ph-1-Ph	97	735h
3-SMe-5-Me	Picrate 164–5	1035
3-SMe-1,5-Ph_2	62	330e
3-SMe-5-Me-1-Ph	(327, 175/10)	330a
3-SMe-4,5-Ph_2	160	1227b
3-SMe-5-Me-1-o-MeC$_6$H$_4$	(135/32)	330a
3-SMe-4,5-Me_2-1-Ph	40	330c
3-SMe-5-Me-1-p-MeC$_6$H$_4$	(151/31)	330a
3-SMe-4-Et-5-Me-1-Ph	(160–5/12)	330c
3-SMe-4-SCH$_2$Ph-5-Ph	132	1227b
3-SMe-5-NH_2-4-CN-1-CONH$_2$	214.5–5.0	1270
3-SEt-5-Me	Picrate 132–3	1035
3-SCH$_2$Ph	Picrate 110–11	1035
3-S(CH$_2$)$_2$CO$_2$H-5-Me	Picrate 134–5	1035
3-SCH$_2$Ph-5-Me	Picrate 145.0–5.5	1035
4-SCH$_2$Ph-1-Ph	84.5–5.0	59a
4-SCH$_2$Ph-3-Me-1,5-Ph_2	108	446, 506a,
5-SMe-1-Ph	(142–3/14)	552b
5-SMe-1-Me-3-Ph	(184/10)	330d
5-SMe-3-Me-1-Ph	(165–6/11)	552b
5-SMe-3,4-Me_2-1-Ph	56	552b
5-SMe-3-Me-1-o-NO$_2$C$_6$H$_4$	61	330h
5-SMe-3-Me-1-m-NO$_2$C$_6$H$_4$	84	330h
5-SMe-3-Ph-1-m-NO$_2$C$_6$H$_4$	106	330e
5-SMe-3-Me-1-p-NO$_2$C$_6$H$_4$	135–6; 139	330h, 552b
5-SMe-3-Me-1-p-NH$_2$C$_6$H$_4$	132	330h
5-SMe-1,3-Me_2	(243)	321j
5-SMe-1,3-Ph_2	(225/11)	330e
5-SMe-3-Me-4-N_2Ph-1-Ph	63	735h
5-SEt-3-Me-1-Ph	(178/19)	552b
5-SEt-3,4-Me_2-1-Ph	(180–5/15)	552b
5-SEt-3-Me-4-N_2Ph-1-Ph	71	735h
5-S-i-Pr-3-Me-1-Ph	(176/16)	552b
5-S-i-Bu-3-Me-1-Ph	(168–75/9)	552b
5-SCH$_2$CH=CH$_2$-3-Me-1-Ph	56–7	552b
5-SPh-3-Me-4-N_2Ph-1-Ph	115	735h
5-SCH$_2$Ph-3-Me-1-Ph	(246/20)	552b
5-SCOPh-3-Me-1-Ph	93	330f, 1354d
5-SCOPh-3-Me-1-p-MeC$_6$H$_4$	114	330f, 1354d
1-Ph-4-S(1'-Phpyrazolyl-4')	153	126c
1,3,5-Me$_3$-4-S(1',3',5'-Me$_3$pyrazolyl-4')	139–40	126c
3,5-Me$_2$-1-Ph-4-S(3',5'-Me$_2$-1'-Phpyrazolyl-4')	132–3	126c
3,5-Me$_2$-1-Pr-4-S(3',5'-Me$_2$-1'-Prpyrazolyl-4')	(225–30/13)	126c
3-Me-1,5-Ph_2-4-S(3'-Me-1',5'-Ph$_2$pyrazolyl-4')	202	446
Sulphonic acids and sulphones		
4-SO$_3$H-3-Me-1,5-Ph_2	Liquid	446
5-SO$_3$H-3-Me-1-Ph	235	330f
3-SO$_2$Me-5-Me-1-Ph	105	330a, 1035
3-SO$_2$Me-5-Me-1-o-MeC$_6$H$_4$	122	330a

Table A.22. Continued

	M.p. (b.p./mmHg)/°C	Ref.
Sulphonic acids and sulphones		
$3\text{-}SO_2Me\text{-}5\text{-}Me\text{-}1\text{-}p\text{-}MeC_6H_4$	85	330a
$3\text{-}SO_2Me\text{-}1,5\text{-}Ph_2$	121	330e
$3\text{-}SO_2Me\text{-}4,5\text{-}Ph_2$	250	1227b
$3\text{-}SO_2Me\text{-}5\text{-}NH_2\text{-}1\text{-}CONH_2\text{-}4\text{-}CN$	190–3	1270
$4\text{-}SO_2Me\text{-}5\text{-}NH_2\text{-}1\text{-}CONH_2\text{-}3\text{-}CN$	203–4	1270
$5\text{-}SO_2Me\text{-}1,3\text{-}Me_2$		321j
$5\text{-}SO_2Me\text{-}3\text{-}Me\text{-}1\text{-}Ph$	89–90	552b
$5\text{-}SO_2Me\text{-}3\text{-}Me\text{-}1\text{-}o\text{-}NO_2C_6H_4$	160	330h
$5\text{-}SO_2Me\text{-}3\text{-}Me\text{-}1\text{-}m\text{-}NO_2C_6H_4$		330h
$5\text{-}SO_2Me\text{-}3\text{-}Me\text{-}1\text{-}p\text{-}NO_2C_6H_4$	154	330h
$5\text{-}SO_2Me\text{-}3,4\text{-}Me_2\text{-}1\text{-}Ph$	137	552b
$5\text{-}SO_2Me\text{-}3,4\text{-}Me_2\text{-}1\text{-}p\text{-}BrC_6H_4$	178	552b
$5\text{-}SO_2Me\text{-}1,3\text{-}Ph_2$	162	330e
$5\text{-}SO_2Me\text{-}3\text{-}Ph\text{-}1\text{-}m\text{-}NO_2C_6H_4$	148	330e
$5\text{-}SO_2Me\text{-}3\text{-}Me\text{-}4\text{-}N_2Ph\text{-}1\text{-}Ph$	156	735h
$5\text{-}SO_2Et\text{-}3\text{-}Me\text{-}1\text{-}Ph$	61–2	552b
$5\text{-}SO_2Et\text{-}3,4\text{-}Me_2\text{-}1\text{-}Ph$	(180–5/15)	552b
$5\text{-}SO_2\text{-}i\text{-}Pr\text{-}3\text{-}Me\text{-}1\text{-}Ph$	(176/16)	552b
$4\text{-}SO_2CH_2Ac\text{-}3\text{-}Me\text{-}1,5\text{-}Ph_2$	110.5–11.5	1256
$4\text{-}SO_2CH_2Ph\text{-}3\text{-}Me\text{-}1,5\text{-}Ph_2$	162	446, 506a
$5\text{-}SO_2CH_2Ph\text{-}3\text{-}Me\text{-}1\text{-}Ph$	92	552b
$4\text{-}SO_2Ph\text{-}1\text{-}Ph$	145.5–6.0	323a
$4\text{-}SO_2Ph\text{-}5\text{-}NH_2\text{-}1\text{-}CONH_2\text{-}3\text{-}CN$	194–5	1270
$5\text{-}SO_2Ph\text{-}3\text{-}Me\text{-}4\text{-}N_2Ph\text{-}1\text{-}Ph$	145	735h
$4\text{-}SO_2C_6H_4\text{-}p\text{-}Me\text{-}1,5\text{-}Ph_2$	164	1266
$4\text{-}SO_2C_6H_4\text{-}p\text{-}Me\text{-}5\text{-}C_6H_4\text{-}p\text{-}NO_2\text{-}1\text{-}Ph$	215.0–16.5	1437
$4\text{-}SO_2C_6H_4\text{-}p\text{-}Me\text{-}5\text{-}NH_2\text{-}1\text{-}CONH_2\text{-}3\text{-}CN$	191.0–2.5	1270
$1\text{-}Ph\text{-}4\text{-}SO_2(1'\text{-}Phpyrazolyl\text{-}4')$	179	126c
$3,5\text{-}Me_2\text{-}1\text{-}Pr\text{-}4\text{-}SO_2(3',5'\text{-}Me_2\text{-}1'\text{-}Prpyrazolyl\text{-}4')$	(202–5/10)	126c
$3,5\text{-}Me_2\text{-}1\text{-}Ph\text{-}4\text{-}SO_2(3',5'\text{-}Me_2\text{-}1'\text{-}Phpyrazolyl\text{-}4')$	184–5	126c
$1,3,5\text{-}Me_3\text{-}4\text{-}SO_2(1',3',5'\text{-}Me_3pyrazolyl\text{-}4')$	99–100	126c
$1,5\text{-}Ph_2\text{-}3,4\text{-}(SO_2C_6H_4\text{-}p\text{-}Me)_2$	233–4	1437
Other S-containing groups		
$1,3,5\text{-}Me_3\text{-}4\text{-}SCl$	61–2	126c
$3\text{-}Me\text{-}1,5\text{-}Ph_2\text{-}4\text{-}SCl$	Gum	1256
$3,5\text{-}Me_2\text{-}4\text{-}SO_2Cl$	98	898l
$3\text{-}Me\text{-}1,5\text{-}Ph_2\text{-}4\text{-}SO_2Cl$	124.0–5.5	446, 1256
$3,5\text{-}Me_2\text{-}4\text{-}SCN$	209	898l
$1\text{-}Ph\text{-}4\text{-}SCN$	60–2	59a
$3\text{-}Me\text{-}1\text{-}Ph\text{-}5\text{-}SSO_2Ph\ (?)$	Oil	330f, 1354d
$3\text{-}Me\text{-}1\text{-}Ph\text{-}5\text{-}SNEt_2$	95–6	1256
Imidazoles		
Thiols and thioethers		
2-SH	225–7	1438a
2-SH-1-Me	143	1439, 1440
2-SH-4-Me	244–5	374a, 1438a
$2\text{-}SH\text{-}1,4\text{-}Me_2$	211–12	1175
$2\text{-}SH\text{-}1,5\text{-}Me_2$	261–2	1175
$2\text{-}SH\text{-}4,5\text{-}Me_2$	270	978a

Table A.22. Continued

	M.p. (b.p./mmHg)/°C	Ref.
Thiols and thioethers		
2-SH-4-i-Pr	150	1441
2-SH-4-i-Bu	188–9	1438b, 1441
2-SH-4-t-Bu	234–5	1441
2-SH-4-C_5H_{11}	114–15	1438b, 1441
2-SH-4-C_6H_{13}	115–16	1441
2-SH-5-Me-4-$CH_2CH=CH_2$	238–9	373b
2-SH-1-Me-5-C_5H_{11}	143–4	1438b
2-SH-1-Ph	181–2	1175, 1440
2-SH-4-Ph	267	507, 1165, 1442
2-SH-1-CH_2Ph	144–5	363a
2-SH-4-CH_2Ph	221–2	1438b, 1441
2-SH-5-Me-4-CH_2Ph	279–80	1443
2-SH-4-Me-1-Ph	190–1	1175
2-SH-4-Me-5-Ph	290–5	1443
2-SH-4,5-Ph_2-1-o-MeC_6H_4	288–9	1444
2-SH-4,5-Ph_2-1-p-MeC_6H_4	319–20	1444
4-SH-1-Bu-2,5-Ph_2	178–9	1446
4-SH-1,5-Ph_2-2-i-Pr	225–7	1446
4-SH-1,2,5-Ph_3	241	1446
4-SH-1-CH_2CO_2Et-2,5-Ph_2	182	1446
4-SH-1-Bu-2-i-Pr-5-C_6H_4-o-Me	173–5	1446
4-SH-1-Bu-2-Ph-5-C_6H_4-o-Me	172–6	1446
4-SH-1-Bu-2-i-Pr-5-C_6H_4-p-Me	216–20	1446
4-SH-1-Bu-2-Ph-5-C_6H_4-p-Me	168–70	1446
4-SH-1-Bu-2-i-Pr-5-C_6H_4-p-OMe	220–6	1446
4-SH-1-Bu-2-Ph-5-C_6H_4-p-OMe	177–9	1446
4-SH-1-Bu-2-i-Pr-5-CH_2CHMe_2	162–4	1446
4-SH-1-Bu-2-i-Pr-5-Me	158–60	1446
4-SH-1-Bu-2-Ph-5-Me	168–71	1446
2-SMe	139	1438a
2-SMe-1-Me	(225)	1440
2-SMe-1-Ph	54	1440
5-SMe-4-NO_2	187–9	544
5-SMe-1-Me-4-NO_2	119–20	544
2-SEt-4-Me	69–71	374a
2-SEt-1-Ph	Picrate 119–20	1175
2-SEt-1,4-Me_2	Picrate 105	1175
5-SEt-4-NO_2	196–8	544
5-SEt-1-Me-4-NO_2	68–70	544
5-SEt-2-i-Pr-4-Pr	121.0–2.5	1445a
2-SPr-4,5-Ph_2	174	1444
5-SPr-4-NO_2	159–61	544
5-SPr-1-Me-4-NO_2	43–4	544
5-SBu-4-NO_2	215–16	544
5-SBu-1-Me-4-NO_2	45–6	544
5-SBu-2-i-Pr-4-Pr	Oil	1445a
2-SCH_2Ph	153	1436
2-SCH_2Ph-4-Pr	183–4	1441
2-SCH_2Ph-4-Ph	176.5–7.5	1442
2-SCH_2Ph-4,5-Ph_2	185–6	1444
2-SCH_2Ph-4,5-Ph_2-1-p-MeC_6H_4	191	1444
4-SMe-1,2-Me_2-5-Ph	121	1446

Table A.22. Continued

	M.p. (b.p./mmHg)/°C	Ref.
Thiols and thioethers		
4-SCH$_2$Ph-1-Bu-2-i-Pr-5-Ph	(200–10/0.5)	1446
4-SCH$_2$Ph-1,2-Me$_2$-5-Ph	(200/0.25)	1446
4-SCOPh-1-Bu-2-i-Pr-5-Ph	(242–4/0.2)	1446
4-SCOPh-1,2-Me$_2$-5-Ph	168–71	1446
5-SCH$_2$Ph-4-NO$_2$	212–13	544
5-SCH$_2$Ph-1-Me-4-NO$_2$	113–15	544
5-SCH$_2$Ph-1-Me-4-NH$_2$.HCl	181–2	544
5-SCH$_2$Ph-2-i-Pr-4-Pr	88–9	1445a
5-SPh-2-i-Pr-4-Pr	155.0–5.5	1445a
5-SPh-2-Pr-4-i-Pr	156.5–7.5	1445a
5-SPh-2,4-Pr$_2$	126.0–6.5	1445a
5-SPh-2,4-i-Pr$_2$	160.0–0.5	1445a
5-SPh-2-Et-4-Pr	119–20	1445a
5-SPh-2-Et-4-i-Pr	125–7	1445a
5-SPh-4-Et-2-i-Pr	181.5–2.5	1445a
5-SPh-4-i-Pr-2-Me	164.5	1445a
5-SPh-4-Et-2-Me	166.5–7.0	1445a
5-SPh-2,4-Et$_2$	159.5–61.5	1445b
5-SPh-4-Me-2-Pr	188.5–9.5	1445b
5-SPh-4-Et-2-Pr	163.5–4.5	1445b
5-SPh-2-i-Pr-4-Me	201.5–3.0	1445b
5-SPh-1-Me-4-NO$_2$	77–8	544
5-SPh-1-Me-4-NH$_2$.HCl	210–11	544
5-SC$_6$H$_4$-o-NH$_2$-1-Me-4-NO$_2$	113–15	544
5-SC$_6$H$_4$-o-Me-2-i-Pr-4-Pr	148.0–8.5	1445a
5-SC$_6$H$_4$-p-Me-2-i-Pr-4-Pr	165–6	1445a
5-SC$_6$H$_4$-p Cl-2-i-Pr-4-Pr	178.5–80.0	1445a
5-SC$_6$H$_4$-p-OMe-2-i-Pr-4-Pr	167.0–7.5	1445a
5-SC$_6$H$_4$-p-NO$_2$-2,4-Pr$_2$	193.5–4.0	1445a
5-S(α-C$_{10}$H$_7$)-2-i-Pr-4-Pr	184.5–5.0	1445a
5-S(β-C$_{10}$H$_7$)-2-i-Pr-4-Pr	182.0–3.5	1445a
Sulphonic acids, sulphones, and sulphonamides		
2-SO$_3$H	303	543
2-SO$_3$H-4-Me	280	374a
2-SO$_3$H-4,5-Ph$_2$	315–18	454
2-SO$_3$H-1,4-Me$_2$-5-NO$_2$	293	412b
4-SO$_3$H	307	543, 1011
4-SO$_3$H-2-Me	279	453a
4-SO$_3$H-5-Me	278	543
4-SO$_3$H-1-CH$_2$CO$_2$H-2-Me-5-Ph	>270	1446
4-SO$_3$H-1-Bu-2-i-Pr-5-Ph	>230	1446
4-SO$_3$H-2-Et-4-Ph	>270	1446
4-SO$_3$H-5-Br	280	374a
4-SO$_3$H-5-Br-2-Me	266	11, 453a
4-SO$_3$H-5-Br-1-Me	284	374b
4-SO$_3$H-5-Ph	>300	1288
4-SO$_3$H-5-NO$_2$	300	374b
4-SO$_3$H-1-Me-5-NO$_2$	299–300	1101a
5-SO$_3$H-1-Me-4-Br	256	374b
5-SO$_3$H-1-Me-4-NO$_2$	254	374b
4,5-(SO$_3$H)$_2$		453a
4,5-(SO$_3$H)$_2$-2-Me	280	453a

Table A.22. Continued

	M.p. (b.p./mmHg)/°C	Ref.
Sulphonic acids, sulphones, and sulphonamides		
2-SO$_2$Me-1-Ph	115–16	1440
2-SO$_2$Me-1-Me-5-NO$_2$	91.0–2.5	607b
5-SO$_2$Me-4-NO$_2$	276–7	544
5-SO$_2$Me-1-Me-4-NO$_2$	160–1	544
5-SO$_2$Me-4-NH$_2$.HCl	193–5	544
5-SO$_2$Et-4-NO$_2$	246.5–7.0	544
5-SO$_2$Et-1-Me-4-NO$_2$	158–60	544
5-SO$_2$Pr-4-NO$_2$	215–17	544
5-SO$_2$Pr-1-Me-4-NO$_2$	165–6	544
5-SO$_2$Bu-1-Me-4-NO$_2$	82–4	544
2-SO$_2$Ph-1-Me	164–5	607b
5-SO$_2$Ph-1-Me-4-NO$_2$	130–3	544
5-SO$_2$CH$_2$Ph-4-NO$_2$	252–6	544
5-SO$_2$CH$_2$Ph-1-Me-4-NO$_2$	161.0–2.5	544
5-SO$_2$CH$_2$Ph-4-NH$_2$.HCl	208–10	544
5-SO$_2$CH$_2$Ph-1-Me-4-NH$_2$.HCl	221–2	554
4-SO$_2$C$_6$H$_4$-p-Me-1,5-Ph$_2$	188.5–9.0	1448
4-SO$_2$C$_6$H$_4$-p-Me-1-Ph-5-C$_6$H$_4$-p-NO$_2$	240	1448
4-SO$_2$C$_6$H$_4$-p-Me-5-Ph-1-C$_6$H$_4$-p-NO$_2$	232	1448
4-SO$_2$C$_6$H$_4$-p-Me-5-Ph-1-C$_6$H$_{11}$	215	1448
4-SO$_2$C$_6$H$_4$-p-Me-5-C$_6$H$_4$-p-NO$_2$-1-C$_6$H$_{11}$	264.5–5.5	1448
2-SO$_2$NH$_2$-4,5-(CO$_2$Et)$_2$	235–6	1101a
2-SO$_2$NH$_2$-4,5-(CONH$_2$)$_2$	>300	1101a
4-SO$_2$NH$_2$-5-Br		1392
4-SO$_2$NH$_2$-5-NH$_2$		1392
4-SO$_2$NH$_2$-5-NO$_2$	261–2	1101a
4-SO$_2$NH$_2$-1-Me-5-NO$_2$	208–10	1101a
5-SO$_2$NH$_2$-1-Me-4-NO$_2$	176–7	1101a
1,2,3-Triazoles		
4-SH-1-Me	124–5	1326d
5-SH-1-Me	102	1326d
4-SH-1-CH$_2$Ph	72–3	1326d
4-SMe-1-Me		1326d
5-SMe-1-Me	Oil	1326e
4-SMe-1-CH$_2$Ph	57–60	1326d
4-SCH$_2$Ph-1-CH$_2$Ph	71	1326e
5-SCH$_2$Ph-1-Me	Oil	1326e
4-SCOPh-1-CH$_2$Ph	153–4	1326e
5-SCOPh-1-Me	152–3	1326e
1,2,4-Triazoles		
Thiols[a] and thioethers		
3-SH	216	1118b, c, 1449
3-SH-1-Me	218–20	1118c
3-SH-5-Me	283–4	1455
3-SH-1-Et	153–5	1118c
3-SH-2-Et	85–6	1118c
3-SH-5-Et	248–51	1450
3-SH-5-NH$_2$	300–2	1198
3-SH-5-NHPh	286–90	1191a
3-SH-5-OH	202	1451e

Table A.22. Continued

	M.p. (b.p./mmHg)/°C	Ref.
Thiols[a] and thioethers		
3-SH-4-Bu	202–3	1118c
3-SH-5-Pr	208–10	1450
3-SH-1-C_6H_{11}	183–5	1118c
3-SH-2-C_6H_{11}	133–5	1118c
3-SH-4-C_6H_{11}	166–7	1118c
3-SH-5-C_6H_{11}	230–2	1118c
3-SH-5-SMe	254	979
3-SH-4-Et-5-Me	139	613c
3-SH-1-Ph	189	1196b
3-SH-1,5-Me$_2$	277–8	1455
3-SH-4,5-Me$_2$	210; 215–16	613c, 1455
3-SH-5-Me-1-Ph	182	613c
3-SH-5-Me-4-Ph	220	613c
3-SH-5-(α-pyridyl)	270–2	1452a
3-SH-4-Me-5-(α-pyridyl)	210–3	1452a
3-SH-5-Ph	257–60	86a, 1191d
3-SH-4-(β-pyridyl)	233–5	1118c
3-SH-4-(γ-pyridyl)	171–2	1118c
3-SH-4-Me-5-Ph	166	1331
3-SH-4,5-Ph$_2$	282	1331
3-SH-1,5-Ph$_2$	188	1451c
3-SH-4-COPh-5-Ph	235	1451c
3-SH-1-CH_2Ph	211–12	1118c
3-SH-2-CH_2Ph	156–7	1118c
3-SH-5-C_6H_4-*p*-Cl	292–6	1191d
3-SH-5-C_6H_4-*p*-OMe	256–9	1191d
5-SH-1,3-Me$_2$	168–9	1455
5-SH-1-Ph	178	1451a, 1454
5-SH-2-Me-3-(α-pyridyl)	163–4	1452a
5-SH-1-Me-3-(α-pyridyl)	255–60	1452a
3,5-(SH)$_2$	196	979
3,5-(SH)$_2$-1-Ph	216	1451b
3-SMe-1-Ph	43 (194–6/15)	613c
3-SMe-5-Me	120–1	1455
3-SMe-5-Me-1-Ph	(149–50/2)	613c
3-SMe-5-NHPh	187–8	1191a
3-SMe-4-Me-5-Ph	138	86b
3-SMe-5-Ph	164	1453
3-SMe-1,5-Me$_2$	66–7	1455
3-SMe-4,5-Me$_2$	55–7	613c, 1455
3-SMe-5-Me-4-Ph	119	613c
3-SMe-5-Me-4-Et	(186–8/10)	613c
3-SMe-4-Me-5-(α-pyridyl)	68–70	1452a
3-SMe-5-(α-pyridyl)	140–1	1452a
3-SMe-5-C_6H_4-*p*-OMe	126	86a
3-SMe-4,5-Ph$_2$	165–6	1453
3-SMe-5-Ph-4-C_6H_4-*o*-Me	130	1453
3-SMe-5-Ph-4-C_6H_4-*p*-Me	176	1453
5-SMe-1,3-Me$_2$	48–9	1455
3-SEt-4,5-Me$_2$	(206/12)	613c
3-SEt-5-Me-1-Ph	(157–9/0.1)	613c
3-SEt-5-Me-4-Ph	106	613c
3-SEt-5-Ph	164	1453

Table A.22. Continued

	M.p. (b.p./mmHg)/°C	Ref.
Thiols[a] and thioethers		
3-SEt-4,5-Ph$_2$	148	1453
3-SEt-5-Ph-4-C$_6$H$_4$-o-Me	107	1453
3-SEt-5-Ph-4-C$_6$H$_4$-p-Me	148	1453
3-SCH$_2$Ph-1-Ph	64	1451a, 1454
3-SCH$_2$Ph-5-Ph	160	1451d
3-SCH$_2$Ph-1,5-Ph$_2$	100	1451c
3-SCH$_2$Ph-5-Ph-4-COPh	174	1451d
3-SCH$_2$Ph-5-NH$_2$	109–11	1198
3-SCH$_2$C$_6$H$_4$-p-Cl-5-NH$_2$	142–4	1198
3-S(5'-NH$_2$-1',2',4'-triazolyl-3')-5-NH$_2$	232–6	1198
3,5-(SMe)$_2$	91	979
3,5-(SCH$_2$Ph)$_2$-4-COPh	91	1451b
Sulphonic acids, sulphones, sulphoxides, sulphonyl chlorides and sulphonamides		
3-SO$_3$H	342	1118b, c
3-SO$_3$H-1-Me	330–2	1118c
3-SO$_3$H-2-Me	276–8	1118c
3-SO$_3$H-4-Me	225–58	1118c
3-SO$_3$H-5-Me	350	1118b
3-SO$_3$H-1-Et	318–20	1118c
3-SO$_3$H-2-Et	175–7	1118c
3-SO$_3$H-4-Et	310–12	1118c
3-SO$_3$H-5-Et	339–41	1118c
3-SO$_3$H-1-C$_6$H$_{11}$	286–7	1118c
3-SO$_3$H-1-Ph	357	1118b
3-SO$_3$H-5-Ph	353	1118b
3-SO$_3$H-1-CH$_2$Ph	307–8	1118c
3-SO$_3$H-5-CH$_2$Ph	264–6	1118c
3-SO$_3$H-4-(α-pyridyl)	228–30	1118c
3-SO$_3$H-4-(β-pyridyl)	259–61	1118c
3-SO$_3$H-4-(γ-pyridyl)	307–8	1118c
3-SO$_3$H-5-(α-pyridyl)	>350	1118c
3-SO$_3$H-5-(β-pyridyl)	>360	1118b
3-SO$_3$H-5-(γ-pyridyl)	>370	1118b
3-SO$_2$Me-5-Ph	160	86a
3-SO$_2$Me-5-NHPh	246–8	1191a
3-SO$_2$CH$_2$Ph	147–9	1118c
3,5-(SO$_2$Me)$_2$-1-Ph	182	1451b
3,5-(SO$_2$Me)$_2$-1-C$_6$H$_4$-p-NO$_2$	234	1451b
3-SOCH$_2$Ph	134–6	1118c
3-SOMe-5-NHPh	240–2	1191a
3-SO$_2$Cl	130	1118b
3-SO$_2$NH$_2$	224.5–5.5	770b
3-SO$_2$NH$_2$-1-Me	182–3	1118c
3-SO$_2$NH$_2$-5-Et	215–17	1118c
3-SO$_2$NH$_2$-1-CH$_2$Ph	182–3	1118c
3-SO$_2$NH$_2$-1-C$_6$H$_{11}$	113–15	1118c
3-SO$_2$NH$_2$-5-CH$_2$Ph	207–8	1118c
3,5-(SO$_2$NH$_2$)$_2$-4-Ph	242.5–3.5	770b
Tetrazoles		
Thiols and thioethers		
5-SH	205	24c

Table A.22. Continued

	M.p. (b.p./mmHg)/°C	Ref.
Thiols and thioethers		
5-SH-1-Me	126	83d, 484, 840
5-SH-1-CH$_2$CH=CH$_2$	69	1093c
5-SH-1-Ph	152	24a, 1093c
5-SH-1-C$_6$H$_4$-p-Me	150–1	1064c
5-SH-1-C$_6$H$_4$-o-Me	129	1064d
5-SH-1-C$_6$H$_3$-o,p-Me$_2$	141	1093c
5-SH-1-(β-naphthyl)	164	1093c
5-SMe	144–6	24c, 1180
5-SMe-1-Ph	80–2	83d, 590b
5-SMe-1-C$_6$H$_4$-p-NO$_2$	147–8	590b
5-SMe-1-C$_6$H$_4$-p-NH$_2$	160–1	590b
5-SEt	86–8	1180
5-S-i-Pr	61–3	1180
5-SCH$_2$CH=CH$_2$	65–7	1180
5-SBu	96–7	1180
5-SCH$_2$Ph	134–5	1180
5-SPh	92–3	1180
5-SC$_6$H$_4$-p-NH$_2$	199–200	1180
5-SC$_6$H$_4$-p-NO$_2$	189–91	1180
5-SCH$_2$Ph-1-Ph	71	83d
Sulphonic acids, sulphones and sulphonamides		
5-SO$_3$H	Syrup	24c
5-SO$_3$H-1-Me	(K$^+$ salt) 321	83d
5-SO$_3$H-1-Ph	(K$^+$ salt) 242	24a
5-SO$_3$H-1-C$_6$H$_4$-o-Me	129	1064d
5-SO$_2$Me	120	24c
5-SO$_2$Me-1-Ph	84	83d
5-SO$_2$NH$_2$-1-Me	139–40	770b
5-SO$_2$NH$_2$-1-Ph	157–8	770b

(a) Compounds listed as thiols may exist as thiones.

Appendix 3
References to work overlooked in the preparation of the main text, or appearing later (up to September 1974)

Chapter 2

Section 2.1. X-Ray crystallographic studies of pyrazole[1470], anti-pyrine[1471], histamine[1472], and amino-mercapto-1,2,4-triazoles[1473].

Dipole moments of 1,2,3-triazole and its two *N*-methyl derivatives in benzene at 25 °C (showing 1,2,3-triazole to contain 83% of 2*H*-1,2,3-triazole)[1474].

Section 2.2.1. Ultraviolet spectroscopy of nitroso- and nitro-pyrazoles [1475], of 4-hydroxypyrazoles in relation to tautomerism[1476a], and of 4-amino-1,2,3-triazoles[1470a,b].

Section 2.2.2. Infrared spectroscopy of 4-hydroxypyrazoles[1476b], 4-amino-1,2,3-triazoles[1477a,b], and 3,5-dinitro-1,2,4-triazole[1478]. Microwave spectrum of 1-deutero-1*H*-1,2,3-triazole[1479].

Section 2.2.3. [1]H-n.m.r. spectroscopy of 4-hydroxypyrazoles[1476a], substituted triazoles and tetrazoles[1480], 4-amino-1,2,3-triazoles[1477a, b], and azole anions[1481].

[13]C-n.m.r. spectroscopy of phenylpyrazoles and 1,2,3-triazoles[1482] and histidine[1483]. Use of ^{13}C-n.m.r. coupling constants in studying azole tautomerism (in CDCl$_3$ at 32 °C 1,2,3-triazole exists as 2 : 3 mixture of 2*H*- and 1*H*-1,2,3-triazole)[1484]. Use in determining site of *N*-methylation[1485].

[14]N-n.m.r. spectroscopy of azoles[1486].

Section 2.3.1. Acidity constants of 4-hydroxypyrazoles (in relation to their tautomerism)[1476b], diazopyrazoles[1487], nitroso- and nitro-pyrazoles[1475], 1-methyltetrazole (from [1]H-n.m.r. spectroscopy pK_a ~ -2.4)[1488], 1-aryl-5-methyltetrazoles[1489], and 4-amino-1,2,3-triazoles[1477a,b].

Section 2.3.2. Copper(I) and silver(I) complexes of pyrazoles[1490]. Imidazole complexes with tetracyanoethylene[1491].

Section 2.4. M.O. calculations, using *ab initio* wave functions, giving charge distributions and moments of azoles[1492].

Chapter 3

Section 3.1. Friedel–Crafts aroylation of pyrazoles[1493].

Section 3.3. Reaction of antipyrine (at C-4) with isoquinoline (at C-1) in presence of benzoyl chloride[1494].

Section 3.4. Pyrazol-5-ones with aldazines give 4,4'-di-(pyrazol-5-one-yl) alkanes[1495].

Section 3.5. The deuteration of imidazole at C-2 has been re-examined [1496] over the range pD = 1 to 16. The rate profile showed important differences at pD < 6 and pD > 14 from that reported by Vaughan *et al.* (see main text). The mechanism involving ylide formation at C-2 from the imidazolium cation by action of OD⁻ uniquely is supported. For deuteration at C-2, C-4 and C-5 in 1-methylimidazole (pD = 0 to 13) this same kind of mechanism is operative. At pD > 13 ylide formation from 1-methylimidazole itself makes a significant contribution to exchange at C-2 and C-4. In strong acids exchange at all three positions proceeds by the S_E mechanism. Results are reported for 1,3-dimethyl-imidazolium. Relative reactivities at C-2, C-4, and C-5 of 1-methyl-imidazole are 54,500 : 1.6 : 1 (ylide mechanism), and 1 : 73 : 120 (S_E reaction of cation).

Hydrogen–deuterium exchange at C-5 in 5-deutero-1-methyltetrazole at 67.6 °C and pH = –2.8 to 7.2 has been studied[1488]. At high pH an ylide is formed from OH⁻ and the substrate, and below pH = 2 the ylide arises from the protonated substrate. At pH = 3 to 5 an acidity independent term of uncertain mechanism is dominant. The catalytic effect of copper(II) and zinc(II) on the ylide mechanism is marked.

Section 3.6. Diazo-coupling at C-4 of 2-alkylimidazoles[1497].

Section 3.7.3. 2-(2'-Hydroxy-2'-propyl)imidazole is di-iodinated at C-4 and C-5[1498].

Kinetics of iodination of 1-phenylpyrazol-5-ones[1499].

Section 3.8. Nuclear and side-chain lithiation of methylpyrazoles[1500].

Section 3.9. Nitration at C-4 of t-butylpyrazoles[1475].

The poly-nitration of imidazole can be achieved[1501]. With hot mixed acid imidazole gave 4-nitro- (30%) and 4,5-dinitro-imidazole (20%) and 2-methylimidazole behaved analogously. Similarly 4-nitro- gave 4,5-dinitro-, and 2,4-dinitro- gave 2,4,5-trinitro-imidazole (26%), but 4,5-dinitroimidazole could not be nitrated. With nitric acid in acetic acid-acetic anhydride, 4-nitro- gave one N,4-dinitroimidazole which in sulphuric acid gave 4,5-dinitroimidazole.

Further kinetic studies of the nitration of 1-phenylpyrazole and its nitro-derivatives and quaternary salts in sulphuric acid[1502], and of 1-phenylpyrazol-5-ones[1503]. Nitration of phenyl-1,2,3-triazoles[1504].

2-(2'-Hydroxy-2'-propyl)-4,5-diiodoimidazole is nitrated at C-4 with displacement of iodine[1498].

Section 3.13.1. Imidazole-4-carboxylic acid and dimethyl sulphate gave the zwitterion from 1,3-dimethylimidazolium 4-carboxylic acid, the natural product norzooanemonin [1505]. Imidazole adds to dicyclohexyl-carbodiimide giving the guanidine [1506].

1,2,3- and 1,2,4-triazoles in acid-catalysed reactions with acetylated ribofuranose give mixtures of *N*-ribofuranosyl derivatives [1507].

Methylation of 3-amino-1,2,4-triazole gives varying proportions of 1-, 2-, and 4-methyl compounds (but no methylamino compound) depending on presence or absence of base [1508]. 5-Aminotetrazole and phenacyl bromide give 5-amino-1-phenacyl-1*H*-1,2,3,4-tetrazole and some 2,4-diphenacyl-5-iminotetrazole [1509].

1,2,3-Triazoles tend to undergo substitutive alkylation at N-1, and their anions (in additive alkylation also) mainly at N-2 [1510].

Section 3.13.2. 1-Acetyl-4-methyl-2-R-imidazole (R = H, Me, Et) (see §3.13.5, below) are quaternised with benzyl bromide. Hydrolysis gives 1-benzyl-5-methyl-2-R-imidazoles [1511].

Section 3.13.3. *N*-Dinitrophenylation of azoles [1512]. *N*-Picrylation of imidazoles [1513]. 2-Aminoimidazole reacted at both cyclic and exocyclic nitrogen atoms. With picryl chloride, 4-amino-1,2,3-triazole gave the 4-picrylamino compound; this with picryl fluoride gave 1-picryl-4-picryl-amino-1*H*-1,2,3-triazole [1514]. Sodium 5-phenyltetrazolate and di-*p*-tolyliodonium bromide gave mixtures of 5-phenyl-*N*-tolyltetrazoles, possibly by benzyne addition [1515].

Section 3.13.4. Addition of pyrazol-5-ones to methyl acrylate [1516] and acrylonitrile [1517]. Addition of 1,2,3-triazole to acetylene [1518], and of 5-aminotetrazole to methyl propiolate [1519] (see §3.14 for reaction of 2-amino-1-methylimidazole with dimethyl acetylenedicarb-oxylate).

Section 3.13.5. *C*-, *O*-, and *N*-Acetylation of 3-methylpyrazol-5-one, involving some thermal acetyl group migrations [1520]. 4-Methyl-2-R-imidazoles give 1-acetyl-4-methyl-2-R-imidazoles (R = H, Me, Et) on acetylation [1511].

3-Amino-1,2,4-triazole with organic isocyanates and isothiocyanates reacts at cyclic or exocyclic nitrogen depending on the conditions. *N*→*N'* acyl migrations occur [1521]. 3-Amino-1,2,4-triazole with acetyl chloride gives 1-acetyl-5-amino-1*H*-1,2,4-triazole, which can rearrange to the 2-acetyl isomer [1522]. *O*- and *N*-Acylation of sodium salt of 5-hydroxy-1-phenyltetrazole [1523].

Section 3.13.6. Kinetics of reactions of arenesulphonyl chlorides with imidazole [1524].

Section 3.14. With dimethyl acetylenedicarboxylate, 2-amino-1-methyl-imidazole gives, as well as the expected imidazopyrimidine derivative, 5% of dimethyl 2-amino-1,3-diazepine-5,6-dicarboxylate by ring-expansion [1525].

Chapter 4

Section 4.2. Kinetics of reaction of piperidine with 5-chloro-1-methyl-3-nitro-1H-1,2,4-triazoles[1526].

Section 4.4. 4-Azidopyrazoles from the diazonium compounds[1527]. 3-Azido-1,2,4-triazoles similarly[1528].

Section 4.7. 1-Benzyl-4-bromo- and 1-benzyl-4-iodo-2-methylpyrazolium are both hydrolysed by caustic soda to a mixture of 1-benzyl-2-methylpyrazol-3- and -5-one (ratio 1.62). 4-Bromo-, 3-bromo- and 5-bromo-1-methyl-2-phenylpyrazolium give on hydrolysis 1-methyl-2-phenylpyrazol-5-, -3- and -5-one, so that in the case of 4-bromo-1-methyl-2-phenylpyrazolium hydrolysis cannot involve bromine migration in an ylide (a conclusion supported by H–D exchange rates), and an aryne mechanism is also excluded. The only likely mechanism is supposed to be one of addition (of OH⁻) followed by elimination (of HBr)[1529]. Reaction of pyrazolium salts with N-bromoacetamide OMe⁻ has been studied[1530].

Section 4.8. 3-Nitropyrazoles[1531] and 2-nitroimidazoles[1532] from the corresponding diazonium salts.

Chapter 5

Section 5.2. 4,5-Dicyano-2-diazo-2H-imidazole and benzotrifluoride give isomeric 4,5-dicyano-2-trifluoromethylphenylimidazoles, and difluoro-(4,5-dicyano-2-fluoro-1-imidazolyl)-phenylmethane[1533].

Section 5.3. 4,5-Dicyano-2-diazo-2H-imidazole and 1,2-dichloroethane give 2-chloro-1-β-chloroethyl-4,5-dicyanoimidazole[1534].

Chapter 6

Section 6.1. Hydride reduction of quaternary pyrazolium salts[1534–5] (with ring-opening in some cases).

Section 6.3. Thermolytic ring-opening of 4-azidopyrazoles[1527]. 5-p-Toluenesulphonylaminotetrazole with p-toluenesulphonyl chloride in pyridine at room temperature gives nitrogen and $MeC_6H_4SO_2N = = \bar{\overset{+}{C}}(NC_5H_5) = = NNHSO_2C_6H_4Me$[1536].

Section 6.6. Mass spectroscopy of trimethylsilylpyrazoles[1537], nitro-imidazoles[1538], amino- and azido-1,2,4-triazoles[1539], and 1,2,4-triazoline-5-thiones[1540].

Chapter 7

Photochemical transformation of 4-aroyl-1-aryl- and 1-aryl-4-formyl-pyrazoles into 1,2-bis(1-arylimidazol-4-yl)ethane-1,2-diols[1541].

Photochemical ring-opening of antipyrine and its derivatives[1542].

Chapter 8

CNDO/2 study of reactivity of ylides formed by de-protonation of azolium ions[1543].

Chapter 9

Section 9.1. Reaction in alkaline solution of 1-phenylimidazole-2-carbaldehyde with reagents such as nitromethane, malonic acid and acetophenone[1544]. Dinitrogen tetroxide converts the oximes of 4-formylimidazoles into the trinitromethyl derivatives[1545].

In the presence of dithionite 1-(*o*-nitrophenyl)imidazole-2-carbaldehyde cyclises to imidazo[1,2-*a*]quinoxaline[1546].

The alkylative decarboxylation of 1-carbalkoxypyrazoles requires a polar aprotic solvent and is subject to catalysis by nucleophiles[1547]. The transition metal catalysed reaction of 1,1′-carbonyldipyrazoles with aldehydes and ketones to form alkylidene dipyrazoles is sensitive to electronic and steric effects[1548].

Structure of 1-methyl-3,4- and 1-methyl-4,5-pyrazole-dicarbaldehydes [1549].

Reaction of 4-acetylpyrazoles with phosphorus pentachloride and sodamide in turn to give acetylenic pyrazoles[1550].

One or both carboxyl functions are replaced by CCl_3 when 4,5-dicarboxyimidazoles react with phosphorus pentachloride in thionyl chloride[1551].

Section 9.2. The 2-methyl group in 1,2-dimethyl-5-nitroimidazole is converted into cyanide by heating with nitrosyl chloride or a nitrogen oxide[1552]. The methyl group is active enough to *N*-alkylate pyridine (Ortoleva–King reaction) and to condense with carbonyl compounds [1553]. Reaction of 1-benzyl-2-methylimidazole with benzoyl chloride in acetonitrile–triethylamine results in the benzoylation of the methyl group. This is a surprising reaction when one considers that the 2-methyl group does not behave as if activated in lithiation[1554].

Section 9.3. Electrochemical reduction of 3-styrylpyrazoles does not produce radical anions. The products isolated are 3-(2-phenylethyl)-pyrazoles[1555]. Allylic rearrangements have been noted with some allyl derivatives of 1,2,4-triazol-5-thiones[1556].

Bromine with 3-(1-phenylpyrazol-4-yl)acrylic acid gives the β-bromo compound. Alkaline treatment converts this into the β-ketoacid by an addition–elimination reaction[1557].

2-Vinylimidazole-4,5-dicarboxylic acid readily polymerises[1558]. Under the influence of radical activators mercaptans add in anti-Markownikoff fashion to *N*-vinylimidazoles[1559].

The Mannich reaction converts ethynylpyrazoles into aminopropynyl-pyrazoles[1560].

Section 9.4. Imidazole diazonium fluoroborates decompose on irradiation to fluoroimidazoles, but thermal decomposition does not proceed in the same way[1561]. The diazonium fluoroborates are very stable compounds which form sublimable meso-ionic products when treated with base[1561]. The reduction of 2,2'-azodiimidazoles to 2-aminoimidazoles has been accomplished[1562]. A study has been made[1563] of the nitrosamines of 1,2,4-triazoles and tetrazoles in relation to structure, reduction to hydrazines, and thermal arylation. In an ESR study of radicals formed by γ-irradiation of 1-substituted 5-nitroamino- and -5-nitrosamino-tetrazoles only the former were found to suffer cleavage of the exocyclic N–N bond (to form NO_2. radicals)[1564].

Section 9.6. Azide–tetrazole tautomerism normally favours the azide in the azole series[1565-6], but in azole anions is displaced towards the tetrazole[1567].

Section 9.7. Although reduction failed using the Stephen reaction, lithium aluminium hydride, or Raney nickel, the catalytic hydrogenation of 5-cyano-1,2,3-triazoles over palladium in hydrochloric acid of a critical concentration gave the aldehydes[1477a].

Section 9.10. Tetrazoles may be alkylated selectively at N-1 when the 5-substituted-2-(tri-n-butylstannyl)tetrazoles are treated with alkylating agents[1567].

Section 9.13. Meso-ionic sulphonamidates have been prepared by base treatment of 4-mesyl- and 4-tosyl-aminopyrazolium methoperchlorates [1568]. Whereas the latter compounds are thermally stable the mesyl compounds rearrange when boiled in benzonitrile. An annular *N*-methyl group migrates to the exocyclic nitrogen.

Reduction of phototetrazolium salts with sodium amalgam, sodium dithionite, or sodium stannite produces tetrazolinyl radicals[1569].

Section 9.15. The kinetics of hydrolysis of 1-tosylimidazole and some quaternary salts have been studied as a function of pH and temperature [1570]. 4-Methylsulphinylpyrazoles are not desulphurised by aluminium amalgam, Raney nickel, or zinc dust–acetic acid treatment; the product is the 4-methylthiopyrazole[1571].

Section 9.16. The chromium(VI) oxide-3,5-dimethylpyrazole complex has been recommended as a reagent for the oxidation of alcohols to carbonyl compounds[1572]. A recently prepared gold(I) derivative of pyrazole appears to exist as a trimer analogous to trimeric pyrazole[1573].

The properties of 4-aryl-5-triphenylphosphonium-1,2,3-triazole ylides and 4-aryl-1,2,3-triazol-5-yltriphenylphosphoranes have been studied[1574]. The ylides are soluble in dipolar, aprotic solvents, are relatively stable, and are excellent nucleophiles.

1-Trimethylsilyl-1*H*-1,2,4-triazoles have been shown by [1]H- and ^{13}C-n.m.r. studies to undergo intermolecular exchange of the trimethylsilyl group (trimethylgermanyl and triethoxysilyl similarly)[1575].

References

1 Knorr, L. (a) *Ber. dtsch. chem. Ges. 16* (1883) 2597; (b) with Blank, A. *18* (1885) 311; (c) *Justus Liebigs Annln Chem. 279* (1894) 188; (d) with Laubman, H. *Ber. dtsch. chem. Ges. 22* (1889) 172; (e) *28* (1895) 696; (f) with Rabe, P. *Justus Liebigs Annln Chem. 293* (1896) 42.
2 Buchner, E. (a) *Ber. dtsch. chem. Ges. 22* (1889) 842; (b) p. 2165; (c) *Justus Liebigs Annln Chem. 273* (1893) 214; (d) with v. der Heide, C. *Ber. dtsch. chem. Ges. 35* (1902) 31; (e) with Hachumian, C. *35* (1902) 37; (f) with Fritsch, M. *Justus Liebigs Annln Chem. 273* (1893) 256.
3 *The Ring Index,* 2nd edition, American Chemical Society, 1960.
4 Laurent, A. *J. prakt. Chem. 35* (1845) 455.
5 Debus, H. *Justus Liebigs Annln Chem. 107* (1858) 199.
6 Wyss, G. (a) *Ber. dtsch. chem. Ges. 9* (1876) 1543; (b) *10* (1877) 1365; (c) *10* (1877) 1369; (d) *10* (1877) 1373.
7 Hantzsch, A. (a) *Justus Liebigs Annln Chem. 249* (1888) 1; (b) with Silberrad, O. *Ber. dtsch. chem. Ges. 33* (1900) 58.
8 Fischer, E. (a) with Troschke, H. *Ber. dtsch. chem. Ges. 13* (1880) 706; (b) with Knoevenagel, O. *Justus Liebigs Annln Chem. 239* (1887) 194.
9 Radziszewski, B. *Ber. dtsch. chem. Ges.* (a) *15* (1882) 1493; (b) p. 2706; (c) *16* (1883) 487; (d) *16* (1883) 747; *17* (1884) 1289.
10 Japp, F. R. and Robinson, H. H. *Ber. dtsch. chem. Ges. 15* (1882) 1268; Japp, F. R. p. 2140; *16* (1883) 284.
11 Light, L. and Pyman, F. L. *J. chem. Soc. 121* (1922) 2626.
12 Marckwald, W. (a) with Wohl, A. *Ber. dtsch. chem. Ges. 22* (1889) 1353; (b) *25* (1892) 2354; (c) with Bott, A. *29* (1896) 2914.
13 Bamberger, E. (a) *Ber. dtsch. chem. Ges. 25* (1892) 274; (b) *Justus Liebigs Annln Chem. 273* (1893) 311; (c) with de Gruyter P. *Ber. dtsch. chem. Ges. 26* (1893) 2385; (d) *Justus Liebigs Annln Chem. 305* (1899) 289.
14 Gabriel, S. (a) with Posner, T. *Ber. dtsch. chem. Ges. 27* (1894) 1037, 1141; (b) *41* (1908) 1926; (c) *Justus Liebigs Annln Chem. 350* (1906) 118.
15 Jänecke, E. *Ber. dtsch. chem. Ges. 32* (1899) 1095.
16 Pyman, F. L. (a) *J. chem. Soc. 97* (1910) 1814; (b) *99* (1911) 2172; (c) *109* (1916) 186; (d) *121* (1922) 2616; (e) *123* (1923) 3359.
17 v. Pechmann, H. (a) *Ber. dtsch. chem. Ges. 21* (1888) 2751; (b) *28* (1895) 1624; (c) with Wedekind, E. *28* (1895) 1693; (d) with Burkard, E. *33* (1900) 3590; (e) with Bauer, W. *42* (1909) 659; (f) *Justus Liebigs Annln Chem. 262* (1891) 265.
18 Bladin, J. A. *Ber. dtsch. chem. Ges.* (a) *18* (1885) 1544; (b) *18* (1885) 2907; (c) *19* (1886) 2598; (d) *23* (1890) 1810, 3785; (e) *25* (1892) 741; (f) *25* (1892) 1411; (g) *25* (1892) 1412; (h) *26* (1893) 545, 2736.
19 Dimroth, O. *Ber. dtsch. chem. Ges.* (a) *35* (1902) 1029; (b) p. 1038; (c) p. 4041; (d) with Taub, L. *39* (1906) 3912; (e) with Aickelin, H.

39 (1906) 4390.

20 Andreocci, A. (a) *Gazz. chim. ital. 19* (1889) 448; *Atti Accad. naz. Lincei* (1890) 209; (b) *Ber. dtsch. chem. Ges. 25* (1892) 225; (c) *Atti Accad. naz. Lincei* (1891) 458.

21 Widman, O. *Ber. dtsch. chem. Ges. 26* (1893) 2617.

22 Pellizzari, G. (a) with Bruzzo, M. *Atti Accad. naz. Lincei 10*I (1901) 363; (b) *10*II (1901) 297; (c) *11*I (1902) 20; (d) *Gazz. chim. ital. 24*II (1894) 222; (e) *32*I (1902) 189; (f) with Soldi, A. *35*I (1905) 373; (g) *41*II (1911) 20; (h) with Cuneo, G. *Ber. dtsch. chem. Ges. 27* (1894) Ref. 407.

23 Wedekind, E. *Ber. dtsch. chem. Ges.* (a) *29* (1896) 1846; (b) *31* (1898) 942.

24 Freund, M. *Ber. dtsch. chem. Ges.* (a) with Hempsel, H. *28* (1895) 74; (b) *29* (1896) 2483; (c) with Paradies, T. *34* (1901) 3110.

25 Thiele, J. *Justus Liebigs Annln Chem.* (a) *270* (1892) 1; (b) with Marais, J. T. *273* (1893) 144; (c) with Ingle, H. *287* (1895) 233.

26 Nasini, R. and Carrara, G. *Gazz. chim. ital. 24*I (1894) 256.

27 Hückel, W. *Theoretische Grundlagen der Organischen Chemie*, Leipzig, 1934.

28 Noe, F. F. and Fowden, L. *Biochem. J. 77* (1960) 543.

29 Schröter, H.-B., Neumann, D., Katritzky, A. R. and Swinbourne, F. J. *Tetrahedron 22* (1966) 2895.

30 Ehrlich, H. W. W. *Acta crystallogr. 9* (1956) 655.

31 Krebs Larsen, F., Lehmann, M. S., Søtofte, I. and Rasmussen, S. E. *Acta chem. scand. 24* (1970) 3248.

32 Nygaard, L., Nielsen, F. T., Kirchheimer, J., Maltesen, G., Rastrup-Andersen, J. and Sørensen, G. O. *J. molec. Spectrosc. 3* (1969) 491.

33 Reimann, C. W., Mighell, A. D. and Mauer, F. A. *Acta crystallogr. 23* (1967) 135.

34 Mighell, A. D., Reimann, C. W. and Santoro, A. *Acta crystallogr. B25* (1969) 595.

35 Vijayan, M. and Viswamitra, M. A. *Acta crystallogr. B24* (1968) 1067.

36 Will, G. *Z. Kristallogr. 129* (1969) 211.

37 Martinez-Carrera, S. *Acta crystallogr. 20* (1966) 783.

38 Visser, G. J. and Vos, A. *Acta crystallogr. B27* (1971) 1802.

39 Garfinkel, D. and Edsall, J. T. *J. Amer. chem. Soc. 80* (1958) 3807.

40 Donohue, J., Lavine, L. R. and Rollett, J. S. *Acta crystallogr. 9* (1956) 655.

41 Bennett, I., Davidson, A. G. H., Harding, M. M. and Morelle, I. *Acta crystallogr. B26* (1970) 1722.

42 Davies, D. R. and Blum, J. J. *Acta crystallogr. 8* (1955) 129.

43 van Remoortere, F. P. and Boer, F. P. *J. chem. Soc. (B)* (1971) 976.

44 Zimmermann, H. (a) *Justus Liebigs Annln Chem. 612* (1958) 193; (b) *Z. Elektrochem. 63* (1959) 601, 608; (c) *Angew. Chem. (int. Edn.) 3* (1964) 157; (d) with Geisenfelder, H. *Z. Elektrochem. 65* (1961) 368.

45 Deuschl, H. *Ber. Bunsen Ges. phys. Chem. 69* (1965) 550.

46 Goldstein, P., Ladell, J. and Abowitz, G. *Acta crystallogr. B25* (1969) 135.

47 Bolton, K., Brown, R. D., Burden, F. R. and Mishra, A. *Chem. Comm.* (1971) 873.

48 Isaacs, N. W. and Kennard, C. H. L. *J. chem. Soc. (B)* (1971) 1270.

49 Bryden, J. H. *Acta crystallogr.* (a) *8* (1955) 211; (b) *9* (1956) 874.

50 Palenik, G. J. *Acta crystallogr. 16* (1963) 596.

51 (a) McEwan, W. S. and Rigg, M. W. *J. Am. chem. Soc. 73* (1951) 4725; (b) Williams, M. M., McEwan, W. S. and Henry, R. A. *J. phys. Chem. 61* (1957) 261.

52 Hückel, W. (a) with Bretschneider, H. *Ber. dtsch. chem. Ges. 70* (1937) 2024; (b) with Jahnentz, W. *75* (1942) 1438; (c) with Datow, J. and Simmersbach, E. *Z. physik. Chem. A186* (1940) 129.

53 Goldschmidt, H. *Ber. dtsch. chem. Ges. 14* (1881) 1844.

54 Wallach, O. (a) *Justus Liebigs Annln Chem. 214* (1882) 309; (b) *235* (1886) 233; (c) *214* (1882) 257; (d) *Ber. dtsch. chem. Ges. 16* (1883) 534;

(e) *15* (1882) 644.

55 Hunter, L. (a) with Hayes, H. T. *J. chem. Soc.* (1941) 1; (b) with Marriott, J. A. p. 777; (c) (1945) 806; (d) in *Progress in Stereochemistry*, vol. 1, edited by W. Klyne, Butterworths, London, 1954.

56 Mingoia, Q. *Gazz. chim. ital. 61* (1931) 449.

57 Bacchetti, T., Salvatori, T. and Palladino, N. *Chim. Ind. (Milan) 46* (1964) 1337.

58 Hofmann, K. *Imidazole and its Derivatives*, part I, Interscience Publishers, New York, 1953.

59 Finar, I. L. *J. chem. Soc.* (a) with Godfrey, K. E. (1954) 2293; (b) (1955) 1205; (c) with Hurlock, R. J. (1957) 3024; (d) with Lord, G. H. (1957) 3314; (e) with Hurlock, R. J. (1958) 3259; (f) with Lord, G. H. (1959) 1819; (g) with Manning, M. (1961) 2733; (h) with Miller, D. B. (1961) 2769; (i) with Foster, T. *(C)* (1967) 1494; (j) with Rackham, D. M. *(C)* (1967) 2650. (k) Finar, I. L. and Mooney, E. F. *Spectrochim. Acta 20* (1964) 1269; (l) *J. chem. Soc. (B)* (1968) 725.

60 Balbiano, L. *Gazz. chim. ital.* (a) *18* (1888) 354; (b) *19* (1889) 128; (c) *19* (1889) 688; (d) with Marchetti, G. *23* (1893) 485. (e) Balbiano, L. *Ber. dtsch. chem. Ges. 23* (1890) 1448.

61 Rupe, H. (a) with Labhardt, H. *Ber. dtsch. chem. Ges. 33* (1900) 233; (b) with Metz, G. *36* (1903) 1092. (c) Rupe, H. and Hubner, A. *Helv. chim. Acta 10* (1927) 846.

62 Rojahn, C. A. *Ber. dtsch. chem. Ges.* (a) *55* (1922) 291; (b) *55* (1922) 2959; (c) *59* (1926) 607; (d) with Fegeler, H. *63* (1930) 2510.

63 v. Auwers, K. *Justus Liebigs Annln Chem.* (a) with Buschmann, W. and Heidenreich, R. *435* (1924) 277; (b) with Kohlhaas, W. *437* (1924) 36; (c) with v. Sass, L. Frhr. and Wittekindt, W. *444* (1925) 195; (d) with Conrad, J., Ernecke, A. and Ottens, B. *469* (1929) 57; (e) with Conrad, J. and Ernecke, A. *469* (1929) 82; (f) with Bergmann, F. *472* (1929) 287; (g) *508* (1934) 51.

64 Wiley, R. H. and Hexner, P. E. *Org. Synth. 31* (1951) 43.

65 Marchetti, G. (a) *Atti accad. naz. Lincei 22*I (1892) 356; (b) *Gazz. chim. ital. 22*II (1892) 368.

66 v. Auwers, K. *Ber. dtsch. chem. Ges.* (a) with Broche, H. *55* (1922) 3880; (b) with Schmidt, W. *58* (1925) 528; (c) with Ottens, B. *58* (1925) 2072; (d) with Hollmann, H. *59* (1926) 601; (e) with Stuhlmann, H. *59* (1926) 1043; (f) with Hollmann, H. *59* (1926) 1282; (g) with Mausolf, C. *60* (1927) 1730; (h) with Mauss, W. *61* (1928) 2411; (i) with Ungemach, O. *66* (1933) 1205; (j) with Ungemach, O. *66* (1933) 1690.

67 Claisen, L. and Roosen, P. (a) *Ber. dtsch. chem. Ges. 24* (1891) 1888; (b) *Justus Liebigs Annln Chem. 278* (1894) 374. (c) Claisen, L. *Justus Liebigs Annln Chem. 297* (1897) 37.

68 Ach, F. *Justus Liebigs Annln Chem. 253* (1889) 44.

69 Stoermer, R. *Ber. dtsch. chem. Ges.* (a) *36* (1903) 3986; (b) *40* (1907) 484; (c) with Johnannsen, D. *40* (1907) 3701. (d) Stoermer, R. and Martinsen, O. *Justus Liebigs Annln Chem. 352* (1907) 322.

70 Parham, W. E. *J. Am. chem. Soc.* (a) with Bleasdale, J. L. *72* (1950) 3843; (b) with Hasek, W. R. *76* (1954) 799. (c) Parham, W. E. and Aldre, I. M. *J. org. Chem. 25* (1960) 1259.

71 Smith, A. and Ransom, J. H. *Justus Liebigs Annln Chem. 289* (1896) 325.

72 Minunni, G. and D'Urso, S. *Gazz. chim. ital. 58* (1928) 691.

73 Bischler, A. *Ber. dtsch. chem. Ges. 26* (1893) 1881.

74 Hüttel, R. *Chem. Ber.* (a) with Jochum, P. *85* (1952) 820; (b) with Büchele, F. and Jochum, P. *88* (1955) 1577; (c) with Büchele, F. *88* (1955) 1586; (d) with Kratzer, J. *92* (1959) 2014; (e) with Schwarz, W. and Wunsch, F. *94* (1961) 2993; (f) with Schwarz, W., Miller, J. and Wunsch, F. *95* (1962) 223; (g) with Authaler, A. *96* (1963) 2879.

75 (a) Heller, G. *Ber. dtsch. chem. Ges. 40* (1907) 114; (b) Heller, G.,

Köhler, W., Gottfried, S., Arnold, H. and Herrmann, H. *J. prakt. Chem.* *120* (1928) 49.

76 Wolff, L. (a) *Justus Liebigs Annln Chem.* *313* (1900) 1; (b) *Ber. dtsch. chem. Ges.* *37* (1904) 2827; (c) *Justus Liebigs Annln Chem.* *394* (1912) 23; (d) with Krüche, R. *394* (1912) 48; (e) with Thielepape, E. *420* (1920) 275.

77 Hüttel, R. *Justus Liebigs Annln Chem.* (a) with Gebhart, A. *558* (1947) 34; (b) with Schäfer, O. and Jochum, P. *593* (1955) 200; (c) with Welzel, G. *593* (1955) 207; (d) with Schäfer, O. and Welzel, G. *598* (1956) 186; (e) with Wagner, E. and Sickenberger, B. *607* (1957) 109; (f) with Wagner, E. and Jochum, P. *593* (1955) 179; (g) with Schön, M. E. *625* (1959) 55.

78 Bertho, A. *Ber. dtsch. chem. Ges.* *58* (1925) 859.

79 Jonas, A. and v. Pechmann, H. *Justus Liebigs Annln Chem.* *262* (1891) 277.

80 Olivieri-Mandalà, E. (a) with Coppola, A. *Gazz. chim. ital.* *40*II (1910) 435; (b) with Alagna, B. *40*II (1910) 441; (c) with Passalacqua, T. *41*II (1911) 430; (d) with Passalacqua, T. *43*II (1913) 465; (e) with Passalacqua, T. *43*II (1913) 468; (f) *43*II (1913) 487; (g) *44*II (1914) 175; (h) *45*I (1915) 302; (i) *46*I (1916) 298. (j) Olivieri-Mandala, E. *Atti Accad. naz. Lincei* *19*I (1910) 228.

81 Muller, E. and Ludsteck, D. *Chem. Ber.* *88* (1955) 921.

82 Ghighi, E. and Pozzo-Balbi, T. *Gazz. chim. ital.* *71* (1941) 228.

83 Stollé, R. (a) *J. prakt. Chem.* *68* (1903) 130; (b) with Münch, E. and Kind, W. *70* (1904) 43; (c) with Thoma, K. *73* (1906) 288; (d) with Henke-Stark, F. *124* (1930) 261; (e) with Ehrmann, E., Rieder, D., Wille H., Winter, H. and Henke-Stark, F. *134* (1932) 282; (f) with Dietrich, W. *139* (1934) 193. (g) Stollé, R. and Adam, G. *Ber. dtsch. chem. Ges.* *57* (1924) 1656; (h) Stollé, R. *62* (1929) 1118.

84 Dedichen, G. *Ber. dtsch. chem. Ges.* (a) *39* (1906) 1831; (b) *39* (1906) 1849.

85 Young, G. and Oates, W. H. *J. chem. Soc.* *79* (1901) 659.

86 Hoggarth, E. *J. chem. Soc.* (a) (1949) 1160; (b) (1949) 1918; (c) (1950) 612.

87 Brunner, K. *Mh. Chem.* *36* (1915) 509.

88 Einhorn, A., Bischkopff, E. and Szelinski, B. *Justus Liebigs Annln Chem.* *343* (1905) 229.

89 Cleve, A. *Ber. dtsch. chem. Ges.* *29* (1896) 2671.

90 Pinner, A. *Ber. dtsch. chem. Ges.* (a) *30* (1897) 1871; (b) *35* (1902) 4131; (c) with Schwarz, R. *35* (1902) 2441. (d) Pinner, A. *Justus Liebigs Annln Chem.* *297* (1897) 221; (e) *298* (1897) 39.

91 Engelhardt, R. *J. prakt. Chem.* *54* (1896) 143, 153.

92 Busch, M. and Schneider, C. *J. prakt Chem.* *89* (1914) 310.

93 Mihina, J. S. and Herbst, R. M. *J. org. Chem.* *15* (1950) 1082.

94 Gryszkiewicz-Trochimowski, O. *C.r. hebd. Séanc. Acad. Sci. Paris* *246* (1958) 2627.

95 Markgraf, J. H., Bachmann, W. T. and Hollis, D. P. *J. org. Chem.* *30* (1965) 3472.

96 v. Braun, J. and Rudolph, W. *Ber. dtsch. chem. Ges.* *74* (1941) 264.

97 Burness, D. M. *J. org. Chem.* *21* (1956) 97.

98 Habraken, C. L. and Moore, J. A. *J. org. Chem.* *30* (1965) 1892.

99 Jacquier, R. and Elguero, J. *Bull. Soc. chim. Fr.* (1961) 199.

100 v. Auwers, K. *J. prakt. Chem.* (a) with Niemeyer, F. *110* (1925) 153; (b) with Mauss, H. *110* (1925) 204; (c) with Daniel, W. *110* (1925) 235; (d) with Bähr, K. *116* (1927) 65; (e) with Mauss, H. *117* (1927) 311; (f) with Cauer, E. *126* (1930) 146; (g) with Cauer, E. *126* (1930) 198; (h) with Dietrich, K. *139* (1934) 65; (i) with Breyhan, T. *143* (1935) 259.

101 Jensen, K. A. and Friediger, A. *Kgl. Danske Videnskab. Selskab., Math.-fys. Medd.* *20* no. 20 (1943) 1.

102 Minkin, V. I., Osipov, O. A., Garnovskii, A. D. and Simonov, A. M. *Russ.*

J. phys. Chem. 36 (1962) 245.

103 Pozharskii, A. F. and Simonov, A. M. *J. Gen. Chem. U. S. S. R. 34* (1964) 224.

104 Kaufman, M. H., Ernsberger, F. M. and McEwan, W. S. *J. Am. chem. Soc. 78* (1956) 4197.

105 Kirchhoff, W. H. *J. Am. chem. Soc. 89* (1967) 1312.

106 Broadus, J. D. and Vaughan, J. D. *J. phys. Chem. 72* (1968) 1005.

107 Giller, S. A., Mazheika, I. B., Grandberg, I. I. and Gorbacheva, L. I. *Khim. geterotsikl Soedinenii* (1967) 130.

108 Sitkina, L. M., Pozharskii, A. F. and Simonov, A. M. *J. gen. Chem. U. S. S. R. 37* (1967) 2103.

109 Mazheika, I. B., Chipen, G. I. and Giller, S. A. *Khim. geterotsikl Soedinenii* (1966) 776. Pevzner, M. S., Federova, E. Y., Shokor, I. N. and Bagal, L. N. (1971) 275.

110 Lounsbury, J. B. *J. phys. Chem. 67* (1963) 721.

111 Kost, A. N. (a) with Gents, F. *Chem. Abstr. 53* (1959) 9197; (b) with Faizova, G. K. and Grandberg, I. I. *J. gen. Chem. U. S. S. R. 33* (1963) 525; (c) with Grandberg, I. I. in *Advances in Heterocyclic Chemistry 6* (1966) 347.

112 McFarlane, W. D. *Biochem. J. 30* (1936) 1199.

113 Mason, S. F. in *Physical Methods in Heterocyclic Chemistry 2* (1963) 1, edited by Katritzky, A. R., Academic Press, New York.

114 Elguero, J. (a) with Jacquier, R., Tarrago, G. and Tien Duc, H. C. N. *Bull. Soc. chim. Fr.* (1966) 293; (b) with Jacquier, R. (1966) 2832; (c) with Jacquier, R. and Tarrago, G. (1966) 2990; (d) with Jacquier, R. and Tien Duc, H. C. N. (1966) 3744; (e) with Guiraud, G., Jacquier, R. and Tien Duc, H. C. N. (1967) 328; (f) with Gonzalez, E. and Jacquier, R. (1967) 2998; (g) with Jacquier, R. and Tarrago, G. (1967) 3772, 3780; (h) with Gonzalez, E. and Jacquier, R. (1968) 707; (i) with Jacquier, R. and Tizane, D. (1970) 1121, 1129; (j) with Imbach, J.-L. and Jacquier, R. *J. Chim. phys. 62* (1965) 643; (k) with Jacquier, R. and Tien Duc, H. C. N. *Bull. Soc. chim. Fr.* (1966) 3727; (l) with Gonzalez, E. and Jacquier, R. (1968) 5009.

115 Yagil, T. *Tetrahedron 23* (1967) 2855.

116 Noyce, D. S., Ryder, E. and Walker, B. H. *J. org. Chem. 20* (1955) 1681.

117 Atkinson, M. R., Parkes, E. A. and Polya, J. B. *J. chem. Soc.* (1954) 4256.

118 Dal Monte, D., Mangini, A. and Passerini, R. *Gazz. chim. ital. 86* (1956) 797.

119 Cohen-Fernandes, P. and Habraken, C. L. *Recl Trav. chim. Pays-Bas Belg. 86* (1967) 1249.

120 Habraken, C. L., van Woerkom, P. C. M., de Wind, H. W. and Kallenberg, C. G. M. *Recl Trav. chim. Pays-Bas Belg. 85* (1966) 1191.

121 Finar, I. L. and Rackham, D. M. *J. chem. Soc. (B)* (1968) 211.

122 Hoffer, M., Toome V. and Brossi, A. *J. heterocycl. Chem. 3* (1966) 454.

123 Henry, R. A., Finnegan, W. G. and Lieber E. *J. Am. chem. Soc. 76* (1954) 2894.

124 Turner, R. A. *J. Am. chem. Soc. 71* (1949) 3472.

125 Tabak, S., Grandberg, I. I. and Kost, A. N. (a) *Chem. Abstr. 63* (1965) 5626; (b) *Tetrahedron 22* (1966) 2703.

126 Grandberg, I. I. *J. gen. Chem. U. S. S. R.* (a) with Kost, A. N. *29* (1959) 1069; (b) with Kost, A. N. *30* (1960) 217; (c) *31* (1961) 505; (d) with Vasina, L. G., Volkova, A. S. and Kost, A. N. *31* (1961) 1765; (e) with Kost, A. N. *31* (1961) 3454; (f) with Kost, A. N. *32* (1962) 2974; (g) *32* (1962) 2978; (h) *33* (1963) 511.

127 Fanshawe, W. J., Bauer, V. J. and Safir, S. R. *J. org. Chem. 29* (1964) 942.

128 Austin, M. W., Blackborow, J. R., Ridd, J. H. and Smith, B. V. *J. chem. Soc.* (1965) 1051.

129 Lynch, B. M. and Hung, Y.-Y. *Can. J. Chem. 42* (1964) 1605.

130 Katritzky, A. R. (a) with Maine, F. W. *Tetrahedron 20* (1964) 299, 315; (b) with Maine, F. W. and Golding, S. *21* (1965) 1693; (c) with Ambler, A. P. in *Physical Methods in Heterocyclic Chemistry 2* (1962) edited by Katritzky, A. R., Academic Press, New York; (d) with Taylor, P. J. *4* (1971) 265, edited by Katritzky, A. R., Academic Press, New York, 1963.
131 O'Brien, D. F. and Gates, J. W. *J. org. Chem. 31* (1966) 1538.
132 Yasuda, H. (a) with Midorikawa, H. *J. org. Chem. 31* (1966) 1722; (b) *Bull. chem. Soc. Japan 40* (1967) 1239.
133 Leandri, G., Mangini, A., Montanari, F. and Passerini, R. *Gazz. chim. ital. 85* (1955) 769.
134 Schubert, H. and Baumann, H. P. *Z. phys. Chem. (Leipzig) 203* (1954) 351.
135 Pozharskii, A. F. *J. gen. Chem. U. S. S. R. 34* (1964) 632.
136 Bredereck, H., Gompper, R. and Reich, F. *Chem. Ber. 93* (1960) 1389.
137 Ames, B. N. and Mitchell, H. K. (a) *J. Am. chem. Soc. 74* (1952) 252; (b) *J. biol. Chem. 212* (1955) 687.
138 Wieland, T. and Schneider, G. *Justus Liebigs Annln Chem. 580* (1953) 159.
139 Barlin, G. B. *J. chem. Soc. (B)* (1967) 641.
140 Gallo, G. G., Pasqualucci, C. R., Radaelli, P. and Lancini, G. C. *J. org. Chem. 29* (1964) 862.
141 Grimison, A., Ridd, J. H. and Smith, B. V. *J. chem. Soc.* (1960) 1352, 1357. Ridd, J. H. and Smith, B. V. *J. chem. Soc.* (1960) 1363.
142 Chua, S. O., Cook, M. J. and Katritzky, A. R. *J. chem. Soc. (B)* (1971) 2350.
143 Heath, H., Lawson, A. and Rimington, C. (a) *J. chem. Soc.* (1951) 2215; (b) p. 2217.
144 Hartzel, L. W. and Benson, F. R. *J. Am. chem. Soc. 76* (1954) 667.
145 Dal Monte D., Mangini, A., Passerini, R. and Zauli, C. *Gazz. chim. ital. 88* (1958) 977.
146 Ruccia, M. (a) *Annali Chim. 49* (1959) 720; (b) *50* (1960) 1367; (c) with Cusmano, S. *Chem. Abstr. 58* (1963) 5661.
147 Hoover, J. R. E. and Day, A. R. *J. Am. chem. Soc. 78* (1956) 5832.
148 Lieber, E. (a) with Patinkin, S. H. and Tao, H. H. *J. Am. chem. Soc. 73* (1951) 1792; (b) with Levering, D. R. and Patterson, L. J. *Analyt. Chem. 23* (1951) 1594; (c) with Chao, T. S. and Rao, C. N. R. *J. org. Chem. 22* (1957) 654; (d) with Rao, C. N. R. and Chao, T. S. *Spectrochim. Acta 10* (1958) 250; (e) with Rao, C. N. R., Chao, T. S. and Rubinstein, H. *Can. J. Chem. 36* (1958) 1441; (f) with Ramachandram, J., Rao, C. N. R. and Pillai, C. N. *Can. J. Chem. 37* (1959) 563; (g) with Enkoji, T. *Inorg. Synth. 6* (1960) 62; (h) with Rao, C. N. R. and Pillai, C. N. *Current Sci. 26* (1957) 167.
149 Albert, A. (a) with Goldacre, R. and Phillips, J. *J. chem. Soc.* (1948) 2240; (b) with Tratt, K. *Chem. Comm.* (1966) 243; (c) *Chem. Comm.* (1967) 684; *J. chem. Soc. (C)* (1968) 2076; (d) in *Physical Methods in Heterocyclic Chemistry 1* (1963) edited by Katritzky, A. R., Academic Press, New York; (e) *J. chem. Soc. (C)* (1970) 230.
150 Staab, H. A. *Chem. Ber.* (a) *89* (1956) 1927; (b) *89* (1956) 2088; (c) *90* (1957) 1326; (d) with Seel, G. *92* (1959) 1302; (e) with Wendel, K. *93* (1960) 2902. Staab, H. A. *Justus Liebigs Annln Chem.* (f) *609* (1957) 75; (g) *609* (1957) 83; (h) *622* (1959) 31; (i) with Benz, W. *648* (1961) 72; (j) with Walther, G. *657* (1962) 98; (k) with Wendel, K. *694* (1966) 86; (l) with Wendel, K. *694* (1966) 91. Staab, H. A. *Angew. Chem. (int. Edn.)* (m) *1* (1962) 351; (n) with Mannschreck, A. *2* (1963) 216. (o) Staab, H. A. and Mannschreck, A. *Tetrahedron Lett.* (1962) 913; (p) with Irngartinger, H., Mannschreck, A. and Mou-Thai Wu *Justus Liebigs Annln Chem. 695* (1966) 55.
151 Garbrecht, W. L. and Herbst, R. M. *J. org. Chem.* (a) *18* (1953) 1003; (b) *18* (1953) 1269.
152 Fallon, F. G. and Herbst, R. M. *J. org. Chem. 22* (1957) 933.

153 Huisgen, R. (a) with Sauer, J., Sturm, H. J. and Markgraf, J. H. *Chem. Ber.*
    93 (1960) 2106; (b) with Sauer, J. and Seidel, M. *94* (1961) 2503;
    (c) with Seidel, M. *94* (1961) 2509; (d) with Koch, H. J. *Justus Liebigs*
    Annln Chem. 591 (1955) 200; (e) with Sauer, J. and Sturm, H. J. *Angew.*
    Chem. 70 (1958) 272.
154 Elpern, B. and Nachod, F. C. *J. Am. chem. Soc. 72* (1950) 3379.
155 Murphy, D. B. and Picard, J. P. *J. org. Chem. 19* (1954) 1807.
156 Horwitz, J. P., Fisher, B. E. and Tomasewski, A. J. *J. Am. chem. Soc. 81*
    (1959) 3076.
157 Bianchi, G., Boulton, A. J., Fletcher, I. J. and Katritzky, A. R. *J. chem.*
    Soc. (B) (1971) 2355.
158 Evans, N. A., Whelan, D. J. and Johns, R. B. *Tetrahedron 21* (1965) 3351.
159 Gagnon, P. E. (a) with Boivin, J. L. and Brown, P. A. *Can. J. Res. 28B*
    (1950) 720; (b) with Boivin, J. L., MacDonald, R. and Yaffe, L. *Can. J.*
    Chem. 32 (1954) 823.
160 Girod, E., Delley, R. and Höfliger, F. *Helv. Chim. Acta 40* (1957) 408.
161 Gordon, A. A., Katritzky, A. R. and Popp, F. D. *Tetrahedron Supp. 7*
    (1966) 213.
162 Lawson, A. and Morley, H. V. *J. chem. Soc.* (1956) 1103.
163 (a) Reynolds, G. A. and van Allen, J. A. *J. org. Chem. 24* (1959) 1478.
    (b) Sandström, J. and Wennerbeck, I. *Acta chem. Scand. 20* (1966) 57.
164 Zecchina, A., Cerruti, L., Coluccia, S. and Borello, E. *J. chem. Soc. (B)*
    (1967) 1363.
165 Perchard, C. and Novak, A. *Spectrochim. Acta 23A* (1967) 1953.
166 Cordes, M. and Walter, J. L. *Spectrochim. Acta 24A* (1968) 237.
167 Bellocq, A. M. (a) with Garrigou-Lagrange, C. *J. Chim. phys. 66* (1969)
    1511; (b) with Perchard, C., Novak, A. and Josien, M. L. *62* (1965) 1334;
    (c) with Garrigou-Lagrange, C. *C. r. hebd. Séanc. Acad. Sci. Paris 268B*
    (1969) 1420; (d) with Garrigou-Lagrange, C. *Spectrochim. Acta 27A*
    (1971) 1091.
168 Borello, E. (a) with Zecchina, A. *Annali Chim. 52* (1962) 1302; (b) with
    Zecchina, A. and Guglielminotti, E. *J. chem. Soc. (B)* (1969) 307; (c) with
    Zecchina, A. *Spectrochim. Acta 19* (1963) 1703; (d) with Zecchina, A.
    and Guglielminotti, E. *J. chem. Soc. (B)* (1966) 1243.
169 Garber, L. L., Sims, L. B. and Brubaker, C. H. *J. Am. chem. Soc. 90* (1968)
    2518.
170 Zerbi, G. and Alberti, C. *Spectrochim. Acta* (a) *18* (1962) 407; (b) *19*
    (1963) 1261.
171 Jacquier, R., Pétrus, C. and Pétrus, F. *Bull. Soc. chim. Fr.* (1966) 2971.
172 Dorn, H. (a) with Hilgetag, G. and Zubek, A. *Chem. Ber. 98* (1965) 3368;
    (b) with Hilgetag, G. and Zubek, A. *99* (1966) 178; (c) with Hilgetag, G.
    and Zubek, A. *99* (1966) 183; (d) with Zubek, A. *101* (1968) 3265;
    (e) with Zubek, A. *Z. Chem. 7* (1967) 343; (f) with Dilcher, H. *Justus*
    Liebigs Annln Chem. 707 (1967) 141.
173 Yepishina, L. V., Slovetski, V. I., Osipov, V. G., Lebedev, O. V., Khmel-
    nitskii, L. I. and Sevostyanova, V. V. *Chem. het. Cpds (U. S. S. R.)* (1967)
    716.
174 Johnson, F. and Nascitavicus, W. A. *J. org. Chem. 29* (1964) 153.
175 Browne, E. J. and Polya, J. B. (a) *J. chem. Soc.* (1962) 575; (b) *J. chem.*
    Soc. (C) (1969) 1056.
176 Rao, C. N. R. and Venkataraghavan, R. *Can. J. Chem. 42* (1964) 43.
177 Kauer, J. C. (a) with Carboni, R. A., Hatchard, W. R. and Harder, R. J.
    J. Am. chem. Soc. 89 (1967) 2626; (b) with Carboni, R. A. *89* (1967)
    2633; (c) with Sheppard, W. A. *J. org. Chem. 32* (1967) 3580.
178 Bonino, G. B. and Manzoni-Ansidei, R. *Atti Accad. naz. Lincei 22* (1935)
    438.
179 Kohlrausch, K. W. F. and Seka, R. *Ber. dtsch. chem. Ges. 71* (1938) 985.
180 King, S. T. *J. phys. Chem. 74* (1970) 2133.

181 Brickmann, J. and Zimmermann, H. *Ber. Bunsenges. phys. Chem.* 70 (1966) 521.
182 Bogunets, N. P., Baturicheva, Z. B. and Naboĭkin, Y. V. *Chem. Abstr.* 53 (1959) 9813.
183 Anderson, D. M. W., Duncan, J. L. and Rossotti, F. J. C. *J. chem. Soc.* (a) (1961) 140; (b) (1961) 2165; (c) (1961) 4201.
184 Vinogradov, S. N. and Kilpatrick, M. *J. phys. Chem.* 68 (1964) 181.
185 Otting, W. *Chem. Ber.* (a) 89 (1956) 1940; (b) 89 (1956) 2887.
186 Lukton, A. *Nature, Lond.* 192 (1961) 422.
187 Geiseler, G., Fruwert, J. and Kürnecke, A. *Z. phys. Chem. (Frankfurt)* 41 (1964) 49.
188 Croatto, U. and Fava, A. *Annali Chim.* 54 (1964) 735, 1340.
189 Ried, W. and Königstein, F.-J. *Justus Liebigs Annln Chem.* (a) 622 (1959) 37; (b) 625 (1959) 53.
190 Birkofer, L. (a) *Chem. Ber.* 76 (1943) 769; (b) with Richter, P. and Ritter, A. 93 (1960) 2804; (c) with Ritter, A. and Richter, P. 96 (1963) 2750; (d) with Wegner, P. 99 (1966) 2512; (e) with Franz, M. 100 (1967) 2681; (f) with Wegner, P. 100 (1967) 3485; (g) with Gilgenberg, W. and Ritter, A. *Angew. Chem.* 73 (1961) 143; (h) with Ritter, A. and Richter, P. *Angew. Chem. (int. Edn.)* 1 (1962) 267.
191 Potts, K. T. *Chem. Rev.* 61 (1961) 87.
192 Taylor, E. C. (a) with Barton, J. W. and Osdene, T. S. *J. Am. chem. Soc.* 80 (1958) 421; (b) with Barton, J. W. 81 (1959) 2448; (c) with Hartke, K. S. 81 (1959) 2452, 2456.
193 Vinokurov, V. G., Troitskaya, V. S. and Grandberg, I. I. *J. gen. Chem. U. S. S. R.* (a) 34 (1964) 657; (b) 35 (1965) 1293; (c) with Pentin, Y. A. 33 (1963) 2531; *Chem. Abstr.* 62 (1965) 2692.
194 Dick, I., Bacaloglu, R. and Maurer, A. *Rev. Romaine de Chim.* 12 (1967) 607.
195 Shchipanov, V. P. (a) *Chem. Abstr.* 65 (1966) 2248a; (b) with Portnova, S. L., Krasnova, V. A., Sheinker, Y. N. and Postovskii, I. Y. 64 (1966) 11056d; (c) with Postovskii, I. Y. 65 (1966) 15368h; (d) with Postovskii, I. Y. 74 (1971) 41715h; (e) with Ryssakova E. N. *Khim. geterotsikl. Soedinenii* (1968) 215.
196 Cogrossi, C. *Spectrochim. Acta* 22 (1966) 1385.
197 Newman, G. A. and Pauwels, P. J. S. *Tetrahedron* (a) 25 (1969) 4605; (b) 26 (1970) 1571; (c) 26 (1970) 3429.
198 Dahn, H. and Rolzler, G. *Helv. chim. Acta* 43 (1960) 1555.
199 Snavely, F. A. and Suydam, F. H. *J. org. Chem.* 24 (1959) 2039.
200 Toda, S. *Chem. Abstr.* 55 (1961) 4150.
201 Pelz, W., Püschel, W., Schellenberger, H. and Loffler, K. *Angew. Chem.* 72 (1960) 967.
202 Jones, R., Ryan, A. J., Sternhell, S. and Wright, S. E. *Tetrahedron* 19 (1963) 1497.
203 Carpino, L. A. *J. Am. chem. Soc.* 80 (1958) 5796.
204 de Stevens, G., Halamandaris, A., Wenk, P. and Dorfman, L. *J. Am. chem. Soc.* 81 (1959) 6292.
205 Refn, S. *Spectrochim. Acta* 17 (1961) 40.
206 Logemann, W., Lauria, F. and Zambari, V. *Chem. Ber.* 88 (1955) 1353; 89 (1956) 620.
207 Cardani, C., Mantegani, A., Cavarelleri, B. and Sianesi, I. L. *Gazz. chim. ital.* 90 (1960) 1746.
208 Gompper, R. (a) *Chem. Ber.* 90 (1957) 382; (b) with Herlinger, H. 89 (1956) 2825; (c) with Hoyer, E. and Herlinger, H. 92 (1959) 550; (d) *Advances in Heterocyclic Chemistry* 2 (1963) 245.
209 Maggio, F., Werber, G. and Lombardo, G. *Annali Chim.* 50 (1960) 491.
210 Howard, J. C. and Burch, H. A. *J. org. Chem.* 26 (1961) 1651.
211 Blackman, A. J. and Polya, J. B. *J. chem. Soc. (C)* (1971) 1016.

212 Imbach, J. L., Jacquier, R. and Romane, A. *J. heterocycl. Chem.* 4 (1967) 451.
213 Stambaugh, J. E. and Manthei, R. W. *Can. Spectrosc.* 13 (1968) 134, 143.
214 Habraken, C. L., Munter, H. J. and Westgeest, J. C. P. *Recl. Trav. chim. Pays-Bas Belg.* 86 (1967) 56.
215 Tensmeyer, L. C. and Ainsworth, C. *J. org. Chem.* 31 (1966) 1878.
216 Barlin, G. B. and Batterham, T. J. *J. chem. Soc. (B)* (1967) 516.
217 Grimmett, M. R., Hartshorn, S. R., Schofield, K. and Weston, J. B. *J. chem. Soc. (Perk. II)* (1972) 1654.
218 Williams, J. K. *J. org. Chem.* 29 (1964) 1377.
219 Mannschreck, A., Seitz, W. and Staab, H. A. *Ber. Bunsenges. phys. Chem.* 67 (1963) 470.
220 Abraham, R. J. and Thomas, W. A *J. chem. Soc. (B)* (1966) 127.
221 Reddy, G. S., Hobgood, R. T. and Goldstein, J. H. *J. Am. chem. Soc.* 84 (1962) 336.
222 Caesar, F. and Overberger, C. G. *J. org. Chem.* 33 (1968) 2971.
223 Potts, K. T. (a) with Crawford, T. H. *J. org. Chem.* 27 (1962) 2631; (b) 28 (1963) 543; (c) with Roy, S. K. and Jones, D. P. 32 (1967) 2245; (d) with Roy, S. K. and Jones, D. P. *J. heterocycl. Chem.* 2 (1965) 105; (e) with Armbruster, R. and Houghton, E. 8 (1971) 773.
224 Freiberg, W., Krüger, C.-F. and Radeglia, R. *Tetrahedron Lett.* (1967) 2109.
225 Jacquier, R., Roumestant, M.-L. and Viallefont, P. (a) *Bull. Soc. chim. Fr.* (1967) 2630; (b) p. 2634.
226 Joop, N. and Zimmermann, H. (a) *Z. Elektrochem.* 66 (1962) 440; (b) p. 541.
227 Bystrov, V. F., Grandberg, I. I. and Sharova, G. I. *Optics and Spectroscopy* 17 (1964) 31.
228 Hung, N. C. *Dissert. Abstr.* 24 (1963) 2287.
229 Creagh, L. T. and Truitt, P. *J. org. Chem.* 33 (1968) 2956.
230 Roumestant, M. L., Viallefont, P., Elguero, J. and Jacquier, R. *Tetrahedron Lett.* (1969) 495.
231 Moore, D. W. and Whittaker, A. G. *J. Am. chem. Soc.* 82 (1960) 5007.
232 Overberger, C. G., Salamone, J. C. and Yaroslavsky, S. *J. org. Chem.* 30 (1965) 3580.
233 Moore, J. A. and Habraken, C. L. *J. Am. chem. Soc.* 86 (1964) 1456.
234 Naidu, M. S. R. and Bensusan, H. B. *J. Am. chem. Soc.* 33 (1968) 1307.
235 Gold, H. *Justus Liebigs Annln Chem.* 688 (1965) 205.
236 Kaul, B. L., Nair, P. M., Rao, A. V. R. and Venkataraman, K. *Tetrahedron Lett.* (1966) 3897.
237 Snavely, F. A. and Yoder, C. H. *J. org. Chem.* 33 (1968) 513.
238 Lestina, G. J. and Regan, T. H. *J. org. Chem.* 34 (1969) 1685.
239 Pugmire, R. J. and Grant, D. M. *J. Am. Chem. Soc.* 90 (1968) 4232.
240 Weigert, F. J. and Roberts, J. D. *J. Am. chem. Soc.* 90 (1968) 3543.
241 Tori, K. and Nakagawa, T. *J. phys. Chem.* 68 (1964) 3163.
242 Rees, R. G. and Green, M. J. *J. chem. Soc. (B)* (1968) 387.
243 Haake, P., Bausher, L. P. and Miller, W. B. *J. Am. chem. Soc.* 91 (1969) 1113.
244 Feeney, J., Newman, G. A. and Pauwels, P. J. S. *J. chem. Soc. (C)* (1970) 1842.
245 Parnell, E. W. *Tetrahedron Lett.* (1970) 3941.
246 Ferguson, I. J. and Schofield, K. Unpublished results.
247 Gonzalez, E. and Jacquier, R. *Bull. Soc. chim. Fr.* (1968) 5006.
248 Hansen, L. D. (a) with Baca, E. J. and Scheiner, P. *J. heterocycl. Chem.* 7 (1970) 991; (b) with West, B. D., Bacca, E. J. and Blank, C. L. *J. Am. chem. Soc.* 90 (1968) 6588.
249 Kochetkov, N. K. and Ambrush, I. *J. gen. Chem. U. S. S. R.* 27 (1957) 2781.
250 Habraken, C. L., Westra, E. C., Bomhoff, G. H. and Heytink, R. A. *Recl*

Trav. chim. Pays-Bas Belg. 85 (1966) 1194.
251 Khan, M. A. (a) with Lynch, B. M. and Hung, Y.-Y. *Can. J. Chem. 41* (1963) 1540; (b) with Polya, J. B. and Lynch, B. M. *46* (1968) 2629.
252 Veibel, S., Eggersen, K. and Linholt, S. *Acta chem. scand. 8* (1954) 768.
253 Pešák, M., Greksáková, O., Kopecky, F. and Čelechovský, J. *Colln Czech. chem. Commun. 32* (1967) 2031.
254 Tanford, C. and Wagner, M. L. *J. Am. chem. Soc. 75* (1953) 434. *Cf.* Edsall, J. T., Felsenfeld, G., Goodman, D. S. and Gurd, F. R. N. *76* (1954) 3054; Li, N. C., Chu, T. L., Fujii, C. T. and White, J. M. *77* (1955) 859.
255 Walba, H. and Isensee, R. W. *J. org. Chem.* (a) *21* (1956) 702; (b) *26* (1961) 2789.
256 Kirby, A. H. M. and Neuberger, A. *Biochem. J. 32* (1938) 1146.
257 Datta, S. P. and Grzybowski, A. K. *J. chem. Soc. (B)* (1966) 136.
258 Hanania, G. I. H., Irvine, D. H. and Abu-Issa, I. *J. chem. Soc.* (1964) 5689.
259 Levy, M. *J. biol. Chem. 109* (1935) 361.
260 Nozaki, Y., Gurd, F. R. N., Chen, R. F. and Edsall, J. T. *J. Am. chem. Soc 79* (1957) 2123.
261 Jencks, W. P. and Carriuolo, J. *J. biol. Chem. 234* (1959) 1272.
262 Cowgill, R. W. and Clark, W. M. *J. biol. Chem. 198* (1952) 33.
263 Storey, B. T., Sullivan, W. W. and Moyer, C. L. *J. org. Chem. 29* (1964) 3118.
264 Laviron E. *Bull. Soc. chim. Fr.* (1963) 2840.
265 Kajfež, F., Kolbak, D., Fajdiga, T., Oklobdžija, M., Slamnik, M. and Sunjić, V. *Croat chem. Acta 39* (1967) 199.
266 Blažević, N., Kajfež, F. and Šunjić, V. *J. heterocycl. Chem. 7* (1970) 227.
267 Taylor, E. G. *Can. J. Res. 20B* (1942) 161.
268 Kröger, C.-F. (a) with Freiberg, W. *Z. Chem. 5* (1965) 381; (b) with Mietchen, R. *7* (1967) 184; (c) with Franck, H. *Angew. Chem. (int. Edn.) 4* (1965) 434, 879; (d) with Schoknecht, G. and Beyer, H. *Chem. Ber. 97* (1964) 398; (e) with Hummel, L., Mietscher, M. and Beyer, H. *98* (1965) 3025; (f) with Mietchen, R. *100* (1967) 2250; (g) with Etzold, G. and Beyer, H. (with Busse, G.) *Justus Liebigs Annln Chem. 664* (1963) 156.
269 Bagal, L. I. (a) with Pevzner, M. S. and Lopyrev, V. A. *Chem. Abstr. 65* (1966) 12204; (b) with Pevzner, M. S. and Lopyrev, V. A. *70* (1969) 77876t; (c) with Pevzner, M. S., Samarenko, V. Y. and Egorov, A. P. *74* (1971) 76376a; (d) with Pevzner, M. S., Samarenko V. Y. and Egorov, A. P. *74* (1971) 99948c.
270 Schmidt, J. and Gehlen, H. *Z. Chem. 5* (1965) 304.
271 Brown, H. C. and Cheng, M. T. *J. org. Chem. 27* (1962) 3243.
272 Rochlin, P., Murphy, D. B. and Helf, S. *J. Am. chem. Soc 76* (1954) 1451.
273 Li, N. C., White, J. M. and Doody, E. *J. Am. chem. Soc. 76* (1954) 6219.
274 Fargher, R. G. (a) *J. chem. Soc. 117* (1920) 668; (b) *119* (1921) 158; (c) with Pyman, F. L. *115* (1919) 217; (d) *115* (1919) 1015.
275 Baltzer, O. and v. Pechmann, H. *Justus Liebigs Annln Chem. 262* (1891) 314.
276 Pedersen, C. *Acta chem. scand.* (a) *12* (1958) 1236; (b) *13* (1959) 888.
277 Benson, F. R. *Chem. Rev. 41* (1947) 1; *Heterocyclic Compounds,* vol. 8, edited by Elderfield, R. C., Wiley and Sons, New York, 1967.
278 Perrin, D. D. *J. chem. Soc.* (1965) 5590.
279 Charton, M. (a) *J. org. Chem. 30* (1965) 3346; (b) *J. chem. Soc. (B)* (1969) 1240.
280 Collis, M. J. and Edwards, G. R. *Chemy Ind.* (1971) 1097.
281 Bruice, T. C. and Schmir, G. L. (a) *J. Am. chem. Soc. 79* (1957) 1663; (b) *80* (1958) 148. (c) Bruice, T. C. and Benkovic, S. J. in *Bio-organic Mechanisms,* vol 1, W. A. Benjamin Inc., New York, 1966.
282 Kuhn, R. and Westphal, O. *Ber. dtsch. chem. Ges. 73* (1940) 1109.
283 Strain, H. H. *J. Am. chem. Soc. 49* (1927) 1995.
284 Oddo, B. and Mingoia, Q. *Gazz. chim. ital. 58* (1928) 584. John, W. *Ber.*

dtsch. chem. Ges. 68 (1935) 2283.
285 Nicholls, D. and Warburton, B. A. *J. inorg. nucl. Chem. 32* (1970) 3871.
286 Daugherty, N. A. (a) with Brubaker, C. H. *J. Am. chem. Soc. 83* (1961) 3779; (b) with Swisher, J. H. *Inorg. Chem. 7* (1968) 1651.
287 Reedijk, J. *Recl Trav. chim. Pays-Bas Belg.* (a) *88* (1969) 1451; (b) *89* (1970) 605; (c) *89* (1970) 993.
288 Snavely, F. A., Sweigart, D. A., Yoder, C. H. and Terzis, A. *Inorg. Chem. 6* (1967) 1831.
289 Gurd, F. R. W. and Goodman, D. S. *J. Am. chem. Soc. 74* (1952) 670.
290 Martin, R. B. and Edsall, J. T. *J. Am. chem. Soc. 80* (1958) 5033.
291 Chakravorty, A. and Colta, F. A. *J. phys. Chem. 67* (1963) 2878.
292 Brubaker, C. H. *J. Am. chem. Soc. 82* (1960) 82. Harris, A. D., Herber, R. H., Jonassen, H. B. and Wertheim, G. K. *85* (1963) 2927.
293 Holm, R. D. and Donnelly, P. L. *J. inorg. nucl. Chem. 28* (1966) 1887.
294 Hanania, G. I. H. and Irvine, D. H. *J. chem. Soc.* (1964) 5694.
295 Andrews, A. C. and Zebolsky, D. M. *J. chem. Soc.* (1965) 742.
296 Popov, A. I. (a) with Holm, R. D. *J. Am. chem. Soc. 81* (1959) 3250; (b) with Bisi, C. C. and Craft, M. *80* (1958) 6513.
297 D'Itri, F. M. and Popov, A. I. (a) *Inorg. Chem. 5* (1966) 1670; (b) *6* (1967) 597; (c) *J. Am. chem. Soc. 90* (1968) 6476.
298 Wehman, T. C. and Popov, A. I. *J. phys. Chem. 70* (1966) 3688.
299 Baenziger, N. C., Nelson, A. D., Tulinsky, A., Bloor, J. H. and Popov, A. I. *J. Am. chem. Soc. 89* (1967) 6463.
300 Easteal, A. J. and Ruthven, M. G. *Chemy Ind.* (1966) 600.
301 Lee, A. G. *J. chem. Soc. (A)* (1971) 880.
302 di Paolini, A. and Goria, E. *Gazz. chim. ital. 63* (1932) 1048. Jarvis, J. A. J. *Acta Cryst. 15* (1962) 964.
303 Hamano, H. and Hameka, H. F. *Tetrahedron 18* (1962) 985.
304 Orgel, L. E., Cottrell, T. L., Dick, W. and Sutton, L. E. *Trans. Faraday Soc. 47* (1951) 113.
305 Owen, A. J. *Tetrahedron 14* (1961) 237.
306 Brown, R. D. and Heffernan, M. L. *Aust. J. Chem. 13* (1960) 49.
307 Dewar, M. J. S. (a) with King, F. E. *J. chem. Soc.* (1945) 114; (b) with Morita, T. *J. Am. chem. Soc. 91* (1969) 796.
308 Adam, W. (a) with Grimison, A. *Theor. Chim. Acta 7* (1967) 342; (b) with Grimison, A. and Rodriguez, G. *Tetrahedron 23* (1967) 2513.
309 Burton, R. E. and Finar, I. L. *J. chem. Soc. (B)* (1970) 1692.
310 Bloor, J. E. and Breen, D. L. *J. Am. chem. Soc. 89* (1967) 6835.
311 Flurry, R. L., Stout, E. W. and Bell, J. J. *Theor. Chim. Acta 8* (1967) 203, 360.
312 Fischer-Hjalmars, I. and Nag-Chaudhuri, J. *Acta chem. Scand. 23* (1969) 2963.
313 Kamiya, M. *Bull chem. Soc. Japan 43* (1970) 3344.
314 Vaughan, J. D. (a) with Lambert, D. G. and Vaughan, V. L. *J. Am. chem. Soc. 86* (1964) 2857; (b) with Fullerton, D. C. and Chin-An Chang *Int. J. quant. Chem. 2* (1968) 205; (c) with O'Donnell, M. O. *Tetrahedron Lett.* (1968) 3727.
315 Bodor, N., Dewar, M. J. S. and Harget, A. J. *J. Am. chem. Soc. 92* (1970) 2929.
316 Deschamps, J., Arriau, J. and Parmentier, P. *Tetrahedron 27* (1971) 5779.
317 Brack, A. *Justus Liebigs Annln Chem. 682* (1965) 165.
318 *U. S. Pat.* (a) 2,066,954 (1937) [*Chem. Abstr. 31* (1937) 1212]; (b) 2,384,369 (1945); (c) 2,399,598 (1946); (d) 2,493,747 (1950); (e) 2,515,878 (1950); (f) 2,526,632 (1950); (g) 2,694,703 (1954); (h) 2,750,379 (1956) [*Chem. Abstr. 51* (1957) 2054]; (i) 3,294,640 (1966) [*Chem. Abstr. 66* (1967) 85786w]; (j) 3,341,548 (1967) [*Chem. Abstr. 68* (1968) 105,198c]; (k) 3,544,585 [*Chem. Abstr. 74* (1971) 76414m].

319 Severini, O. *Atti Accad. naz. Lincei* [47] 7 ii (1891) 377.
320 Ochiai, E. (a) *J. pharm. Soc. Japan* 60 (1940) 164; (b) with Sibata, M. 59 (1939) 185; (c) with Utahashi, K. 60 (1940) 104; (d) with Utahashi, K. 60 (1940) 312.
321 Michaelis, A. *Ber. dtsch. chem. Ges.* (a) with Burmeister, R. 25 (1892) 1502; (b) with Pasternak, R. 32 (1899) 2398; (c) with Behn, H. 33 (1900) 2595; (d) with Schwabe, G. 33 (1900) 2607; (e) with Sudendorf, T. 33 (1900) 2615; (f) 38 (1905) 154; (g) with Schlecht, H. 39 (1906) 1954; (h) with Engelhardt, F. 41 (1908) 2668; (i) with Schenk, K. 40 (1907) 3568; (j) with Lachwitz, A. 43 (1910) 2106; (k) with Kirstein, E. 46 (1913) 3603; (l) with Rojahn, C. A. 50 (1917) 537.
322 Butler, D. E. and DeWald, H. A. *J. org. Chem.* 36 (1971) 2542.
323 Grandberg, I. I. (a) with Tabak, S. V., Bobrova, N. I., Kost, A. N. and Vasina, L. G. *Khim. Geterotsikl Soedinenii, Akad, Nauk Latv. S. S. R.* (1965) 407; (b) with Kost, A. N. and Zheltikova, N. N. *J. gen. Chem. U. S. S. R.* 30 (1960) 2907; (c) with Khyuchko, G. V. 32 (1962) 1876; (d) *Chem. Abstr.* 63 (1965) 8339; (e) with Tabak, S. V. *Chem. Abstr.* 63 (1965) 5625; (f) with Wei-P'i Ting and Kost, A. N. *J. gen. Chem. U. S. S. R.* 31 (1961) 2153; (g) with Din Wei-P'i, Shchegoleva, V. N. and Kost, A. N. 31 (1961) 1771; (h) with Vasina, L. G. and Kost, A. N. *J. gen. Chem. U. S. S. R.* 30 (1960) 3292; (i) 31 (1961) 2149; (j) *Chem. Abstr.* 64 (1966) 9704.
324 *Ger. Pat.* (a) 90,959 (1896); (b) 111,724 (1899); (c) 254,487 (1911); (d) 270,487 (1914); (e) 489,363 (1925); (f) 579,818 (1933); (g) 611,003 (1935); (h) 617,360 (1935); (i) 701,135 [*Chem. Abstr.* 35 (1941) 7422]; (j) 703,844 (1930); (k) 708,262 (1941); (l) 745,625 (1943) [*Chem. Abstr.* 40 (1946) 1543]; (m) 909,342 (1954); (n) 1,136,342 (1962) [*Chem. Abstr.* 58 (1963) 5724]; (o) 1,138,058 (1962); (p) 1,956,711 (1971) [*Chem. Abstr.* 75 (1971) 49086v].
325 Cornforth, J. and Huang, H. T. *J. chem. Soc.* (1948) 731.
326 Schönherr, H.-J. and Wanzlick, H.-W. *Chem. Ber.* 103 (1970) 1037.
327 Fusco, R., D'Alo, F. and Masserini, A. *Gazz. chim. ital.* 96 (1966) 1084.
328 Rodionov, V. M. and Fedorova, A. M. *Chem. Abstr.* 48 (1954) 671.
329 Stolz, F. (a) *Ber. dtsch. chem. Ges.* 27 (1894) 407; (b) 28 (1895) 623; (c) 33 (1900) 262; (d) 36 (1903) 3279; (e) *J. prakt. Chem.* (2) 55 (1897) 145.
330 Michaelis, A. *et al. Justus Liebigs Annln Chem.* (a) 338 (1905) 267; (b) 339 (1905) 117; (c) 350 (1906) 288; (d) 352 (1907) 152; (e) 358 (1908) 127; (f) 361 (1908) 251; (g) 373 (1910) 129; (h) 378 (1911) 293; (i) 385 (1911) 44; (j) 397 (1913) 119; (k) 407 (1915) 229.
331 Ghosh, T. N. *J. Indian chem. Soc.* 14 (1937) 123.
332 Losco, G. *Gazz. chim. ital.* (a) 69 (1939) 639; (b) 70 (1940) 284; (c) with Passerini, M. 70 (1940) 410.
333 Perroncito, G. (a) *Gazz. chim. ital.* 67 (1937) 158; (b) *Chem. Abstr.* 34 (1940) 2842.
334 Schiedt, B. *J. prakt. Chem.* 157 (1941) 203.
335 Ridi, M. (a) *Gazz. chim. ital.* 71 (1941) 542; (b) with Papini, P. 77 (1947) 99; (c) with Papini, P. 78 (1948) 3; (d) 82 (1952) 746; (e) with Checchi, S. 83 (1953) 36; (f) with Papini, P. and Checchi, S. 91 (1961) 973; (g) with Checchi, S. *Annali chim.* 43 (1953) 816.
336 Dains, F. B. and Brown, E. W. *J. Am. chem. Soc.* 31 (1909) 1148. Dains, F. B. and Daily, A. E. *Chem. Abstr.* 26 (1932) 427.
337 Mansberg, E. and Shaw, G. *J. chem. Soc.* (1953) 3467.
338 Wislicenus, W., Elvert, H. and Kurtz, P. *Ber. dtsch. chem. Ges.* 46 (1913) 3395.
339 *Brit. Pat.* (a) 426,718 (1933); (b) 544,647 (1942); (c) 556,266 (1943); (d) 583,367 (1946); (e) 614,471 (1948); (f) 732,820 (1955).
340 Hünig, S. *Justus Liebigs Annln Chem.* 574 (1951) 99.

341 Ogata T., Tanno, R. and Nishida, K. *Chem. Abstr. 47* (1953) 5284.
342 Schultz, G. and Rohde, G. *J. prakt. Chem. (2) 87* (1931) 119.
343 Chatterjee, S. C. and Das, A. K. *J. Am. chem. Soc. 41* (1919) 707.
344 Passerini, M. *Gazz. chim. ital.* (a) with Casini, V. *67* (1937) 332; (b) with Losco, G. *69* (1939) 658.
345 Kira, M. A. and Bruckner-Wilhelms, A. *Acta Chimica Acad. Sci. Hung. 56* (1968) 47.
346 Kocwa, A. *Chem. Abstr.* (a) *31* (1937) 1803; (b) p. 1804.
347 Ito, I. *J. pharm. Soc. Japan 76* (1956) 167.
348 Mach-Phuoc Sinh and Buu-Hoi, N. P. *Bull. Soc. chim. Fr.* (1967) 802.
349 Benary, E. and Schmidt, A. *Ber. dtsch. chem. Ges. 57* (1924) 517.
350 Kaufmann, H. P. (a) with Liepe, J. *Ber. dtsch. chem. Ges. 56* (1923) 2514; (b) with Küchler, K. *67* (1934) 944; (c) with Huang, L. S. *75* (1942) 1214, 1236; (d) with Steinhoff, F. *Arch. Pharm., Berl. 278* (1940) 437.
351 Klosa, J. *Arch. Pharm., Berl. 288* (1955) 217.
352 Asher, T. *Ber. dtsch. chem. Ges. 30* (1897) 1018.
353 Biltz, H. (a) *Ber. dtsch. chem. Ges. 40* (1907) 2630; (b) *40* (1907) 4799; (c) *41* (1908) 1754, 1761; (d) *Justus Liebigs Annln Chem. 368* (1909) 156, 262.
354 Duschinsky, R. and Dolan, L. A. *J. Am. chem. Soc. 67* (1945) 2079; *68* (1946) 2350; *70* (1948) 657.
355 Cusack, N. J., Shaw, G. and Litchfield, G. J. *J. chem. Soc. (C)* (1971) 1501.
356 Diels, O. and Alder, K. *Justus Liebigs Annln Chem. 498* (1932) 1.
357 Ginzburg, O. F. (a) with Terushkin, V. R. *Zh. obshch. Khim. 23* (1953) 1049; (b) p. 1890; (c) *Chem. Abstr. 49* (1955) 1049.
358 Chrzczonowicz, S., Zwierzak, A. and Achmatowicz, O. *Chem. Abstr. 49* (1955) 14742.
359 *Spanish Pat.* 219,066 (1955) [*Chem. Abstr. 50* (1956) 9449].
360 Meyer, A. and Bouchet, G. *C. r. hebd. Séanc. Acad. Sci Paris 227* (1948) 345.
361 Colonna, M., Bruni, P. and Guerra, A. M. *Gazz. chim. ital. 96* (1966) 1410.
362 Hamana, M. and Noda, H. *Chem. pharm. Bull., Tokyo 15* (1967) 1380.
363 Jones, R. G. (a) *J. Am. chem. Soc. 71* (1949) 383; (b) *71* (1949) 3994; (c) with Ainsworth, C. *77* (1955) 1538; (d) with Mann, M. J. and McLaughlin, K. C. *J. org. Chem. 19* (1954) 1428.
364 Dvoretzky, J. and Richter, G. H. *J. org. Chem. 15* (1950) 1285.
365 Bachman, G. B. and Heisey, L. V. *J. Am. chem. Soc. 68* (1946) 2496.
366 Ainsworth, C. *J. Am. chem. Soc. 77* (1955) 621.
367 Eichenberger, K., Stuber, F. A. and Schmidt, P. *Angew. Chem. (int. Edn.) 6* (1967) 967.
368 (a) Windaus, A. *Ber. dtsch. chem. Ges. 42* (1909) 758. (b) Ewins, A. J. *J. chem. Soc. 99* (1911) 2052; Erlenmeyer, H., Waldi, D. and Sorkin, E. *Helv. chim. Acta 31* (1948) 32.
369 Grindley, R. and Pyman, F. L. *J. chem. Soc.* (1927) 3128.
370 Sonn, A., Hotes, E. and Sieg, H. *Ber. dtsch. chem. Ges. 57* (1924) 953, 2134.
371 Jocelyn, P. C. *J. chem. Soc.* (1957) 3305.
372 Roe, A. M. *J. chem. Soc.* (1963) 2195.
373 Sarasin, J. (a) *Helv. chim. Acta 6* (1923) 370; (b) p. 377; (c) with Wegmann, E. *7* (1924) 713, 720.
374 Balaban, I. E. (a) with Pyman, F. L. *J. chem. Soc. 121* (1922) 947; (b) with Pyman, F. L. *125* (1924) 1564; (c) with King, H. (1927) 1858; (d) (1930) 268; (e) with Pyman, F. L. (1930) 397; (f) (1932) 2423.
375 Fournari, P., de Cointet, P. and Laviron E. *Bull. Soc. chim. Fr.* (1968) 2438.
376 Simonov, A. M. and Sitkina, L. M. *U. S. S. R. Pat.* 178,384 (1966).
377 (a) *U. S. Pat.* 3,290,328 (1966) [*Chem. Abstr. 66* (1967) 55488]; (b) *Netherlands Pat.* 6,609,552 (1967) [*Chem. Abstr. 67* (1967) 54123u].

378 Fränkel, S. and Zeimer, K. *Biochem. Z. 110* (1920) 234; Wellisch, J. *49* (1913) 173; *Swiss Pat.* 92,297 [*Chem. Abstr. 17* (1923) 2119].
379 Stocker, F. B., Kurtz, J. L., Gilman, B. L. and Forsyth, D. A. *J. org. Chem. 35* (1970) 883.
380 Liccari, J. J., Hartzel, L. W., Dougherty, G. and Benson, F. R. *J. Am. chem. Soc. 77* (1955) 5386.
381 Gerngross, O. *Ber. dtsch. chem. Ges.* (a) *42* (1909) 398; (b) *45* (1912) 509; (c) *46* (1913) 1908; (d) p. 1913.
382 Bader, H., Downer, J. D. and Driver, P. *J. chem. Soc.* (1950) 2775.
383 Amâl, H. (a) *Chem. Abstr. 37* (1943) 3091; (b) *44* (1950) 6853; (c) *45* (1951) 610; (d) with Kapuano, L. *Pharm. Acta Helv. 26* (1951) 379; (e) with Ozger, A. *Chem. Abstr. 46* (1952) 4534.
384 Pathak, B. and Ghosh, T. N. *J. Indian chem. Soc. 26* (1949) 371.
385 Mustafa, A., Asker, W., Harhash, A. H., Foda, K. M., Jahme, H. H. and Kassab, N. A. *Tetrahedron 20* (1964) 531.
386 Mannich, C. and Krösche, W. *Arch. Pharm. Berl. 250* (1912) 647.
387 Hellmann, H. and Opitz, G. *Chem. Ber. 89* (1956) 81.
388 Bodendorf, K. (a) with Mildner, J. and Lehmann, T. *Justus Liebigs Annln Chem. 563* (1949) 1; (b) with Raaf, H. *592* (1955) 26; (c) with Ziegler, W. *Chem. Ber. 88* (1955) 1197.
389 Poppelsdorf, F. and Holt, S. J. *J. chem. Soc.* (1954) 1124.
390 Pyl, T. (a) with Lahmer, H. and Beyer, H. *Chem. Ber. 94* (1961) 3217; (b) with Melde, S. and Beyer, H. *Justus Liebigs Annln Chem. 663* (1963) 108.
391 Behal, A. and Choay, E. *Annls Chim. Phys. (6) 27* (1892) 330.
392 Knorr, L. (a) *Ber. dtsch. chem. Ges. 17* (1884) 546; (b) *17* (1884) 2032; (c) with Klotz, C. *20* (1887) 2545; (d) with Laubmann, H. *21* (1888) 1205; (e) with Duden, P. *25* (1892) 759; (f) *28* (1895) 706; (g) *28* (1895) 714; (h) *29* (1896) 249; (i) *30* (1897) 909, 915; (j) with Köhler, A. *39* (1906) 3257; (k) *Justus Liebigs Annln Chem. 238* (1887) 137; (l) *293* (1896) 3 (footnote); (m) with Stolz, F. *293* (1896) 58.
393 Lachowicz, B. *Mh. Chem. 17* (1896) 343.
394 Cocker, W. and Turner, D. G. *J. chem. Soc.* (1940) 57. Anker, R. M. and Cook, A. H. (1944) 489.
395 Sawdey, G. W., Ruoff, M. K. and Vittum, P. W. *J. Am. chem. Soc. 72* (1950) 4947.
396 Schiemann, G. and Winkelmuller, W. *Ber. dtsch. chem. Ges. 66* (1933) 727.
397 Mundici, C. M. *Gazz. chim. ital. 39(ii)* (1909) 123. Porai-Koshits, A. E. and Dinaburg, M. S. *Zh. obshch. Khim. 24* (1954) 635.
398 Finger, H. *J. prakt. Chem. (2) 76* (1907) 93; with Zeh, W. *(2) 82* (1910) 50.
399 Wheeler, H. L. and Hoffman, C. *Am. chem. J. 45* (1911) 368. Wheeler, H. L. and Brautlecht, C. p. 446.
400 Elguero, J., Jacquier, R., Pellegrin, V. and Tabacik, V. *Bull. Soc. chim. Fr.* (1970) 1974.
401 Gillespie, R. J., Grimison, A., Ridd, J. H. and White, R. F. M. *J. chem. Soc.* (1958) 3228.
402 Dizabo, P. and Belloc, J. *Bull. Soc. chim. Fr.* (1969) 230.
403 Harris, T. M. and Randall, J. C. *Chemy. Ind.* (1965) 1728.
404 Vaughan, J. D., Mughrabi, Z. and Chung Wu, E. *J. org. Chem. 35* (1970) 1141.
405 Chung Wu, E. and Vaughan, J. D. *J. org. Chem. 35* (1970) 1146.
406 Olofson, R. A. (a) with Landesberg, J. M., Houk, K. N. and Michelman, J. S. *J. Am. chem. Soc. 88* (1966) 4265; (b) with Thompson, W. R. and Michelman, J. S. *86* (1964) 1865.
407 Norris, W. P. and Henry, R. A. (a) *Tetrahedron Lett.* (1965) 1213; (b) *J. org. Chem. 29* (1964) 650.
408 Rung, F. and Behrend, M. *Justus Liebigs Annln Chem. 271* (1892) 28.

409 Burian, R. *Ber. dtsch. chem. Ges.* 37 (1904) 696.
410 Pauly, H. (a) *Hoppe-Seyler's Z. physiol. Chem.* 42 (1904) 508; 94 (1915) 284; (b) with Arauner, E. *J. prakt. Chem. (2) 118* (1928) 33; (c) with Gundermann, K. *Ber. dtsch. chem. Ges.* 41 (1908) 3999; (d) 43 (1910) 2243.
411 Simonov, A. M. (a) with Garnovskiĭ, A. D. *J. gen. Chem. U. S. S. R.* 31 (1961) 106; (b) with Garnovskiĭ, A. D., Sheinker, Y. N., Khritisch, B. I. and Trofimova, S. S. 33 (1963) 565; (c) with Sitkina, L. M. *Chem. Abstr.* 65 (1966) 13686; (d) with Sitkina, L. M. and Pozharskii, A. F. *Chemy Ind.* (1967) 1454.
412 Pyman, F. L. (a) with Ravald, L. A. *J. chem. Soc. 117* (1920) 1429; (b) with Timmis, L. B. *123* (1923) 494; (c) with Stanley, E. *125* (1924) 2484; (d) with Timmis, L. B. *J. Soc. Dyers Colour.* 38 (1922) 269.
413 Dent, C. E. *Biochem. J.* 43 (1948) 169. Urbach, K. F. *Proc. Soc. Exp. Biol. Med.* 68 (1948) 430. Huebner, C. F. *J. Am. chem. Soc.* 73 (1951) 4667.
414 Ruccia, M. (a) with Santostasi, M. L. and Cusmano, S. *Annali Chim.* 50 (1960) 335; (b) with Natale, G. and Cusmano, S. *Gazz. chim. ital.* 90 (1960) 831; (c) with Natale, G. p. 1047; (d) with Werber, G. p. 1140.
415 Brown, R. D., Duffin, H. C., Maynard, J. C. and Ridd, J. H. *J. chem. Soc.* (1953) 3937.
416 Grimison, A. and Ridd, J. H. *Proc. chem. Soc.* (1958) 256; *J. chem. Soc.* (1959) 3019.
417 Guarneri, M. and Duda, L. (a) *Annali Chim.* 49 (1959) 958; (b) 51 (1961) 446.
418 Hunter, G. (a) with Hlynka, I. *Biochem. J.* 31 (1937) 488; (b) with Nelson, J. A. *Can. J. Res. 19B* (1941) 296.
419 Partridge, M. W. and Stevens, M. F. G. *J. chem. Soc. (C)* (a) (1966) 1127; (b) (1967) 1828.
420 Verkade, P. E. and Dhont, J. *Recl Trav. chim. Pays-Bas Belg.* 64 (1945) 165.
421 v. Rothenburg, R. (a) *Ber. dtsch. chem. Ges.* 25 (1892) 3441; (b) 26 (1893) 2053; (c) *J. prakt. Chem. (2) 52* (1895) 23.
422 Fichter, F., Enzenauer, J. and Uellenberg, E. *Ber. dtsch. chem. Ges.* 33 (1900) 494.
423 Venkataraman, K. *The Chemistry of Synthetic Dyes,* vol. 1, p. 607, Academic Press, New York, 1952. Wiley, R. H. and Wiley, P. *Pyrazolones, Pyrazolidones, and Derivatives,* Interscience Publishers, New York, 1964.
424 Ziegler, J. H. and Locher, M. *Ber. dtsch. chem. Ges.* 20 (1887) 840.
425 Anschütz, R. (a) *Justus Liebigs Annln Chem.* 294 (1897) 219; (b) with Schwickerath, K. 284 (1895) 9; (c) with Montfort, W. F. 284 (1895) 8, 19.
426 Musante, C. (a) *Gazz. chim. ital.* 75 (1945) 109; (b) 75 (1945) 121; (c) 78 (1948) 178; (d) with Fabbrini, L. 84 (1954) 595; (e) *Chem. Abstr.* 45 (1951) 5879.
427 Ioffe, I. S. and Khavin, Z. Y. *Chem. Abstr.* 40 (1946) 2847; 42 (1948) 903; 42 (1948) 1933.
428 Curtius, T. (a) with Thompson, J. *Ber. dtsch. chem. Ges.* 39 (1906) 1388; (b) with Thompson, J. p. 4140; (c) *J. prakt. Chem.* 50 (1894) 508; (d) *et al.* 85 (1912) 37.
429 Dimroth, O. (a) *Justus Liebigs Annln Chem.* 335 (1904) 1; (b) with Eberhardt, E. 335 (1904) 86; (c) 364 (1909) 183; (d) with Aickelin, H. and Merckle, E. 373 (1910) 352.
430 Manchot, W. (a) *Ber. dtsch. chem. Ges.* 31 (1898) 2444; (b) 43 (1910) 1312; (c) with Noll, R. *Justus Liebigs Annln Chem.* 343 (1905) 1.
431 Morgan, G. T. (a) with Reilly, J. *J. chem. Soc. 105* (1914) 435; (b) with Reilly, J. *109* (1916) 155; (c) with Ackerman, I. *123* (1923) 1308.
432 Reimlinger, H. (a) *Chem. Ber.* 92 (1959) 970; (b) 93 (1960) 1857; (c) with van Overstraeten, A. and Viehe, H. G. 94 (1961) 1036; (d) with Oth, J. F. M. 97 (1964) 331; (e) with Mousselois, C. H. 98 (1965) 1805;

(f) with Noels, A., Jadot, J. and van Overstraeten, A. *103* (1970) 1942; (g) with Noels, A. and Jadot, J. *103* (1970) 1949; (h) with Merényi, R. *103* (1970) 3284.

433 Mazzara, G. and Borgo, A. *Gazz. chim. ital.* *36*II (1906) 348; *Atti. Accad. naz. Lincei 15*I (1906) 704.

434 Lutz, A. W. and DeLorenzo, S. *J. heterocycl. Chem. 4* (1967) 399.

435 Severini, O. *Gazz. chim. ital. 23*I (1893) 284.

436 Elguero, J., Jacquier, R. and Tien Duc, H. C. N. *Bull. Soc. chim. Fr.* (1967) 2617.

437 Worrall, D. E. (a) *J. Am. chem. Soc. 59* (1937) 933; *60* (1938) 1198; (b) with Lavin, E. *61* (1939) 104; (c) with Lerner, M. and Washnock, J. *61* (1939) 105.

438 Ware, E. *Chem. Rev. 46* (1950) 403.

439 Westöö, G. *Acta chem. scand. 6* (1952) 1499.

440 Chattaway, F. D. (a) with Irving, H. *J. chem. Soc.* (1931) 786; (b) with Ashworth, D. R. (1933) 475, 1389, 1624; (c) with Ashworth, D. R. and Grimwade, M. (1935) 117.

441 Kitamura, R. and Sunagawa, G. (a) *J. pharm. Soc. Japan 60* (1940) 60; (b) *60* (1940) 65; (c) *61* (1941) 26.

442 (a) Sonn, A. and Litten, W. *Ber. dtsch. chem. Ges. 66* (1933) 1512. (b) Leulier, A. *J. pharm. Chim. (vii) 29* (1924) 447; (c) with Cohen, R. *(viii) 29* (1939) 245.

443 (a) Mukherjee, S. L., Gupta, P. R., Laskar, S. L. and Raymahassay, S. *J. Indian chem. Soc. 30* (1953) 841. (b) *Indian Pat.* 49,798; 50,925 (1955). (c) Franchi, G. *Chem. Abstr. 50* (1956) 8611. (d) Jowett H. A. D. and Potter, C. E. *J. chem. Soc. 83* (1903) 464.

444 Atherton, J. H. and Fields, R. *J. chem. Soc. (C)* (1968) 1507.

445 Closs, G. L. and Heyn, H. *Tetrahedron 22* (1966) 463.

446 Barry, W. J. *J. chem. Soc.* (1961) 3851.

447 Brain, E. G. and Finar, I. L. *J. chem. Soc.* (1958) 2435.

448 Schmir, G. L. (a) with Bruice, T. C. *J. Am. chem. Soc. 80* (1958) 1173; (b) with Cohen, L. A. *Biochemistry 4* (1965) 533; (c) with Jones, W. M and Cohen, L. A. *4* (1965) 539.

449 Langenbeck, W. *J. prakt. Chem. (2) 119* (1928) 77.

450 Linda P. *Tetrahedron 25* (1969) 3297.

451 Tertov B A., Burykin, V. V. and Sadekov, I. D. *Chem. Abstr. 71* (1969) 124328y.

452 Weidenhagen, R. (a) with Herrmann, R. *Ber. dtsch. chem. Ges. 68* (1935) 2205; (b) with Herrmann, R. and Wegner, H. *70* (1937) 570.

453 Forsyth, W. G. (a) with Moore, J. A. and Pyman, F. L. *J. chem. Soc. 125* (1924) 919; (b) with Pyman, F. L. *127* (1925) 573; (c) with Nimkar, V. K. and Pyman, F. L. (1926) 800; (d) with Pyman, F. L. (1930) 397.

454 Lamb, I. D. and Pyman, F. L. *J. chem. Soc. 125* (1924) 706.

455 El Khadem, H., Mansour, H. A. R. and Meshreki, M. H. *J. chem. Soc. (C)* (1968) 1329.

456 Lynch, B. M. (a) with Tze-Lock Chan *Can. J. Chem. 41* (1963) 274; (b) *41* (1963) 2380.

457 Henseke, G. and Schmeisky, I. *J. prakt. Chem. 33* (1966) 256.

458 Boyer, J. H. (a) with Straw, D. *J. Am. chem. Soc. 74* (1952) 4506; (b) *74* (1952) 6274; (c) with Selvarajan, R. *Tetrahedron Lett.* (1969) 47.

459 Settepani, J. A. and Stokes, J. B. *J. org. Chem. 33* (1968) 2606.

460 Smith, L. (a) *Chem. Abstr. 44* (1950) 1490; (b) with Merits, I. and Norlöv, B. *49* (1955) 15863.

461 Kröger, C.-F., Miethchen, R., Frank, H., Siemer M. and Pilz, S. *Chem. Ber. 102* (1969) 755.

462 Möllenhoff, C. *Ber. dtsch. chem. Ges. 25* (1892) 1941.

463 Muckermann, E. *Ber. dtsch. chem. Ges. 42* (1909) 3449.

464 Janssen, R. and Ruysschaert, H. *Bull. Soc. chim. Belg. 67* (1957) 270.

465 De, S. C. *Qu. J. Indian chem. Soc. 3* (1926) 30.
466 Nef, J. U. *Justus Liebigs Annln Chem. 266* (1891) 52.
467 Ledrut, J. and Combes, G. (a) *C. r. hebd. Séanc. Acad. Sci. Paris 231* (1950) 1513; (b) with de Graef, H. *Bull. Soc. chim. Belges 61* (1952) 331.
468 Greenberg, H., van Es, T. and Backeberg, O. G. *J. org. Chem. 31* (1966) 3951.
469 (a) Johnson, T. B. and Hoffman, C. *Am. chem. J. 47* (1912) 20; (b) Dakin, H. D. *Biochem. J. 13* (1919) 398; (c) Hepner, B. and Frenkenberg, S. *J. prakt. Chem. 134* (1932) 249; (d) Nye, J. F. and Mitchell, H. K. *J. Am. chem. Soc. 69* (1947) 1382.
470 Clementi, S., Forsythe, P. P., Johnson, C. D. and Katritzky, A. R. *J. chem. Soc. (Perk. 2)* (1973) 1675.
471 Burton, A. G., Forsythe, P. P., Johnson, C. D. and Katritzky, A. R. *J. chem. Soc. (B)* (1971) 2365.
472 Boulton, B. E. and Coller, B. A. W. *Aust. J. Chem. 24* (1971) 1413.
473 Giles, D., Parnell, E. W. and Renwick, J. D. *J. chem. Soc. (C)* (1966) 1179.
474 Ridd, J. H. (a) *J. chem. Soc.* (1955) 1238; (b) in *Physical Methods in Heterocyclic Chemistry,* vol. 4, edited by Katritzky, A. R., London, 1971.
475 Schutte, L. (a) with Kluit, P. P. and Havinga, E. *Tetrahedron Supplement no. 7* (1966) 295; (b) with Havinga, E. *Recl. Trav. chim. Pays-Bas 86* (1967) 385; (c) with Havinga, E. *Tetrahedron 26* (1970) 2297.
476 Lambert, D. G. and Jones, M. M. *J. Am. chem. Soc. 88* (1966) 5537.
477 de la Mare, P. B. and Ridd, J. H. *Aromatic Substitution: Nitration and Halogenation,* Butterworths, London, 1959.
478 Vaughan, J. D., Jewett, G. L. and Vaughan, V. L. *J. Am. chem. Soc. 89* (1967) 6218.
479 Ragno, M. *Gazz. chim. ital. 68* (1938) 741.
480 Alley, P. W. and Shirley, D. A. (a) *J. Am. chem. Soc. 79* (1957) 4922; (b) *80* (1958) 6271.
481 Snyder, H. R., Ferbane, F. and Bright, D. B. *J. Am. chem. Soc. 74* (1952) 3246.
482 Micetich, R. G. *Can. J. Chem. 48* (1970) 2006.
483 Iverson, P. E. and Lund, H. *Acta chem. scand. 20* (1966) 2649.
484 Raap, R. *Can. J. Chem. 49* (1971) 1792.
485 Schrauth, W. and Bauerschmidt, H. *Ber. dtsch. chem. Ges. 47* (1914) 2736.
486 Patel. H. P. and Tedder, J. M. (a) *J. chem. Soc.* (1963) 4589; (b) with Webster, B. *Chemy. Ind.* (1961) 1163.
487 *Brit. Pat.* (a) 15,759 (1912); (b) 514,203 (1939); (c) 938,726 (1963); (d) 971,606 (1964); (e) 991,644 (1965).
488 *U. S. Pat.* 3,102,890 (1963).
489 Janssen, J. W. A. M. (a) with Habraken, C. L. *J. org. Chem. 36* (1971) 3081; (b) with Koerners, H. J., Kruse, C. G. and Habraken, C. L. *38* (1973) 1777.
490 Coburn, M. D. (a) with Jackson, T. E. *J. het. Chem. 5* (1968) 199; (b) *7* (1970) 345; (c) *7* (1970) 707; (d) *8* (1971) 153.
491 Habraken, C. L., Cohen-Fernandes, P., Balian, S. and van Erk, K. C. *Tetrahedron Lett.* (1970) 479.
492 Behrend, R. and Schmitz, J. *Justus Liebigs Annln Chem. 277* (1893) 310.
493 Matsuura, T., Banba, A. and Ogura, K. *Tetrahedron 27* (1971) 1211.
494 Kochergin, P. M. (a) *Chem. Abstr. 64* (1966) 9709e; (b) with Verenikina, S. G. and Bushueva, K. S. *64* (1966) 9709g; (c) *66* (1967) 104954t; (d) with Bashkir, E. A. *66* (1967) 104955u.
495 Trout, G. E. and Levy, P. R. *Recl Trav. chim. Pays-Bas Belg.* (a) *84* (1965) 1257; (b) *85* (1966) 765.
496 Cox, J. S. G., Fitzmaurice, C., Katritzky, A. R. and Tiddy, G. J. T. *J. chem. Soc. (B)* (1967) 1251.
497 Hazeldine, C. E., Pyman, F. L. and Winchester, J. *J. chem. Soc. 125* (1924) 1431.

498 Bhagwat, V. K. and Pyman, F. L. *J. chem. Soc. 127* (1925) 1832.
499 Dal Monte Casoni, D. (a) *Annali Chim. 48* II (1958) 783; (b) *Gazz. chim. ital. 89* (1959) 1539.
500 *Beilsteins Handbuch der Organischen Chemie*, vol. 25, p. 120, 4th edition, 1936.
501 Eastman, R. H. and Detert, F. L. *J. Am. chem. Soc. 70* (1948) 962.
502 Stagno D'Alcontres, G. *Gazz. chim. ital. 80* (1950) 446.
503 Parrini, V. *Annali Chim. 47* (1957) 929.
504 Lund, H. *J. chem. Soc.* (1935) 418.
505 Behaghel, W. and Buchner, E. *Ber. dtsch. chem. Ges. 35* (1902) 34.
506 Barry, W. J. (a) with Finar, I. L. and Simmonds, A. B. *J. chem. Soc.* (1956) 4974; (b) with Birkett, P. and Finar, I. L. *J. chem. Soc. (C)* (1969) 1328.
507 Grant, R. L. and Pyman, F. L. *J. chem. Soc. 119* (1921) 1893.
508 Ellis, G. P., Epstein, C., Fitzmaurice, C., Goldberg, L. and Lord, G. H. *J. Pharm. Pharmacol. 16* (1964) (a) p. 400; (b) p. 801.
509 Tröger, J. and Thomas, H. *J. prakt. Chem. 110* (1925) 42.
510 Riebsomer, J. L. (a) *J. org. Chem. 13* (1948) 415; (b) with Stauffer, D. A. *16* (1951) 1643.
511 Bishop, C. T. *Science 117* (1953) 715.
512 Lossen, W. (a) with Lossen, C. *Justus Liebigs Annln Chem. 263* (1890) 103; (b) with Statius, F. *298* (1897) 91.
513 McManus, J. M. and Herbst R. M. (a) *J. org. Chem. 24* (1959) 1044; (b) p. 1387.
514 Chipen, G. I. and Grinshtein, V. Y. (a) with Preiman, R. P. *J. gen. Chem. U. S. S. R. 32* (1962) 447; (b) *32* (1962) 452; (c) with Tiltin, M. *Chem. Abstr. 59* (1963) 6227; (d) p. 12790; (e) with Grinvalde, A. p. 12791.
515 Herbst, R. M. (a) with Roberts, C. W. and Harvill, E. J. *J. org. Chem. 16* (1951) 139; (b) with Garrison J. A. *18* (1953) 941; (c) with Percival, D. F. *19* (1954) 439; (d) with Garrison, J. A. *22* (1957) 278; (e) with Percival, D. F. *22* (1957) 925; (f) with Froberger, C. F. *22* (1957) 1050; (g) with Wilson, K. R. and Haak, W. J. *24* (1959) 1046.
516 Backer, H. J. (a) with Mulder, C. H. K. *Recl Trav. chim. Pays-Bas Belg. 44* (1925) 1113; (b) with Meyer, W. *45* (1926) 82.
517 Krohs, W. *Chem. Ber. 88* (1955) 866.
518 Franchimont, A. P. N. and Klobbie, E. A. *Recl Trav. chim. Pays-Bas Belg. 7* (1888) 12, 236; *8* (1889) 283; Franchimont, A. P. N. and van Erp, H. *15* (1896) 165. Harries, C. *Justus Liebigs Annln Chem. 361* (1908) 71.
519 Stuckey, R. E. *J. chem. Soc.* (1947) 331.
520 Zimmermann, W. and Cuthbertson, D. P. *Hoppe-Seyler's Z. physiol. Chem. 205* (1932) 38.
521 Matthes, H. and Rammstedt, O. *Arch. Pharm. Berl. 245* (1907) 112. Hougounenq, L., Florence, G. and Couture, E. *Bull. Soc. Chim. biol. 7* (1925) 58.
522 Iseki, T., Suguira, T., Yasunaga, S. and Nakasina. M. *Ber. dtsch. chem. Ges. 74* (1941) 1420.
523 Altschul, J. *Ber. dtsch. chem. Ges. 25* (1892) 1842.
524 Lederer, L. *J. prakt. Chem. (2) 45* (1892) 83.
525 Thoms, H. and Schnupp, J. *Justus Liebigs Annln Chem. 434* (1923) 296.
526 Grünanger, P., Finzi, P. V. and Fabbri. E. *Gazz. chim. ital. 90* (1960) 413.
527 Bülow, C. and Haas, K. (a) *Ber. dtsch. chem. Ges. 42* (1909) 4638; (b) *43* (1910) 2647.
528 Ferguson, I. J., Grimmett, M. R. and Schofield, K. *Tetrahedron Lett.* (1972) 2771.
529 Barnett, J. W., Ferguson, I. J. and Schofield, K., unpublished results.
530 Birkofer, L. and Franz, M. *Chem. Ber. 104* (1971) 3062.
531 Dorn, H. and Dilcher, H. *Justus Liebigs Annln Chem. 717* (1968) 118.
532 Cusmano, S. (a) *Gazz. chim. ital. 69* (1939) 594; (b) *69* (1939) 621; (c) with Didonna, G. *85* (1955) 208; (d) with Ruccia. M. *88* (1958) 463.

533 Pesson, M. and Polmanss, G. *C. r. hebd. Séanc. Acad. Sci., Paris 247* (1958) 787.
534 Checchi, S., Papini, P. and Ridi, M. (a) *Gazz. chim. ital. 85* (1955) 1160; (b) *85* (1955) 1558; (c) *86* (1956) 631.
535 Betti, M. *Gazz. chim. ital. 34 (i)* (1904) 179.
536 Dutt, S. and Dharam, I. N. *Chem. Abstr. 34* (1940) 425.
537 Mohr, E. *J. prakt. Chem. (2) 79* (1909) 1.
538 (a) Ajello, T. *Gazz. chim. ital. 70* (1940) 401; (b) Amorosa, M. *Chem. Abstr. 34* (1940) 7910; (c) Efros, L. G. and Davidenkov, L. R. *40* (1952) 8100.
539 (a) Klebanskiĭ, A. L. and Lemke, A. L. *Chem. Abstr. 29* (1935) 6891; (b) Brockmühl, M. *31* (1937) 5796; (c) *Jap. Pat.* 130 (1954), *Chem. Abstr. 49* (1955) 11020.
540 *U. S. Pat.* (a) 307,399 (1884); (b) 579,412 (1897); (c) 2,005,505 (1935); (d) 2,068,790 (1937); (e) 2,234,866 (1941); (f) 2,315,836 (1943); (g) 2,476,525 (1949); (h) 2,476,549 (1949); (i) 2,883,373 (1959); (j) 3,066,137 (1962).
541 (a) Heymons, A. and Rohland, W. *Ber. dtsch. chem. Ges. 66* (1933) 1654; (b) *Can. Pat.* 374,822 (1938).
542 Rondestvedt, C. S. and Chang, P. K. *J. Amer. Chem. Soc. 77* (1955) 6532.
543 Barnes, G. R. and Pyman, F. L. *J. chem. Soc.* (1927) 2711.
544 Bennett, L. L. and Baker, H. T. *J. Am. chem. Soc. 79* (1957) 2188.
545 Fisher, M. H., Nicholson, W. H. and Stuart, R. S. *Can. J. Chem. 39* (1961) 1336.
546 Barry, W. J., Finar, I. L. and Khatkhate, G. V. *J. chem. Soc. (C)* (1968) 1120.
547 Sanna, G. *Chem. Abstr. 37* (1943) 2356; with Sollai, V. *Gazz. chim. ital. 72* (1942) 313.
548 Crippa, G. B. and Guarneri, M. *Gazz. chim. ital. 85* (1955) 199.
549 Otto, H. H. *Arch. Pharm. (Weinheim) 304* (1971) 504.
550 Pesin, V. G., Khaletskiĭ, A. M. and Thyum-Syan Den *J. Gen. Chem. U. S. S. R. 28* (1958) 2841.
551 Michaelis, A. with (a) Röhmer, H. *Ber. dtsch. chem. Ges. 31* (1898) 3003; (b) Bindewald, H. *33* (1900) 2873; (c) Gunkel, E. *34* (1901) 723; (d) Bender, F. *36* (1903) 523; (e) Schmidt, O. *43* (1910) 2116; (f) Stau, B. *46* (1913) 3612.
552 Michaelis, A. (a) *et al. Justus Liebigs Annln Chem. 320* (1902) 1; (b) *et al. 331* (1904) 197; (c) with Klopstock, H. *354* (1907) 102; (d) *385* (1911) 1; (e) with Bressel, H. *407* (1915) 274.
553 Bauer, L., Dhawan, D. and Mahajanshetti, C. S. *J. org. Chem. 31* (1966) 2491.
554 Büchi, J. with (a) Ursprung, R. and Lauener, G. *Helv. chim. Acta 32* (1949) 984; (b) Meyer, H. R., Hirt, R., Hunziker, F., Eichenberger, E. and Lieberherz, R. *38* (1955) 670.
555 Wright, J. B., Dulin, W. E. and Markillie, J. H. *J. med. Chem. 7* (1964) 102.
556 Witiak, D. T. with (a) Lu, M. C. *J. org. Chem. 33* (1968) 4451; (b) Sinha, B. K. *35* (1970) 501.
557 Arndt, F. G. in *Organic Analysis,* vol. 1, edited by Mitchell, J., Kolthoff, I. M., Proskauer, E. S. and Weissberger, A. New York, 1953.
558 Kyrides, L. P., Zientz, F. B., Steahly, G. W. and Morrill, H. L. *J. org. Chem. 12* (1947) 577.
559 Hubball, W. and Pyman, F. L. *J. chem. Soc.* (1928) 21.
560 Davis, S. B. and Ross, W. F. *J. Am. chem. Soc. 69* (1947) 1177.
561 Haring, M. *Helv. chim. Açta 42* (1959) 1845.
562 Townsend, L. B. (a) *Chem. Rev. 67* (1967) 533; (b) with Rousseau, R. J. and Robins, R. K. *J. heterocycl. Chem. 4* (1967) 311.
563 Dankova, T. F., Genkin, E. I. and Preobrazhenskiĭ, N. A. *Zh. obshch. Khim. 15* (1945) 189.

564 Allsebrook, W. E., Gulland, J. M. and Story, L. F. *J. chem. Soc.* (1942) 232.
565 Baddiley, J., Buchanan, J. G., Hardy, F. E. and Stewart, J. *J. chem. Soc.* (1959) 2893.
566 Gulland, J. M. and Macrae, T. F. *J. chem. Soc.* (1933) 662.
567 Bergmann, E. and Heinhold, H. *J. chem. Soc.* (1936) 505.
568 Howard, G. A., McLean, A. C., Newbold, G. T., Spring, F. S. and Todd, A. R. *J. chem. Soc.* (1949) 232.
569 Giesemann, H. (a) with Hälschke, G. *Chem. Ber.* *92* (1959) 92; (b) with Lettau, H. and Mannsfeldt, H.-G. *93* (1960) 570; (c) with Oelschlägel, A. and Pfau, H. *93* (1960) 576.
570 Baxter, R. A. and Spring, F. S. *J. chem. Soc.* (1945) 232.
571 Jowett, H. A. D. *J. chem. Soc.* *87* (1905) 405.
572 Asinger, F., Offermanns, H. and Krings, P. *Justus Liebigs Annln Chem.* *719* (1968) 145.
573 Sakami, W. and Wilson, D. N. *J. biol. Chem.* *154* (1944) 215.
574 Hudson, R. F. and Withey, R. J. *J. chem. Soc.* (1964) 3513.
575 Oliver, J. E. and Stokes, J. B. *J. heterocycl. Chem.* *7* (1970) 961.
576 Begtrup, M. (a) with Pedersen, C. *Acta chem. Scand.* *19* (1965) 2022; (b) with Pedersen, C. *20* (1966) 1555; (c) with Pedersen, C. *21* (1967) 633; (d) with Pedersen, C. and Hansen, K. *21* (1967) 1235; (e) *23* (1969) 2025; (f) with Kristensen, P. A. *23* (1969) 2733; (g) *24* (1970) 1819; (h) *25* (1971) 249; (i) *25* (1971) 795; (j) *25* (1971) 803; (k) with Poulsen, K. V. *25* (1971) 2087.
577 Atkinson, M. R. and Polya, J. B. (a) *J. chem. Soc.* (1954) 141; (b) (1954) 3319; (c) *Chemy Ind.* (1954) 462.
578 Carpenter, W. R. *J. org. Chem.* *27* (1962) 2085.
579 Finnegan, W. G. (a) with Henry, R. A. and Lieber, E. *J. org. Chem.* *18* (1953) 779; (b) with Smith, S. R. *J. Chromatography* *5* (1961) 461.
580 Fries, K. and Saftien, K. *Ber. dtsch. chem. Ges.* *59* (1926) 1246.
581 Henry, R. A. (a) with Dehn, W. M. *J. Am. chem. Soc.* *71* (1949) 2297; (b) *73* (1951) 4470; (c) with Finnegan W. G. *76* (1954) 923; (d) with Finnegan, W. G. and Lieber, E. *76* (1954) 88; (e) with Finnegan, G. and Lieber, E. *77* (1955) 2264.
582 Elpern, B. *J. Am. chem. Soc.* *75* (1953) 661.
583 Hattori, K., Lieber, E. and Horwitz, J. P. *J. Am. chem. Soc.* *78* (1956) 411.
584 Raap. R. and Howard, J. *Can. J. Chem.* *47* (1969) 813.
585 Fraser. R. R. and Haque, K. E. *Can. J. Chem.* *46* (1968) 2855.
586 Kaiser, D. W. and Peters, G. A. *J. org. Chem.* *18* (1953) 196.
587 Bryden, J. H., Henry, R. A., Finnegan, W. G., Boschan, R. H., McEwan, W. S. and van Dolah, R. W. *J. Am. chem. Soc.* *75* (1953) 4863.
588 von Braun, J. (a) with Keller, W. *Ber. dtsch. chem. Ges.* *65* (1932) 1677; (b) *Justus Liebigs Annln Chem.* *507* (1933) 14.
589 Huff, L., Forkey, D. W. and Henry, R. A. *J. org. Chem.* *35* (1970) 2074; Ansell, G. B. *Chem. Comm.* (1970) 684.
590 Scott. F. L. (a) with O'Donovan, D. G. and Reilly, J. *J. appl. Chem. Lond.* *2* (1952) 368; (b) with Britten, F. C. and Reilly, J. *J. org. Chem.* *21* (1956) 1191; (c) with Butler, R. N. *J. chem. Soc.* (1966) 1202; (d) with Tobin, J. C. *J. chem. Soc. (C)* (1971) 703.
591 Butler, R. N. and Scott, F. L. (a) *J. org. Chem.* *31* (1966) 3182; (b) *32* (1967) 1224; (c) with Feeney, J. *J. chem. Soc. (B)* (1967) 919; (d) *J. chem. Soc. (C)* (1967) 239; (e) *J. chem. Soc. (C)* (1968) 1711.
592 Fosse, R., Hieulle, A. and Bass, L.-W. *C. r. hebd. Séanc. Acad. Sci. Paris* *178* (1924) 811.
593 *Jap. Pat.* (a) 1717 (1951); (b) 2779 (1952); (c) 1289, 6281, 7984 (1962); (d) 12633, 12634 (1965).
594 Beyer, H. and Stehwein, D. *Arch. Pharm., Berl.* *286* (1953) 13.
595 (a) Rodionov, V. *Bull. Soc. chim. Fr.* *39* (1926) 305; (b) Valyashko, N. A.

and Bliznyukov, V. I. *Chem. Abstr.* 24 (1930) 5751; (c) *Encyclopedia of Chemical Technology*, vol. 1, p. 858, Interscience Publishers, New York, 1947; (d) p. 859.

596 Sonn, A. and Litten, W. *Ber. dtsch. chem. Ges.* 66 (1933) 1582.
597 Grandmougin, E., Havas, E. and Guyot, G. *Chemikerzeitung 37* (1913) 812.
598 Elwood, J. K. and Gates, J. W. *J. org. Chem. 32* (1967) 2956.
599 McGee, M. A., Newbold, G. T., Redpath, J. and Spring, F. S. *J. chem. Soc.* ·(1960) 2536.
600 Arndt. F. (a) with Loewe, L. and Tarlan-Akön, A. *Chem. Abstr. 42* (1948) 8190; (b) with Loewe, L. and Ergener, L. *43* (1949) 579.
601 *Swiss Pat.* (a) 266,236 (1950); (b) 269,087 (1950); (c) 269,088 (1950); (d) 373,840 (1964).
602 Hilbert, G. E. *J. Am. chem. Soc. 54* (1932) 3413.
603 Biltz, H. and Slotta, K. *J. prakt. Chem. 113* (1926) 233.
604 Scarpati, R., Sica, D. and Lionetti, A. *Gazz. chim. ital. 93* (1963) 90.
605 Daeniker, H. H. and Druey, J. *Helv. chim. Acta 45* (1962) 2441.
606 Chipen, G. I. and Bolkadere, R. P. *Chem. Abstr. 70* (1969) 115078u.
607 *Netherlands Pat.* (a) 6,406,077 (1964); (b) 6,409,120 (1965); (c) 6,501,015 (1965); (d) 6,501,053 (1965); (e) 6,504,121 (1965); (f) 6,510,485 (1966); (g) 301,383 (1965) [*Chem. Abstr. 64* (1966) 8196].
608 *Belg. Pat.* (a) 611,906 (1962); (b) 621,842 (1963); (c) 630,494 (1963); (d) 671,402 (1966).
609 Acree, S. F. (a) *Ber. dtsch. chem. Ges. 36* (1903) 3139. (b) *Am. chem. J. 32* (1904) 609; (c) *37* (1907) 71; (d) *38* (1907) 1.
610 Cohen, A., King, H. and Strangeways, W. I. *J. chem. Soc.* (1931) 3043.
611 Kjellin, G. and Sandström, J. *Acta chem. scand. 23* (1969) 2879.
612 Hünig, S. and Oette, K.-H. *Justus Liebigs Annln Chem. 641* (1961) 94.
613 Duffin, G. F., Kendall, J. D. and Waddington, H. R. J. (a) *Chemy Ind.* (1954) 1458; (b) (1955) 1355; (c) *J. chem. Soc.* (1959) 3799.
614 Nirenburg, V. L. and Postovskiĭ, I. Y. (a) *J. gen. Chem. U. S. S. R. 34* (1964) 2540; (b) *34* (1964) 3245; (c) *Chem. Abstr. 63* (1965) 11544.
615 Knorr, L. (a) *Ber. dtsch. chem. Ges. 37* (1904) 3520; (b) with Weidel, A. *42* (1909) 3523; (c) *Justus Liebigs Annln Chem. 293* (1896) 1.
616 Shepard, E. R. and Shonle, H. A. *J. Am. chem. Soc. 69* (1947) 2269.
617 *Brit. Pat.* (a) 521,821 (1938); (b) 626,475 (1951); (c) 766,380 (1957); (d) 785,334 (1957); (e) 837,471 (1960); (f) 844,419 (1960); (g) 863,060 (1961); (h) 891,515 (1962); (i) 972,956 (1964); (j) 974,523 (1964); (k) 999,381 (1965); (l) 1,070,243 (1967); (m) 1.157,256 (1969).
618 Boekelheide, V. and Fedoruk, N. A. *J. Am. chem. Soc. 90* (1968) 3830.
619 Pozharskii, A. F., Chegolya, T. N. and Simonov, A. M. *Chem. Abstr. 69* (1968) 95825j.
620 Chegolya, T. N. *Chem. Abstr. 71* (1969) 12411p.
621 Olofson, R. A. and Kendall, R. V. *J. org. Chem. 35* (1970) 2246.
622 Seiber, J. N. and Tolkmith, H. *Tetrahedron Lett.* (1967) 3333.
623 Wiley, R. H. (a) with Smith, N. R., Johnson, D. M. and Moffat, J. *J. Am. chem. Soc. 76* (1954) 4933; (b) with Moffat, J. *77* (1955) 1703; (c) with Smith, N. R., Johnson, D. M. and Moffat J. *77* (1955) 2572.
624 Gabel, W. and Schmidegg, O. *Mh. Chem. 47* (1926) 748.
625 Hernler, F. *Mh. Chem. 48* (1927) 392.
626 Balli, H. and Kersting, F. *Justus Liebigs Annln Chem. 647* (1961) 1.
627 Voltz, J. *Chimia (Switz.) 15* (1961) 168.
628 Benson, F. R., Hartzel, L. W. and Savell, W. L. *J. Am. chem. Soc. 73* (1951) 4457.
629 Harvill E. K., Roberts, C. W. and Herbst, R. M. *J. org. Chem. 15* (1950) 58.
630 Becker, H. G. O. and Boettcher, H. (a) with Roethling, T. and Timpe, H. J. *Chem. Abstr. 64* (1966) 19596; (b) *Tetrahedron 24* (1968) 2687.

631 Ogilvie, J. W. and Corwin, A. H. *J. Am. chem. Soc. 83* (1961) 5023.
632 Khan, M. A. and Polya, J. B. *J. chem. Soc. (C)* (1970) 85.
633 Zahn, H. and Pfannmüller, H. *Biochem. Z. 330* (1958) 97.
634 Imbach, J.-L. and Jacquier, R. *C. r. hebd. Séanc. Acad. Sci. Paris 257* (1963) 2683.
635 Hubert, A. J. and Reimlinger, H. *Chem. Ber. 103* (1970) 3811.
636 Chiriac, C. and Zugrǎvescu, I. *Rev. Roum. Chem. 15* (1970) 1201.
637 Andreani, F., Andrisano, R., Della Casa, C. and Tramontini, M. *Tetrahedron Lett.* (1968) 1059.
638 Acheson, R. M. (a) with Poulter, P. W. *J. chem. Soc.* (1960) 2138; (b) with Vernon, J. M. (1962) 1148; (c) with Foxton, M. W., Abbott, P. J. and Mills, K. R. *J. chem. Soc. (C)* (1967) 882; (d) with Foxton, M. W. (1968) 389; (e) *Advances heterocycl. Chem. 1* (1963) 125.
639 Crabtree, A. and Johnson, A. W. *J. chem. Soc.* (1962) 1510.
640 Renn, D. W. and Herbst, R. M. *J. org. Chem. 24* (1959) 473.
641 Schwan, T. J. *J. heterocycl. Chem. 5* (1968) 199.
642 Ried, W. (a) with Pfaender, P. *Justus Liebigs Annln Chem. 640* (1961) 111; (b) with Deuschel, G. and Kotelko, A. *642* (1961) 121; (c) with Beck, B. M. *646* (1961) 96; (d) with Köcher, E.-U. *647* (1961) 116; (e) with Lohwasser, H. *699* (1966) 88.
643 Profft, E. and George, W. *Justus Liebigs Annln Chem. 643* (1961) 136.
644 Franke, W. and Kraft, R. *Chem. Ber. 86* (1953) 797.
645 Fox, H. H. and Gibas, J. T. *J. org. Chem. 20* (1955) 60.
646 Seidel, F., Thier, W., Uber, A. and Dittmer, J. *Ber. dtsch. chem. Ges. 68* (1935) 1913.
647 Grigat, E. and Putter, R. *Chem. Ber. 97* (1964) 3027.
648 Barthel, J. and Schmeer, G. *Justus Liebigs Annln Chem. 738* (1970) 195.
649 Walter, W. and Radke, M. (a) *Ang. Chem. (int. Edn.) 7* (1968) 302; (b) *Justus Liebigs Annln Chem. 739* (1970) 201.
650 Ruggli, P. *Helv. chim. Acta 3* (1920) 559.
651 Bergmann, M. and Zervas, L. *Hoppe-Seyler's Z. physiol. Chem. 175* (1928) 145.
652 Caplow, M. and Jencks, W. P. *Biochemistry 1* (1962) 883.
653 Fife, T. H. *J. Am. chem. Soc. 87* (1965) 4597.
654 Fee, J. A. and Fife, T. H. *J. org. Chem. 31* (1966) 2343.
655 Klinman, J. P. and Thornton, E. R. *J. Am. chem. Soc. 90* (1968) 4390.
656 Anderson, G. W. and Paul, R. *J. Am. chem. Soc. 80* (1958) 4423.
657 Rivett, D. E. and Wilshire, J. F. K. *Aust. J. Chem. 19* (1966) 165.
658 Reddy, G. S. (a) with Mandell, L. and Goldstein, J. H. *J. chem. Soc.* (1963) 1414; (b) with Gehring, D. G. *J. org. Chem. 32* (1967) 2291.
659 Bunton, C. A. *J. chem. Soc.* (1963) 6045.
660 Pullakat, T. J. and Urry, G. *Tetrahedron Lett.* (1967) 1953.
661 Stadtman, E. R. and White, F. H. *J. Am. chem. Soc. 75* (1953) 2022; *Chem. Abstr. 48* (1954) 8275.
662 Bender, M. L. and Turnquest, B. W. *J. Am. chem. Soc. 79* (1957) 1652.
663 Oakenfull, D. G. *J. chem. Soc. (B)* (1970) 197.
664 Jencks, W. P. *Catalysis in Chemistry and Enzymology,* McGraw-Hill, New York, 1969. Bender, M. L. *Mechanisms of Homogeneous Catalysis from Protons to Proteins,* Wiley-Interscience, New York, 1971.
665 Blyth, C. A. and Knowles, J. R. *J. Am. chem. Soc. 93* (1971) 3021.
666 Overberger, C. G. and Yuen, P. S. *J. Am. chem. Soc. 92* (1970) 1667.
667 Mollica, J. A. and Connors, K. A. *J. Am. chem. Soc. 89* (1967) 308.
668 Brunner, K. and Medweth, J. *Mh. Chem. 47* (1926) 741.
669 (a) v. Meyer, A. *J. prakt. Chem. (2) 90* (1914) 1. (b) Anderson, E. L., Casey, J. E., Greene, L. C., Lafferty, J. J. and Reiff, H. E. *J. med. Chem. 7* (1964) 259. (c) Dymek, W., Janik, B. and Zimon, R. *Chem. Abstr. 63* (1965) 18066.
670 Makisumi, Y. (a) *Chem. pharm. Bull., Tokyo 9* (1961) 801, 808, 878;

(b) *10* (1962) 612; (c) *10* (1962) 620; (d) with Kano, H. *11* (1963) 67; (e) with Watanabe, H. and Tori, K. *12* (1964) 204.

671 Takamizawa, A. (a) with Hayashi, S. *Chem. Abstr. 59* (1963) 5147; (b) with Hamajima, Y. *65* (1966) 16983, 16984; (c) with Hayashi, S. *65* (1966) 20144; (d) with Hamashima, Y. *66* (1967) 10951, 10952, 10957.

672 Beyer, H., Pyl, T. and Wünsch, K.-H. *Chem. Ber. 93* (1960) 2209.

673 Burtles, R. and Pyman, F. L. *J. chem. Soc. 127* (1925) 2012.

674 Cook, A. H. (a) with Downer, J. D. and Heilbron, Sir I. *J. chem. Soc.* (1948) 1262; (b) with Davis, A. C., Heilbron, Sir I. and Thomas, G. H. (1949) 1071.

675 van den Bos, B. G. (a) with Koopmans, M. J. and Huisman, H. O. *Recl Trav. chim. Pays-Bas 79* (1960) 807; (b) *79* (1960) 836; (c) with Schipperheyn, A. and van Deursen, F. W. *85* (1966) 429.

676 Gehlen, H. (a) with Blankenstein, G. *Justus Liebigs Annln Chem. 651* (1962) 137; (b) with Roebisch, G. *660* (1962) 148; (c) with Schade, W. *675* (1964) 180; (d) with Winzer, V. *681* (1965) 100; (e) with Schmidt, J. *682* (1965) 123.

677 Levin, Y. A. (a) with Fedotova, A. P., Rakova, N. F., Savicheva, G. A. and Kukhtin, V. A. *J. gen. Chem. U. S. S. R. 33* (1963) 1279; (b) with Gul'kina, N. A. and Kukhtin, V. A. *33* (1963) 2603; (c) with Kukhtin, V. A. *33* (1963) 2608; (d) with Fedotova, A. P. and Kukhtin, V. A. *34* (1964) 501; (e) with Platonova, R. N. and Kukhtin, V. A. *Chem. Abstr. 64* (1966) 15878.

678 Ponomarev, A. A. and Lipanova, M. D. *Chem. Abstr. 56* (1962) 2440.

679 Papini, P. and Manganelli, M. *Gazz. chim. ital. 80* (1950) 855.

680 Torrey, H. A. and Zanetti J. E. *Am. chem. J. 44* (1910) 391.

681 Weissberger, A. and Porter, H. D. *J. Am. chem. Soc. 65* (1943) 1495.

682 Jassmann, E. and Schulz, H. *Pharmazie 18* (1963) 461.

683 Lawson, A. *J. chem. Soc.* (1957) 1443.

684 Waugh, R. C., Ekeley, J. B. and Ronzio, A. R. *J. Am. chem. Soc. 64* (1942) 2028. Cornforth, J. W. *The Chemistry of Penicillin,* Princeton University Press, 1949, p. 688.

685 Ginsburg, S. *J. org. Chem. 27* (1962) 4062.

686 Korte, F. and Störiko, K. *Chem. Ber. 94* (1961) 1956.

687 Begtrup, M. and Pedersen, C. *Acta chem. Scand. 23* (1969) 1091.

688 Thiele, J. (a) with Schleussner, K. *Justus Liebigs Annln Chem. 295* (1897) 129; (b) with Manchot, W. *303* (1898) 33.

689 Busch, M. and Heinrichs, C. *Ber. dtsch. chem. Ges. 34* (1901) 2331.

690 Padwa, A. *J. org. Chem. 30* (1965) 1274.

691 Bertrand, M., Elguero, J., Jacquier, R. and Le Gras, J. *C. r. hebd. Séanc. Acad. Sci. Paris 262C* (1966) 782. Le Gras, J. *263C* (1966) 1460.

692 Jozefczak, C., Bram, G. and Vilkas, M. *C. r. hebd. Séanc. Acad. Sci. Paris 271C* (1970) 553.

693 *U. S. Pat.* (a) 2,444,605-6-7 [*Chem. Abstr. 42* (1948) 7178, 7179, 7180]; (b) 2,575,182 (1951) [*Chem. Abstr. 46* (1952) 1904]; (c) 2,931,814 (1960) [*Chem. Abstr. 54* (1960) 19716]; (d) 3,047,582 (1962) [*Chem. Abstr. 59* (1963) 15419]; (e) 3,056,780 (1962) [*Chem. Abstr. 58* (1963) 5703]; (f) 3,096,329 [*Chem. Abstr. 59* (1963) 14004]; (g) 3,121,092 (1964) [*Chem. Abstr. 60* (1964) 12020d]; (h) 3,190,886 (1965) [*Chem. Abstr. 63* (1965) 11749e]; (i) 3,202,512 (1965); (j) 3,293,259 (1966) [*Chem. Abstr. 66* (1967) P55489j].

694 Veldstra, H. and Wiardi, P. W. *Recl Trav. chim. Pays-Bas Belg. 61* (1942) 627.

695 Anderson, G. W., Faith, H. E., Marson, H. W., Winne', ℩. S. and Robbin, R. O. *J. Am. chem. Soc. 64* (1942) 2902.

696 Hultquist, M. E., Germann, R. P., Webb, J. S., Wright, W. B., Roth, B., Smith, J. M. and Subba Row, Y. *J. Am. chem. Soc. 73* (1951) 2558.

697 Jensen, K. A. (a) *Chem. Abstr. 36* (1942) 5793; (b) with Pedersen, C.

Acta chem. Scand. 15 (1961) 991; (c) with Holm, A. and Rachlin, S. *20* (1966) 2795.

698 Himmelbauer, R. *J. prakt. Chem. (2) 54* (1896) 177.
699 Boyle, F. T. and Jones, R. A. Y. *J. chem. Soc. (Perk. II)* (1973) 164.
700 Coburn, M. D. *J. het. Chem. 7* (1970) 455.
701 Botwinnik, M. M. and Prokofiev, M. A. *J. prakt. Chem. (2) 148* (1937) 191.
702 Müller, E. and Meier, H. *Justus Liebigs Annln Chem. 716* (1968) 11.
703 Luijten, J. G. A., Janssen, M. J. and van der Kerk, G. J. M. *Recl Trav. chim. Pays-Bas Belg. 81* (1962) 202.
704 Trofimenko, S. (a) *J. Am. chem. Soc. 88* (1966) 1842; *89* (1967) 3165 4948; *91* (1969) 588; (b) *88* (1966) 5588.
705 Seel, F. and Sperber, V. *Angew. Chem. (int. Edn.) 7* (1968) 70.
706 Mayers, D. F. and Kaiser, E. T. *J. Am. chem. Soc. 90* (1968) 6192.
707 Dorn, H. and Arndt, D. *Justus Liebigs Annln Chem. 750* (1971) 39.
708 Rosenberg, T. *Arch Biochem. Biophys. 105* (1964) 315.
709 Beard, L. N. and Lenhert, P. G. *Acta crystallogr. 24B* (1968) 1529.
710 Jampel, E., Wakselman, M. and Vilkas, M. *Tetrahedron Lett.* (1968) 3533.
711 Blakeley, R., Kerst, F. and Westheimer, F. H. *J. Am. chem. Soc. 88* (1966) 112.
712 Reimlinger, H. with (a) Noels, A., Jadot, J. and van Overstraeten, A. *Chem. Ber. 103* (1970) 1954; (b) Peiren, M. A. and Merényi, R. *103* (1970) 3252.
713 Hart, C. V. *J. Am. chem. Soc. 50* (1928) 1922.
714 Dorn, H. and Zubek, A. *Angew. Chem. (int. Edn.) 6* (1967) 958.
715 Dornow, A. (a) with Siebrecht, M. *Chem. Ber. 93* (1960) 1106; (b) with Menzel, H. and Marx, P. *97* (1964) 2185; (c) with Dehmer, K. *100* (1967) 2577.
716 Schmidt, P., Meier, K. and Druey, J. *Angew. Chem. 70* (1958) 344.
717 Hayashi, S. *Chem. Abstr. 63* (1965) 5644.
718 Solomons, T. W. G. and Voigt, C. F. *J. Am. chem. Soc. 87* (1965) 5256.
719 Allen, C. F. H. (a) with Beilfuss, H. R., Burness, D. M., Reynolds, G. A., Tinker, J. F. and Van Allan, J. A. *J. org. Chem. 24* (1959) 779; (b) p. 787; (c) p. 793; (d) p. 796; (e) with Reynolds, G. A., Tinker, J. F. and Williams, L. A. *25* (1960) 361.
720 de Cat, A. and van Dormael, A. *Bull. Soc. chim. Belg. 59* (1950) 573; *60* (1951) 69.
721 Prokof'ev, M. A. and Shvochkin, Yu. P. *Chem. Abstr. 49* (1955) 9661; *50* (1956) 3458.
722 Birr, E. J. *Z. wiss. Phot. 47* (1952) 2.
723 Sutherland, D. R. (a) with Tennant, G. *Chem. Commun.* (1969) 1070; (b) with Tennant, G. *J. chem. Soc. (C)* (1971) 706; (c) with Tennant, G. p. 2156; (d) with Tennant, G. and Vevers, R. J. S. *J. chem. Soc. (Perk. I)* (1973) 943.
724 Bülow, C. (a) *Ber. dtsch. chem. Ges. 42* (1909) 2208; (b) p. 2594; (c) p. 4429; (d) *43* (1910) 3401.
725 Libermann, D. and Jacquier, R. *Bull. Soc. chim. Fr.* (1962) 355.
726 Steck, E. A. and Brundage, R. P. *J. Am. chem. Soc. 81* (1959) 6289.
727 Giuliano, R. and Leonardi, G. *Chem. Abstr. 53* (1959) 10248. Williams, L. A. *J. chem. Soc.* (1961) 3046; (1962) 2222. Tenor, E. and Kröger, C. F. *Chem. Ber. 97* (1964) 1373. Kreutzberger, A. *Chem. Ber. 99* (1966) 2237. Spickett, R. G. W. and Wright, S. H. B. *J. chem. Soc. (C)* (1967) 503.
728 Paudler, W. W. and Helmick, L. S. *J. het. Chem. 3* (1966) 269.
729 Reimlinger, H. (a) *Chem. Ber. 103* (1970) 1900; (b) with Peiren, M. A. *Chem. Ber. 103* (1970) 3266; (c) with Jacquier, R. and Daunis, J. *104* (1971) 2702.
730 Potts, K. T. and Hirsch, C. *J. org. Chem. 33* (1968) 143.
731 Brady, L. and Herbst, R. M. *J. org. Chem. 24* (1959) 922. Powell, H. *Dissert. Abstr. 22* (1962) 4189.

732 Godt H. C. and Blicke, F. F. *J. org. Chem. 34* (1969) 2008.
733 Sunjić, V., Fajdiga, T., Japelj, M. and Rems, P. *J. heterocycl. Chem. 6* (1969) 53.
734 Zauhar, J. and Ladouceur, B. F. *Can. J. Chem. 46* (1968) 1079.
735 Michaelis, A. with (a) Röhmer, H. *Ber. dtsch. chem. Ges. 31* (1898) 2907; (b) Röhmer, H. *31* (1898) 3193; (c) Voss, U. and Greuss, M. *34* (1901) 1300; (d) Hepner, E. *36* (1903) 3271; (e) Leonhardt, R. *36* (1903) 3597; (f) Eisenschmidt, C. *37* (1904) 2228; (g) Kobert, K. *42* (1909) 2765; (h) Leonhardt, R. and Wahle, K. *Justus Liebigs Annln Chem. 338* (1905) 183.
736 Thielepape, E. and Spreckelsen, O. *Ber. dtsch. chem. Ges. 55* (1922) 2929.
737 Kochergin, P. M., Tsyganova, A. M., Shlikunova, V. S. and Klykov, M. A. *Chem. Abstr. 76* (1972) 126867a.
738 Prümenko, B. A. and Kochergin, P. M. *Chem. Abstr. 76* (1972) 25174s.
739 *Fr. Pat.* 1,389,779 [*Chem. Abstr. 63* (1965) 7019].
740 Gorbacheva, L. I., Grandberg, I. I. and Kost, A. N. *J. gen. Chem. U. S. S. R. 34* (1964) 652.
741 Kauffmann, T., Nurnberg, R., Schulz, J. and Stabba R. *Tetrahedron Lett.* (1967) 4273.
742 de Bie, D. A., van der Plas, H. C. and Geurtsen, G. *Recl Trav. chim. Pays-Bas Belg. 90* (1971) 594.
743 Duquénois, P. and Amal. H. *Bull. Soc. chim. Fr. 9* (1942) 718.
744 Mustafa, A., Asker, W., Shalaby, A. F. A. and Selim, Z. *J. org. Chem. 26* (1961) 1779.
745 Smith, P. A. S. (a) with Krbechek, L. O. and Resemann, W. *J. Am. chem. Soc. 86* (1964) 2025; (b) with Wirth, J. G. *J. org. Chem. 33* (1968) 1144; (c) with Breen, G. J. W., Hajek, M. K. and Awang, D. V. C. *35* (1970) 2215.
746 Durrant, G. J., Foottit, M. E., Ganellin, C. R., Loynes, J. M., Pepper, E. S. and Roe, A. M. *Chem. Commun.* (1968) 108.
747 O'Donovan, D. G., Carroll, A. P. and Reilly, J. *Nature, Lond. 185* (1960) 460.
748 Montequi, F. *Chem. Abstr. 21* (1927) 3353.
749 Mann, F. G. and Porter, J. W. G. *J. chem. Soc.* (1945) 751.
750 Rousseau, R. J., Townsend, L. B. and Robins, R. K. *Chem. Commun.* (1966) (1966) 265.
751 Taylor, E. C. and Loeffler, P. K. *J. Am. chem. Soc. 82* (1960) 3147.
752 Lund, H. *J. chem. Soc.* (1933) 686.
753 King, H. and Murch, W. O. *J. chem. Soc. 123* (1923) 621.
754 Bell, F. *J. chem. Soc.* (1941) 285.
755 Mayer, K. *Ber. dtsch. chem. Ges. 36* (1903) 717.
756 Hill, H. B. and Black, O. F. *Am. chem. J. 33* (1905) 292.
757 Clemo, G. R. and Holmes, T. *J. chem. Soc.* (1934) 1739.
758 Björkquist, B., Helgstrand, E. and Stjernström, N. E. *Acta chem. Scand. 21* (1967) 2295.
759 Kitamura R. *Chem. Abstr. 35* (1941) 4770.
760 Jacobsen, P. and Jost, H. *Justus Liebigs Annln Chem. 400* (1913) 195.
761 Kashparov, I. S. and Pozharskii, A. F. *Chem. Abstr. 75* (1971) 35922c.
762 Freeman, J. P. and Gannon, J. J. *J. org. Chem. 34* (1969) 194.
763 Moureu, C. and Lazennec, I. *C. r. hebd. Séanc. Acad. Sci. Paris 143* (1906) 1239.
764 Jagerspacher, C. *Ber. dtsch. chem. Ges. 28* (1895) 1283.
765 Shimada, K., Kuriyama, S., Kanazawa, T., Satok, M. and Toyoshima, S. *Chem. Abstr. 74* (1971) 141631w.
766 Witkowski, J. T. and Robins, R. K. *J. org. Chem. 35* (1970) 2635.
767 Musliner, W. J. and Gates, J. W. *J. Am. chem. Soc. 88* (1966) 4271.
768 Barfknecht, C. F., Smith, R. V. and Reif, V. D. *Can. J. Chem. 48* (1970) 2138.

769 Chapman, D. D., Jones, E. T., Wilgus, H. S., Nelander, D. H. and Gates, J. W. *J. org. Chem. 30* (1965) 1520.
770 Roblin, R. O. (a) with Williams, J. H., Winnek, P. S. and English, J. P. *J. Am. chem. Soc. 62* (1940) 2002; (b) with Clapp, J. W. *72* (1950) 4890.
771 Cheng, C. C. *J. heterocycl. Chem. 5* (1968) 195.
772 Lancini G. C. and Lazzari, E. *Experientia 21* (1965) 83.
773 Bearman, A. G. and Tautz, W. *J. Am. chem. Soc. 87* (1965) 389.
774 Angelini, C. and Martani, A. *Chem. Abstr. 50* (1956) 1649, 3416.
775 *East Ger. Pat.* 59,288 (1967) [*Chem. Abstr. 70* (1969) 28922w].
776 Grimmett, M. R. Private communication; Begg, C. G., Grimmett, M. R. and Yu-Man, L. *Aust. J. Chem. 26* (1973) 415.
777 Lynch, B. M. (a) with Kahn, M. A. *Can. J. Chem. 41* (1963) 2086; (b) with Chang, H. S. *Tetrahedron Lett.* (1964) 617; (c) (1964) 2965.
778 Dou, H. J. M. (a) with Lynch, B. M. *Tetrahedron Lett.* (1965) 897; (b) with Lynch, B. M. *Bull. Soc. chim. Fr.* (1966) 3815; (c) with Lynch, B. M. (1966) 3820; (d) with Vernin, G. and Metzger, J. (1972) 1173.
779 Lettau, H. and Heckel H. J. *Z. Chem. 11* (1971) 62.
780 v. Walther, R. and Litter, H. *J. prakt. Chem. 83* (1911) 171.
781 Darapsky, A. *Ber. dtsch. chem. Ges. 46* (1913) 863.
782 van Alphen, J. *Recl Trav. chim. Pays-Bas 62* (1943) 485.
783 Bouchet, P., Elguero, J. and Jacquier, R. *Tetrahedron 22* (1966) 2461.
784 (a) Aubagnac, J.-L., Elguero, J., Jacquier, R. and Tizané, D. *Tetrahedron Lett.* (1967) 3705. (b) Elguero, J., Jacquier, R. and Tizané, D. *Tetrahedron 27* (1971) 133.
785 Hinman, R. L., Elefson, R. D. and Campbell, R. D. *J. Am. chem. Soc. 82* (1960) 3988.
786 Waser, E. and Gratsos, A. *Helv. chim. Acta 11* (1928) 944.
787 Schubert, H. (a) with Hofmann, S. *J. prakt. Chem. 7* (1959) 119; (b) with Selisko, L. *16* (1962) 1; (c) with Hagen, E. and Lehmann, G. *17* (1962) 173.
788 Godefroi, E. F. *J. org. Chem. 33* (1968) 860.
789 LaForge, R. A., Cosgrove, C. E. and D'Adamo, A. *J. org. Chem. 21* (1956) 988.
790 Kuhn, R. and Jerchel, D. *Ber. dtsch. chem. Ges. 74* (1941) 941, 949.
791 Nineham, A. W. *Chem. Rev. 55* (1955) 355. Hooper, W. D. *Rev. Pure and Appl. Chem. 19* (1969) 221.
792 Grimshaw, J. and Trocha-Grimshaw, J. *J. chem. Soc. (Perk. I)* (1973) 1275.
793 Parham W. E. (a) with Serres, C. and O'Connor, P. R. *J. Am. chem. Soc. 80* (1958) 588; (b) with Braxton, H. G. and O'Connor, P. R. *J. org. Chem. 26* (1961) 1805.
794 Loudon, J. D. and Young, L. B. *J. chem. Soc.* (1963) 5496.
795 Wibaut, J. P. and Boon, J. W. P. *Helv. chim. Acta 44* (1961) 1171.
796 Searles, S. and Hine, W. R. *J. Am. chem. Soc. 79* (1957) 3175.
797 Fichter, F. and de Montmollin, H. *Helv. chim. Acta 5* (1922) 256.
798 Gillis, B. T. and Weinkam, R. *J. org. Chem. 32* (1967) 3321.
799 Rees, C. W. and Yelland, M. *J. chem. Soc. (Perk. I)* (1973) 221.
800 Freeman, J. P., Surbey, D. L. and Kassner. J. E. *Tetrahedron Lett.* (1970) 3797.
801 Veibel, S. *Acta chem. Scand. 26* (1972) 3685.
802 Jowett, H. A. D. *J. chem. Soc. 83* (1903) 438.
803 Schubert, H. and Rudorf, H.-D. *Angew. Chem. (int. Edn.) 5* (1966) 674.
804 van Meeteren, H. W. and van der Plas, H. C. *Recl Trav. chim. Pays-Bas Belg. 88* (1969) 204.
805 Takayama, Y. and Oeda, H. *Chem. Abstr. 27* (1933) 2099; *28* (1934) 6438.
806 Baumgärtel, H. and Zimmermann, H. *Chem. Ber.* (a) *99* (1966) 843; (b) *102* (1969) 1755.
807 Radziszewski, R. *Ber. dtsch. chem. Ges. 10* (1877) 70.

808 Dufraisse, C., Etienne, A. and Martel, J. *C. r. hebd. Séanc. Acad. Sci. Paris* *244* (1957) 970; Dufraisse, C. and Martel, J. *244* (1957) 3106.

809 White, E. H. and Harding, M. J. C. *J. Am. chem. Soc. 86* (1964) 5686; *Photochem. Photobiol. 4* (1965) 1129.

810 Sonnenberg, J. and White, D. M. *J. Am. chem. Soc. 86* (1964) 5685.

811 White, D. M. and Sonnenberg, J. *J. Am. chem. Soc. 88* (1966) 3825.

812 Wasserman, H. H., Stiller, K. and Floyd, M. B. *Tetrahedron Lett.* (1968) 3277.

813 Sakai, K. and Anselme, J.-P. (a) *Tetrahedron Lett.* (1970) 3851; (b) *Bull. chem. Soc. Japan 45* (1972) 306.

814 Gelus, M. and Bonnier, J.-M. *J. Chim. phys. 65* (1968) 253.

815 Wentrup, C. and Crow, W. D. *Tetrahedron 27* (1971) 361.

816 Dimroth, O. *Justus Liebigs Annln Chem.* (a) *338* (1905) 143; (b) *364* (1908) 183; (c) *373* (1910) 336; (d) *377* (1910) 127.

817 Dimroth, O. and Taub, L. *Ber. dtsch. chem. Ges. 39* (1906) 3912.

818 Brown, D. J. in *Mechanisms of Molecular Migrations*, vol. 1, edited by Thyagarajan, B. S. Interscience, New York, 1968.

819 Dimroth, O. and Michaelis, W. *Justus Liebigs Annln Chem. 459* (1927) 39.

820 Lieber, E. (a) with Henry, R. A. and Finnegan, W. G. *J. Am. chem. Soc. 75* (1953) 2023; (b) with Rao, C. N. R. and Chao, T. S. *79* (1957) 5962.

821 Brown, B. R., Hammick, D. L. and Heritage, G. G. *J. chem. Soc.* (1953) 3820.

822 Hermes, M. E. and Marsh, F. D. *J. Am. chem. Soc. 89* (1967) 4760.

823 Anderson, D. J., Gilchrist, T. L., Gymer, G. E. and Rees, C. W. *Chem. Commun.* (1971) 1518. Gilchrist, T. L., Gymer, G. E. and Rees, C. W. *Chem. Commun.* (1971) 1519; *J. chem. Soc. (Perk. I)* (1973) 555.

824 Selvarajan, S. and Boyer, J. H. *J. heterocycl. Chem. 9* (1972) 87.

825 Garbrecht, W. L. and Herbst, R. M. *J. org. Chem. 18* (1953) 1269, 1283.

826 Herbst, R. M. (a) with Klingbeck, J. E. *J. org. Chem. 23* (1958) 1912; (b) *26* (1961) 2372.

827 Hantzsch, A. and Vagt, A. *Justus Liebigs Annln Chem. 314* (1901) 339.

828 Huisgen, R. (a) with Seidel, M., Sauer, J., McFarland, J. W. and Wallbillich, G. *J. org. Chem. 24* (1959) 892; (b) with Grashey, R., Seidel, M., Wallbillich, G., Knupfer, H. and Schmidt, R. *Justus Liebigs Annln Chem. 653* (1962) 105; (c) with Sauer, J. and Seidel, M. *654* (1962) 146; (d) with Sauer, J. and Seidel, M. *Chem. Ber. 93* (1960) 2885; (e) with Sturm, H. J. and Seidel, M. *94* (1961) 1555; (f) with Axen, C. and Seidel, H. *98* (1965) 2966; (g) *Angew. Chem. 72* (1960) 359.

829 Markgraf, J. H., Brown, S. H., Kaplinsky, M. W. and Peterson, R. G. *J. org. Chem. 29* (1964) 2629.

830 Behringer, H. (a) with Fischer, H. J. *Chem. Ber. 95* (1962) 2546; (b) with Matner, M. *Tetrahedron Lett.* (1966) 1663.

831 Smith, P. A. S. *J. Am. chem. Soc. 76* (1954) 436; Smith, P. A. S. and Leon, E. *80* (1958) 4647; Vaughan, J. and Smith, P. A. S. *J. org. Chem. 23* (1958) 1909.

832 Boyer, J. H. and Miller, E. J. *J. Am. chem. Soc. 81* (1959) 4671. Temple, C. and Montgomery, J. A. *86* (1964) 2946; *J. org. Chem. 30* (1965) 826, 829. Temple, C., Coburn, W. C., Thorpe, M. C. and Montgomery, J. A. *30* (1965) 2395. Smirnova, N. B., Postovskii, I. Y., Vereshchagina, N. N., Lundina, I. B. and Mudretsova, I. I. *Chem. Abstr. 69* (1968) 106647f. Scott, F. L., Cronin, D. A. and O'Halloran, J. K. *J. chem. Soc. (C)* (1971) 2769.

833 Soon-Yung Hong and Baldwin, J. E. *Tetrahedron 24* (1968) 3787.

834 Pinner, A. *Justus Liebigs Annln Chem. 298* (1897) 1.

835 Sauer, J., Huisgen, R. and Sturm, H. J. *Tetrahedron 11* (1960) 241.

836 Arnold, C. and Thatcher, D. N. *J. org. Chem. 34* (1969) 1141.

837 Brown, H. C. and Kassal, R. J. *J. org. Chem. 32* (1967) 1871.

838 Nagy, H. K., Tomson, A. J. and Horwitz, J. P. *J. Am. chem. Soc. 82* (1960)

1609.
839 Hagedorn, I. and Winkelmann, H.-D. *Chem. Ber.* 99 (1960) 850.
840 Raap, R. *Can. J. Chem.* 49 (1971) 2139.
841 Scott, F. L., Murphy, C. M. B. and Reilly, J. *Nature, Lond.* 167 (1951) 1037.
842 Knorr, L. *Ber. dtsch. chem. Ges.* 39 (1906) 3265.
843 Fusco, R., Rosnati, V. and Pagani, G. *Tetrahedron Lett.* (a) (1966) 1739; (b) (1967) 4541.
844 Grandberg, I. I. and Bobrova, N. I. *Chem. Abstr.* 64 (1966) 3516.
845 Cusmano, S. and Ruccia, M. (a) *Gazz. chim. ital.* 85 (1955) 1686; (b) 86 (1956) 187.
846 Garnovskii, A. D., Simonov, A. M., Minkin, V. I. and Dionis'ev, V. D. *Chem. Abstr.* 62 (1965) 8530.
847 Schönherr, H.-J. and Wanzlick, H.-W. *Justus Liebigs Annln Chem.* 731 (1970) 176.
848 Bamberger, E. and Berlé, B. *Justus Liebigs Annln Chem.* 273 (1893) 342.
849 Ruggli, P., Ratti, R. and Henzi, E. *Helv. chim. Acta* 12 (1929) 332. Ruggli, P. and Henzi, E., p. 362.
850 Windaus, A. (a) with Knoop, F. *Ber. dtsch. chem. Ges.* 38 (1905) 1166; (b) with Vogt, W. 40 (1907) 3691; (c) 43 (1910) 499; (d) with Dörries, W. and Jensen, H. 54 (1921) 2745; (e) with Langenbeck, W. 55 (1922) 3706.
851 Stensiö, K.-E. *Acta chem. Scand.* 25 (1971) 1483.
852 Ashley, J. W. and Harrington, C. R. *J. chem. Soc.* (1930) 2586.
853 Heller, G. *Ber. dtsch. chem. Ges.* 37 (1904) 3112.
854 Babad, E. and Ben-Ishai, D. *J. heterocycl. Chem.* 6 (1969) 235.
855 Hirch, B. and Bassl, A. *Chem. Abstr.* 59 (1963) 15410.
856 Fusco, R. and Rossi, S. *Tetrahedron* 3 (1958) 209.
857 Wright, J. B. *J. org. Chem.* 34 (1969) 2474.
858 Jennings, A. L. and Boggs, J. E. *J. org. Chem.* 29 (1964) 2065.
859 Nishiwaki, T. (a) *J. chem. Soc. (B)* 9 (1967) 885; (b) *Bull. chem. Soc. Japan* 42 (1969) 3024.
860 Kheml'nitskii, R. A. and Krasnoshcheck, A. P. (a) with Polyakova, A. A. and Grandberg, I. I. *Chem. Abstr.* 68 (1968) 34199r; (b) with Vysotskii, V. I. 72 (1970) 99729r.
861 Krasnoshcheck, A. P. (a) with Khmel'nitskii, R. A., Polyakova, A. A. and Grandberg, I. I. *Chem. Abstr.* 69 (1968) 2346r; (b) with Minkin, V. I. 70 (1969) 19432j.
862 van Thuijl, J., Klebe, K. J. and van Houte, J. J. (a) *Org. Mass Spectrom.* 3 (1970) 1549; (b) 5 (1971) 1101; (c) *J. heterocycl. Chem.* 8 (1971) 311.
863 Simons, B. K., Kallury, R. K. M. R. and Bowie, J. H. *Org. Mass Spectrom.* 2 (1969) 739.
864 Desmarchelier, J. M. and Johns, R. B. *Org. Mass Spectrom.* 2 (1969) 697.
865 Finar, I. L. and Millard, B. J. *J. chem. Soc. (C)* (1969) 2497.
866 Bowie, J. H. (a) with Cooks, R. G., Lawesson, S. O. and Schroll, G. *Aust. J. Chem.* 20 (1967) 1613; (b) with Donaghue, P. F., Rodda, H. J. and Simons, B. K. *Tetrahedron* 24 (1968) 3965.
867 Hodges, R. and Grimmett, M. R. *Aust. J. Chem.* 21 (1968) 1085.
868 Seiler, N., Schneider, H. and Sonnenberg, D. *Z. anal. Chem.* 252 (1970) 127.
869 Ikekawa, N. and Honma, Y. *Tetrahedron Lett.* (1967) 1197.
870 Lord, G. H., Millard, B. J. and Memel, J. *J. chem. Soc. (Perk. I)* (1973) 572.
871 Compernolle, F. and Dekeirel, M. *Org. Mass Spectrom.* 5 (1971) 427.
872 Briggs, P. R., Parker, W. L. and Shannon, T. W. *Chem. Comm.* (1968) 727.
873 Becker, H. G. O., Beyer, D. and Timpe, H. J. *J. prakt. Chem.* 312 (1970) 869.
874 Brady, L. E. *J. heterocycl. Chem.* 7 (1970) 1223.
875 Baker, F. S., Busby, R. E., Iqbal, M., Parrick, J. and Shaw, C. J. G. *Chemy. Ind.* (1969) 1344.

876 Jones, R. L. and Rees, C. W. *J. chem. Soc. (C)* (1969) (a) p. 2251; (b) p. 2255.
877 Takimoto, H. H. and Denault, G. C. (a) *Tetrahedron Lett.* (1966) 5369; (b) with Hatta, S. *J. heterocycl. Chem. 3* (1966) 119.
878 Kuhn, R., Neugebauer, F. A. and Trischmann, H. *Mh. Chem. 97* (1966) 846.
879 Tiefenthaler, H., Dörscheln, W., Göth, H. and Schmid, H. *Helv. chim. Acta 50* (1967) 2244.
880 Beak, P. (a) with Miesel, J. L. and Messer, W. R. *Tetrahedron Lett.* (1967) 5315; (b) with Messer, W. *Tetrahedron 25* (1969) 3287.
881 Eğe, S. N. *J. chem. Soc. (C)* (1969) 2624.
882 Reisch, J. and Fitzek, A. *Tetrahedron Lett.* (1969) 271.
883 Bouchet, P., Coquelet, C., Elguero, J. and Jacquier, R. *Tetrahedron Lett.* (1973) 891.
884 Burgess, E. M., Carithers, R. and McCullagh, L. *J. Am. chem. Soc. 90* (1968) 1923.
885 Cantrell, T. S. and Haller, W. S. *Chem. Comm.* (1968) 977.
886 Willey, F. G. *Angew. Chem. (int. Edn.) 3* (1964) 138.
887 Kato, H., Shiba, T. and Miki, Y. *Chem. Comm.* (1972) 498.
888 Moriarty, R. M. and Kliegman, J. M. (a) with Shovlin, C. *J. Am. chem. Soc. 89* (1967) 5958; (b) p. 5959.
889 Chae, Y. B., Chang, K. S. and Kim, S. S. *Chem. Abstr. 70* (1969) 20031j.
890 Scheiner, P. (a) *J. org. Chem. 34* (1969) 199; (b) with Dinda, J. F. *Tetrahedron 26* (1970) 2619; (c) with Litchman, W. M. *Chem. Comm.* (1972) 781.
891 Angadiyavar, C. S. and George, M. V. *J. org. Chem. 36* (1971) 1589.
892 Hobbs, P. D. and Magnus, P. D. *J. chem. Soc. (Perk. I)* (1973) 469.
893 Bassett, J. M. and Brown, R. D. *J. chem. Soc.* (1954) 2701. Brown, R. D. and Heffernan, M. L. *Aust. J. Chem. 12* (1959) 543.
894 Adam, W. and Grimison, A. *Tetrahedron 22* (1966) 835.
895 Hüttel, R. (a) *Ber. dtsch. chem. Ges. 74* (1941) 1680; (b) with Riedl, J., Martin, H. and Franke, K. *93* (1960) 1425; (c) with Franke, K., Martin, H. and Riedl, J. *93* (1960) 1433; (d) with Schneiderhan, T., Hertwig, H., Leuchs, A., Reinecke, V. and Miller, J. *Justus Liebigs Annln Chem. 585* (1954) 115.
896 Mingoia, Q. (a) *Gazz. chim. ital. 63* (1933) 242; (b) with Ingraffia, F. *64* (1934) 279.
897 Ried, W. (a) with Schleimer, B. *Justus Liebigs Annln Chem. 619* (1958) 43; (b) with Königstein, F. *622* (1959) 37; (c) with Schleimer, B. *626* (1959) 98; (d) with Beck, B. M. *657* (1962) 98; (e) with Schmidt, E. *676* (1964) 114; (f) with Mengler, H. *678* (1964) 95; (g) with Schleimer, B. *Angew. Chem. 70* (1958) 164.
898 Grandberg, I. I. (a) with Kost, A. N. and Sibiryakova, D. V. *J. gen. Chem. U. S. S. R. 30* (1960) 2896; (b) with Kost, A. N. *30* (1960) 2916; (c) with Py, D. V., Shchegolova, V. I. and Kost, A. N. *31* (1961) 1770; (d) with Kost, A. N. *31* (1961) 3454; (e) with Kost, A. N. *32* (1962) 866; (f) with Kost, A. N. *32* (1962) 1542; (g) with Vinokurov, V. G., Troitskaya, V. S. and Sharova, G. I. *32* (1962) 3515; (h) with Gorbacheva, L. I. and Kost, A. N. *33* (1963) 503; (i) with Gorbacheva, L. I., Kost, A. N. and Sibiryakova-Fedotova, D. V. *33* (1963) 508; (j) with Krashashchek, A. P., Kost, A. N. and Faizova, G. K. *33* (1963) 2521; (k) with Sharova, G. I. *Khim. geterotsikl. Soedin.* (1968) 325; (l) with Milovanova, S. N., Kost, A. N. and Nette, I. T. *Chem. Abstr. 56* (1962) 9368.
899 Comrie, A. M. *J. chem. Soc. (Perk. I)* (1972) 1193.
900 v. Auwers, K. (a) with Cauer, E. *J. prakt. Chem. 126* (1930) 177; (b) with Wiegand, C. *134* (1932) 82; (c) with Dersch, F. *Justus Liebigs Annln Chem. 426* (1928) 104; (d) with Cauer, E. *470* (1929) 284; (e) with König, F. *496* (1932) 27; (f) with Mauss, W. *Ber. dtsch. chem. Ges. 59* (1926) 611;

(g) with Cauer, E. *61* (1928) 2402.
901 Fusco, R. in *The Chemistry of Heterocyclic Compounds, Pyrazoles, Pyr-azolones, Pyrazolidines, Indazoles and Condensed Rings,* edited by R. H. Wiley, p. 684, Wiley-Interscience, New York, 1967.
902 Elguero, J. (a) with Jacquier, R. *C. r. hebd. Séanc. Acad. Sci. Paris 260* (1965) 606; (b) with Gelin, R., Gelin, S. and Tarrago, G. *Bull. Soc. chim. Fr.* (1970) 231; (c) with Jacquier, R. and Mignonac-Mondon, S. (1970) 1346; (d) with Jacquier, R. and Mignonac-Mondon, S. (1972) 2807.
903 Henry, R. A. *J. Am. Chem. Soc. 72* (1950) 5343.
904 Staab, H. A. (a) with Otting, W. and Ueberle, A. *Z. Elektrochem. 61* (1957) 1000; (b) with Lüking, M. and Dürr, F. H. *Chem. Ber. 95* (1962) 1275; (c) with Mannschreck, A. *95* (1962) 1284; (d) with Walther, G. *95* (1962) 2070; (e) with Walther, G. and Rohr, W. *95* (1962) 2073; (f) with Rohr, W. and Graf, F. *98* (1965) 1122; (g) with Graf, F. and Rohr, W. *98* (1965) 1128; (h) with Malek, G. *99* (1966) 2955; (i) *Angew. Chem. 71* (1959) 194; (j) with Rohr, W. and Mannschreck, A. *73* (1961) 143; (k) with Datta, A. P. *Angew. Chem. (int. Edn.) 3* (1964) 132; (l) with Braenling, H. *Justus Liebigs Annln Chem. 654* (1962) 119; (m) with Jost, E. *655* (1962) 90; (n) with Wendel, K. and Datta, A. P. *694* (1966) 78.
905 Folli, U. and Iarossi, D. *Bull. Sci. Fac. Chim. Ind. Bologna 26* (1968) 61.
906 Birkofer, L. (a) with Richter, P. and Ritter, A. *Chem. Ber. 92* (1959) 1302; (b) with Plath, D. and Ritter, A. *96* (1963) 2090; (c) with Franz, *105* (1972) 1759; (d) with Ritter, A. and Richter, P. *Tetrahedron Lett.* (1962) 195; (e) with Ritter, A. in *Newer Methods of Preparative Organic Chemistry,* edited by G. W. Foerst, vol. 5, p. 211, New York, 1968.
907 Ykman, P., L'abbe, G. and Smets, G. (a) *Tetrahedron Lett.* (1970) 5225; (b) *Tetrahedron 27* (1971) 5623; (c) *Chemy. Ind.* (1972) 886.
908 Van den Bos, B. G. (a) *Recl Trav. chim. Pays-Bas, Belg. 79* (1960) 1129; (b) with Schoot, C. J., Koopmans, M. J. and Meltzer, J. *80* (1961) 1040.
909 *Neth. Appl.* (a) 6,407,401 (1965) [*Chem. Abstr. 62* (1965) 15362]; (b) 6,415,324 (1965) [*Chem. Abstr. 64* (1966) 2092]; (c) 6,609,596 (1967) [*Chem. Abstr. 67* (1967) 64397].
910 Cipens, G. and Grinsteins, V. *Chem. Abstr. 63* (1965) 13243.
911 Hirata, T. (a) with Twanmoh, L., Wood, H. B., Goldin, A. and Driscoll, J. S. *J. heterocycl. Chem. 9* (1972) 99; (b) with Wood, H. B., and Driscoll, J. S. *J. chem. Soc. (Perk. I)* (1973) 1209.
912 Coburn, M. D. (a) with Loughran, E. D. and Smith, L. C. *J. heterocycl. Chem. 7* (1970) 1149; (b) with Neumann, P. N. *7* (1970) 1391.
913 Gehlen, H. (a) with Röbisch, G. *Justus Liebigs Annln Chem. 663* (1963) 119; (b) with Dost, J. *665* (1963) 144.
914 Taylor, E. C. (a) with Hendess, R. W. *J. Am. Chem. Soc. 87* (1965) 1980; (b) with Loeffler, P. K. *J. org. Chem. 24* (1959) 2035.
915 Huisgen, R. (a) with Grashey, R., Gotthardt, H. and Schmidt, R. *Angew. Chem. (int. Edn.) 1* (1962) 48; (b) with Gotthardt, H. and Grashey, R. *Angew. Chem. 74* (1962) 30; (c) with Funke, E., Schäfer, F. C., Gotthardt, H. and Brunn, E. *Tetrahedron Lett.* (1967) 1809; (d) with Grashey, R., Aufderhaar, E. and Kunz, R. *Chem. Ber. 98* (1965) 642; (e) with Knorr, R., Moebius, L. and Szemies, G. *98* (1965) 4014; (f) with Seidel, M., Wallbillich, G. and Knupfer, H. *Tetrahedron 17* (1962) 3.
916 Scott, F. L. (a) *Chimia 11* (1957) 163; (b) with O'Mahony, T. A. F. *Tetrahedron Lett.* (1970) 1841; (c) with Lambe, T. M. and Tobin, J. C. *Chem. Commun.* (1971) 411; (d) with O'Sullivan, D. A. and Reilly, J. *Chemy. Ind.* (1952) 782.
917 Jencks, W. P. and Salvesen, K. *J. Am. Chem. Soc. 93* (1971) 1419.
918 Guibe-Jampel, E. and Wakselman, M. (a) *Bull. Soc. chim. Fr.* (1971) 2554; (b) with Vilkas, M. (1971) 1308.
919 Page, M. I. and Jencks, W. P. (a) *J. Am. Chem. Soc. 94* (1972) 3263; (b) *94* (1972) 8818, 8828.

920 Oakenfull, D. G. and Jencks, W. P. (a) *J. Am. Chem. Soc. 93* (1971) 178;
(b) *93* (1971) 188.
921 Gerstein, J. and Jencks, W. P. *J. Am. Chem. Soc. 86* (1964) 4655.
922 Kirsch, J. F. and Jencks, W. P. *J. Am. Chem. Soc. 86* (1964) 837.
923 Carlsson, L. O. and Sandström, J. *Acta chem. scand. 24* (1970) 299.
924 Gompper, R. (a) with Herlinger, H. *Chem. Ber. 89* (1956) 2816; (b) with
Ruhle, H. *Justus Liebigs Annln Chem. 626* (1960) 83.
925 Zalikin, A. A., Nikitenkova, L. P. and Strepikheev, Yu. A. *J. gen. Chem.
U. S. S. R. 41* (1971) 1940.
926 Bergman, J. *Tetrahedron Lett.* (1972) 4723.
927 Paul, R. and Anderson, G. W. (a) *J. Am. Chem. Soc. 82* (1960) 4596;
(b) *J. org. Chem. 27* (1962) 2094.
928 Weygand, F., Prox, A., Schmidhammer, L. and König, W. *Angew. Chem.
(int. Edn.) 2* (1963) 183.
929 Bestmann, H. J., Sommer, N. and Staab, H. A. *Angew. Chem. (int. Edn.) 1*
(1962) 270.
930 Hecht, F. and Rüchardt, C. *Chem. Ber. 96* (1963) 1281.
931 Otting, W. and Staab, H. A. *Justus Liebigs Annln Chem. 622* (1959) 23.
932 Appel, R. and Hauss, A. *Chem. Ber. 93* (1960) 405.
933 Staab, H. A. (a) with Schaller, H. and Cramer, F. *Angew. Chem. 71* (1959)
736; (b) with Wendel, K. *72* (1960) 708; (c) with Wendel, K. *73* (1961)
26.
934 Cramer, F. and Schaller, H. (a) *Chem. Ber. 94* (1961) 1634; (b) with Staab,
H. A. *94* (1961) 1612.
935 Schaller, H., Staab, H. A. and Cramer, F. *Chem. Ber. 94* (1961) 1621.
936 Floret, A. *Bull. Soc. chim. Fr.* (1971) 1109.
937 Corey, E. J. (a) with Winter, R. A. E. *J. Am. Chem. Soc. 85* (1963) 2677;
(b) with Corey, R. A. and Winter, R. A. E. *87* (1965) 934.
938 Horton, D. and Turner, W. N. *Tetrahedron Lett.* (1964) 2531.
939 Walter, W. and Radke, M. *Justus Liebigs Annln Chem.* (1973) 676.
940 Markley, L. D. and Dorman, L. C. *Tetrahedron Lett.* (1970) 1787.
941 Thé, K. I. and Peterson, L. K. *Chem. Commun.* (1972) 841.
942 Kosuge, T., Okeda, H., Aburatani, M., Ito, H. and Kosake, S. *J. pharm. Soc.
Japan 74* (1954) 1086.
943 Rojahn, C. A. (a) with Kühling, H. E. *Arch. Pharm. Berl. 264* (1926) 337;
(b) with Fahr, K. *Justus Liebigs Annln Chem. 434* (1923) 252; (c) with
Seitz, A. *437* (1924) 297; (d) with Trieloff, H. *445* (1925) 296.
944 Greco, C. B. and Pellegrini, F. (a) *J. chem. Soc. (Perk. I)* (1972) 720;
(b) with Pesce, M. A. (1972) 1623; (c) with Pesce, M. A. *J. heterocycl.
Chem. 9* (1972) 967.
945 Finar, I. L. (a) with Utting, K. *J. chem. Soc.* (1959) 4015; (b) with Walter,
B. H. (1960) 1588; (c) with Saunders, R. J. (1963) 3967; (d) with
Saunders, H. E. *J. chem. Soc. (C)* (1969) 1495.
946 Wijnberger, C. and Habraken, C. L. *J. heterocycl. Chem. 6* (1969) 545.
947 *Brit. Pat.* (a) 785,185 (1957) [*Chem. Abstr. 52* (1958) 5478]; (b) 797,144
(1958) [*Chem. Abstr. 53* (1959) 4983]; (c) 837,838 (1960) [*Chem. Abstr.
54* (1960) 24804]; (d) 884,851 (1961) [*Chem. Abstr. 61* (1964) 3116];
(e) 923,734 (1963) [*Chem. Abstr. 59* (1963) 11705]; (f) 1,026,631
(1966) [*Chem. Abstr. 64* (1966) 19630]; (g) 1,114,154 (1970) [*Chem.
Abstr. 75* (1971) 140848f]; (h) 1,119,638 (1968) [*Chem. Abstr. 69*
(1968) 106705y].
948 Henkel, K. and Weygand, F. *Ber. dtsch. chem. Ges. 76* (1943) 812.
949 Nesmeyanov, A. N. and Rybinskaya, M. I. (a) *Dokl. Akad. Nauk, S. S. S. R.
158* (1964) 408; (b) *167* (1966) 109; (c) with Kochetkov, N. K. *Izv.
Akad. Nauk. S. S. S. R. Otd. Khim. nauk.* (1950) 350; *Chem. Abstr. 45*
(1951) 1585.
950 Zbiral, E. and Bauer, E. *Tetrahedron 28* (1972) 4189.
951 Klages, A. and Rönnenburg, A. *Ber. dtsch. chem. Ges. 36* (1903) 1128.

952 Castellana, V. *Gazz. chim. ital. 36* (1906) 48.
953 Cadogan, J. I. G., Mitchell, J. R. and Sharp, J. T. *J. chem. Soc. (Perk. I)* (1972) 1304.
954 Brain, E. G. and Finar, I. L. *J. chem. Soc.* (1957) 2356.
955 Birkinshaw, J. H., Oxford, A. E. and Raistrick, H. *Biochem. J. 30* (1936) 394.
956 Bowden, K. and Jones, E. R. H. *J. chem. Soc.* (1946) 953.
957 Farnum, D. G. and Yates, P. (a) *J. Am. Chem. Soc. 84* (1962) 1399; (b) *Chemy. Ind.* (1960) 659.
958 Russel, P. B. *J. Am. Chem. Soc. 75* (1953) 5315.
959 Borsche, W. and Hahn, H. (a) *Justus Liebigs Annln Chem. 537* (1939) 219; (b) with Wagner-Roemmich, M. *554* (1943) 15.
960 Panizzi, L. (a) *Gazz. chim. ital. 73* (1943) 13; (b) with Benati, O. *76* (1946) 66; (c) *77* (1947) 283; (d) with Monti, E. *77* (1947) 556.
961 Dains, F. B. (a) with Harger, R. N. *J. Am. Chem. Soc. 40* (1918) 562; (b) with Long, W. S. *43* (1921) 1200.
962 Fusco, R. (a) with Justoni, R. *Gazz. chim. ital. 67* (1937) 3; (b) *69* (1939) 344; (c) *69* (1939) 353; (d) *72* (1942) 411; (e) with Romani, R. *78* (1948) 332; (f) with Bianchetti, G. and Pocar, D. *91* (1961) 1233; (g) (g) with Dalla-Croce, P. *99* (1969) 69; (h) with Bianchetti, G., Pocar, D. and Ugo, R. *Chem. Ber. 96* (1963) 802; (i) with Rossi, S. *Ann. Chim. (Rome) 50* (1960) 277; (j) with Rossi, S. *Rend. Ist. Lomb. Sci. e Lett., Cl. Sci. 93* (1959) 334.
963 Kochetkov, N. K., Ambrush, I. and Ambrush, T. I. *Zh. obshch. Khim. 29* (1959) 2964.
964 Wolff, L. (a) with Lüttringhaus, A. *Justus Liebigs Annln Chem. 313* (1900) 1; (b) with Fertig, F. *313* (1900) 12; (c) *325* (1902) 177; (d) with Hall, A. A. *Ber. dtsch. chem. Ges. 36* (1903) 3612.
965 Balbiano, L. (a) *Ber. dtsch. chem. Ges. 19* (1889) 128; (b) *23* (1890) 1103; (c) with Severini, O. *Gazz. chim. ital. 23* (1893) 309; (d) with Marchetti, G. *23* (1893) 490.
966 Smith, L. I. and Howard, K. L. *J. Am. Chem. Soc. 65* (1943) 159.
967 Rateb, L. (a) with Soliman, G. *J. chem. Soc.* (1960) 1426; (b) with Mokhtar, H. *J. chem. Soc. (C)* (1968) 1845.
968 Terent'ev, A. P., Grandberg, I. I., Sibiryakova, D. V. and Kost, A. N. *Zh. obshch. Khim. 30* (1960) 2925; *Chem. Abstr. 55* (1961) 16518.
969 Schubert, H. (a) *J. prakt. Chem. 8* (1959) 333; (b) with Heydenhauss, D. *22* (1963) 304; (c) with Ruehberg, B. and Friedrich, G. *32* (1966) 249; (d) with Simon, H. and Jumar, A. *Z. Chem. 8* (1968) 62; (e) with Rudorf, H. D. *11* (1971) 175.
970 Iversen, P. E. (a) with Lund, H. *Acta chem. scand. 21* (1967) 279; (b) *24* (1970) 2459.
971 Turner, R. A. (a) *J. Am. Chem. Soc. 71* (1949) 3476; (b) with Huebner, C. F. and Scholz, C. R. *71* (1949) 2801.
972 Hubball, W. and Pyman, F. L. *J. chem. Soc.* (1928) 21.
973 Cornforth, J. W. (a) with Huang, H. T. *J. chem. Soc.* (1948) 1960; (b) with Huang, H. T. (1948) 731; (c) with Fawaz, E., Goldsworthy, L. G. and Robinson, R. (1949) 1549; (d) with Cookson, E. (1952) 1085; (e) with Cornforth, R. H. (1953) 93.
974 Jones, R. G. (a) *J. Am. chem. Soc. 71* (1949) 644; (b) with McLaughlin, K. C. *71* (1949) 2444; (c) *74* (1952) 1085; (d) *74* (1952) 4889; (e) with Mann, M. J. *75* (1953) 4048; (f) with Whitehead, C. W. *J. org. Chem. 20* (1955) 1342.
975 Bredereck, H. (a) with Sell, R. and Effenberger, F. *Chem. Ber. 97* (1964) 3407; (b) with Effenberger, F., Hofmann, A. and Hajek, M. *Angew. Chem. (int. Edn.) 2* (1963) 655.
976 Godefroi, E. F. (a) with Corvers, A. and de Groot, A. *Tetrahedron Lett.* (1972) 2173; (b) with Van der Eycken, C. A. M. and Janssen, P. A. J.

J. org. Chem. 32 (1967) 1259.
977 Grimmett, M. R. (a) *Advan. heterocycl. Chem. 12* (1970) 103; (b) with Richards, E. L. *J. chem. Soc.* (1965) 3751; (c) with Hodges, R. and Richards, E. L. *Aust. J. Chem. 21* (1968) 505.
978 Ochiai, E. (a) with Ikuma, S. *Ber. dtsch. chem. Ges. 69* (1936) 1147; (b) with Tamamushi, Y. and Nagasawa, H. *73* (1940) 28.
979 Arndt, F. and Milde, E. *Ber. dtsch. chem. Ges. 54* (1921) 2089.
980 Pinkerton, F. H. and Thames, S. F. *J. heterocycl. Chem. 9* (1972) 67.
981 Begg, C. G. and Grimmett, M. R. (a) with Lee, Y. M. *Aust. J. Chem. 26* (1973) 415; (b) with Wethey, P. D. *26* (1973) 2435; (c) unpublished results.
982 Sonn, A. (a) with Hotes, E. and Sieg, H. *Ber. dtsch. chem. Ges. 57* (1924) 2134; (b) with Greif, P. *66* (1933) 1900.
983 Brocklehurst, K. and Griffiths, J. R. *Tetrahedron 24* (1968) 2407.
984 Sheehan, J. C. and Robinson, C. A. *J. Am. Chem. Soc.* (a) *71* (1949) 1436; (b) *73* (1951) 1207.
985 Wiley, R. H. (a) with Smith, N. R., Johnson, D. M. and Moffat, J. *J. Am. Chem. Soc. 77* (1955) 3412; (b) with Hussung, K. F. and Moffat, J. *J. org. Chem. 21* (1956) 190.
986 Karabinos, J. V. *Euclides (Madrid) 16* (1956) 279; *Chem. Abstr. 51* (1957) 15508.
987 Haskins, W. T., Hann, R. M. and Hudson, C. S. *J. Am. Chem. Soc. 69* ·(1947) 1461.
988 El Khadem, H. (a) with Kolkaila, A. M. and Meshreki, M. H. *J. chem. Soc.* (1963) 2957, 3531; (b) with Rateb, L. and Mokhtar, H. *J. chem. Soc. (C)* (1968) 1845; (c) with Shaban, M. A. E. and Nassr, M. A. (1970) 2167.
989 Anderson, L. and Aronson, J. N. *J. org. Chem. 24* (1959) 1812.
990 Garg, H. G. (a) *Indian J. Chem. 4* (1966) 200; (b) *J. org. Chem. 26* (1961) 948.
991 Quilico, A. and Musante, C. *Gazz. chim. ital. 71* (1941) 327.
992 Wittig, G., Bangert, F. and Kleine, H. *Ber. dtsch. chem. Ges. 61* (1928) 1140.
993 Fournier, J. O. and Miller, J. B. *J. heterocycl. Chem. 2* (1965) 488.
994 Ajello, T. (a) with Cusmano, S. *Gazz. chim. ital. 70* (1940) 770; (b) with Tornetta, B. *77* (1947) 332.
995 Browne, E. J. (a) *Aust. J. Chem. 22* (1969) 2251; (b) *24* (1971) 393; (c) *Tetrahedron Lett.* (1970) 943; (d) with Polya, J. B. *J. chem. Soc.* (1962) 5149; (e) with Polya, J. B. *J. chem. Soc. (C)* (1968) 824.
996 Frericks, G. and Beckurts, A. *Arch. Pharm. 237* (1899) 346; *Br. Abstr. 76* (1899) 808.
997 Vereshchagina, N. N. and Postovskii, I. Ya. *Nauchn. Dokl. Vysshei Shkoly Khim. i Khim. Tecknol.* (1959) 341; *Chem. Abstr. 54* (1960) 510.
998 Bamberger, E. *Justus Liebigs Annln Chem. 273* (1893) 267; (b) with Witter, H. *Ber. dtsch. chem. Ges. 26* (1893) 2786; (c) with de Gruyter, P. *26* (1893) 2783.
999 Gryszkiewics-Trochimowski, O. *C. r. hebd. Séanc. Acad. Sci. Paris 246* (1958) 2627.
1000 Fisher, B. E., Tomson, A. J. and Horwitz, J. P. *J. org. Chem. 24* (1959) 1650.
1001 Moderhaek, D. *Justus Liebigs Annln Chem. 758* (1972) 29.
1002 Yates, P. and Farnum, D. G. *Tetrahedron Lett.* (1970) 22.
1003 Jordaan, A. and Arndt, R. R. *J. heterocycl. Chem. 5* (1968) 723.
1004 Vinokurov, V. G., Troitskaya, V. S., Solokhina, N. D. and Grandberg, I. I. *Zh. obshch. Khim. 33* (1963) 506; *Chem. Abstr. 59* (1963) 1615.
1005 Michaelis, A. (a) with Blume, R. *Justus Liebigs Annln Chem. 339* (1905) 165; (b) *385* (1911) 1.
1006 Jensen, B. S. *Acta chem. scand. 13* (1959) 1668.
1007 Potts, K. T. and Husain, S. *J. org. Chem. 36* (1971) 3368.

1008 Hensel, H. R. *Chem. Ber.* *98* (1965) 1325; *99* (1966) 868.
1009 Del Corona, L., Massaroli, G. G., Signorelli, G. and Musa, G. *Boll. Chim. Farm. 110* (1971) 645; *Chem. Abstr. 77* (1972) 5402.
1010 Smith, P. A. S., Clegg, J. M. and Lakritz, J. *J. org. Chem. 23* (1958) 1595.
1011 Pyman, F. L. *J. chem. Soc. 125* (1924) 919.
1012 Rufer, C., Kessler, H. J. and Schroeder, E. *Chim. Ther. 7* (1972) 5.
1013 Sheinker, Yu. N., Ambrush, I. and Kochetkov, N. K. *Dokl. Akad. Nauk. S. S. S. R. 123* (1958) 709; *Chem. Abstr. 53* (1959) 7145.
1014 Burger, A. and Bernabe, M. *J. med. Chem. 14* (1971) 883.
1015 Rozhkov, A. M., Kolmakova, U. G. and Mamaev, V. P. *Izv. Sib. Otd. Akad. Nauk. S. S. S. R. Ser. Khim. Nauk.* (1972) 144; *Chem. Abstr. 77* (1972) 139895.
1016 Behringer, H. (a) with Kohl, K. *Chem. Ber. 89* (1956) 2648; (b) with Türck, U. *99* (1966) 1815.
1017 Binte, H. J. and Henseke, G. *Z. Chem. 5* (1965) 268.
1018 Van Thielen, J., Van Thien, T. and De Schryver, F. C. *Tetrahedron Lett.* (1971) 3031.
1019 Reimlinger, H. (a) *Chem. Ber. 97* (1964) 3493; (b) with van Overstraeten, A. *99* (1966) 3350; (c) with King, G. S. D. and Peiren, M. A. *103* (1970) 2821; (d) *103* (1970) 3284; (e) with de Ruiter, E. and Peiren, M. A. *104* (1971) 3961; (f) with Peiren, M. A. and Merényi, R. *105* (1972) 103; (g) *Justus Liebigs Annln Chem. 713* (1968) 113.
1020 Musante, C. (a) *Gazz. chim. ital. 67* (1937) 682; (b) *69* (1939) 523; (c) *72* (1942) 537; (d) *73* (1943) 355; (e) with Pino, P. *77* (1947) 199; (f) with Berretti, R. *79* (1949) 683; (g) *Il Farmaco 6* (1951) 32.
1021 *U. S. Pat.* (a) 2,655,506 (1953) [*Chem. Abstr. 48* (1954) 11496]; (b) 2,872,371 (1959) [*Chem. Abstr. 53* (1959) 13179]; (c) 2,965,643 (1960) [*Chem. Abstr. 57* (1962) 11211]; (d) 2,997,467 (1958) [*Chem. Abstr. 56* (1962) 5934]; (e) 2,998,425 (1959) [*Chem. Abstr. 56* (1962) 5974]; (f) 2,998,426 (1959) [*Chem. Abstr. 56* (1962) 5974]; (g) 3,004,959 (1961) [*Chem. Abstr. 56* (1962) 15518]; (h) 3,021,336 (1962) [*Chem. Abstr. 56* (1962) 15518]; (i) 3,036,086 (1962) [*Chem. Abstr. 57* (1962) 6147].
1022 von Pechmann, H. and Burkard, E. *Ber. dtsch. chem. Ges. 33* (1900) 3594.
1023 Alberti, C. (a) *Gazz. chim. ital. 89* (1959) 1017; (b) with Zerbi, G. *Il Farmaco 16* (1961) 527; (c) with Tironi, C. *Farmaco (Pavia) ed. Sci. 19* (1964) 618.
1024 Kohler, E. P. and Steele, L. L. *J. Am. Chem. Soc. 41* (1919) 1093.
1025 Knorr, L. (a) with Jödiche, F. *Ber. dtsch. chem. Ges. 18* (1885) 2256; (b) *20* (1887) 1096; (c) *28* (1895) 688; (d) *28* (1895) 699; (e) with Macdonald, J. *Justus Liebigs Annln Chem. 279* (1894) 217; (f) *279* (1894) 232; (g) *279* (1894) 273.
1026 Buchner, E. (a) with Dessauer, H. *Ber. dtsch. chem. Ges. 26* (1893) 258; (b) *27* (1894) 3247; (c) with Behagel, W. *35* (1902) 34; (d) with Lehmann, L. *35* (1902) 35; (e) with Papendieck, A. *Justus Liebigs Annln Chem. 273* (1893) 214, 232.
1027 Parham, W. E. and Bleasdale, J. L. *J. Am. Chem. Soc. 73* (1951) 4664.
1028 Soliman, G. and Rateb, L. *J. chem. Soc.* (1956) 3663.
1029 Claisen, L. (a) *Justus Liebigs Annln Chem. 278* (1893) 261; (b) *295* (1897) 301, 311.
1030 Cusmano, S. (a) *Gazz. chim. ital. 70* (1940) 227; (b) with Ruccia, M. *85* (1955) 1329, 1339; (c) with Tiberio, T. *86* (1956) 507; (d) with Ruccia, M. *88* (1958) 463.
1031 Corsano, S., Capito, L. and Bonamico, M. *Ann. Chim. (Rome) 48* (1958) 140; *Chem. Abstr. 52* (1958) 16339.
1032 Wislicenus, W. (a) with Bindemann, W. *Justus Liebigs Annln Chem. 316* (1901) 18; (b) with Breit, E. *356* (1907) 32.
1033 Rosengärten, G. D. *Justus Liebigs Annln Chem. 279* (1894) 237.
1034 Bouveault, L. and Bongert, A. *Bull. Soc. chim. Fr. 27* (1902) 1095.

1035 Sandström, J. *Ark. Kemi 8* (1955) 523; *Chem. Abstr. 50* (1956) 12029.
1036 Zincke, T. and Kegel, O. *Ber. dtsch. chem. Ges. 22* (1889) 1478.
1037 Maki, Y. and Takaya, M. *Chem. pharm. Bull., Tokyo 20* (1972) 747.
1038 Richter, R. *Helv. chim. Acta 35* (1952) 478.
1039 Benary, E. (a) *Ber. dtsch. chem. Ges. 42* (1909) 3912; (b) *43* (1910) 1065;
     (c) with Soenderop, H. and Bennewitz, E. *50* (1917) 65; (d) with Silber-
     ström, L. *52* (1919) 1605; (e) with Schmidt, A. *54* (1921) 2157; (f) with
     Hosenfeld, M. *55* (1922) 3417; (g) with Schwoch, G. *57* (1924) 332;
     (h) with Bitter, G. A. *61* (1928) 1057.
1040 Pascual, V. J. and Serratosa, F. *Chem. Ber. 85* (1952) 686.
1041 El-Sayed, A. A. and Ohta, M. *Bull. chem. Soc., Japan 46* (1973) 947.
1042 Alemagna, A., Bachetti, T. and Rossi, S. *Gazz. chim. ital. 93* (1963) 748.
1043 Eidebenz, E. and Koulen, K. *Arch. Pharm. 281* (1943) 171.
1044 Feist, F. *Justus Liebigs Annln Chem. 345* (1906) 110.
1045 Owen, L. N. and Somade, H. M. B. *J. chem. Soc.* (1947) 1030.
1046 Falco, E. A. and Hitchings, G. H. *J. Am. Chem. Soc. 78* (1956) 3143.
1047 Bülow, C. (a) with Schlesinger, A. *Ber. dtsch. chem. Ges. 32* (1899) 2880;
     (b) with Schlesinger, A. *33* (1900) 3362; (c) *42* (1909) 4424.
1048 Bianchetti, G. and Pocar, D. (a) *Gazz. chim. ital. 92* (1962) 799; (b) with
     Dalla Croce, P. *94* (1964) 340.
1049 Curtius, T. *J. prakt. Chem. 91* (1915) 39.
1050 Haake, P. and Bausher, L. P. (a) *J. phys. Chem. 72* (1968) 2213; (b) with
     McNeal, J. P. *J. Am. Chem. Soc. 93* (1971) 7045.
1051 Onishchuk, A. E. *Zh. obshch. Khim. 25* (1955) 984.
1052 Martin, P. K., Matthews, H. R., Rapoport, H. and Thyagarajan, G. *J. org.
     Chem. 33* (1968) 3758.
1053 Cook, A. H. (a) in *The Chemistry of Penicillin,* edited by K. T. Clarke,
     J. R. Johnson and R. Robinson, Princeton University Press, 1949, p. 38;
     (b) with Jones, D. G. *J. chem. Soc.* (1941) 278.
1054 Le Quang Toan and Tefas, D. *Farmacia 10* (1962) 19; *Chem. Abstr. 58*
     (1963) 3409.
1055 *French Pat.* (a) 1,184,709 (1959) [*Chem. Abstr. 54* (1960) 21135];
     (b) 1,389,363 (1965) [*Chem. Abstr. 62* (1965) 13270]; (c) 1,389,779
     (1965) [*Chem. Abstr. 63* (1965) 7019]; (d) 1,403,372 (1965) [*Chem.
     Abstr. 63* (1965) 14871]; (e) 1,528,151 (1968) [*Chem. Abstr. 71* (1969)
     38958x]; (f) 2,070,695 (1971) [*Chem. Abstr. 77* (1972) 34525e].
1056 Janssen, P. A., van der Eycken, C. A. M. and van Heertum, A. H. M. T.
     J. med. Chem. 8 (1965) 220.
1057 Vinogradova, N. B. and Khromov-Borisov, N. V. *Zh. obshch. Khim. 31*
     (1961) 1466.
1058 Tamamushi, Y. *J. pharm. Soc., Japan* (a) *53* (1933) 580; (b) *55* (1935)
     1053; (c) *57* (1937) 1023.
1059 Bauer, L. (a) with Nambury, C. N. V. and Dhawan, D. *J. heterocycl. Chem.
     1* (1964) 275; (b) with Mahajanshetti, C. S. *4* (1967) 325.
1060 Carbon, J. A. *J. Am. Chem. Soc. 80* (1958) 6083.
1061 Peratoner, A. (a) with Azzarello, E. *Atti accad. Lincei 16* (1908) 318;
     Chem. Abstr. 2 (1908) 1271; (b) with Palazzo, F. C. *Gazz. chim. ital. 38*I
     (1908) 76.
1062 Meltzer, R. I., Lewis, A. D., McMillan, F. H., Genzer, J. D., Leonard, F.
     and King, J. A. *J. Am. pharm. Assoc. 42* (1953) 594.
1063 Yamada, Y. (a) with Mizoguchi, T. and Ayata, A. *Yakugaku Zasshi 77*
     (1957) 452; *Chem. Abstr. 51* (1957) 14697; (b) with Kumashiro I. and
     Takenishi, T. *Bull. chem. Soc., Japan 41* (1968) 241.
1064 Olivieri-Mandala E. (a) *Gazz. chim. ital. 40*I (1910) 117, 123; (b) *41*I
     (1911) 59; (c) *44*I (1914) 670; (d) *51*II (1921) 195; (e) with Coppola,
     A. *Atti acad. Lincei 19*I (1909) 563; *Chem. Abstr. 4* (1910) 2455.
1065 Boulton, A. J. and Katritzky, A. R. *J. chem. Soc.* (1962) 2083.
1066 Dimroth, O. (a) *Ber. dtsch. chem. Ges. 35* (1902) 1029, 4041; (b) with

Letsche, E. *35* (1902) 4056; (c) *Justus Liebigs Annln Chem. 364* (1909) 212.
1067 Mugnaini, E. and Grünanger, P. *Atti accad. naz. Lincei, Rend., Classe sci. fís., mat. e. nat. 14* (1953) 275; *Chem. Abstr. 49* (1955) 3948.
1068 Charrier, G. and Gallotti, M. *Gazz. chim. ital. 55* (1925) 7.
1069 Bishay, B. B., El Khadem, H., El-Shafei, Z. M. and Meshreki, M. H. *J. chem. Soc.* (1962) 3154.
1070 Riebsomer, J. L. and Sumrell, G. *J. org. Chem. 13* (1948) 807.
1071 Makabe, O., Fukatsu, S. and Umezawa, S. *Bull. chem. Soc., Japan 45* (1972) 2577.
1072 Novikov, S. S. (a) with Brusnikina, V. M. and Rudenko, V. A. *Chem. Abstr. 55* (1961) 23504; (b) with Rudenko, V. A. and Brusnikina, V. M. *Chem. Abstr. 55* (1961) 27282; (c) with Khmel'nitskii, L. I., Lebedev, O. V., Epishina, L. V. and Sevost'yanova, V. V. *Chem. Abstr. 73* (1970) 56028; (d) with Khmel'nitskii, L. I., Lebedev, O. V., Sevost'yanova, V. V. and Epishina, L. V. *Chem. Abstr. 73* (1970) 66491.
1073 Bojarska-Dahlig, H. and Nantka-Namirski, P. *Congr. Sci. Farm., Conf. Commun. 21, Pisa* (1961) 203; *Chem. Abstr. 59* (1963) 6408.
1074 Ford, M. C. and Mackay, D. *J. chem. Soc.* (1958) 1290.
1075 Akimova, G. S., Chistokletov, V. N. and Petrov, A. A. *Zh. organ. Khim. 1* (1965) 2077; *Chem. Abstr. 64* (1966) 9713.
1076 Gallotti, M., Barro, G. and Salto, L. *Gazz. chim. ital. 60* (1930) 866.
1077 Kuraishi, T. *Pharm. Bull., Tokyo 4* (1956) 382; *Chem. Abstr. 51* (1957) 13854.
1078 Baddiley, J., Buchanan, J. G. and Osborne, G. O. *J. chem. Soc.* (1958) 1651.
1079 Looker, J. J. *J. org. Chem. 30* (1965) 638.
1080 Plant, G. W. E. *J. Am. Chem. Soc. 76* (1954) 5801.
1081 Beretta, A. *Gazz. chim. ital. 57* (1927) 173.
1082 Schneider, F. and Schaeg, W. *Z. physiol. Chem. 327* (1962) 74.
1083 Kano, H. and Yamazaki, E. *Tetrahedron 20* (1964) 159.
1084 Bladin, J. A. *Ber. dtsch. chem. Ges. 23* (1890) 1814.
1085 Okamoto, T., Hirobe, M. and Yabe, E. *Chem. pharm. Bull., Tokyo 14* (1966) 523.
1086 Tornetta, B. *Ann. Chim. (Rome) 53* (1963) 244, 253.
1087 Ruccia, M. (a) with Vivona, N. *Ann. Chim. (Rome) 57* (1967) 671; (b) with Cusmano, S. *Chem. Abstr. 62* (1965) 13148.
1088 Regitz, M. and Eistert, B. *Chem. Ber. 96* (1963) 3120.
1089 Spasov, A. and Demirov, G. *Chem. Ber. 102* (1969) 2530.
1090 Sopyrev, V. A. and Vereshchagina, T. N. *USSR Patent* 232977; *Chem. Abstr. 70* (1969) 87817.
1091 Hellmann, H. (a) with Schwiersch, G. *Chem. Ber. 94* (1961) 1868; (b) with Elser, W. *95* (1962) 1955.
1092 Brown, M. and Benson, R. E. *J. org. Chem. 31* (1966) 3849.
1093 Stollé, R. (a) *J. prakt. Chem. 75* (1907) 416; (c) with Nieland, H. and Merkle, M. *117* (1927) 185; (c) with Strittmatter, A. *133* (1932) 60; (d) *227* (1928) 275; (e) *Ber. dtsch. chem. Ges. 38* (1905) 3023; (f) with Orth, O. *58* (1925) 2100; (g) with Schick, E., Henke-Stark, F. and Krauss, L. *62* (1929) 1118.
1094 Basu, U. *J. Ind. chem. Soc. 8* (1931) 119.
1095 Pino, P. *Ann. Chim. (Rome) 40* (1950) 575.
1096 Saper, R. P. *Chem. Abstr. 59* (1963) 1196.
1097 Sprinzl, M., Farkas, J. and Sorm, F. *Tetrahedron Lett.* (1969) 289.
1098 Gloria, T. S. *Chem. Abstr. 77* (1972) 88381.
1099 Klinsberg, E. *Synthesis* (1972) 475.
1100 Weidenhagen, R. (a) with Wegner, H. *Ber. dtsch. chem. Ges. 70* (1937) 2309; (b) with Wegner, H. *Chem. Abstr. 32* (1938) 8416.
1101 Fisher, M. H., Nicholson, W. H. and Stuart, R. S. *Can. J. Chem.* (a) *39*

(1961) 501; (b) *39* (1961) 1336.
1102 Mitsuhashi, K., Takahashi, K., Zaima, T. and Asahara, T. *Kogyo Kagaku Zasshi 74* (1971) 316; *Chem. Abstr. 74* (1971) 125566.
1103 Litchfield, G. J. and Shaw, G. (a) *Chem. Commun.* (1965) 563; (b) *J. Chem. Soc. (B)* (1971) 1474.
1104 Schroeter, G. and Finck, E. *Ber. dtsch. chem. Ges. 71* (1938) 671.
1105 Farina, F. *Chem. Abstr. 48* (1954) 4524.
1106 Albert, A. *Chem. Commun.* (1970) 858.
1107 Shealy, Y. F., Struck, R. F., Holum, L. B. and Montgomery, J. A. *J. org. Chem. 26* (1961) 2396.
1108 Cheng, C. C., Robins, R. K., Cheng, K. C. and Lin, D. C. *J. pharm. Sci. 57* (1968) 1044.
1109 Giesemann, H. *J. prakt. Chem. 1* (1955) 345.
1110 Sunjic, V., Fajdiga, T. and Japelj, M. *J. heterocycl. Chem. 7* (1970) 211.
1111 Isida, T. (a) with Fujimori, S., Nabika, K., Sisido, K. and Kozima, S. *Bull. chem. Soc. Japan 45* (1972) 1246; (b) with Kozima, S., Fujimori, S. and Sisido, K. *45* (1972) 1471; (c) with Kozima, S., Nabika, K. and Sisido, S. *J. org. Chem. 36* (1971) 3807.
1112 Butler, D. E. and Alexander, S. M. *J. org. Chem. 37* (1972) 215.
1113 Stock, A. M., Donahue, W. E. and Amstutz, E. D. *J. org. Chem. 23* (1958) 1840.
1114 *U. S. Pat.* (a) 2,510,724 (1950) [*Chem. Abstr. 44* (1950) 8375];
(b) 2,519,310 (1950) [*Chem. Abstr. 45* (1951) 668]; (c) 2,681,915 (1954) [*Chem. Abstr. 49* (1955) 11020]; (d) 2,710,870 (1955) [*Chem. Abstr. 50* (1956) 6514]; (e) 2,751,395 (1956) [*Chem. Abstr. 51* (1957) 2054]; (f) 2,827,415 (1958) [*Chem. Abstr. 52* (1958) 14957];
(g) 2,888,462 (1959) [*Chem. Abstr. 53* (1959) 17412]; (h) 3,014,916 (1961) [*Chem. Abstr. 56* (1962) 10159]; (i) 3,050,520 (1962) [*Chem. Abstr. 57* (1963) 15120]; (j) 3,054,800 (1962) [*Chem. Abstr. 58* (1963) 10220]; (k) 3,177,223 (1965) [*Chem. Abstr. 63* (1965) 1795];
(l) 3,207,763 (1965) [*Chem. Abstr. 63* (1965) 18096]; (m) 3,255,201 (1966) [*Chem. Abstr. 65* (1966) 13724]; (n) 3,391,156 (1968) [*Chem. Abstr. 69* (1968) 96718p]; (o) 3,658,835 (1972) [*Chem. Abstr. 77* (1972) 48474z].
1115 Windaus, A. (a) *Ber. dtsch. chem. Ges. 39* (1906) 3886; (b) with Langenbeck, W. *56* (1923) 683.
1116 Habraken, C. L. (a) with Cohen-Fernandes, P. *Chem. Commun.* (1972) 37; (b) with Beenakker, C. I. M. and Brusse, J. *J. heterocycl. Chem. 9* (1972) 939.
1117 Patterson, J. M., de Haan, J. W., Boyd, M. R. and Ferry, J. D. *J. Am. Chem. Soc. 94* (1972) 2487.
1118 Blackman, A. J. (a) with Bowie, J. H. *Org. mass Spect. 7* (1973) 57;
(b) with Browne, E. J. and Polya, J. B. *J. chem. Soc. (C)* (1967) 661;
(c) with Polya, J. B. (1970) 2403.
1119 Kametani, T., Kigasawa, K., Ikari, N., Iwata, T. and Saito, M. *Tetrahedron Lett.* (1966) 4849.
1120 Fernandez-Bolanos, J. (a) with Mendenez-Gallego, M. *Chem. Abstr. 67* (1967) 90719; (b) with Martin Lomas, M., Martinez Ruiz, D. and Pradera, M. A. *69* (1968) 5243u.
1121 Komoto, M. *J. agric. Chem. Soc. Japan 36* (1962) 407, 461.
1122 Shimada, K., Kuriyama, S., Kanazawa, T. and Satoh, M. *Yakugaku Zasshi 91* (1971) 231; *Chem. Abstr. 74* (1971) 141633.
1123 Rohr, W. and Staab, H. A. *Angew. Chem. (int. Edn.) 4* (1965) 1073.
1124 Becker, H. G. O. (a) with Boettcher, H., Fischer, G., Rückauf, H. and Sephon, S. *J. prakt. Chem. 312* (1970) 586; (b) with Görmer, G. and Timpe, H.-J. *312* (1970) 610; (c) with Görmer, G., Haufe, H. and Timpe, H.-J. *314* (1972) 101.
1125 Jacobson, C. R. and Amstutz, E. D. *J. org. Chem.* (a) *18* (1953) 1183;

(b) *19* (1954) 1652.
1126 Gutsche, C. D. and Voges, H. W. *J. org. Chem. 32* (1967) 2685.
1127 Van der Plas, H. C. and Jongejan, H. *Recl. Trav. chim. Pays-Bas Belg.*
(a) *91* (1972) 133; (b) *91* (1972) 336.
1128 Trofimenko, S. *J. org. Chem. 35* (1970) 3459.
1129 Paul, H. and Kausmann, A. *Chem. Ber. 101* (1968) 3700.
1130 Sokolov, L. B., Porfireva, I. I. and Petrov, A. A. *Zh. organ. Khim. 1* (1965)
610.
1131 Dornow, A. (a) with Bartsch, W. *Justus Liebigs Annln Chem. 602* (1957)
23; (b) with Hell, H. *Chem. Ber. 93* (1960) 1998.
1132 Light, R. J. and Hauser, C. R. *J. org. Chem. 26* (1961) 1716.
1133 Vasilevskii, S. F., Shvartsberg, M. S. and Kotlyarevskii, I. L. *Izv. Akad. Nauk
S. S. S. R., Ser. Khim.* (1971) 1764; *Chem. Abstr. 75* (1971) 151724.
1134 *Ger. Pat.* (a) 854,955 (1952) [*Chem. Abstr. 52* (1958) 15592];
(b) 1,168,437 (1964) [*Chem. Abstr. 61* (1964) 1873]; (c) 1,215,164
(1966) [*Chem. Abstr. 65* (1966) 8926]; (d) 1,923,643 (1970) [*Chem.
Abstr. 74* (1971) 22838j]; (e) 2,212,080 (1972) [*Chem. Abstr. 78* (1973)
4247u].
1135 *East Ger. Pat.* 36,137 (1965) [*Chem. Abstr. 63* (1965) 11574].
1136 Reppe, W. *Justus Liebigs Annln Chem. 601* (1956) 128.
1137 Overberger, C. G. and Vorchheimer, N. *J. Am. Chem. Soc. 85* (1963) 951.
1138 Ford, M. F. and Watson, D. A. M. *Synthesis* (1973) 47.
1139 Murahashi, S., Nozakura, S., Umehara, A. and Obata, K. *Chem. Abstr. 62*
(1965) 7878.
1140 Padwa, A., Smolanoff, J. and Wetmore, S. I. *Chem. Commun.* (1972) 409.
1141 *Jap. Pat.* (a) 2338 (1962) [*Chem. Abstr. 58* (1963) 7948]; (b) 2346
(1962) [*Chem. Abstr. 58* (1963) 7952]; (c) 5568 (1959) [*Chem. Abstr.
54* (1960) 14271]; (d) 10468 (1959) [*Chem. Abstr. 54* (1960) 18555].
1142 Van Meeteren, H. W. and Van der Plas, H. C. *Tetrahedron Lett.* (1966)
4517.
1143 Jones, J. B. and Hysert, D. W. *Can. J. Chem. 49* (1971) 325.
1144 Shvartsberg, M. S. and Bizhan, L. N. (a) with Kotlyarevskii, I. L. *Bull.
Acad. Sci. U. S. S. R. 20* (1971) 1429; (b) with Zaev, E. E. and
Kotlyarevskii, I. L. *21* (1972) 426.
1145 Cavalleri, B., Ballotta, R. and Lancini, G. C. *J. heterocycl. Chem. 9* (1972)
979.
1146 Simonov, A. M. (a) with Popov, I. I. *Khim. geterotsikl. Soedin 7* (1971)
139; (b) with Vitkevich, N. D. *J. gen. Chem. U. S. S. R. 29* (1959) 2369;
(c) with Vitkevich, N. D. and Zheltonozhko, S. Ya. *30* (1960) 2667;
(d) with Mendelevich, F. A. *Zh. obshch. Khim. 23* (1953) 1387.
1147 Pocar, D., Bianchetti, G. and Dalla-Croce, P. *Chem. Ber. 97* (1964) 1225.
1148 L'abbé, G. and Hassner, A. (a) *J. heterocycl. Chem. 7* (1970) 361;
(b) with Galle, J. E. *Tetrahedron Lett.* (1970) 303.
1149 Hopff, H. and Lippay, M. *J. makromol. Chem. 66* (1963) 157.
1150 *U. S. Pat.* 3,055,911 (1962) [*Chem. Abstr. 58* (1963) 5705].
1151 Finnegan, W. M. and Henry, R. A. *J. org. Chem. 25* (1959) 1565.
1152 Lykkeberg, J. and Klitgaard, N. A. *Acta chem. scand. 26* (1972) 266.
1153 Stephan, E., Vo-Quang, L. and Vo-Quang, Yen, *C. r. hebd. Séanc. Acad.
Sci., Paris 272* (1971) 1731.
1154 Skvortsova, G. G. and Domnina, E. S. (a) with Glaskova, N. P., Chipanina,
N. N. and Shergina, N. I. *J. gen. Chem. U. S. S. R. 41* (1971) 620;
(b) with Makhno, L. P., Frolov, Yu L., Voronov, V. K., Chipanina, N. N.
and Shergina, N. I. *Bull. Acad. Sci. U. S. S. R.* (1970) 2570.
1155 Domnina, E. S., Ivlev, Y. N., Shergina, N. I., Chipanina, N. N., Belousova,
L. V., Frolov, Yu. L. and Skvortsova, G. G. *J. gen. Chem. U. S. S. R. 41*
(1971) 1106.
1156 Cockerill, A. F., Harden, R. C. and Mallen, D. N. B. *J. chem. Soc. (Perk. 2)*
(1972) 1428.

1157 Dürr, H. (a) with Schrader, L. *Z. Naturforsch.* 24b (1969) 536; (b) with Schrader, L. *Chem. Ber.* 103 (1970) 1334; (c) with Sergio, R. and Gombler, W. *Angew. Chem. (int. Edn.)* 3 (1964) 748.
1158 Hori, I. and Igarashi, M. *Bull. chem. Soc. Japan* 44 (1971) 2856.
1159 Dorn, H. (a) with Hilgetag, G. and Zubek, A. *Angew. Chem. (int. Edn.)* 3 (1964) 748; (b) with Dilcher, H. *Z. Chem.* 8 (1968) 420; (c) with Zubek, A. *J. prakt. Chem.* 313 (1971) 1118; (d) with Dieter, A. 313 (1971) 1173.
1160 Bertho, A. and Nüssel, H. *Justus Liebigs Annln Chem.* 457 (1927) 278.
1161 Stafford, W. H., Los, M. and Thomson, N. *Chemy Ind.* (1956) 1277.
1162 Khan, M. A. and Lynch, B. M. *Can. J. Chem.* 49 (1971) 3566.
1163 Huebner, C. F. and Link, K. P. *J. Am. Chem. Soc.* 72 (1950) 4812.
1164 *Swiss Pat.* (a) 344,413 (1960) [*Chem. Abstr.* 57 (1962) 16624]; (b) 520,090 (1972) [*Chem. Abstr.* 77 (1972) 101609].
1165 Clemo, G. R., Holmes, T. and Leitch, G. C. *J. chem. Soc.* (1938) 753.
1166 Torf, S. F., Kudryashova, N. I., Khromov-Borisov, N. V. and Mikhailova, T. A. *Zh. obshch. Khim.* 32 (1962) 1740.
1167 Morgan, G. T. and Reilly, J. *J. chem. Soc.* 103 (1913) 808.
1168 Fabbrini, L. *Ann. Chim. (Rome)* 45 (1955) 728.
1169 Broser, W. and Bollert, U. *Chem. Ber.* 99 (1966) 1767.
1170 Walter, R. and Schickler, P. G. *J. prakt. Chem.* 55II (1897) 305.
1171 Biffin, M. E. C. and Brown, D. J. (a) *Tetrahedron Lett.* (1967) 2029; (b) with Porter, Q. N. *J. chem. Soc. (C)* (1968) 2159.
1172 Mohr, E. *J. prakt. Chem.* 90 (1914) 223, 509.
1173 Baba, H., Hori, I., Hayashi, T. and Midorikawa, H. *Bull. chem. Soc. Japan* 42 (1969) 1653.
1174 Guarneri, M., Giori, P. and Benassi, C. A. *Tetrahedron Lett.* (1971) 665.
1175 Burtles, R., Pyman, F. L. and Reylance, J. *J. chem. Soc.* 127 (1925) 581.
1176 Shaw, G. (a) with Warrener, R. N., Butler, D. N. and Ralph, R. K. *J. chem. Soc.* (1959) 1648; (b) with Butler, D. N. (1959) 4040; (c) with Wilson, D. V. (1963) 1077.
1177 Sarasin, J. *Helv. chim. Acta* 6 (1923) 620.
1178 Kleinfeller, H. and Bönig, G. *J. prakt. Chem.* 132 (1931) 175.
1179 Thiele, J. and Strange, O. *Justus Liebigs Annln Chem.* 283 (1894) 41.
1180 Lieber, E. and Enkoji, T. *J. org. Chem.* 26 (1961) 4472.
1181 Glover, E. E., Rowbottom, K. T. and Bishop, D. C. *J. chem. Soc. (Perk I)* (1972) 2927.
1182 Tertov, B. A. and Burykin, V. V. (a) *Khim. geterotsikl. Soedin* (1970) 1554; *Chem. Abstr.* 74 (1971) 76466; (b) with Koblik, A. V. *Khim. geterotsikl. Soedin.* (1972) 1552; *Chem. Abstr.* 78 (1973) 58308.
1183 Hetzheim, A., Peters, O. and Beyer, H. *Chem. Ber.* 100 (1967) 3418.
1184 Julia, M. and Tam, H. D. *Bull. Soc. chim. Fr.* (1971) 1303.
1185 Leese, C. L. and Timmis, G. H. *J. chem. Soc.* (1961) 3816.
1186 Taguchi, H. *Tetrahedron Lett.* (1973) 1137.
1187 Garcia-Muñoz, G., Madroñero, R., Rico, M. and Saldaña, M. C. *J. heterocycl. Chem.* 6 (1969) 921.
1188 Yoshida, M., Matsumoto, A. and Simamura, O. *Bull. chem. Soc. Japan* 43 (1970) 3587.
1189 Ramart-Lucas, P., Hoch, J. and Grumez, M. *Bull. Soc. chim. Fr.* (1949) 447.
1190 Manchot, W. *Ber. dtsch. chem. Ges.* 43 (1910) 1312.
1191 Kurzer, F. (a) with Douraghi-Zadeh, K. *J. chem. Soc.* (1965) 3912; (b) with Douraghi-Zadeh, K. (1965) 4448; (c) with Douraghi-Zadeh, K. *J. chem. Soc. (C)* (1966) 6; (d) (1970) 1813.
1192 Polya, J. B. *Chemy. Ind.* (1965) 812.
1193 Reilly, B. J. (a) with Madden, D. *J. chem. Soc.* 127 (1925) 2936; (b) with Drumm, P. (1926) 1729; (c) with Bastible, H. E. *Sci. Proc. R. Dublin Soc.* 18 (1926) 343; (d) with Teegan, J. P. and Carey, M. F. 24 (1948) 349; (e) with McSweeney, D. T. *Proc. R. Irish Acad.* 39B (1930) 497.

1194 Garbrecht, W. L. and Herbst, R. M. *J. org. Chem.* (a) *18* (1953) 1003; (b) *18* (1953) 1014.
1195 Bowie, R. A., Gardner, M. D., Neilson, D. G., Watson, K. M., Mahmood, S. and Ridd, V. *J. chem. Soc. (Perk. 1)* (1972) 2395.
1196 Pellizzari, G. (a) *Gazz. chim. ital.* 24I (1894) 481; (b) with Ferro, A. A. *28* (1898) 541; (c) with Repetto, A. *37*II (1907) 317.
1197 McBride, W. R., Henry, R. A. and Skolnik, S. *Analyt. Chem. 25* (1953) 1042.
1198 Godfrey, L. E. A. and Kurzer, F. *J. chem. Soc.* (1960) 3437.
1199 Busch, M. *Ber. dtsch. chem. Ges. 40* (1907) 2093.
1200 Jensen, K. A. and Hansen, O. R. *Acta chem. scand. 6* (1952) 195.
1201 O'Mahony, T. A. F., Butler, R. N. and Scott, F. L. *J. chem. Soc. (Perk. 2)* (1972) 1319.
1202 Butler, R. N. (a) *Chem. Commun.* (1969) 405; (b) (1970) 1096; (c) *J. chem. Soc. (B)* (1970) 138; (d) with Catton, R. C. and Symons, M. C. R. (1970) 378; (e) with Lambe, T. and Scott, F. L. *Chemy Ind.* (1970) 628.
1203 Shchipanov, V. P. (a) *Zh. organ. Khim. 2* (1966) 356; (b) with Postovskii, I. Ya. *2* (1966) 350, 360, 1108
1204 Bianchi, G. and Katritzky, A. R. *Chem. Commun.* (1971) 846.
1205 Fletcher, I. J. and Katritzky, A. R. *Chem. Commun.* (1970) 706.
1206 Gagnon, P. E., Boivin, J. L. and Tremblay, M. *Can. J. Chem. 31* (1953) 673.
1207 Hunter, G. and Hlynka, I. *Can. J. Research 19B* (1941) 305.
1208 Dymek, W., Janik, B. and Samson, O. *Chem. Abstr. 62* (1965) 10428.
1209 Chinone, A., Sato, S. and Ohta, M. *Bull. chem. Soc. Japan 44* (1971) 826.
1210 Meyer, V. *Ber. dtsch. chem. Ges. 21* (1888) 11.
1211 Tabak, S. V., Grandberg, I. I. and Kost, A. N. *J. gen. Chem. U. S. S. R. 34* (1964) 2778.
1212 Kröger, C. F. (a) with Etzold, G. and Beyer, H. *Justus Liebigs Annln Chem. 664* (1963) 146; (b) with Miethchen, R. *Z. Chem. 9* (1969) 378.
1213 Takimoto, H. H., Denault, G. C. and Hotta, S. (a) *J. org. Chem. 30* (1965) 711; (b) *J. heterocycl. Chem. 3* (1966) 119.
1214 Kitaev, Y. P., Savin, V. I., Zverev, V. V. and Popova, G. V. *Khim. geterotsikl. Soedin. 7* (1971) 559.
1215 Müller, E., Beutler, R. and Zeeh, B. *Justus Liebigs Annln Chem. 719* (1969) 72.
1216 Mayer, K. K., Schröppel, F. and Sauer, J. *Tetrahedron Lett.* (1972) 2899.
1217 Duffin, G. F. and Kendall, J. D. *J. chem. Soc.* (1954) 408.
1218 De Mendoza Sans, J. *Chem. Abstr. 76* (1972) 153663.
1219 Yamauchi, O., Tanaka, H. and Uno, T. *Chem. pharm. Bull. Tokyo 14* (1966) 948.
1220 Beaman, A. G., Tautz, W., Gabriel, T. and Duschinsky, R. *J. Am. Chem. Soc. 87* (1965) 389.
1221 *Belg. Pat.* 660,836 (1965) [*Chem. Abstr. 63* (1965) 18097].
1222 *Rom. Pat.* 50786 (1968) [*Chem. Abstr. 70* (1969) 4114h].
1223 Grimmel, H. W. and Morgan, J. F. *J. Am. Chem. Soc. 70* (1948) 1750.
1224 Timpe, H. J. (a) with du Bois, M., Fürstenberg, B. and Herrmann, W. *Z. Chem. 11* (1971) 258; (b) with Becker, H. G. O. *J. prakt. Chem. 314* (1972) 325.
1225 Pesson, M. (a) with Dupin, S. *C. r. hebd. Séanc. Acad. Sci., Paris 252* (1961) 3830; (b) with Dupin, S. and Antoine, M. *253* (1961) 285; (c) with Dupin, S. and Antoine, M. *253* (1961) 992; (d) with Dupin, S. *Bull. Soc. chim. Fr.* (1962) 250; (e) with Dupin, S. and Antoine, M. (1962) 1364.
1226 Tobin, J. C. (a) with Butler, R. N. and Scott, F. L. *Chem. Commun.* (1970) 112; (b) with Hegarty, A. F. and Scott, F. L. *J. chem. Soc. (B)* (1971) 2198.
1227 Beyer, H. (a) *Z. Chem. 10* (1970) 406; (b) with Bulka, E. and Beckhaus, F. W. *Chem. Ber. 92* (1959) 2593.

1228 Heep. U. *Justus Liebigs Annln Chem.* (1973) 578.
1229 Buchanan, J. in *Chemistry and Biology of the Purines,* Boston, Massachusetts, 1957.
1230 Richter, E., Loeffler, J. E. and Taylor, E. C. *J. Am. Chem. Soc. 82* (1960) 3144.
1231 Naylor, R. N., Shaw, G., Wilson, D. V. and Butler, D. N. *J. chem. Soc.* (1961) 4845.
1232 Kreutzberger, A. *J. org. Chem. 27* (1962) 886.
1233 Lancini, G. C. (a) with Lazzari, E., Arioli, V. and Bellani, P. *J. med. Chem. 12* (1969) 775; (b) with Maggi, N. and Sensi, P. *Chem. Abstr. 59* (1963) 10032; (c) with Lazzari, E. and Pallanza, K. *Chem. Abstr. 65* (1966) 700.
1234 Lukens, L. N. and Buchanan, J. M. *J. biol. Chem. 234* (1959) 1791.
1235 Gots, J. S. and Gollub, E. G. *Proc. nat. Acad. Sci. U. S. A. 43* (1957) 826.
1236 Shevlin, P. B. *J. Am. Chem. Soc. 94* (1972) 1379.
1237 Horowitz, J. P. and Grakauskas, V. A. *J. Am. Chem. Soc. 80* (1958) 926.
1238 Pan'kov, A. K., Pevzner, M. S. and Bagal, L. I. *Khim. geterotsikl. Soedin.* (1972) 713; *Chem. Abstr. 77* (1972) 126523.
1239 Fukata, G., Kawazoe, Y. and Taguchi, T. *Tetrahedron Lett.* (1973) 1199.
1240 Sheppard, W. A. and Webster, O. W. *J. Am. Chem. Soc. 95* (1973) 2695.
1241 Hofmann, K. A. and Hoch, H. *Ber. dtsch. chem. Ges. 43* (1910) 1866.
1242 von Auwers, K. (a) *Justus Liebigs Annln Chem. 378* (1911) 218; (b) with Bähr, K. and Frese, E. *441* (1925) 54; (c) with Heimke, P. *458* (1927) 186; (d) with Ungemach, O. *Ber. dtsch. chem. Ges. 66* (1933) 1198.
1243 Michaelis, A. (a) *Justus Liebigs Annln Chem. 338* (1905) 183; (b) with Röhmer, H. *Ber. dtsch. chem. Ges. 31* (1898) 3193; (c) with Zilg, A. *39* (1906) 370.
1244 Garnovskii, A. D., Kuznetsova, L. I., Andreichikov, Yu. P., Osipov, O. A., Kolodyazhnyi, Yu. V., Minkin, V. I., Bren, V. A., Simonov, A. M. and Avdyunina, N. I. *J. gen. Chem. U. S. S. R. 41* (1971) 1838.
1245 O'Callaghan, C. N. and Twomey, D. *Proc. R. Irish Acad. Sect. B. 64* (1965) 187; *Chem. Abstr. 65* (1966) 12191.
1246 Perrin, D. D. in *Dissociation Constants of Organic Bases in Aqueous Solution,* Butterworths, London, 1965.
1247 Hillers, S., Mazeika, I. and Grandberg, I. I. *Khim. geterotsikl. Soedin.* (1965) 103; *Chem. Abstr. 63* (1965) 5506.
1248 Bouchet, P., Elguero, J. and Jacquier, R. *Tetrahedron Lett.* (1964) 3317.
1249 Katritzky, A. R. and Lagowski, J. M. *Adv. in heterocycl. Chem. 2* (1963) 27.
1250 Gauss, W., Heitzer, H. and Petersen, S. *Justus Liebigs Annln Chem. 764* (1972) 131.
1251 Gever, G. *J. Am. Chem. Soc. 76* (1954) 1283.
1252 Reiford, L. C. and Peterson, W. I. *J. org. Chem. 1* (1937) 544.
1253 Marchetti, G. (a) *Gazz. chim. ital. 22*II (1892) 351, 368; (b) *Atti. R. Accad. Lincei 5*I (1892) 337.
1254 Biltz, H. (a) *Justus Liebigs Annln Chem. 391* (1912) 169; (b) with Krebs, P. *391* (1912) 210; (c) with Bülow, H. *457* (1927) 103; (d) with Edlefsen, H. *Ber. dtsch. chem. Ges. 40* (1907) 2630.
1255 Winans, C. F. and Adkins, H. (a) *J. Am. Chem. Soc. 55* (1933) 2051; (b) *55* (1933) 4167.
1256 Alabaster, R. J. and Barry, W. J. *J. chem. Soc. (C)* (1970) 78.
1257 Rossi, S., Maiorana, S. and Bianchetti, G. *Gazz. chim. ital. 94* (1964) 210.
1258 Chiriac, C. and Zugravescu, I. *Rev. roum. Chem. 15* (1970) 1065.
1259 Crocker, H. P. and Hall, R. H. *J. chem. Soc.* (1955) 4489.
1260 Shvachkin, Yu. P., Ryabtsev, M. N. and Krymov, A. P. *J. org. Chem. U. S. S. R. 8* (1972) 665.
1261 Henning, G. (a) with Wolff, F. *Z. Chem. 11* (1971) 153; (b) with Zimmermann, H. *Ber. Bunsenges. phys. Chem. 72* (1968) 630.
1262 Kauffmann, T., Legler, J., Ludorff, E. and Fischer, H. *Angew. Chem. (int.*

Edn.) 11 (1972) 846.
1263 Siegrist, A. *Helv. chim. Acta 50* (1967) 906.
1264 Franck-Neumann, M. and Buchecker, C. *Tetrahedron Lett.* (1972) 937.
1265 Takamizawa, A. and Hayashi, S. (a) *Yakugaku Zasshi 79* (1959) 334; (b) *83* (1963) 373; *Chem. Abstr. 59* (1963) 10022; (c) with Sato, M. *85* (1965) 158; *Chem. Abstr. 62* (1965) 13137.
1266 Grothaus, C. E. and Dains, F. B. *J. Am. Chem. Soc. 58* (1936) 1334.
1267 Sprio, V. and Fabra, I. (a) *Gazz. chim. ital. 86* (1956) 1059; (b) *Chem. Abstr. 59* (1963) 590.
1268 Burns, P. S. *J. prakt. Chem. 47* (1893) 105.
1269 Seidel, J. *J. prakt. Chem. 58* (1898) 134.
1270 Dickinson, C. L., Williams, J. K. and McKusick, B. C. *J. org. Chem. 29* (1964) 1915.
1271 Sutcliffe, E. Y., Zee-Cheng, K. Y., Cheng, C. C. and Robins, R. K. *J. med. pharm. Chem. 5* (1962) 588.
1272 Martin, D., Schwarz, K. H., Rackow, S., Reich, P. and Grundemann, E. *Chem. Ber. 99* (1966) 2302.
1273 Lont, P. J., Van der Plas, H. C. and Koudjis, A. *Recl. Trav. chim. Pays-Bas, Belg. 90* (1970) 207.
1274 de Vries, L. *J. org. Chem. 36* (1971) 3442.
1275 Wakamatsu, H., Saito, T., Kumashiro, I. and Takenishi, T. *J. org. Chem. 31* (1966) 2035.
1276 Matsuda, K. and Morin, L. T. *J. org. Chem. 26* (1961) 3783.
1277 Protopopova, T. V., Klimko, V. T. and Skoldinov, A. P. *Khim. Nauka i Prom. 4* (1959) 805; *Chem. Abstr. 54* (1960) 11037.
1278 Dieckmann, W. and Platz, L. *Ber. dtsch. chem. Ges. 37* (1904) 4638.
1279 Barry, W. J. *J. chem. Soc.* (1958) 1171.
1280 D'Yakonov, I. A. *Zh. obshch. Khim. 17* (1947) 67; *Chem. Abstr. 42* (1948) 902.
1281 Brooklyn, R. J. and Finar, I. L. (a) *J. chem. Soc. (C)* (1968) 466; (b) (1969) 1515.
1282 Lespieau, R. *C. r. hebd. Séanc. Acad. Sci. Paris 133* (1901) 538.
1283 Drumm, P. J. *Proc. roy. Irish Acad. 40B* (1931) 106.
1284 Severini, O. *Atti. Accad. Lincei ΠI* (1892) 391; *Brit. Abstr. 1* (1894) 145.
1285 Delaby, R., Danton, S. and Chabrier, P. *Bull. Soc. chim. Fr.* (1961) 2061.
1286 Roedig, A. and Becker, H. J. *Justus Liebigs Annln Chem. 597* (1955) 214.
1287 Cohen, L. A. and Kirk, K. L. *J. Am. Chem. Soc. 93* (1971) 3060.
1288 Langenbeck, W., Hutschenreuter, R. and Röttig, W. *Ber. dtsch. chem. Ges. 65* (1932) 1750.
1289 Kochergin, P. M. (a) *Khim. geterotsikl. Soedin.* (1965) 398; (b) (1965) 754; (c) (1965) 761; (d) with Verenikina, S. G. and Bushueva, K. S. (1965) 765; (e) with Tsyganova, A. M., Viktorova, L. M. and Peresleni, E. M. (1967) 126; (f) with Klykov, M. A. and Mikhailova, I. S. (1972) 820; (g) *Zh. obshch. Khim. 25* (1956) 2493, 2916; (h) with Shchukina, M. N. *25* (1956) 2905; (i) *34* (1964) 2735; (j) *34* (1964) 3402; (k) with Tsyganova, A. M. and Shlikunova, V. S. *Khim. Farm. Zh. 2* (1968) 22; *Chem. Abstr. 70* (1969) 37714; (l) with Tsyganova, A. M., Shlikunova, V. S. and Klykov, M. A. *7* (1971) 689; *Chem. Abstr. 76* (1972) 126867a.
1290 Gränacher, C. (a) with Schelling, V. and Schlatter, E. *Helv. chim. Acta 8* (1925) 873; (b) with Gulbas, G. *10* (1927) 819.
1291 Karrer, P. and Gränacher, C. *Helv. chim. Acta 7* (1924) 763.
1292 Caló, W., Ciminale, F., Lopez, L., Naso, F. and Todesco, P. E. *J. chem. Soc. (Perk 1)* (1972) 2567.
1293 Goldschmidt, S. and Steigerwald, C. *Ber. dtsch. chem. Ges. 58* (1925) 1346.
1294 Haruki, E., Izumita, S. and Imoto, E. *Kagaku Zasshi 86* (1965) 942; *Chem. Abstr. 65* (1966) 13688.
1295 Laviron, E. and Fournari, P. *Bull. Soc. chim. Fr.* (1966) 518.
1296 McNeill, R., and Weiss, D. E. and Willis, D. *Aust. J. Chem. 18* (1965) 477.

1297 Tamburello, A. and Milazzo, A. *Atti accad. Lincei (Palermo) 16* (1908) 412; *Chem. Abstr. 2* (1908) 1271.
1298 Carnelly, H. W. and Dutt, P. K. *J. chem. Soc. 125* (1924) 2476.
1299 Bachmann, G. B. and Heisey, L. V. *J. Am. Chem. Soc. 71* (1949) 1985.
1300 Baker, W., Ollis, W. D. and Poole, V. D. *J. chem. Soc.* (1950) 3389.
1301 Baumgärtel, H. and Zimmermann, H. *Z. Natur. 18b* (1963) 406; *Chem. Abstr. 59* (1963) 6382.
1302 Gosztonyi, T., Carnmalm, B. and Sjoberg, B. *Acta chem. scand. 24* (1970) 3078.
1303 Snyder, H. R., Verbanac, F. and Bright, D. B. *J. Am. Chem. Soc. 74* (1952) 3243.
1304 Derbyshire, D. H. and Waters, W. A. *J. chem. Soc.* (1950) 3694.
1305 Sio, T. H. and Grimmett, M. R. unpublished data.
1306 Brunings, K. J. *J. Am. Chem. Soc. 69* (1947) 205.
1307 Hayashi, T. and Maeda, K. *Bull. chem. Soc. Japan* (a) *33* (1960) 565; (b) *35* (1962) 2057.
1308 Nakamura, N., Kishida, Y. and Ishida, N. *Chem. pharm. Bull. Tokyo 19* (1971) 1389.
1309 O'Brien, D. H. and Hrung, C. P. *J. organometallic Chem. 27* (1971) 185.
1310 Sachs, F. (a) with Romer, A. *Ber. dtsch. chem. Ges. 35* (1902) 3307; (b) with Alsleben, P. *40* (1907) 664.
1311 Nye, M. J. and Tang, W. P. (a) *Tetrahedron 28* (1972) 455; (b) *28* (1972) 460; (c) *Can. J. Chem. 48* (1970) 3563.
1312 Dittli, C., Elguero, J. and Jacquier, R. *Bull. Soc. chim. Fr.* (1971) 1038.
1313 Freeman, J. P. (a) with Gannon, J. J. and Surbey, D. L. *J. org. Chem. 34* (1969) 187; (b) with Surbey, D. L. *Tetrahedron Lett.* (1967) 4917.
1314 Chattaway, F. D. and Coulson, E. A. *J. chem. Soc.* (1928) 1361.
1315 Westöö, G. *Acta chem. scand. 7* (1953) 360.
1316 Backer, H. J. and Meijer, W. *Recl Trav. chim. Pays-Bas Belg. 45* (1926) 428.
1317 Biquard, D. and Grammaticakis, P. *Bull. Soc. chim. Fr.* (1941) 246.
1318 Grard, J. *Arch. Chem. 13* (1930) 336.
1319 Walker, C. *Am. chem. J. 14* (1892) 576.
1320 Grünanger, P. and Finzi, P. B. *Chem. Abstr. 58* (1963) 516.
1321 Groen, S. H. and Arens, J. F. *Recl Trav. chim. Pays-Bas Belg. 80* (1961) 879.
1322 Boyle, F. T. and Jones, R. A. Y. (a) *J. chem. Soc. (Perk. 1)* (1973) 167; (b) (1973) 170.
1323 Allan, F. J. and Allan, G. G. *Chemy. Ind.* (1964) 1837.
1324 Akagane, K., Allan, F. J., Allan, G. G., Friberg, T., Muircheartaigh, S. O. and Thomson, J. B. *Bull. chem. Soc. Japan 42* (1969) 3204.
1325 Volkamer, M. and Zimmermann, H. (a) *Chem. Ber. 102* (1969) 4177; (b) *103* (1970) 296; (c) with Baumgärtel, H. *Angew. Chem. (int. Edn.) 6* (1967) 947.
1326 Begtrup, M. (a) with Pedersen, C. *Acta chem. scand. 18* (1964) 1333; (b) *25* (1971) 3500; (c) *26* (1972) 715; (d) *26* (1972) 1243; (e) *Tetrahedron Lett.* (1971) 1577.
1327 Hadacek, J. *Chem. Abstr. 55* (1961) 18756.
1328 Widman, O. *Ber. dtsch. chem. Ges. 26* (1893) 2612; *29* (1896) 1946.
1329 Young, G. and Witham, E. *J. chem. Soc. 77* (1900) 224.
1330 Baccar, B. G. and Mathis, F. *C. r. hebd. Séanc. Acad. Sci. Paris 261* (1965) 174.
1331 Hoggarth, E. *J. chem. Soc.* (1949) 1163.
1332 Nakai, R., Sugii, M. and Nakao, H. *Pharm. Bull. Tokyo 5* (1957) 576.
1333 Girard, M. *Ann. Chim. 16* (1941) 326.
1334 Arcus, C. L. and Prydal, B. S. *J. chem. Soc.* (1957) 1091.
1335 Srinivasan, V., Ramachander, G. and Naqui, S. *Arch. Pharm. 295* (1962) 405.

1336 Hohenlohe-Oeringen, K. *Monatsh. 89* (1958) 557.
1337 Rupe, H. *Ber. dtsch. chem. Ges. 28* (1895) 251.
1338 Logemann, W. *Chem. Ber. 91* (1958) 2578.
1339 Bailey, J. R. and Moore, H. N. *J. Am. Chem. Soc. 39* (1917) 279.
1340 Ponzio, G. (a) *Gazz. chim. ital. 29* (1899) 277; (b) *29* (1899) 349; (c) *30* (1900) 459; (d) *31* (1901) 413; (e) with Torres, M. *59* (1929) 461; (f) *J. prakt. Chem. 57* (1898) 160.
1341 Cuneo, G. *Ann. Chim. Farm. 26* (1897) 481; *Brit. Abstr. 76* (1899) 9.
1342 Pinner, A. *Ber. dtsch. chem. Ges. 21* (1888) 1219.
1343 Palazzo, F. C. *Chem. Abstr. 4* (1910) 2455.
1344 Foster, M. O. *J. chem. Soc. 95* (1909) 184.
1345 Iyengar, D. S., Prasad, K. K. and Venkataratnam, R. V. *Tetrahedron Lett.* (1972) 3937.
1346 Dittmer, K., Ferger, M. F. and du Vigneaud, V. *J. biol. Chem. 164* (1946) 19.
1347 Bodendorf, K. and Towliati, H. *Arch. Pharm. 298* (1965) 293; *Chem. Abstr. 63* (1965) 5629.
1348 Brown, B. R. and Hammick, D. L. *J. chem. Soc.* (1947) 1384.
1349 Wieland, H. (a) with Bauer, H. *Ber. dtsch. chem. Ges. 40* (1907) 1680; (b) *42* (1909) 4199.
1350 Kost, A. N., Sagitullin, R. S. and Sun, Y.-S. *Zh. obshch. Khim. 31* (1961) 3280.
1351 McKennis, H. and du Vigneaud, V. *J. Am. Chem. Soc. 68* (1946) 832.
1352 Duschinsky, R., Dolan, L. A., Randall, L. O. and Lehmann, G. *J. Am. Chem. Soc. 69* (1947) 3150.
1353 Daunis, J., Guindo, Y., Jacquier, R. and Viallefont, P. *Bull. Soc. chim. Fr.* (1971) 3296.
1354 Michaelis, A. (a) *Justus Liebigs Annln Chem. 338* (1905) 183; (b) with Klopstock, H. *354* (1907) 102; (c) with Duntze, E. *414* (1914) 21; (d) with Pander, R. *Ber. dtsch. chem. Ges. 37* (1904) 2774.
1355 Porai-Koshits, A. E. and Muravich, Kh. L. *Zh. obshch. Khim. 23* (1953) 1583.
1356 Buzykin, B. I. and Lonshchakova, T. I. *Bull. Acad. Sci. U. S. S. R.* (1971) 2224.
1357 Huntress, E. H. and Olsen, R. T. *J. Am. Chem. Soc. 70* (1948) 2856.
1358 Hukki, J., Laitinen, P. and Alberty, J. E. *Chem. Abstr. 76* (1972) 59517.
1359 Kato, H., Shiba, T., Yoshida, H. and Fujimori, S. *Chem. Commun.* (1970) 1591.
1360 Bach, F. L., Karliner, J. and Van Lear, G. E. *Chem. Commun.* (1969) 1110.
1361 Sunagawa, G. and Watatani, W. *Chem. pharm. Bull. Tokyo 16* (1968) 1300, 1308.
1362 Krivoozheiko, K. M. and El'tsov, A. V. *Zh. organ. Khim. 4* (1968) 1114.
1363 Noels, A., Jadot, J. and van Overstraeten, A. *Chem. Ber. 103* (1970) 1954.
1364 Oddo, B. and Mingoia, Q. *Gazz. chim. ital. 61* (1931) 446.
1365 Schubert, J., Lind, E. L., Westfall, W. M., Pfleger, R. and Li, N. C. *J. Am. Chem. Soc. 80* (1958) 4799.
1366 Eilbeck, W. J., Holmes, F., Taylor, C. E., and Underhill, A. E. *J. chem. Soc. (A)* (1968) 128.
1367 Holmes, F. and Jones, F. *J. chem. Soc.* (1960) 2398.
1368 Brooks, P. and Davidson, N. *J. Am. Chem. Soc. 82* (1960) 2118.
1369 Driver, R. and Walker, W. R. *Aust. J. Chem. 21* (1968) 671.
1370 Sandmark, C. and Branden, C. I. *Acta chem. scand. 21* (1967) 993.
1371 Beetlestone, J. G., Epega, A. A. and Irvine, D. H. *J. chem. Soc. (A)* (1968) 1346.
1372 Ivanoff, M. D. *Bull. Soc. chim. Fr. 39* (1926) 47.
1373 Bast, S. and Andersen, K. K. *J. org. Chem. 33* (1968) 846.
1374 Mirone, P. *Ann. Chim. (Rome) 46* (1956) 39.
1375 Oettinger, B. *Justus Liebigs Annln Chem. 279* (1894) 245.

1376 Caradonna, C. and Stein, M. L. *Ann. Chim. (Rome) 54* (1964) 539.
1377 Ferguson, I. J. and Grimmett, M. R. unpublished data.
1378 Padeiskaya, E. N., Grandberg, I. I., Pershin, G. N., Kost, A. N., Ovseneva, L. G. and Wei-P'i Ting, *Vestn. Mosk. Univ., Ser. II, Khim. 18* (1963) 69; *Chem. Abstr. 60* (1964) 515.
1379 Nakamura, S. *Pharm. Bull. (Japan) 3* (1955) 379.
1380 Lehmstedt, K. (a) *Justus Liebigs Annln Chem. 456* (1927) 253; (b) *507* (1933) 213; (c) with Rolker, B. *Ber. dtsch. chem. Ges. 76* (1943) 879.
1381 Coscar, C., Crisan, C., Horclois, R., Jacob, R. R. M., Robert, J., Tchelitcheff, S. and Vaupre, R. *Arzneimittel-Forsch. 16* (1966) 23; *Chem. Abstr. 66* (1967) 2512.
1382 Sawa, N., Okamura, S. and Hoda, M. *Chem. Abstr. 70* (1969) 77865.
1383 Kajfez, M., Blazevic, N. and Sunjic, V. *Chem. Abstr. 71* (1969) 70534.
1384 Aleshina, G. A., Gireva, R. N., Boldyreva, T. A., Petrova, O. I. and Malyuga, A. V. *Chem. Abstr. 77* (1972) 139984.
1385 Butler, K., Howes, H. L., Lynch, J. E. and Pirie, D. K. *J. med. Chem. 10* (1967) 891.
1386 *Span. Pat.* 305073 (1965) [*Chem. Abstr. 63* (1965) 2981].
1387 Epishina, L. V., Slovetskii, V. I., Osipov, V. G., Lebedev, O. V., Khmel'nitskii, L. I., Sevost'yanova, V. V. and Novikova, T. S. *Khim. geterotsikl. Soedin. 3* (1967) 716 (Engl. 570).
1388 *S. African Pat.* 6800778 [*Chem. Abstr. 71* (1969) 3387].
1389 Pevzner, M. S. (a) with Fedorova, E. Ya., Shokhor, I. N. and Bagal, L. N. *Khim. geterotsikl. Soedin.* (1971) 275; (b) with Samarenko, V. Ya. and Bagal, L. N. (1972) 848; *Chem. Abstr. 77* (1972) 87354.
1390 Mel'nikov, V. V., Stolpakova, V. V., Khor'kova, L. F., Pevzner, M. S. and Mel'nikova, N. N. *Khim. geterotsikl. Soedin.* (1972) 120; *Chem. Abstr. 77* (1972) 18820.
1391 Matthews, H. R. and Rapoport, H. *J. Am. Chem. Soc. 95* (1973) 2297.
1392 Mokrushin, V. S., Nifontov, V. I., Pushkareva, Z. V. and Ofitserov, V. I. *Khim. geterotsikl. Soedin. 7* (1971) 1421.
1393 Neumann, P. N. *J. heterocycl. Chem. 7* (1970) 1159.
1394 Phillips, A. P. *J. org. Chem. 12* (1947) 333.
1395 Meyer, A. *Ann. Chim. (Paris) 1* (1914) 290.
1396 Volodarsky, L. B., Lisack, A. N. and Koptyug, V. A. *Tetrahedron Lett.* (1965) 1565.
1397 Wright, J. B. *J. org. Chem. 29* (1964) 1620.
1398 Franchetti, P., Grifantini, M., Lucarelli, C. and Stein, M. L. *Chem. Abstr. 76* (1972) 85750.
1399 Lettau, H. (a) *Z. Chem. 10* (1970) 431; (b) *11* (1971) 10.
1400 Maffei, S. and Bettinetti, G. F. *Ann. Chim. (Rome) 46* (1956) 812; *Chem. Abstr. 51* (1957) 5759.
1401 Elguero, J., Jacquier, R. and Tizané, D. *Bull. Soc. chim. Fr.* (1969) 1687.
1402 Hünig, S. and Boes, O. *Justus Liebigs Annln Chem. 579* (1953) 28.
1403 Schiele, C. *Angew. Chem. (int. Edn.) 5* (1966) 681.
1404 Curphey, T. J. and Prasad, K. S. *J. org. Chem. 37* (1972) 2259.
1405 Borne, R. F., Aboul-Enein, H. Y. and Baker, J. K. *Spektrochim. Acta 28A* (1972) 393.
1406 Breslow, R. *J. Am. Chem. Soc. 80* (1958) 3719.
1407 Sieber, J. N. and Tolkmith, H. *Tetrahedron Lett.* (1967) 3333.
1408 Boyd, G. V. and Summers, A. J. H. *J. chem. Soc. (B)* (1971) 1648.
1409 Smith, C., Rasmussen, R. and Ballard, S. *J. Am. Chem. Soc. 71* (1949) 1082.
1410 Jerchel, D. and Kuhn, R. *Justus Liebigs Annln Chem. 568* (1950) 185.
1411 Kuhn, R. and Weitz, H. M. *Chem. Ber. 86* (1953) 1199.
1412 Neugebauer, F. A. (a) *Tetrahedron 26* (1970) 4843; (b) with Russell, G. A. *J. org. Chem. 33* (1968) 2744.
1413 Duguchi, Y. and Takagi, Y. *Tetrahedron Lett.* (1967) 3179.

1414 Lee, L. A., Evans, R. and Wheeler, J. W. *J. org. Chem. 37* (1972) 343.
1415 Isida, T., Akiyama, T., Mihara, N., Kozima, S. and Sisido, K. *Bull. chem. Soc. Japan 46* (1973) 1250.
1416 Maender, O. W. and Russell, G. A. *J. org. Chem. 31* (1966) 442.
1417 Cyr, N., Wilks, M. A. J. and Willis, M. R. *J. chem. Soc. (B)* (1971) 404.
1418 Wilks, M. A. J. and Willis, M. R. *J. phys. Chem. 72* (1968) 4717.
1419 Kasai, P. H. and McLeod, D. *J. Am. Chem. Soc. 95* (1973) 27.
1420 Tanino, H., Kondo, T., Okada, K. and Goto, T. *Bull. chem. Soc. Japan 45* (1972) 1474.
1421 Maeda, K. and Hayashi, T. *Bull. chem. Soc. Japan 42* (1969) 3509; *43* (1970) 42.
1422 Philbrook, G. E. and Maxwell, M. A. *Tetrahedron Lett.* (1964) 1111.
1423 Cescon, L., Coraor, G. R., Dessauer, R., Silversmith, E. F. and Urban, E. J. (a) *J. org. Chem. 36* (1971) 2262; (b) with Deutsch, A. S., Jackson, H. L., MacLachlan, A., Marcali, K., Potrafke, E. M. and Read, R. E. *36* (1971) 2267.
1424 Nicholson, I. and Poretz, R. *J. chem. Soc.* (1965) 3067.
1425 Riem, R. H., MacLachlan, A., Coraor, G. R. and Urban, E. J. *J. org. Chem. 36* (1971) 2272.
1426 MacLachlan, A. and Riem, R. H. *J. org. Chem. 36* (1971) 2275.
1427 Cohen, R. L. *J. org. Chem. 36* (1971) 2280.
1428 Bardina, A. A. and Tanaseichuk, B. S. (a) with Nekaeva, I. M. *Khim. geterotsikl. Soedin.* (1972) 1552; (b) with Khomenko, A. A. *J. org. Chem. U. S. S. R. 7* (1971) 1307; (c) with Tsapenkov, A. I. *Chem. Abstr. 78* (1973) 43365.
1429 Tanaseichuk, B. S. and Yartseva, S. V. (a) *Chem. Abstr. 78* (1973) 43366; (b) with Rezepova, L. G. *J. org. Chem. U. S. S. R. 6* (1970) 1068; (c) 7 (1971) 1299.
1430 Tikhonova, L. G. and Tanaseichuk, B. S. *Khim. geterotsikl Soedin.* (1972) 1676.
1431 Bailey, P. S., Bath, S. S., Thomsen, W. F., Nelson, H. H. and Kawas, E. E. *J. org. Chem. 21* (1956) 297.
1432 Effenberger, F. and Mack, K. E. *Tetrahedron Lett.* (1970) 3947.
1433 Hilgetag, G., Kraft, R. and Paul, H. *Z. Chem. 5* (1965) 17.
1434 Mengelberg, M. *Chem. Ber. 93* (1960) 2230.
1435 Wieland, T. and Vogeler, K. *Angew. Chem. 73* (1961) 435.
1436 Harvey, G. R. *J. org. Chem. 31* (1966) 1587.
1437 Panzeri, A. *Rend. 1st Lomb. Sci. e Lett., Classe di Sci. 95* (1961) 432.
1438 Akabori, S. (a) *Ber. dtsch. chem. Ges. 66* (1933) 151; (b) *66* (1933) 159.
1439 Easson, A. P. T. and Pyman, F. L. *J. chem. Soc.* (1932) 1806.
1440 Wohl, A. and Marckwald, W. *Ber. dtsch. chem. Ges. 22* (1889) 568.
1441 Jackman, M., Klenk, M., Fishburn, B., Tullar, B. F. and Archer, S. *J. Am. Chem. Soc. 70* (1948) 2884.
1442 Dodson, R. M. *J. Am. Chem. Soc. 70* (1948) 2753.
1443 Dakin, H. D. and West, R. *J. biol. Chem. 78* (1928) 745.
1444 Bhatt, M. V., Iyer, B. H. and Guha, P. C. *Current Sci. (India) 17* (1948) 184; *Chem. Abstr. 42* (1948) 8799.
1445 Masui, M., Suda, K., and Yamauchi, M. (a) with Yijima, C. *J. chem. Soc. (Perk. 1)* (1972) 1955; (b) with Miyata, H. (1972) 1960.
1446 Asinger, F., Saus, A., Offermanns, H., Krings, P. and Andrée, H. *Justus Liebigs Annln Chem. 744* (1971) 51.
1447 Heath, H., Lawson, A. and Rimington, C. *J. chem. Soc.* (1951) 2223.
1448 van Leusen, A. M. and Oldenziel, O. H. *Tetrahedron Lett.* (1972) 2373.
1449 Kanaoka, M. *J. pharm. Soc. Japan 75* (1955) 1149.
1450 Wojahn, H. *Arch. Pharm. 285* (1952) 122.
1451 Fromm, E. (a) *Justus Liebigs Annln Chem. 361* (1908) 302; (b) with Kayser, E., Brieglef, K. and Fohrenbach, E. *426* (1922) 313; (c) with Trnka, A. *442* (1925) 150; (d) with Kapeller, R., Feniger, M., Krauss, P.,

Schwanenfeld, M. and Wetternik, L. *447* (1926) 294; (e) with Nehring, E. *Ber. dtsch. chem. Ges.* *56* (1923) 1374.
1452 Kubota, S. and Uda, M. (a) with Ohtsuka, M. *Chem. pharm. Bull. Tokyo 19* (1971) 2331; (b) *20* (1972) 2096.
1453 De, S. C. and Chakraverty, T. K. *J. Indian Chem. Soc.* *7* (1930) 875.
1454 Rolla, L. *Gazz. chim. ital.* *38* (1908) 327.
1455 Barascut, J. L., Daunis, J. and Jacquier, R. *Bull. Soc. chim. Fr.* (1973) 323.
1456 Nuhn, P. and Wagner, G. *J. prakt. Chem.* *312* (1970) 90.
1457 Johnson, T. B. and Edens, C. O. *J. Am. Chem. Soc.* *64* (1942) 2706.
1458 Goerdler, J. and Gnad, G. *Chem. Ber.* *99* (1966) 1618.
1459 Jackson, A. O. and Marvell, C. S. *J. biol. Chem.* *103* (1933) 191.
1460 Clark, J., Grantham, R. K. and Lydiate, J. *J. chem. Soc. (C)* (1968) 1122.
1461 Postovskii, I. Ya. and Nirenburg, V. L. *Chem. Abstr.* *64* (1966) 19590.
1462 Ioffe, I. S. and Khavin, Z. Ya. *J. gen. Chem. U. S. S. R.* *14* (1944) 822.
1463 Tsurkan, A. A., Efremenko, V. I. and Groshev, V. V. *Farm. Zh. (Kiev)* *27* (1972) 72; *Chem. Abstr.* *78* (1973) 43372.
1464 Himbert, G. and Regitz, M. *Chem. Ber.* *105* (1972) 2975.
1465 Buechel, K. H., Draber, W., Regel, E. and Plempel, M. *Arzneimittel-Forsch.* *22* (1972) 1260; *Chem. Abstr.* *77* (1972) 152065.
1466 Pump, J. and Wannagat, U. *Justus Liebigs Annln Chem.* *652* (1962) 21.
1467 Agawa, T., Yasui, M. and Matsui, M. *Agr. biol. Chem.* *36* (1972) 1441; *Chem. Abstr.* *77* (1972) 151168.
1468 Goldman, L., Marsico, J. W. and Anderson, G. W. *J. Am. Chem. Soc.* *82* (1960) 2969.
1469 Knorr, L. and Duden, P. *Ber. dtsch. chem. Ges.* *26* (1893) 111.
1470 La Cour, T. and Rasmussen, S. E. *Acta chem. scand.* *27* (1973) 1845.
1471 Singh, T. P. and Vijayen, M. *Acta crystallogr. B29* (1973) 714.
1472 Bonnet, J. J. and Ibers, J. A. *J. Am. Chem. Soc.* *95* (1973) 4829.
1473 Seccombe, R. C. and Kennard, C. H. L. *J. chem. Soc. (Perk. 2)* (1973) 1, 4, 9; Seccombe, R. C., Tillack, J. V. and Kennard, C. H. L. p. 6.
1474 Mauret, P., Fayet, J. P., Fabre, M., Elguero, J. and del C. Pardo, M. *J. Chim. Phys. Physicochim. Biol.* *70* (1973) 1483.
1475 Habraken, C. L., Beenakker, C. I. M. and Brussee, J. *J. heterocycl. Chem.* *9* (1972) 939; Janssen, J. W. A. M., Kruse, C. G., Koeners, H. J. and Habraken, C. L. *10* (1973) 1055.
1476 Nye, M. J. and Tang, W. P. *Tetrahedron* (a) *28* (1972) 455; (b) p. 463.
1477 (a) Albert, A. and Taguchi, H. *J. chem. Soc. (Perk. 1)* (1973) 1629; (b) Albert, A. p. 1634.
1478 Mel'nikov, V. V., Stolpakova, V. V., Pevzner, M. S. and Gidaspov, B. V. *Chem. Abstr.* *80* (1974) 2730p.
1479 Sprensen, G. O., Nygaard, L. and Begtrup, M. *Chem. Commun.* (1974) 605.
1480 Lee, L. A. and Wheeler, J. W. *J. org. Chem.* *37* (1972) 348.
1481 Bradamante, S., Pagani, G. and Marchesini, A. *J. chem. Soc. (Perk. 2)* (1973) 568.
1482 Begtrup, M. *Acta chem. scand.* *27* (1973) 3101.
1483 Reynolds, W. F., Peat, I. R., Freedman, M. H. and Lyerla, J. R. *J. Am. Chem. Soc.* *95* (1973) 328, 6510.
1484 Begtrup, M. *Chem. Commun.* (1974) 702.
1485 Thurber, T. C., Pugmire, R. J. and Townsend, L. B. *J. heterocycl. Chem.* *11* (1974) 645.
1486 Witanowski, M., Stefaniak, L., Januszewski, H., Grabowski, Z. and Webb, G. A. *Tetrahedron 28* (1972) 637.
1487 Vilarrasa, J., Meléndez, E. and Elguero, J. *Tetrahedron Lett.* (1974) 1609.
1488 Kohn, H., Benkovic, S. J. and Olofson, R. A. *J. Am. Chem. Soc.* *94* (1972) 5759.
1489 Agibalov, N. D., Enin, A. S., Koldobskii, G. I., Gidaspov, B. V. and Tomofeeva, T. N. *Zh. Org. Khim.* *8* (1972) 2414.
1490 Okkersen, H., Groeneveld, N. L. and Reedijk, J. *Rec. trav. chim.* *92* (1973)

945.
1491 Kouno, K. and Ueda, Y. *Chem. Pharm. Bull. 19* (1971) 2278.
1492 Palmer, M. H., Findlay, R. H. and Gaskell, A. J. *J. chem. Soc. (Perk. 2)* (1974) 420.
1493 Nakanishi, M. and Kobayashi, R. *Chem. Abstr. 80* (1974) 59940m.
1494 Sheinkman, A. K., Deikalo, A. A., Stupnikova, T. V. and Baranov, S. N. *Chem. Abstr. 76* (1972) 126863w.
1495 Buzykin, B. I. and Lonshchakova, T. I. *Chem. Abstr. 76* (1972) 113126h.
1496 Wong, J. L. and Keck, J. H. *J. org. Chem. 39* (1974) 2398.
1497 *Ger. Pat.* 2 129 524 (1973) [*Chem. Abstr. 78* (1973) 97647n].
1498 *Canadian Pat.* 914 196 (1972) [*Chem. Abstr. 81* (1974) 3943t].
1499 Orban, M., Koros, E. and Krudy-Sandov, K. *Chem. Abstr. 80* (1974) 14212j.
1500 Butler, D. E. and Alexander, S. M. *J. org. Chem. 37* (1972) 215.
1501 Novikov, S. S., Khmel'nitskii, L. I., Lebedev, O. V., Sevast'yanova, V. V. and Epishina, L. V. *Chem. heterocycl. Compounds 6* (1970) 465.
1502 Burton, A. G., Katritzky, A. R., Konya, M. and Tarhan, H. O. *J. chem. Soc. (Perk. 2)* (1974) 389.
1503 Burton, A. G., Dereli, M., Katritzky, A. R. and Tarhan, H. O. *J. chem. Soc. (Perk. 2)* (1974) 382.
1504 Neuman, P. *J. heterocycl. Chem. 8* (1971) 51.
1505 Weinheimer, A. J., Metzner, E. K. and Mole, L. M. *Tetrahedron 29* (1973) 3125.
1506 Rink, H. and Riniker, B. *Helv. chim. Acta 57* (1974) 831.
1507 Lehmkuhl, F. A., Witkowski, J. T. and Robins, R. K. *J. heterocycl. Chem. 9* (1972) 1195; Naik, S. R., Witkowski, J. T. and Robins, R. K. *11* (1974) 57.
1508 Barascut, J.-L., Claramunt, R.-M. and Elguero, J. *Bull. Soc. chim. Fr.* (1973) 1849.
1509 Claramunt, R.-M., Granados, R. and Pedroso, E. *Bull. Soc. chim. Fr.* (1973) 1854.
1510 Tanaka, Y. and Miller, S. I. *Tetrahedron 29* (1973) 3285.
1511 Godefroi, E. F. and Mentjens, J. H. F. M. *Rec. trav. chim. 93* (1974) 56.
1512 Khan, M. A. and Lynch, B. M. *J. heterocycl. Chem. 7* (1970) 1237.
1513 Coburn, M. D. and Neuman, P. N. *J. heterocycl. Chem. 7* (1970) 1391.
1514 Neuman, P. N. *J. heterocycl. Chem. 7* (1970) 1159.
1515 Akiyama, T., Imasaki, Y. and Kawanisi, M. *Chem. Lett.* (1974) 229.
1516 *Ger. Pat.* 1 542 789 (1973) [*Chem. Abstr. 80* (1974) 3513g].
1517 Elnagoli, M. H. and Ohta, M. *Bull. chem. Soc. Japan 46* (1973) 3818.
1518 Voishcheva, O. V., Galkin, V. D., Mikhant'ev, B. I. and Shatalov, G. V. *Chem. Abstr. 80* (1974) 82834p.
1519 Reimlinger, H., Peiren, M. A. and Merényi, R. *Chem. Ber. 105* (1972) 103.
1520 Arakawa, K., Miyasaka, T. and Ochi, H. *Chem. pharm. Bull. 22* (1974) 207.
1521 Hirata, T., Twannoh, L., Wood, H. B., Goldin, A. and Driscoll, J. S. *J. heterocycl. Chem. 9* (1972) 99; Hirata, T., Wood, H. B. and Driscoll, J. S. *J. chem. Soc. (Perk. 1)* (1973) 1209.
1522 Coburn, M. D., Loughran, E. D. and Smith, L. C. *J. heterocycl. Chem. 7* (1970) 1149.
1523 Lippmann, E., Widera, R. and Kleinpeter, E. *Z. Chem. 13* (1973) 429.
1524 Rogne, O. *J. chem. Soc. (Perk. 2)* (1973) 823.
1525 Troxler, F., Weber, H. P., Jaunin, A. and Loosli, H.-R. *Helv. chim. Acta 57* (1974) 750.
1526 Mel'nikova, N. N., Pevzner, M. S., Malysheva, N. M. and Bagal, L. I. *Zh. org. Khim. 9* (1973) 2535.
1527 Smith, P. A. S. and Dounchis, H. *J. org. Chem. 38* (1973) 2958.
1528 Heitke, B. T. and McCarty, C. G. *J. org. Chem. 39* (1974) 1522.
1529 Begtrup, M. *Acta chem. scand. 27* (1973) 2051.
1530 Begtrup, M., Conradsen, N. and Olsen, J. H. *Acta chem. scand. 27* (1973)

2930.
1531 Jones, R. G. and Terando, N. H. *Chem. Abstr. 78* (1973) 4247u.
1532 Cavalleri, B., Ballotta, R. and Lancini, G. C. *J. heterocycl. Chem. 9* (1972) 979.
1533 *Ger. Pat.* 2 317 453 (1973) [*Chem. Abstr. 80* (1974) 14926v].
1534 Elguéro, J., Jacquier, R. and Mignonac-Mondon, S. *Bull. Soc. chim. Fr.* (1972) 2807.
1535 Aubagnac, J.-L., Elguéro, J. and Gilles, J.-L. *Bull. Soc. chim. Fr.* (1973) 288.
1536 Jensen, K. A. and Christopherson, C. *Acta chem. scand. B 28* (1974) 1.
1537 Birkofer, L., Franz, M. and Schmidtberg, G. *Org. Mass. Spectrom. 8* (1974) 347.
1538 Kajfez, F., Sunjic, V., Klasinc, L. and Marsel, J. *Chem. Abstr. 78* (1973) 110004x.
1539 Heitke, B. T. and McCarty, C. G. *Can. J. Chem. 52* (1974) 2861.
1540 Blackman, A. J. and Bowie, J. H. *Aust. J. Chem. 25* (1972) 335.
1541 Nishiwaki, T., Fujiyama, F. and Minamisono, E. *J. chem. Soc. (Perk. 1)* (1974) 1871.
1542 Reisch, J. and Ossenkop, W. F. *Chem. Ber. 106* (1973) 2070; Reisch, J. and Fitzek, A. *Arch. Pharm. 307* (1974) 211.
1543 Schroeder, M. A. and Makino, R. C. *Tetrahedron 29* (1973) 3469.
1544 Sitkina, L. M. and Simonov, A. M. *Chem. heterocycl. Compounds U. S. S. R. 6* (1970) 380.
1545 Novikov, S. S., Khmel'nitskii, L. I., Novikova, T. S., Lebedev, O. V. and Epishina, L. V. *Chem. heterocycl. Compounds U. S. S. R. 6* (1970) 619.
1546 Simonov, A. M. and Uryukina, I. G. *Chem. heterocycl. Compounds U. S. S. R. 7* (1971) 536.
1547 Wilczynski, J. J. and Johnson, H. W. *J. org. Chem. 39* (1974) 1909.
1548 Thé, K. I., Peterson, L. K. and Kiehlmann, E. *Can. J. Chem. 51* (1973) 2448; with Sanger, A. R. *52* (1974) 2367.
1549 Winter, W. and Müller, E. *Chem. Ber. 107* (1974) 2127.
1550 Shvartsberg, M. S., Vasilevskii, S. F. and Kotlyarevskii, I. L. *Chem. Abstr. 79* (1973) 42402.
1551 Mitsuhashi, K. and Zaima, T. *Chem. Abstr. 79* (1973) 31986.
1552 *U. S. Pat.* 3 740 000 [*Chem. Abstr. 79* (1973) 42505e].
1553 Albright, J. D. and Shepherd, R. G. *J. heterocycl. Chem. 10* (1973) 899.
1554 Bastiaansen, L. A. M., Macco, A. A. and Godefroi, E. F. *Chem. Commun.* (1974) 36.
1555 Grimshaw, J. and Trocha-Grimshaw, J. *J. chem. Soc. (Perk. 1)* (1973) 1275.
1556 Bany, T. and Dobosz, M. *Chem. Abstr. 79* (1973) 5298.
1557 Finar, I. L. and Okoh, E. *J. chem. Soc. (Perk. 1)* (1973) 2008.
1558 Krieg, B., Schlegel, R. and Manecke, G. *Chem. Ber. 107* (1974) 168.
1559 Skvortsova, G. G., Glaskova, N. P., Domnina, E. S. and Voronov, V. K. *Chem. heterocycl. Compounds U. S. S. R. 6* (1970) 153.
1560 Vasilevskii, S. F., Slabuka, P. A., Izyumov, E. G., Shvartsberg, M. S. and Kotlyarenskii, I. L. *Bull. Acad. Sci. U. S. S. R. 21* (1972) 2453.
1561 Kirk, K. L. and Cohen, L. A. *J. Am. chem. Soc. 95* (1973) 4619.
1562 Kreutzberger, A. and Schuecker, R. *Arch. Pharm. 306* (1973) 169.
1563 Butler, R. N., Lambe, T. M., Tobin, J. C. and Scott, F. L. *J. chem. Soc. (Perk. 1)* (1973) 1357.
1564 Calton, R. C. and Butler, R. N. *Can. J. Chem. 52* (1974) 1248.
1565 Alcalde, E., de Mendoza, J. and Elguero, J. *Chem. Commun.* (1974) 411.
1566 Heitke, B. T. and McCarty, C. G. *J. org. Chem. 39* (1974) 1522.
1567 Isida, T., Akiyama, T., Nabika, K., Sisido, K. and Kozima, S. *Bull. chem. Soc. Japan 46* (1973) 2176.
1568 Boyd, G. V. and Norris, T. *J. chem. Soc. (Perk. 1)* (1974) 1028.
1569 Neugebauer, F. A. *Angew. Chem. (int. Edn.) 12* (1973) 455.
1570 Monjoint, P. and Laloi-Diard, M. *Bull. Soc. chim. Fr.* (1973) 2357.

1571 Bravo, P. and Ponti, P. P. *J. heterocycl. Chem. 10* (1973) 669.
1572 Corey, E. J. and Fleet, G. W. J. *Tetrahedron Lett.* (1973) 4499.
1573 Bonati, F., Minghetti, G. and Banditelli, G. *Chem. Commun.* (1974) 88.
1574 Tanaka, Y. and Miller, S. I. *J. org. Chem. 38* (1973) 2708.
1575 Torocheshnikov, V. N., Sergeev, N. M., Viktorov, N. A., Gol'din, G. S., Poddubnyi, V. G. and Koltsnova, A. V. *Chem. Abstr. 81* (1974) 24779j; Gol'din, G. S., Maksakova, M. V., Poddubnyi, V. G., Kol'tsova, A. N., Kisin, A. V., Torocheshnikov, V. N. and Simonova, A. A. *J. gen. Chem. U. S. S. R. 44* (1974) 113.
1576 Begg, C. G. and Grimmett, M. R. *J. Chromat. 73* (1972) 238.

Index